An Integrated Approach to Ecology

An Integrated Approach to Ecology

Editor: Lawrence Poole

www.callistoreference.com

Callisto Reference,
118-35 Queens Blvd., Suite 400,
Forest Hills, NY 11375, USA

Visit us on the World Wide Web at:
www.callistoreference.com

ISBN: 978-1-64116-169-5 (Hardback)

Cataloging-in-Publication Data

An integrated approach to ecology / edited by Lawrence Poole.
    p. cm.
Includes bibliographical references and index.
ISBN 978-1-64116-169-5
1. Ecology. 2. Environmental sciences. I. Poole, Lawrence.
QH541 .I58 2019
577--dc23

Table of Contents

Preface

The branch of biology that studies the interactions between organisms and their environment is known as ecology. Biodiversity, biomass, populations of organisms, distribution, competition and cooperation between different species are the major topics that fall under this field of study. The study of ecology can be structured by organizing the biological world into a nested hierarchy, that ranges in scale from genes to cells, organs, organisms, species and so on, up to the level of the biosphere. The practical applications of this field can be found in natural resource management, conservation biology, city planning, and wetland management, among many others. This book unfolds the innovative aspects of ecology that will be crucial for the progress of this field in the future. It consists of contributions made by international experts. Also included herein is a detailed explanation of the various concepts and applications of this field. Those in search of information to further their knowledge will be greatly assisted by this book.

The information shared in this book is based on empirical researches made by veterans in this field of study. The elaborative information provided in this book will help the readers further their scope of knowledge leading to advancements in this field.

Finally, I would like to thank my fellow researchers who gave constructive feedback and my family members who supported me at every step of my research.

Editor

Skeletal light-scattering accelerates bleaching response in reef-building corals

Timothy D. Swain[1,2], Emily DuBois[1,2], Andrew Gomes[3], Valentina P. Stoyneva[3], Andrew J. Radosevich[3], Jillian Henss[1,2], Michelle E. Wagner[1,2], Justin Derbas[3], Hannah W. Grooms[1], Elizabeth M. Velazquez[1], Joshua Traub[1], Brian J. Kennedy[1], Arabela A. Grigorescu[4], Mark W. Westneat[2], Kevin Sanborn[5], Shoshana Levine[5], Mark Schick[5], George Parsons[5], Brendan C. Biggs[6], Jeremy D. Rogers[3], Vadim Backman[3] and Luisa A. Marcelino[1,2]*

Abstract

Background: At the forefront of ecosystems adversely affected by climate change, coral reefs are sensitive to anomalously high temperatures which disassociate (bleaching) photosynthetic symbionts (*Symbiodinium*) from coral hosts and cause increasingly frequent and severe mass mortality events. Susceptibility to bleaching and mortality is variable among corals, and is determined by unknown proportions of environmental history and the synergy of *Symbiodinium*- and coral-specific properties. *Symbiodinium* live within host tissues overlaying the coral skeleton, which increases light availability through multiple light-scattering, forming one of the most efficient biological collectors of solar radiation. Light-transport in the upper ~200 μm layer of corals skeletons (measured as 'microscopic' reduced-scattering coefficient, $\mu'_{S,m}$), has been identified as a determinant of excess light increase during bleaching and is therefore a potential determinant of the differential rate and severity of bleaching response among coral species.

Results: Here we experimentally demonstrate (in ten coral species) that, under thermal stress alone or combined thermal and light stress, low-$\mu'_{S,m}$ corals bleach at higher rate and severity than high-$\mu'_{S,m}$ corals and the *Symbiodinium* associated with low-$\mu'_{S,m}$ corals experience twice the decrease in photochemical efficiency. We further modelled the light absorbed by *Symbiodinium* due to skeletal-scattering and show that the estimated skeleton-dependent light absorbed by *Symbiodinium* (per unit of photosynthetic pigment) and the temporal rate of increase in absorbed light during bleaching are several fold higher in low-$\mu'_{S,m}$ corals.

Conclusions: While symbionts associated with low-$\mu'_{S,m}$ corals receive less total light from the skeleton, they experience a higher rate of light increase once bleaching is initiated and absorbing bodies are lost; further precipitating the bleaching response. Because microscopic skeletal light-scattering is a robust predictor of light-dependent bleaching among the corals assessed here, this work establishes $\mu'_{S,m}$ as one of the key determinants of differential bleaching response.

Keywords: Global climate change, Optical scattering, Coral bleaching, Photosynthesis, Symbiosis

Background

At the forefront of ecosystems adversely affected by climate change, coral reefs are sensitive to anomalously high temperatures which disassociate (bleaching) photosynthetic symbionts (*Symbiodinium*) from coral hosts and cause increasingly frequent and severe mass mortality events [1–4]. Susceptibility to bleaching and mortality is variable among corals [2, 5–8], and is partially determined (at unknown proportions) by a combination of environmental history [9, 10] and the interaction of *Symbiodinium*- [2, 11–14] and coral-specific [8, 15–19] properties (reviewed in [20]).

*Correspondence: l-marcelino@northwestern.edu
[2] Department of Zoology, Field Museum of Natural History, 1400 South Lake Shore Drive, Chicago, IL 60605, USA
Full list of author information is available at the end of the article

As photosynthetic performance of *in hospite Symbiodinium* is often impaired during thermally-induced bleaching (e.g., [21–23]), the interaction of temperature and irradiance exacerbate the bleaching response (reviewed in [4, 20, 24, 25]). Corals under thermal stress experience greater damage to the *Symbiodinium* photosynthetic apparatus (chronic photoinhibition of PSII) and elevated bleaching response when exposed to supraoptimal solar irradiances, indicating that temperature reduces the light intensity threshold for photoinhibition [4, 21, 25, 26].

Symbiodinium live within host tissues overlaying the coral skeleton, which can significantly increase light availability to symbionts through multiple scattering [15–18, 27], and together with within-tissue scatter and dynamic light redistribution (due to tissue contraction and scattering or absorption by host fluorescent pigments) [19, 28] form one of the most efficient biological collectors of solar radiation [15, 29]. This increase in light-availability is dependent on density and absorption properties of symbiont and host pigments and on diffuse reflectance of light from coral skeleton (R_S) and tissue, which is mainly reliant on light scattering and absorption in the skeleton and tissue as well as overall coral morphology [15, 17–19, 27–33]. Scattering in skeletons (characterized by the reduced scattering coefficient, bulk-μ'_S or μ'_S: inverse of the distance a photon travels before randomization) is mainly due to light interaction with skeletal microstructures throughout the entire skeleton (from 50 to 200 nm $CaCO_3$ nanograins to 1–5 μm fiber bundles; [34, 35]) and larger length-scale structures (hundreds of micron size septa to millimeter size corallites; [15, 27, 36]). Furthermore, scattering in the superficial layer of coral skeletons (measured as microscopic-μ'_S or $\mu'_{S,m}$: the inverse distance a short-path length photon travels before randomization [18, 37]) governs light-transport at sub-diffusion path lengths (~100 μm) and is affected by skeletal microstructures, but not larger length-scale structures [18]. Thus $\mu'_{S,m}$ can be described as μ'_S of the skeletal material itself, within the top ~100 μm of the skeleton without voids [18]. Although R_S includes the effect of $\mu'_{S,m}$, it is primarily determined by μ'_S, absorption, and overall coral morphology [15, 18, 27, 29–31].

Greater total skeletal reflectance, associated with higher μ'_S, has been demonstrated to increase light-absorption by at least six times for symbionts *in hospite* and in simulations compared to those in vitro [15, 17]. By estimating absorption efficiency in differentially bleached corals and skeletal models (e.g., polished-laminae), it has been shown that skeletal light amplification (excess light available to the symbiont) is inversely related to symbiont concentration, leading to the prediction that skeletal μ'_S could exacerbate the feedback of increasing

photodamage for remaining *Symbiodinium* as symbiont densities diminish during bleaching (positive feedback-loop hypothesis) [15, 17, 29]. However, the rate of excess light increase as symbiont densities decrease has been demonstrated in models to be highly variable among corals, with high rates of excess light increase inversely correlated with skeletal $\mu'_{S,m}$ [18]. Low skeletal $\mu'_{S,m}$ values were significantly correlated with heightened bleaching susceptibility in a retrospective analysis of global bleaching events for 94 coral taxa, leading to the prediction that $\mu'_{S,m}$ (as the optical property responsible for the rate of feedback) is a potential determinant of the severity of bleaching response for this mechanism [18]. In this previous study, neither μ'_S nor R_S were correlated with historical bleaching response [18].

To consolidate previous findings and provide predictions about the bleaching process that can be experimentally assessed, we propose the optical feedback hypothesis based on the effect of short-path light-transport. Although skeletal contribution to the endosymbiotic light microenvironment is normally small [38], skeletal optical properties become increasingly important as symbionts are lost and the skeleton becomes more exposed to light [18]. As densities of light absorbers (*Symbiodinium* cells and/or their photosynthetic pigments) decrease during the bleaching response, the coral skeleton becomes progressively exposed to downwelling light and dynamically becomes an increasingly significant source of excess light to remaining symbionts, compounding stress on *Symbiodinium* and provoking a more rapid and severe bleaching response. This feedback loop may proceed at differential rates that are determined by the rate at which the skeleton increases excess light to symbionts, as *Symbiodinium* and pigment concentrations decline [18]. As the optical property that is predictive of the rate of excess light increase as a function of pigment density, $\mu'_{S,m}$ affects the rate of feedback and may therefore be a determinate of bleaching severity [18]. We therefore predict that, depending on skeletal $\mu'_{S,m}$, corals that are bleaching should be differentially exposed to stress, and low-$\mu'_{S,m}$ corals should experience: (1) increased rates and severities of bleaching response, with *Symbiodinium* remaining *in hospite* showing increased rates and severities of light stress, and (2) increased skeleton-dependent light absorption by remaining *Symbiodinium*. Furthermore, (3) skeletal $\mu'_{S,m}$ should be a good predictor of the light-dependent bleaching effect but a poor predictor of temperature-dependent bleaching. These predictions of the optical feedback hypothesis have not been experimentally demonstrated among corals with diverse skeletal optical properties ($\mu'_{S,m}$ and Rs); which due to the dynamic nature of feedback, must be assessed as corals undergo bleaching.

Here we describe a heat- and light-stress experiment that demonstrates the effect of skeletal $\mu'_{S,m}$ on bleaching response using ten coral species selected for diversity of bleaching susceptibilities, skeletal optical properties, and *Symbiodinium* thermotolerances. By following the dynamics of holobiont response to stress directly, and developing a novel empirical model of skeleton-dependent light-absorption for *in hospite Symbiodinium*, we assessed the general predictions for coral bleaching under the optical feedback mechanism detailed above. The combined experimental and empirical modeling substantiates the predictions of the optical feedback hypothesis by establishing a connection between the dynamics of skeletal light amplification, bleaching response, *in hospite Symbiodinium* light absorption, and photophysiology among a diverse group of corals.

Results

Skeletal and holobiont optical characteristics

Microscopic scattering, $\mu'_{S,m}$, varied between 1.53 and 5.8 mm^{-1} (Table 1), with low-$\mu'_{S,m}$ corals (defined as below the mean of the ten species assessed: *Merulina* sp., *Pocillopora damicornis*, *Seriatopora hystrix*, and *Stylophora pistillata*) averaging 2.01 $\pm$ 0.27 mm^{-1} (mean $\pm$ std error) and high-$\mu'_{S,m}$ corals (*Diploria labyrinthiformis*, *Goniopora* sp., *Favia favus*, *Montipora foliosa*, and *Montipora digitata*) averaging 4.58 $\pm$ 0.34 mm^{-1}. Consistent with the imperfectly-white coloration of the skeletons, R_S varied between 0.24 and 0.71 (relative to white standard, Table 1). Holobiont reflectance, R_H, varied between 0.02 and 0.26 prior to the initiation of stress (Fig. 1a, e).

Although corals are highly complex structures, the variability detected in repeated measurements of $\mu'_{S,m}$, R_S, and R_H is sufficiently small that we assume colonies can be characterized by mean values. The variability due to irregular surfaces and varying instrument positions is small, as is the coefficient of variation (COV), compared to the observed change in reflectance during bleaching. The average standard error of mean for R_H is <12 % (n = 10 measurements per ramet), and its COV is 38 % (standard deviation relative to mean) while the observed change in reflectance during bleaching increases as much as 300 % (Additional file 1: Figure S1a, f). This level of

Table 1 Optical, tissue, bleaching, and genetic data for individual corals

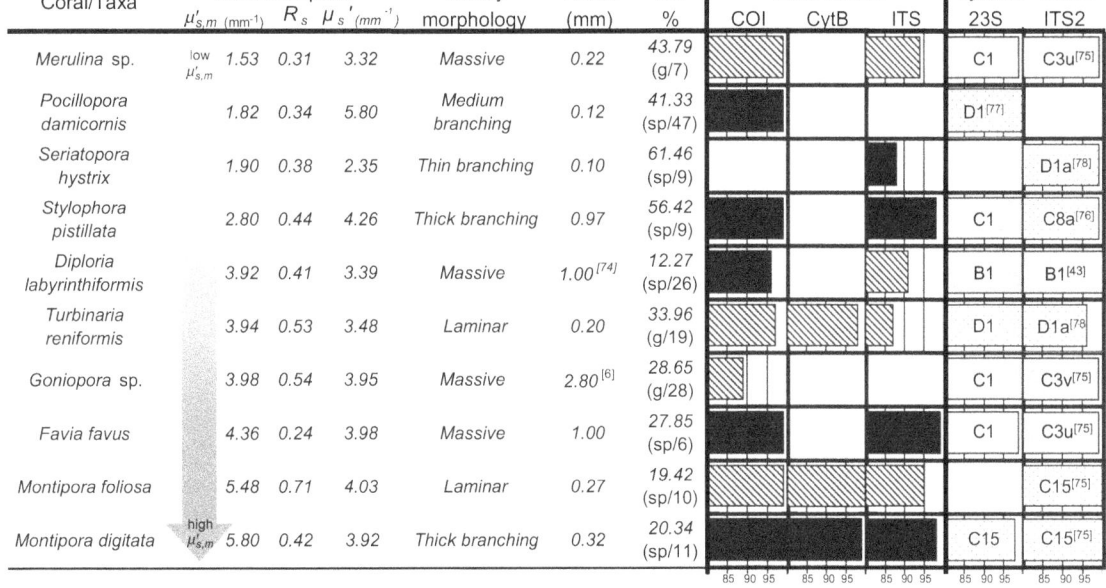

Coral/Taxa	Skeletal/Optics $\mu'_{S,m}$ (mm⁻¹)	R_s	μ_s' (mm⁻¹)	Colony morphology	Tissue (mm)	BRI %	Coral Genes COI	CytB	ITS	*Symbio.* Genes 23S	ITS2
Merulina sp.	low $\mu_{s,m}$ 1.53	0.31	3.32	Massive	0.22	43.79 (g/7)				C1	C3u[75]
Pocillopora damicornis	1.82	0.34	5.80	Medium branching	0.12	41.33 (sp/47)				D1[77]	
Seriatopora hystrix	1.90	0.38	2.35	Thin branching	0.10	61.46 (sp/9)					D1a[78]
Stylophora pistillata	2.80	0.44	4.26	Thick branching	0.97	56.42 (sp/9)				C1	C8a[76]
Diploria labyrinthiformis	3.92	0.41	3.39	Massive	1.00[74]	12.27 (sp/26)				B1	B1[43]
Turbinaria reniformis	3.94	0.53	3.48	Laminar	0.20	33.96 (g/19)				D1	D1a[78]
Goniopora sp.	3.98	0.54	3.95	Massive	2.80[6]	28.65 (g/28)				C1	C3v[75]
Favia favus	4.36	0.24	3.98	Massive	1.00	27.85 (sp/6)				C1	C3u[75]
Montipora foliosa	5.48	0.71	4.03	Laminar	0.27	19.42 (sp/10)					C15[75]
Montipora digitata	high $\mu_{s,m}$ 5.80	0.42	3.92	Thick branching	0.32	20.34 (sp/11)				C15	C15[75]

% Match (85 90 95 for each gene column)

Skeletal optical properties [skeletal scattering ($\mu'_{S,m}$), skeletal reflectance (R_S)], and bulk scattering (μ_s')], tissue thickness (all measured directly, except those annotated with citations [6, 74]), bleaching response index [BRI or the percent coral cover bleached and/or killed during mass bleaching events [18] used here as expected bleaching response for each taxon; parenthetical notation refers to genus- (g) or species-level (sp) estimations and the number of records that estimation is based upon], and genetic identity of corals and *Symbiodinium* assessed in experiment. Nucleotide sequences compared with Genbank (last accessed August 15, 2013) and reported as percent match (bar graphs) with accessions for coral mitochondrial cytochrome oxidase I (COI), cytochrome b (CytB), and nuclear internal transcribed spacer (ITS) genes; and *Symbiodinium* nuclear internal transcribed spacer region 2 (ITS2) and chloroplast 23S ribosomal (23S) genes. Shading of bars indicate the presence (solid black) or absence (diagonal lines) of the target species in Genbank, and low- (solid gray) or high-thermotolerance (stippled) of *Symbiodinium* [as reported in the literature (assuming C3u and C3v are similar to C3) [43, 75–78] and indicated by parenthetical superscript number on the phylotype used to categorize thermotolerance]

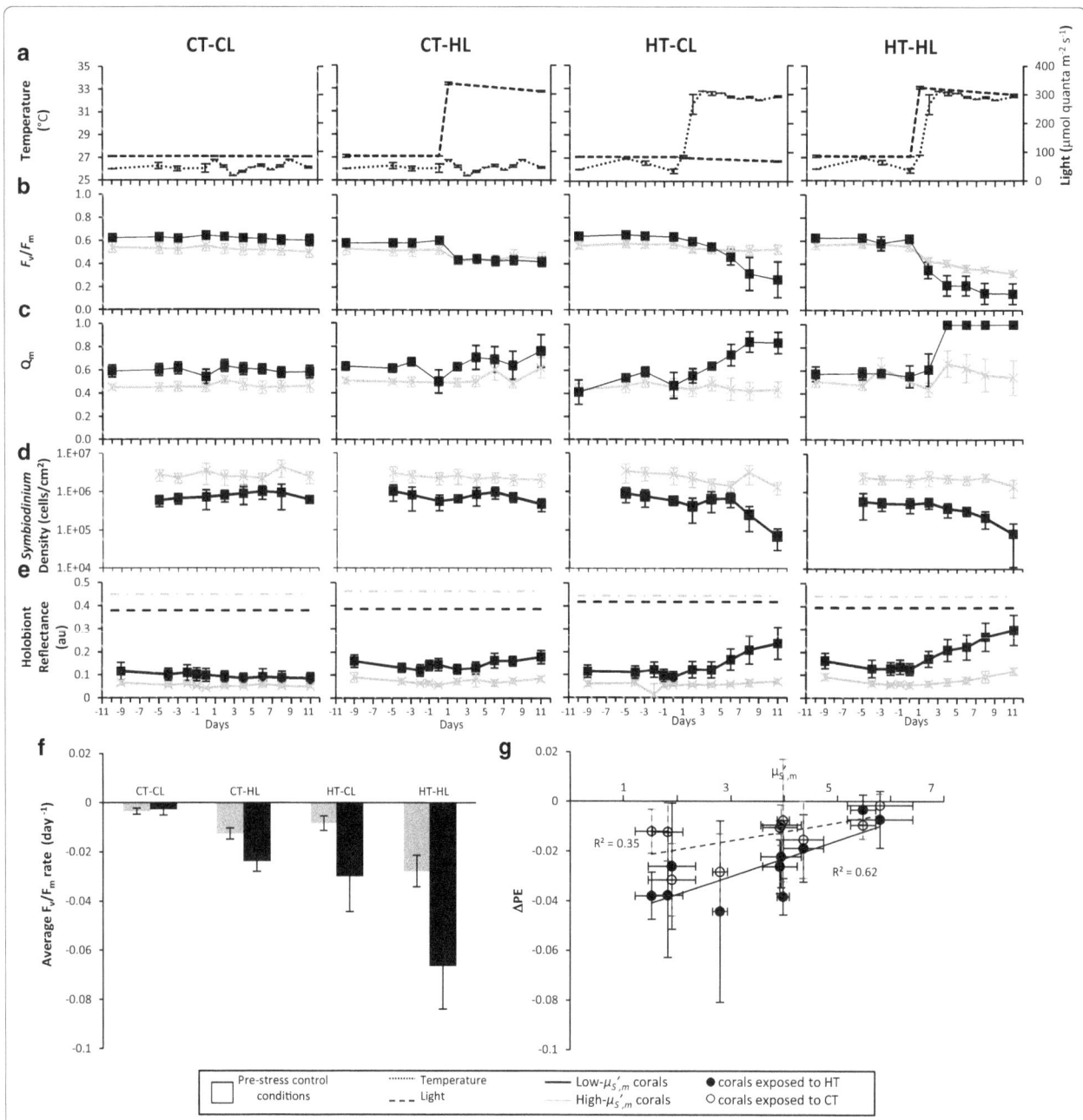

Fig. 1 Dynamics of bleaching response variables. High- and low-$\mu'_{S,m}$ corals (means in *gray* and *black* respectively in **b–f**) responded differentially to experimental light (*broken line* in **a**) and temperature (*dotted line* in **a**) conditions (CT-CL: control temperature [26 °C] and light [83 µmol quanta m^{-2} s^{-1}], CT-HL: control temperature and high light [328 µmol quanta m^{-2} s^{-1}], HT-CL: high temperature [32 °C] and control light, and HT-HL: high temperature and high light; shaded areas are control). Under temperature stress (HT-CL and HT-HL), *Symbiodinium in hospite* of low-$\mu'_{S,m}$ corals experienced suppressed photosynthetic performance (**b**, **c**) and reduced cell density (**d**), and holobiont reflectance (**e**) of low-$\mu'_{S,m}$ corals approached the level of bare skeleton (*dashed lines* in **e** are post-experiment skeletal reflectance). Low-$\mu'_{S,m}$ corals experienced progressively greater average rates of photochemical efficiency loss (CT-CL $p = 0.755$, CT-HL $p = 0.032$, HT-CL $p = 0.112$, and HT-HL $p = 0.042$) as heat and light stress were combined (**f**). Isolating the effect of light from temperature on photochemical efficiency (**g**), $\mu'_{S,m}$ is correlated with the temporal rate of F_v/F_m change $\left(\Delta PE \sim \Delta^2(F_V/F_M)/(\Delta t \Delta I) \right)$ expressed as the difference between CL and HL (Eq. 2) for corals exposed to HT (*filled circles*; $p = 0.007$) or CT (*open circles*; $p = 0.07$). All error bars are standard error of the mean

signal variability is sufficiently low to resolve changes in R_H as small as ~24 %. The COV of $\mu'_{S,m}$ for coral skeletons has been previously determined to be similarly small, at 12 % within a colony (assessing four areas from each of seven colonies) and 29.3 % within a species (assessing 4–8 colonies representing each of seven species) [18].

Low $\mu'_{S,m}$ corals experience increased rates and severities of bleaching and remaining *Symbiodinium* experience increased rates and severities of light stress

Corals in high temperature treatments (high temperature-control light: HT-CL, or high temperature-high light: HT-HL) experienced responses consistent with bleaching, with low-$\mu'_{S,m}$ corals bleaching at greater rates and severities. Under the application of temperature (HT-CL) or light and temperature (HT-HL) stress all corals experienced significant (ANOVA, $p < 0.05$) reductions in *Symbiodinium* cell densities (ρ) and increases in R_H, with the most severe responses among low-$\mu'_{S,m}$ corals (Fig. 1; Additional file 2: Figure S2). Additionally, low-$\mu'_{S,m}$ corals experienced significantly (ANOVA, $p < 0.001$) greater decreases in *Symbiodinium* chlorophyll *a* densities (Chl *a*), with the greatest response occurring under the HT-HL treatment (Additional file 2: Figure S2). Exemplar R_H spectra over the visible (400–700 nm) and near infra-red (>700–800 nm) regions are shown in (Additional file 1) Figure S1 for *S. pistillata* (low-$\mu'_{S,m}$) and *M. digitata* (high-$\mu'_{S,m}$) before and after combined thermal- and light-stress was applied. As symbionts are lost during bleaching of *S. pistallata*, values of R_H approached the values of R_S (Fig. 1e; Additional file 2: Figure S1). Corals in the high light treatment alone (CT-HL) did not experience responses consistent with bleaching and observed differences in the dynamics of R_H, ρ, or Chl *a* between low- and high-$\mu'_{S,m}$ corals (Fig. 1; Additional file 2: Figure S2) are insignificant.

Symbiodinium that remained *in hospite* during bleaching experienced responses consistent with increasing light stress (i.e., corals under HT-CL, HT-HL), however *Symbiodinium* of low-$\mu'_{S,m}$ corals experienced greater rates and severities of light stress (Fig. 1; Additional file 2:

Figure S2). *Symbiodinium* associated with low-$\mu'_{S,m}$ corals experienced significantly suppressed photochemical efficiency (F_v/F_m, linear mixed models, LMM, analysis) and elevated maximum-excitation pressure over PSII (Q_m) (Fig. 1b, c, f). Specifically, the rate of reduction in photosynthetic performance [$\Delta(F_V/F_m)/\Delta t$ and $\Delta Q_m/\Delta t$] was significantly greater for *Symbiodinium* of low-$\mu'_{S,m}$ corals (clustered longitudinal analysis, $\mu'_{S,m}$-group × day interaction term $p = 0.016$ and 0.013, respectively: Fig. 1b, c; Table 2) and photosynthetic function diverged between low- and high-$\mu'_{S,m}$ corals at four and 2 days (for F_v/F_m and Q_m respectively) after stress initiation (marginal analysis, $p = 0.013$ and 0.012, respectively, Fig. 1b, c; Tables 3). Although non-photochemical quenching (Φ_{NPQ}) increased on average by 1.8-fold for low-$\mu'_{S,m}$ and 1.2-fold for high-$\mu'_{S,m}$ corals during bleaching, the dissipation of excess energy through non-photochemical mechanisms was not significantly different across high- and low-$\mu'_{S,m}$ corals (Additional file 2: Figure S2g).

Symbiodinium of low $\mu'_{S,m}$ corals experience increased rates of light absorption

We developed an empirical model of light absorption by *Symbiodinium in hospite* by considering symbiont light-absorption (I_a) as the sum of skeleton-independent absorption (I_{a1}) of downwelling light and skeleton-dependent absorption (I_{a2}) of reflected light (downwelling light not absorbed on the first pass and reflected by the skeleton back into coral tissue) [15–17]. The model relates I_{a1} and I_{a2} with parameters that were experimentally measured: skeletal reflectance, R_S, of the clean skeleton and holobiont reflectance, R_H, measured at different time points throughout the bleaching experiment.

The results of the model of *Symbiodinium* light absorption indicate that the estimated skeleton-dependent light absorbed per unit pigment (I_{a2}/ρ) and its rate ($\Delta(I_{a2}/\rho)/\Delta t$) were several fold higher in low-$\mu'_{S,m}$ corals (Fig. 2a–c, where average ρ for low- and high-$\mu'_{S,m}$ corals are concentrations of Chl *a* in µg/cm^2, Additional file 2: Figure S2). This pattern remained (Fig. 2c) when the effect of downwelling light was isolated (subtracting I_{a2}/ρ determined

Table 2 Hierarchical linear mixed models (LMM) analysis of photosynthetic performance

Metric of bleaching response	$\mu'_{S,m}$ Cluster	Rate (day^{-1})	p value, rate	CLA $\mu'_{S,m}$—day interaction term p value
F_v/F_m	Low-$\mu'_{S,m}$	−0.0319	<0.001	0.016
	High-$\mu'_{S,m}$	−0.0144	0.002	
Q_m	Low-$\mu'_{S,m}$	0.043	<0.001	0.013
	High-$\mu'_{S,m}$	0.011	0.19	

Results of clustered longitudinal analysis (CLA) of high- and low-$\mu'_{S,m}$ corals. Marginal analysis of F_v/F_m performed with values normalized to initial because the dynamic inversion of values (seen at day 4 in Fig. 1b; Additional file 3: Figure S3) makes marginal analysis insensitive to absolute differences over time

Table 3 Hierarchical linear mixed models (LMM) analysis of photosynthetic performance

Metric of bleaching response	Day after application of stress	Difference between high- and low-$\mu'_{S,m}$ groups	p value
F_v/F_m (normalized to initial values)	0	0.0034	0.92
	2	0.054	0.074
	4	0.10	0.013
	6	0.15	0.011
Q_m	0	−0.057	0.22
	2	−0.12	0.012
	4	−0.19	0.003
	6	−0.25	0.002

Results of marginal analysis of the photosynthetic performance (F_v/F_m and Q_m) of high- and low-$\mu'_{S,m}$ corals. Marginal analysis of F_v/F_m performed with values normalized to initial because the dynamic inversion of values (seen at day 4 in Fig. 1b; Additional file 3: Figure S3) makes marginal analysis insensitive to absolute differences over time

under CL from the HL treatment using Taylor expansion, Eq. 2 using I_{a2}/ρ as a metric instead of change in photochemical efficiency). As symbiont densities decrease, I_{a2}/ρ increases at a rate of $-\Delta(I_{a2}/\rho|_{HTHL} - I_{a2}/\rho|_{HTCL})/\Delta\rho$, which follows an inverse-power law function of $\mu'_{S,m}$ ($r^2 = 0.79$), consistent with previously published data on flat-coral models [18]. Parameters chosen are valid at high per-cell pigment concentration, and I_{a2}/ρ significantly underestimates actual values as ρ decreases. Because ρ is reduced in low-$\mu'_{S,m}$ corals during bleaching (Fig. 1d), our estimation of I_{a2} is conservative, and feedback effect is expected to be even more pronounced.

Light and temperature dependent bleaching effects

The light- or temperature-dependent bleaching effects were evaluated for one parameter in particular; the rate of reduction in photochemical efficiency of *Symbiodinium* with bleaching (ΔPE). In the case of light-dependent bleaching effect, ΔPE for corals exposed to CL were subtracted from those exposed to HL for either control (i.e., CT-HL−CT-CL) or high (i.e., HT-HL−HT-CL) temperature (Eq. 2). Thereby, the effect of light on bleaching was determined by calculating the increased light stress [ΔPE (HL−CL)] in the absence and presence of thermal stress. The rate of light-induced reduction in photosynthetic efficiency ΔPE is positively correlated with $\mu'_{S,m}$, approaching 0 (no loss of F_v/F_m with time) at the highest values of $\mu'_{S,m}$, under high ($r^2 = 0.62, p = 0.007$) or control ($r^2 = 0.35, p = 0.07$) temperature (Fig. 1g). Taking a similar approach to isolate the effect of temperature on the rate of reduction in photosynthetic efficiency, ΔPE of corals exposed to CT were subtracted from those exposed to HT for either control (i.e., HT-CL−CT-CL) or high (i.e., HT-HL−CT-HL) light (Eq. 3). Temperature-induced loss

of F_v/F_m over time, ΔPE, is not significantly correlated with $\mu'_{S,m}$ ($r^2 = 0.18$, $p = 0.23$, Additional file 3: Figure S3a). Although all corals experienced some reduction in F_v/F_m (during the 11 days of the experiment) under single stressor treatments (CT-HL and HT-CL), larger reductions were observed under combined heat and light stress with the greatest decline among low-$\mu'_{S,m}$ corals (Fig. 1b).

Factors that did not influence bleaching response

The diversity of corals and symbionts included in these experiments permitted examination of the effects of several factors that have been previously described as determinants of bleaching response (R_S, bulk-μ'_S, coral tissue thickness, colony morphology, *Symbiodinium* thermotolerance) and confounding factors of $\mu'_{S,m}$ (i.e., parameters that correlated with $\mu'_{S,m}$: a priori physiological differences observed among the targeted species during baseline pre-stress measurements, including *Symbiodinium* and Chl a densities, and photochemical efficiency). None of these factors were significantly correlated with the changes in photosynthetic performance observed in bleaching corals.

Corals examined included substantial diversity in R_S, bulk-μ'_S, coral tissue thickness, colony morphology, and *Symbiodinium* thermotolerances (Table 1). Skeletal reflectance was not significantly associated with changes in F_v/F_m or Q_m (Fig. 3b, c, f; Additional file 4: Figure S4; LMM, $p > 0.15$). Bulk-μ'_S (Table 1) was not significantly associated with the rate of reduction in photosynthetic efficiency ΔPE ($r^2 = 0.02$, $p > 0.5$). The experimental corals included thin (*S. hystrix*), medium (*P. damicornis*), and thick branching (*S. pistillata* and *M. digitata*) colony morphologies, as well as laminar (*M. foliosa* and *T. reniformis*) and massive (*Merulina* sp., *D. labyrinthiformis*, *Goniopora* sp., and *F. favus*) forms; however colony morphology was not significantly associated with light- ($r^2 = 0.001$, $p > 0.5$) nor temperature- ($r^2 = 0.02$, $p > 0.5$) dependent ΔPE. Coral tissue thickness varied between 0.1 and 2.8 mm (Table 1), but was not significantly associated with light- ($r^2 = 0.12$, $p > 0.5$) nor temperature- ($r^2 = 0.05$, $p > 0.5$) dependent ΔPE. Experimental corals hosted some of the highest (C8a, C15, D1 and D1a) or lowest (B1 and, assuming similar to C3, C3u and C3v) thermotolerance phylotypes known (Table 1). However *Symbiodinium* thermotolerance was not significantly associated with F_v/F_m or Q_m (LMM, $p > 0.05$), and the observed trends have greater losses of photosynthetic performance among high-thermotolerance physiotypes (Fig. 3d, e, g; Additional file 5: Figure S5).

Physiological differences between low- and high-$\mu'_{S,m}$ corals were detected in the absence of stress: low-$\mu'_{S,m}$ corals had higher baselines for F_v/F_m (Fig. 1b) and Chl a (Additional file 3: Figure S3c) and lower baselines for

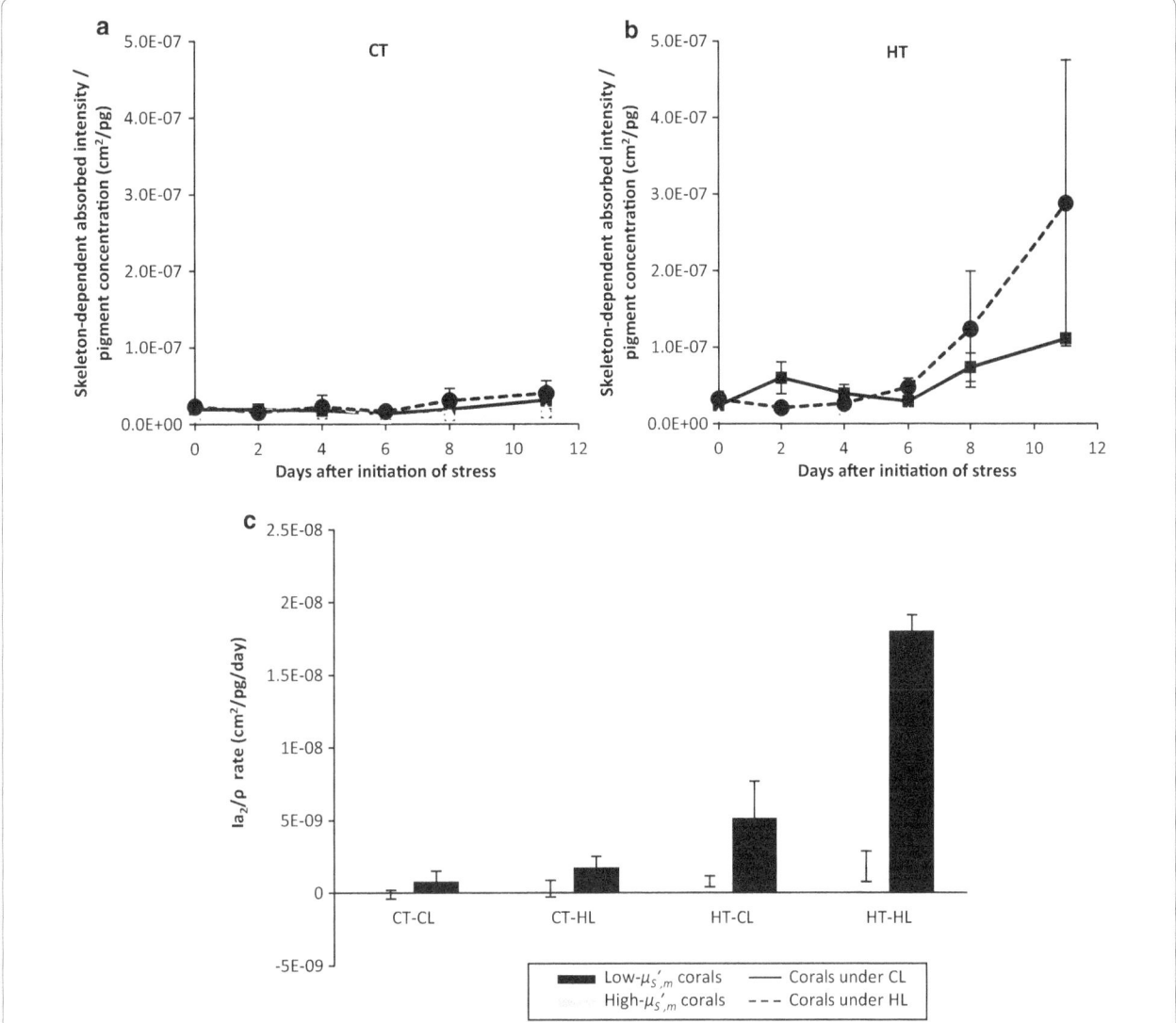

Fig. 2 Dynamics of modeled *Symbiodinium* light absorption *in hospite* due to skeletal backscattering ($\mu'_{S,m}$). *Symbiodinium in hospite* of high- (*gray line*) and low-$\mu'_{S,m}$ (*black line*) corals are (conservatively) predicted by an empirical model to have differential skeleton-dependent light absorption per unit pigment (I_{a2}/ρ). Under **a** CT, the absorption of light in high- and low-$\mu'_{S,m}$ corals is similar when exposed to CL (*solid line*) and HL (*broken line*). Under **b** HT, the absorption of light in low-$\mu'_{S,m}$ corals is several times larger under either light condition, but the increase under HL is dramatic. Additionally, the increase in (conservatively) estimated temporal rates of light absorbed per unit pigment ($\Delta(I_{a2}/\rho)/\Delta t$) in low-$\mu'_{S,m}$ corals (*black bars*) is progressively greater as heat and light stress were combined (**c**). All abbreviations follow Fig. 1 and error bars are standard error of the mean

Symbiodinium density, ρ, (Fig. 1d) during monitoring prior to experimental manipulation (t test, all $p < 0.001$). Only baseline-F_v/F_m had a significant correlation with F_v/F_m (LMM, $p = 0.01$), and also correlated with ΔPE under HT ($r^2 = 0.45$). However, this correlation was unstable and primarily caused by a single datapoint (*M. digitata*), without which r^2 dropped to 0.12. Baseline-F_v/F_m could not predict ΔPE under CT ($r^2 < 0.07$), and the difference between mean ΔPE of baseline-low- and baseline-high-F_v/F_m was not significant ($p > 0.25$ versus 0.007 for $\mu'_{S,m}$ as the explanatory variable).

Discussion

Results of the bleaching experiment and empirical light-absorption model are consistent with predictions of the optical feedback hypothesis. Bleaching corals with skeletal nanostructures that scatter light at relatively low $\mu'_{S,m}$ experienced increased rates and severities of bleaching response (ΔR_H, ρ, Chl a; Fig. 1d, e; Additional files 1 and 2: Figures S1, S2), light stress on retained *Symbiodinium* ($\Delta F_v/F_m$, Q_m; Fig. 1), and amounts and rates of skeleton-dependent light absorption by remaining *Symbiodinium* [(I_{a2}/ρ) and ($\Delta(I_{a2}/\rho)/\Delta t$); Fig. 2] relative to corals with

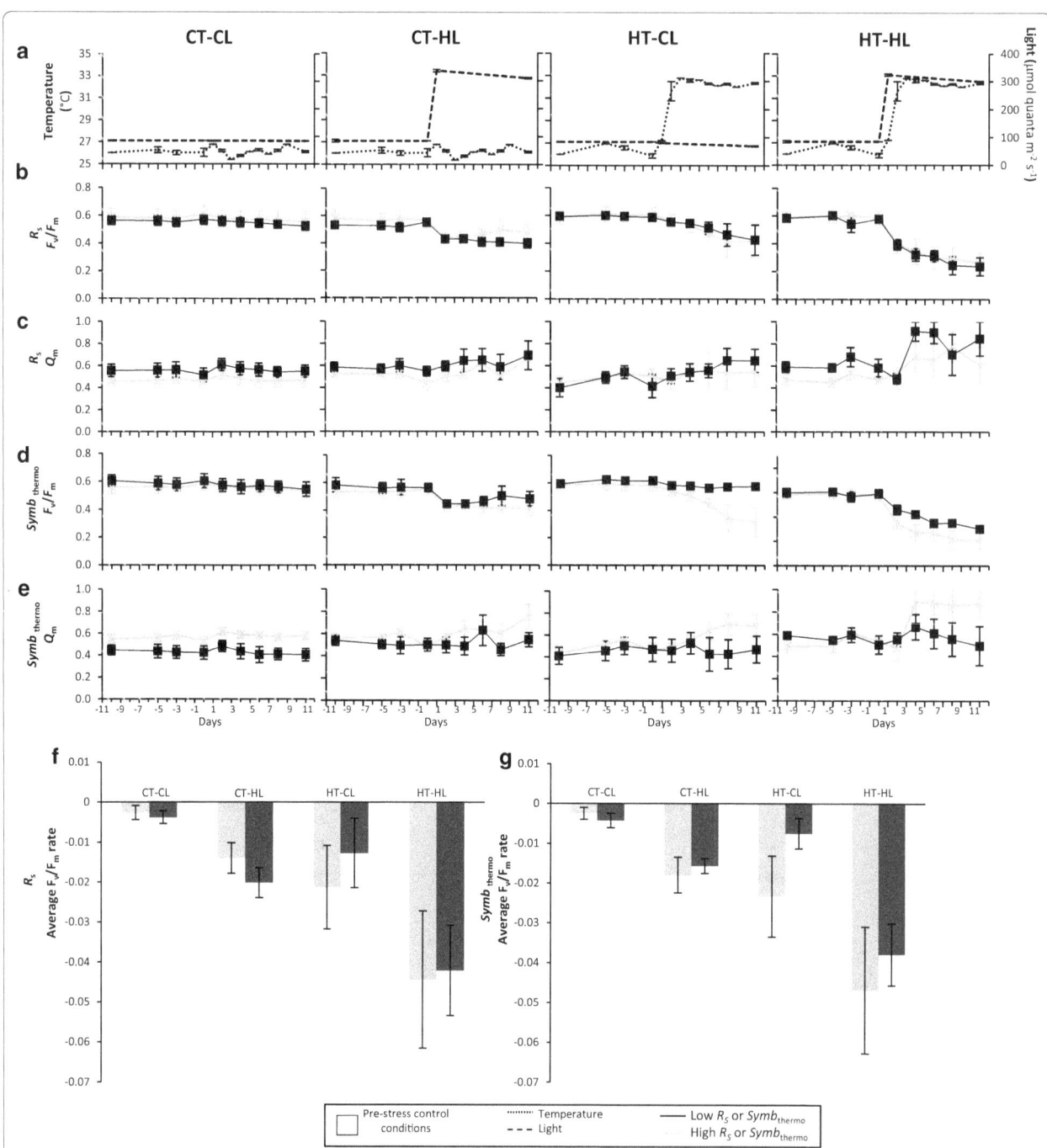

Fig. 3 Effects of skeletal reflectance (R_S) and *Symbiodinium* thermotolerance ($Symb_{thermo}$) on photosynthetic performance dynamics. High- and low-(means in *gray* and *black* respectively in **b–g**) R_S and $Symb_{thermo}$ corals responded similarly to experimental light (*broken line* in **a**) and temperature (*dotted line* in **a**) conditions (described in Fig. 1). Photosynthetic performance was similarly suppressed under increased stress in corals grouped by R_S (**b, c**) and was modestly (but non-significantly) more suppressed for corals hosting high-thermotolerance *Symbiodinium* (**d, e**). Both low- and high-R_S corals experienced a progressively greater average rate of photochemical efficiency loss (CLL $p = 0.64$; CHL $p = 0.28$; TLL $p = 0.55$ and THL $p = 0.91$) as heat and light stress were increased (**f**), and both low and high-$Symb_{thermo}$ corals experienced a progressively greater average rate of photochemical efficiency loss (CLL $p = 0.47$; CHL $p = 0.70$; TLL $p = 0.26$ and THL $p = 0.68$) as heat and light stress were increased (**g**). All *error bars* are *standard error* of the mean

skeletal nanostructures that scatter light at relatively high $\mu'_{S,m}$.

Low $\mu'_{S,m}$ corals experience increased rates and severities of bleaching and remaining *Symbiodinium* experience increased rates and severities of light stress

Although all corals experienced some response to increased temperature, differentially increased bleaching was detected among low-$\mu'_{S,m}$ corals as early as day 2 (under HL-HT) and no later than day 6 (under CL-HT) after initiation of stress (Fig. 1; Additional file 2: Figure S2). Similarly, differentially decreased photosynthetic performance of retained *Symbiodinium* was nearly simultaneous with bleaching (within the sampling periods of the experimental design) and was detected among low-$\mu'_{S,m}$ corals as early as day 2 (under HL-HT) and no later than day 6 (under CL-HT) of the experiment (Fig. 1; Additional file 2: Figure S2).

Change in photosynthetic performance was evaluated by measuring changes in F_v/F_m and Q_m of all corals before and after the application of stress. F_v/F_m indicates the proportion of potentially active PSII reaction centers under dark-adapted conditions [39] and significant decreases in F_v/F_m over time under light- and heat-stress have been measured in bleaching corals (e.g., [4, 24, 25]). Q_m [40, 41] is sensitive to effective quantum yield (Φ_{PSII}) oscillations as a result of the induction of multiple photoprotective pathways that compete for energy dissipation when light absorption exceeds photochemistry and indicates the proportion of active (or open) PSII reaction centers under peak irradiance [40]. Values approximating 0 indicate light-limitation with most reaction centers open, ≈ 1 indicate photoinhibition with most reaction centers closed, and photoacclimation is indicated when Q_m remains unchanged during suppressed photochemical efficiency [11, 12, 40]. Differential rates of divergence of photosynthetic performance (at day 4 for F_v/F_m and day 2 for Q_m; Fig. 1) indicate that *Symbiodinium* associated with high-$\mu'_{S,m}$ corals were experiencing photoacclimation (Q_m remains unchanged while F_v/F_m decreases modestly) while those associated with low-$\mu'_{S,m}$ corals were experiencing photoinhibition (Q_m approaches one, while F_v/F_m decreases significantly); consistent with observations of bleaching corals [11, 40].

All corals dissipated excess energy through Φ_{NPQ} at similar levels (increase of 1.2-fold to 1.8-fold after thermal stress, Additional file 2: Figure S2g). This finding may seem unexpected as Φ_{NPQ} is mainly affected by photoprotective pathways (downregulation of PSII antenna pigments and the xanthophyll cycle) [42], and given the increased light stress experienced by low-$\mu'_{S,m}$ corals, a greater increase in Φ_{NPQ} would be expected compared to high-$\mu'_{S,m}$ corals. However, while suppressed F_v/F_m and

elevated Q_m are often associated with severe bleaching response (e.g., [11, 12]), there is no consensus [43–46] that variation in NPQ is indicative of resistance [47, 48] or sensitivity [44] to thermal stress and photoinhibition.

Symbiodinium associated with low-$\mu'_{S,m}$ corals absorb light at higher rates and amounts

We developed an empirical model of light-absorption for *in hospite Symbiodinium* to test the assumption that the susceptibility of low-$\mu'_{S,m}$ corals is driven by a feedback-loop between absorber loss (decrease in ρ) and the rate of light amplification increase, which exposes remaining symbionts to rapidly increasing light. The rate of light amplification increase is modeled as: $-\frac{d(I_a/I_{a1})}{d\rho}$; where I_a is the fraction of incident light absorbed in tissue and I_{a1} is the fraction incident absorbed on its first pass through tissue. Change in the rate of light amplification increase is a consequence of a higher rate of light absorption per pigment due to skeletal reflectance, which is modeled as: $\frac{d(I_{a2}/\rho)}{d\rho}$; where $I_{a2} = I_a - I_{a1}$ is the fraction of incident light not absorbed on the first pass, scattered by the skeleton back into the tissue and subsequently absorbed.

The empirical model of light-absorption for *in hospite Symbiodinium* is a generalization of prior models [15, 17], however it differentiates between downwelling (skeleton-independent) and reflected (skeleton-dependent) light-absorption so that the effect of skeletal optical properties on light intensity experienced by symbionts is explicitly estimated and repeated passes of light between tissue and skeleton can be accounted for. The model expresses I_a, I_{a1} and I_{a2} through experimentally determined values for R_S, R_H, and three model parameters describing light transport properties of the holobiont (α, β, γ; see "Methods"). Downwelling light that is not absorbed during the first pass can be returned to tissues by the skeleton, lost to absorption, or diffusely scattered out of the colony [17, 19, 27, 28] and may repeatedly pass between skeleton and tissue (i.e., aided by skeletal morphology; [19, 30]). Thus, I_{a2} may be the result of multiple passes of light through tissue caused by multiple reflections of the skeleton [15, 17]. For a flat coral model (no multiple passes through tissue), and neglecting absorption of light reflected by the skeleton in tissue, our model (Eqs. 5, 6 and 7) converges to the approximate solution used to estimate the absorption of light based on holobiont and skeletal reflectance values [15, 29, 49].

The estimated *Symbiodinium* light absorption indicates that the effect of $\mu'_{S,m}$ on light absorption by *Symbiodinium*, I_a, is substantial. Skeleton-dependent light absorbed per unit pigment (I_{a2}/ρ) and its rate ($\Delta(I_{a2}/\rho)/\Delta t$) were several fold higher in low-$\mu'_{S,m}$ corals (Fig. 2a–c). This pattern was even more pronounced for combined light and temperature stress (Fig. 2b) and remained when the effect of downwelling light was isolated (Fig. 2c) (subtracting I_{a2}/ρ determined under CL from the HL treatment). Parameters (α, β, γ) chosen are valid at high per-cell pigment concentration and (I_{a2}/ρ) significantly underestimates actual values as ρ decreases. Because ρ is greatly reduced in low-$\mu'_{S,m}$ corals during bleaching compared to high-$\mu'_{S,m}$ (Fig. 1d), these calculations are expected to underestimate I_{a2}, and the feedback effect is expected to be even more pronounced.

$\mu'_{S,m}$ is a robust predictor of light-dependent bleaching, but not of temperature-dependent bleaching

By mathematically isolating the effect of light on bleaching from temperature and other confounding factors, including those unknown (light-dependent bleaching effect), we found that the rate of reduction in photochemical efficiency during bleaching (ΔPE) is associated with $\mu'_{S,m}$, indicating that $\mu'_{S,m}$ is one of the determinants of light-dependent bleaching severity. The rate of light-induced loss of F_v/F_m is much more pronounced in low-$\mu'_{S,m}$ corals; high-$\mu'_{S,m}$ corals are nearly invariable under HT or CT conditions (ΔPE approached 0; Fig. 1f, g). While $\mu'_{S,m}$ was a robust predictor of light-dependent bleaching as it explained 62 % of the variance in ΔPE for HT ($r^2 = 0.62$, $p = 0.007$, Fig. 1g), it was not a robust predictor of the temperature-dependent bleaching as $\mu'_{S,m}$ explained only 18 % of the variance in ΔPE for HL ($r^2 = 0.18$, $p = 0.23$, Additional file 3: Figure S3a).

Heat and light stress have a compounding effect on bleaching response; differential sensitivity to light is amplified by temperature (Fig. 1f, g) as excess light generated by skeletal scattering may overwhelm photosystems impaired by thermal stress. Heat reduces the ability of *Symbiodinium* to utilize light in photosynthesis [4, 23, 25, 50] and can uncouple energy absorption from photochemistry [23, 50]; resulting in excess energy independent of light increase. Therefore, *Symbiodinium* may perceive heat stress as an increase in excitation pressure over photosystem II [23, 50] and experience an increase in excess light as a result of an increase in temperature. In the absence of increased temperature stress, the effect of light-transport in the surface of the coral skeleton seems low, but once temperature increases and bleaching is initiated, the effect of light stress becomes remarkable, in particular for low-$\mu'_{S,m}$ corals (r^2 for $\Delta PE(\mu'_{S,m})$ is two times lower for CT than HT; 0.35 and 0.62, respectively, Fig. 1g). $\mu'_{S,m}$ explained 35 % of light-and temperature-dependent bleaching variance ΔPE for HL and HT ($r^2 = 0.35$, $p = 0.07$, Additional file 3: Figure S3b). The ecological relevance of high- and low-$\mu'_{S,m}$ remains to be fully understood, but current evidence points to very distinct ecological strategies. Skeleton deposited by corals is

made of calcium carbonate nanograins (about 50–200 nm diameter) (e.g., [35]) that govern the scattering properties of the skeleton and present a fractal micro-morphology (i.e., structures between 30 and 1000 nm that have a similar degree of compactness [18]) likely reflective of their growth strategy and skeletogenesis. Corals with higher rates of linear extension, rather than skeletal infilling (typical of branching species), often have the lowest $\mu'_{S,m}$ values and are typically thin branching, as opposed to corals with high-$\mu'_{S,m}$ which often have higher skeletal density and are massive or thick branching [18]. A prior study of light scattering and skeletal fractality in 150 coral skeletons representing 94 coral taxa demonstrated that high and low-$\mu'_{S,m}$ corals are important species in a variety of ecosystems. For example, *S. hystrix* and *S. pistillata*, two representatives of the Pocilloporidae family with low-$\mu'_{S,m}$, can be frequently found in Central, Eastern, and Western Indo-Pacific reefs, while *Porites lobata* and *Orbicella annularis* of the Poritidae and Merulinidae families with high-$\mu'_{S,m}$ are important species in Eastern Indo-Pacific and Caribbean reefs, respectively.

This study focused on the light scattering within skeleton and light absorbed by *Symbiodinium in hospite*, but did not evaluate light scattering within coral tissue which has been shown to significantly modulate light availability to symbionts. Light scattering causes lateral redistribution within tissue and increases light availability to symbionts [19, 28] while host fluorescent pigments [33, 51] or tissue contraction [19, 52] may reduce light stress by regulating light exposure and travel within tissue. Direct evidence for the optical feedback hypothesis would require in vivo measurements of *Symbiodinium* light-absorption rates as the coral undergoes bleaching and separation of skeleton-dependent effects, which has proven to be a technical challenge. However, combining the model of light absorbed by *Symbiodinium in hospite* developed in this study and light available to *Symbiodinium* within the coral tissue measured with light microsensors [19, 28, 32] will improve models of the optics of intact corals. In fact, integrating within-tissue light scattering with skeletal scattering will allow for a comprehensive evaluation of the mechanisms of light scattering by skeleton and tissue in modulating light to symbionts and their role in bleaching response.

Factors that did not influence bleaching response

Neither R_S (Fig. 3b, c, f; Additional file 4: Figure S4; LMM $p > 0.15$) nor μ'_S ($r^2 = 0.02$ for ΔPE $p > 0.5$) were significantly correlated with the severity of bleaching response. Light reflectance in coral skeletons is a complex process, and an important distinction must be made between $\mu'_{S,m}$, which governs short-path light transport in the superficial skeletal layer, and the reduced scattering coefficient of the entire skeletal material, μ'_S. Short-path transport is primarily driven by scattering of nanograins and fiber bundles of the top ~100 μm and is less influenced by larger structures such as overall morphology of corallites, optical properties of deeper skeletal material, or absorption. Although R_S includes the effect of short-path light-transport, it is primarily determined by μ'_S, absorption, and overall coral morphology (see Additional file 6: Text S1.1). In agreement with this, μ'_S assessed for the ten coral species in the present study was not a good predictor of bleaching response. This difference between $\mu'_{S,m}$ and μ'_S was also observed for 22 coral taxa [18]; modeling of the effect of $\mu'_{S,m}$ on bleaching showed that the rate of increase of light enhancement with decrease of absorbers (microspheres modeling symbiont pigments) is inversely dependent on $\mu'_{S,m}$. Although this model couldn't be applied to test the effect of μ'_S on bleaching in thin (1–2 mm) polished skeletal laminae, integrating sphere measurements of μ'_S for 22 coral taxa showed no correlation with their bleaching susceptibility, further supporting observations of the current study [18].

Skeletal R_S and $\mu'_{S,m}$ affect coral physiology through two opposing light-modulation pathways: $\mu'_{S,m}$ is inversely related to the *rate* of light amplification increase [18], R_S is directly related to total light amplification [15, 17, 30]. Both $\mu'_{S,m}$ and R_S have the potential to increase light availability to symbionts [15–18, 29] and exacerbate the bleaching response [15, 18]. While our results identified a connection between $\mu'_{S,m}$ and bleaching response, no correlation between R_S or μ'_S and F_v/F_m was detected. Parallel to the hypothesis that the threshold for bleaching is determined by temperature increase rate [53], the threshold for light-enhanced bleaching appears to be determined by light-increase rate (associated with $\mu'_{S,m}$) rather than the total light (associated with R_S).

Even though *Symbiodinium* thermotolerance (physiotype) has been shown to increase holobiont thermotolerance (1–2 °C [54]) in a pattern that dominates current theory explaining differential bleaching susceptibility [2, 11–14], it was not associated with bleaching response in these experiments. While three associations had similar tolerances and susceptibilities, the most thermotolerant symbionts (D1, D1a, and C8a) were hosted by the most bleaching susceptible corals (*P. damicornis, S. hystrix and S. pistillata* [5, 6, 18]), and the most thermosensitive symbionts (B1, C3v, and C3u) were hosted by the most bleaching resistant corals (*D. labyrinthiformis, Goniopora* sp., and *F. favus* [5, 6, 18]); providing an opportunity to detect effects of symbiont physiotypes. Similar to recent evidence that differential bleaching susceptibility cannot be explained by symbiont thermotolerance alone [10, 55, 56], no positive correlation between *Symbiodinium* thermotolerance and F_v/F_m or Q_m was detected (LMM,

p > 0.5 and 0.05, respectively, Fig. 3d, e, g; Additional file 5: Figure S5). While thermotolerance is demonstrable within a single life-stage of an individual species [11] or in isolation [43], it is generally context-dependent within the physiological and physical properties of the coral host [8, 12, 55, 57] and environment [12, 58].

We evaluated potential confounding factors of $\mu'_{S,m}$: in the absence of stress, low-$\mu'_{S,m}$ corals had lower *Symbiodinium* density, higher chlorophyll, and higher F_v/F_m (t test, all $p < 0.001$), but these factors were not found to significantly associate with differential bleaching severity among the ten studied coral species. While this study cannot rule out the existence of other unknown potential confounders that may correlate with $\mu'_{S,m}$ and better explain the differential bleaching severity among these species, we have proposed a mechanism that explains the association of $\mu'_{S,m}$ with differential bleaching severity.

Conclusions

Skeletal scattering was predictive of beaching susceptibility in these experiments and, if these results are representative of wider patterns, then they indicate that skeletal scattering is one of the key determinants of differential bleaching susceptibility. While symbionts associated with low-$\mu'_{S,m}$ corals may receive less total light from their skeletons, they are predicted to experience a higher rate of (skeletally-derived) light increase once bleaching is initiated and absorbing bodies are lost; further precipitating the bleaching response. While $\mu'_{S,m}$ explained 62 % of the light-dependent variance in bleaching response, it was a poor predictor of the temperature-dependent variance and it explained 35 % of the light- and temperature-dependent bleaching variance. Therefore, the remaining variance must be explained by other determinants of bleaching susceptibility. Symbiont phylotype can affect host physiology, holobiont fitness, and bleaching susceptibility [12, 54, 59]; higher symbiont densities per coral cell increase the risk of coral bleaching [55]; coral morphological and physiological properties modulate available light to the symbiont, determine early stress responses, and regulate symbiont photosynthetic demand for CO_2 [8, 30, 57]; within-tissue light scattering increases light availability to symbionts [19, 28] and may reduce the threshold for bleaching. The challenge now is to discern the contribution of the key determinants of bleaching susceptibility in order to identify the most effective management and remediation strategies to protect the remaining diversity of coral-*Symbiodinium* associations in a changing climate.

Methods

The predictions of the optical feedback hypothesis were experimentally assessed by monitoring the effects of differential $\mu'_{S,m}$ on the dynamics of bleaching response for a diverse set of 10 corals and modelling skeleton-dependent light absorption by *Symbiodinium* from experimentally measured values of coral reflectance (R_H during bleaching and R_S of bare skeletons). Low-$\mu'_{S,m}$ corals should experience increased rates and severities of bleaching-response as indicated by dynamically decreased density of *Symbiodinium* ($\Delta\rho$) and/or photosynthetic pigments per *Symbiodinium* cell (ΔChl a) and increased skeletal exposure (ΔR_H), increased rates and severities of light stress on the *Symbiodinium* which remain *in hospite* as indicated by photosynthetic performance ($\Delta F_v/F_m$ and ΔQ_m) and increased light absorption ($\Delta I_{a2}/\rho$). Because of the diversity of corals employed in this study, we assessed alternative factors (known and hypothesized) for their contribution to experimental bleaching responses, including physical properties of the host (skeletal reflectance and coral tissue thickness), and differences in *Symbiodinium* phylotype thermotolerance known from the historical record.

Coral host and *Symbiodinium* types

Colonies were prescreened for diversity of $\mu'_{S,m}$, R_S, and *Symbiodinium* thermotolerance (Table 1). Coral were selected from live collections of Shedd Aquarium, Chicago, IL, USA (*P. damicornis*, *S. hystrix*, *S. pistillata*, *T. reniformis*, *M. foliosa*, and *M. digitata* originating from the Indo-pacific; and *D. labyrinthiformis* originating from Key West, Florida, USA) or obtained through A&M Aquatics, Lansing, MI, USA (*Goniopora* sp., *F. favus*, and *Merulina* sp. originating from Jakarta, Indonesia or Fiji). All corals were property of Shedd Aquarium, who granted research approval through their institutional review board; none of the coral species are listed as endangered or threatened by the US Endangered Species Act. All colonies were acclimated under control conditions (26 °C and 83.1 ± 1 µmol quanta m^{-2} s^{-1} on a 10/14 h light/dark cycle) 2–4 weeks prior to fragmentation and recovered 3–5 weeks under the same conditions. Ramets were created by cutting parent colonies into 32 ~ 1.5 cm^2 explants with a wet tile-saw primed with artificial sea water (37 $^0/_{00}$ salinity) and mounted to natural stone tiles using aquarium epoxy or ethyl 2-cyanoacrylate. Mounted corals where evenly distributed among four sectors in two aquaria. The tissue thickness of eight of the ten colonies were measured directly (reported as the mean of ten measurements) from size-standardized digital photos (using ImageJ version 1.47; NIH) of live colonies when cut in cross section, while the tissue thickness of *D. labyrinthiformis* and *Goniopora* sp. were estimated from published measurements (Table 1).

Holobiont tissue was scraped from skeletons and nucleic acids were extracted using standard protocols

[60]. Identification markers [*Symbiodinium* nuclear internal transcribed spacer region 2 (ITS2) and chloroplast 23S ribosomal DNA (23S rDNA), and Scleractinia mitochondrial cytochrome oxidase I (COI), cytochrome b (CytB), and nuclear ITS] were selectively amplified by polymerase chain reaction (PCR) using standard reagents and the primers and annealing temperatures listed in Additional file 7: Table S1a and Additional file 6: Text S1.2. PCR products were separated by gel electrophoresis and directly sequenced using the amplification primers and identified by similarity (i.e., BLAST search) with GenBank accessions (Table 1). All DNA sequences created in this study are accessioned in GenBank as documentation of identity (Additional file 8: Table S1b). Morphological identification [61] was used for coral taxa novel to Genbank (Table 1). Thermotolerance of *Symbiodinium* phylotypes was designated following previous research (Table 1).

Microscopic reduced light-scattering coefficient, $\mu'_{S,m}$

Microscopic-skeletal scattering ($\mu'_{S,m}$) was measured using low-coherence enhanced backscattering spectroscopy (LEBS) on corals cleaned with pressurized artificial seawater, soaked for <12 h in 3 % sodium hypochlorite, rinsed, and dried. We focused on short-propagating photons from the upper ~100–200 microns of skeletons to reduce the effects of 'bulk-scattering' properties [18]. The LEBS instrument has been previously described [62–64], and its application to coral ecology demonstrated [18]; but briefly, this method uses constructive interference of photons observed as an angular intensity cone centered in the backscattering direction to measure microscopic-scattering through broadband partial spatial coherence illumination. The LEBS instrument uses linearly polarized collimated broadband illumination directed at the surface of a coral skeleton at 15° angle of incidence, and light backscattered by the coral is collected using a lens, a polarizer, and an imaging spectrograph coupled with a CCD camera. The camera records a matrix of light-scattering intensities, $I_{LEBS}(\theta, \lambda)$, as a function of wavelength λ (450–700 nm) and backscattering angle θ (−5 to 5 degrees). The spatial coherence length of illumination, *Lsc*, was fixed at ~57 microns at 600 nm illumination. The reduced scattering coefficient of μ'_S was measured on cleaned coral skeletons using the enhanced backscattering spectroscopy (EBS) method as previously described [64–66].

Skeletal and holobiont reflectance (R_s and R_H)

Holobiont reflectance, R_H, is used to quantify bleaching: as *Symbiodinium* cell and photopigment density decrease, the skeleton becomes increasingly visible through host tissues and R_H increases [15, 16, 27, 29]. To prepare corals for R_S measurements, tissue was removed from ramets with their skeletons remaining attached to their tiles so that they could be returned to the same location and orientation as they were during the collection of R_H measurements. Preservation of the experimental conditions during measurement of R_S insured that the intensity and direction of downwelling incident light was maintained and that R_S would be comparable to R_H. Tissue was removed (by pressurized water), and preserved for *Symbiodinium* and pigment density analysis, and cleaned (as above) prior to measurement of R_S.

Reflectance, R_H and R_S, were measured as spectral reflectance using an optical fiber (Thorlabs SFS200/220Y) attached to a spectrometer (Ocean Optics USB4000). This method uses the Lambertian nature of the diffusely reflected light to enable hand-held measurement. Radiant flux is independent of angle and distance for a flat Lambertian scattering surface, however coral surfaces are irregular and small signal variations occur in different fiber positions. To account for this variation, ten measurements were collected randomly across the geometry of the ramet for each time point and specimen. The fiber was held at a distance of 1–2 cm from the upper surfaces of the ramet, near normal to the illumination source, while simultaneously avoiding shading the interrogation spot. The aperture of the fiber and refractive index of the water determine the acceptance angle of light, therefore this method interrogates a 3–6 mm diameter spot which will include signal from polyp and coenosarc. Measurements were normalized to a white reflectance standard (PTFE, Ocean Optics) adjacent to each ramet. The raw spectral reflectance for R_S and R_H was not further processed (e.g., by applying low-pass filters that smooth signal averages of high frequencies, making the spectra appear less variable), as the signal to noise ratio is sufficiently high to distinguish changes in R_H during bleaching (Additional file 6: Text S1.1).

Experimental design
The two experimental aquaria are 420 L (~25 cm depth) recirculating unidirectional (2.5–4 cm/s) baffled flumes, with the corals at ~15 cm depth. The illuminating arrays (high color temperature that approximates sunlight) are divided by suspended shades to allow independent control of light conditions in each half of each aquarium. Explants were assigned to light sectors (8 ramets of each coral species) and randomly distributed within a sector to acclimate. See Additional file 9: Figure S6, and Additional file 6: Text S1.3 for details.

Stress was induced in three treatments (control remained static) by increasing the temperature to 32.3 ± 0.5 °C (over 2 days) in one aquarium and light levels to 328.1 ± 4.3 µmol photons m^2/s in half of both aquaria (dynamic photoinhibition has been observed at

200–400 µmol quanta m^{-2} s^{-1} [67] and a trial experiment showed chronic photoinhibition of these corals at >400 umol quanta m^{-2} s^{-1} with no increase in temperature). This established four conditions: (1) control temperature and control light (CT-CL: 26.2 ± 1 °C at 83 ± 1 µmol photons m^2/s), (2) control temperature and high light (CT-HL: 26.2 ± 1 °C at 328 ± 4.3 µmol photons m^2/s), (3) high temperature and control light (HT-CL: 32.3 ± 0.5 °C at 83 ± 1 µmol photons m^2/s), and (4) high temperature and high light (HT-HL: 32.3 ± 0.5 °C at 328 ± 4.3 µmol photons m^2/s) (Additional file 9: Figure S6). Ramets were assessed every second day for 10 days prior to stress induction and 11 days thereafter (Additional file 6: Text S1.3). Any ramets with necrotic tissue (1.3 % of replicates) were removed from the experiment. Bleaching response was evaluated by the dynamics of *Symbiodinium* and photopigment density, holobiont reflectance, and *Symbiodinium* photosynthetic performance.

Symbiodinium photophysiology
Symbiodinium photosynthetic performance was assessed through pulse-amplitude modulation (PAM) chlorophyll fluorometry with a 1.5 mm diameter optical fiber and the following instrument settings: measuring intensity 6, saturation intensity 12, saturation width 0.6, and actinic light intensity 9. Induction curves were collected with the F_0'-mode (far-red light) activated and a delay of 40 s, a width of 20 s, and a length of 13 cycles (Additional file 6: Text S1.4). Dark-adapted yield of photosystem II (PSII) was measured (where $F_v/F_m = F_m - F_0/F_m$) at 07:20–08:00 h (prior to sunrise) and induction curve analyses were performed at 09:00–13:00 h (at peak irradiance). Data for induction curves were collected through the steady state of F' and F_m' and effective quantum yield ($\Phi_{PSII} = F_m' - F'/F_m'$), non-photochemical quenching ($\Phi_{NPQ} = F'/F_m' - F'/F_m$) and non-regulated heat dissipation ($\Phi_{NO} = F'/F_m$) were calculated from steady state measurements where $\Phi_{PSII} + \Phi_{NPQ} + \Phi_{NO} = 1$ [42]. Photochemical efficiency, F_v/F_m, was used as a metric of bleaching response and has repeatedly been shown to decrease during bleaching [25, 43]. *Symbiodinium* exhibit Φ_{PSII} oscillations when light absorption exceeds photochemistry [40], which is measured here as maximum excitation pressure over photosystem II, or $Q_m = 1 - [(\Phi_{PSII\ at\ peak\ light})/(F_v/F_m\ at\ dawn)]$ [40, 41] (Additional file 6: Text S1.5).

Symbiodinium and photosynthetic pigment density
Symbiodinium cells were collected using pressurized seawater and the resulting slurry was concentrated by centrifugation before being divided into aliquots for hemocytometer cell counts (Additional file 6: Text S1.6) and high-performance liquid chromatography (HPLC)

analysis of photosynthetic pigment identities and concentrations (Additional file 6: Text S1.7) using established procedures and gradients [68]. Surface area estimation of skeletons (for normalizing cell counts) were estimation using the single-dip wax method [69].

Statistical analysis

General linear model ANOVAs were performed in Minitab to test the effect of μ'_{Sm} on change in *Symbiodinium* cell and photosynthetic pigments density, ΔR_H, $\Delta F_v/F_m$, or ΔQ_m. Hierarchical linear mixed models (LMM) were applied in Stata 11.2 to account for the repeated-measures design [70] to assess the overall effect of treatment (time, light, and temperature) on bleaching response in the 11-day experiment (Additional file 6: Text S1.8). These analyses focused on the effect of potential explanatory variables (μ'_{Sm}, R_S, and *Symbiodinum* thermotolerance) on photophysiological response (F_v/F_m and Q_m).

Determining light-dependent and temperature-dependent bleaching effects

To determine the effect of light and temperature on bleaching separately, we used Taylor Series Expansion to mathematically isolate factors of interest (effect of light or temperature on temporal rates of F_v/F_m decrease) and cancel out known and unknown confounders across conditions because the physical conditions of the live animal experiment cannot be made precisely identical across all ramets. For example, potential confounders such as differential tissue thickness and localized morphology-induced flow diversity among explants of the same colony could alter mass transfer across the diffuse boundary layer and affect bleaching response [16, 71–73]; these factors cannot be fully controlled among such a large number of ramets. However, they can be mathematically cancelled out from all conditions by subtracting the difference between temporal rates of F_v/F_m decrease (*PE*) under control and stress conditions for each environmental factor (light or temperature).

To determine the light-dependent bleaching effect, we examined the difference between *PE* under control and high light conditions. For a given ramet i, the temporal rate of F_v/F_m decrease, $PE_i = \frac{\Delta(F_V/F_m)_i}{\Delta t}$, where t is time after the initiation of bleaching, was expressed as the first order Taylor expansion over temperature, light intensity, and potential confounding (including unknown) factors:

$$PE_i(T, I) = X_i + \Delta T \left.\frac{\partial PE_i}{\partial T}\right|_{T_1, I_1} + \Delta I \left.\frac{\partial PE_i}{\partial I}\right|_{T_1, I_1}, \quad (1)$$

where ΔT is the difference between experimental temperature T and control T_1, ΔI is the difference between experimental light intensity I and control I_1, and X_i

accounts for all other conditions (e.g., localized flow rates, without assuming that they are identical across ramets). To mathematically isolate the effect of light from temperature and confounding factors, *PE* values for corals exposed to CL ($I = I_1 \equiv I_{CL}$) were subtracted from corals exposed to HL ($I = I_2 \equiv I_{HL}$) for either control ($T = T_1$) or high ($T = T_2$) temperature:

$$\Delta PE_i(T_j) = PE_i(T_j, I_{HL}) - PE_i(T_j, I_{CL})$$
$$= (I_{HL} - I_{CL}) \left.\frac{\partial PE_i}{\partial I}\right|_{T_j, I_{CL}} \propto \left.\frac{\partial^2 (F_V/F_m)_i}{\partial t \partial I}\right|_{T_j},$$
$$(2)$$

where index j indicates either high ($j = 2$) or control ($j = 1$) temperature environment. In the first order approximation, this differential quantity ΔPE_i is independent of factors not directly related to illumination.

Similarly, to mathematically isolate the effect of temperature from light and confounding factors (temperature-dependent bleaching effect), *PE* values for corals exposed to CT were subtracted from corals exposed to HT:

$$\Delta PE_i(I_j) = PE_i(T_{HT}, I_j) - PE_i(T_{CT}, I_j)$$
$$= (T_{HT} - T_{CT}) \left.\frac{\partial PE_i}{\partial T}\right|_{T_{CT}, I_j} \propto \left.\frac{\partial^2 (F_V/F_m)_i}{\partial t \partial T}\right|_{I_j},$$
$$(3)$$

where index j indicates either HL ($j = 2$) or CL ($j = 1$) environment.

ΔPE was analyzed as a function of potential explanatory variables (potential determinants of bleaching response; μ'_{Sm}, R_S, tissue thickness, and *Symbiodinium* thermotolerance) and confounders of μ'_{Sm} (initial F_v/F_m, initial *Symbiodinium* and chl *a* density), thereby removing differences in bleaching response that are not explicitly related to light.

Skeleton-dependent light absorption model

We developed a novel model of *Symbiodinium* light absorption, which, in comparison to existing models, accounts for skeleton-driven absorption and multiple reentry effects. Incident light absorption by *Symbiodinium* (fraction I_a) can be viewed as the result of absorption of downwelling light (fraction I_{a1} of the incident light) and skeleton-dependent absorption (fraction $I_{a2} = I_a - I_{a1}$) of light reflected by the skeleton [15–17]. Light that is not absorbed in the first pass ($1 - I_{a1}$) can be reflected by the skeleton back into the tissue, lost to skeletal absorption, or diffusely scattered out of the colony [17, 19, 27, 28]. This process may involve multiple passes of light through tissue due to multiple reentries of unabsorbed light back into the skeleton and subsequent reflections by the skeleton [15,

17]. Because direct quantification of light absorption by pigments in live corals is not currently possible, we developed an empirical model relating I_{a1} and I_{a2} with experimentally measurable parameters R_S and R_H.

Starting with balance equations for R_H and I_a, we solve for I_{a1} and I_{a2} [see Additional file 6: Text S1.9 for detailed derivation using equations (4) through (7)]:

$$R_H = R_1(1 - I_{a1})(1 - a_2), \tag{4}$$

$$I_{a1} = \frac{1}{2\alpha}\left(1 + \alpha - \sqrt{(1+\alpha)^2 - 4\alpha(1 - \beta R')}\right), \tag{5}$$

$$I_a = I_{a1} + (1 - I_{a1})R_1 a_2 + (1 - I_{a1})\gamma(R_S - R_1), \tag{6}$$

$$I_{a2} = I_a - I_{a1} = (1 - I_{a1})\left(\frac{\alpha}{\beta}I_{a1} + \gamma\frac{\beta-1}{\beta}\right)R_S, \tag{7}$$

where $R' = R_H/R_S$, $\beta = R_S/R_1$, $\alpha = a_2/I_{a1}$ with R_1 the fraction of unabsorbed light that is leaving the holobiont after being reflected by the skeleton back into tissue including all reentries and a_2 the fraction of this reflected light that is absorbed by the pigments in the tissue, and γ is the fraction of light that is absorbed by tissue through processes other than I_{a1} or a_2 divided by $(R_S - R_1)$.

Coefficients α, β, and γ depend on coral morphology, its optical properties, and the concentration of absorbing pigments in tissue (see Additional file 6: Text S1.9 for detailed explanation). Coefficient $\alpha(>1)$ describes the amplification of light absorption due to elongation of light paths through the tissue caused by diffuse skeletal reflection of unabsorbed downwelling light, which is why α increases as the concentration of absorbing pigments decreases. Coefficients β and γ are related to the non-flatness of the skeleton and account for the reentry effect. In the special case of no reentry (flat coral model), $\beta = \gamma = 1$ and $1 < \alpha \leq 2$. Non-flat skeletons can create $\alpha > 2$ due to multiple reentry [15] and $\beta > 1$ and $\gamma < 1$ for non-flat geometries. If reentry is neglected, Eq. 7 for I_a converges to the solution that has been conventionally used to estimate the light absorption based on holobiont and skeletal reflectances [15, 29, 49], $I_a \approx 1 - R'$, if one of the following two conditions is satisfied: I_{a2} can be neglected (most of the absorption is due to the downwelling light) or $R_S = 1$. Even though α, β, and γ depend on concentration and the optical properties of the skeletons, the model can still be used to estimate the range of I_{a1} and I_{a2}. Indeed, I_{a2} increases with α (e.g., as symbionts leave). Thus, we can obtain the lower bound on I_{a2} by using Eqs. (5) and (7) with $\alpha = \beta = \gamma = 1$.

Additional files

Additional file 1: Figure S1. Dynamics of holobiont reflectance (R_H). Panels a–f are aligned into columns defined by light (broken line in a) and temperature (dotted line in a) conditions (described in Figure S1). Response of an exemplar low-$\mu'_{S,m}$ coral (S. pistillata) through (b) time series photos of explants, (c) spectral R_H, and (f) means (black line) and standard errors of the 10 random measurements collected to estimate R_H normalized to its skeleton reflectance at 675 nm. Response of an exemplar high-$\mu'_{S,m}$ coral (M. digitata) through (d) time series photos of coral explants, (e) spectral R_H and (f) means (gray line) and standard errors of the 10 random measurements collected to estimate R_H normalized to its skeleton reflectance at 675 nm. Spectral skeletal reflectance (R_S) in panels c and e shown to contextualize R_H with the limit of R_S values in the visible spectrum where photopigments have substantial absorption (e.g., 675 nm, chlorophyll a absorption peak); for wavelengths > 700 nm, the limit of R_H may be greater than R_S. As corals bleached and less than 10 % of symbionts remained associated with the host, R_H approached the values of R_S.

Additional file 2: Figure S2. Dynamics of bleaching response variables for corals grouped by $\mu'_{S,m}$. Panels (a–h) aligned into columns defined by experimental conditions (CT-CL: control temperature [26 °C] and light [83 µmol quanta m^{-2} s^{-1}], CT-HL: control temperature and high light [328 µmol quanta m^{-2} s^{-1}], HT-CL: high temperature [32 °C] and control light, and HT-HL: high temperature and high light; shaded areas are control). Responses of high- (gray line) and low-$\mu'_{S,m}$ (black line) corals for (a) holobiont reflectance (dashed lines are the corresponding post-experiment skeletal reflectance), (b) Symbiodinium cell density, (c) chlorophyll a density per Symbiodinium cell, (d) maximal photosynthetic efficiency, (e) effective quantum yield of photosystem II, (f) excitation pressure over photosystem II, (g) non-photochemical quenching, and (h) non-regulated heat dissipation. All error bars are standard error.

Additional file 3: Figure S3. $\mu'_{S,m}$ and temperature- and light-induced bleaching response. $\mu'_{S,m}$-specific temporal rate of F_v/F_m change ($\Delta PE \sim \Delta^2(F_v/F_M)/(\Delta t \Delta I)$) after stress-initiation is expressed as (a) the difference between CT and HT conditions (Eq. 3) for corals exposed to HL (filled circles; $p = 0.22$) and CL (open circles; $p = 0.44$), isolating the effect of temperature on bleaching response, and (b) ΔPE for HL and HT conditions ($p = 0.07$), where both temperature- and light-dependent bleaching response is evaluated. Although $\mu'_{S,m}$ predicts light-dependent bleaching ($r^2 = 62.3$ and $p = 0.007$, Fig. 1g), it is a weak predictor of temperature-dependent bleaching and light- and temperature-dependent bleaching.

Additional file 4: Figure S4. Dynamics of bleaching response variables for corals grouped by skeletal reflectance (R_S). Panels (a–h) aligned into columns defined by experimental conditions (described in Figure S3). Responses of high- (gray line) and low-R_S (black line) corals for (a) holobiont reflectance (dashed lines are the corresponding post-experiment skeletal reflectance), (b) Symbiodinium cell density, (c) chlorophyll a density per Symbiodinium cell, (d) maximal photosynthetic efficiency, (e) effective quantum yield of photosystem II, (f) excitation pressure over photosystem II, (g) non-photochemical quenching, and (h) non-regulated heat dissipation. All error bars are standard error.

Additional file 5: Figure S5. Dynamics of bleaching response variables for corals grouped by Symbiodinium thermotolerance ($Symb_{thermo}$). Panels (a–h) aligned into columns defined by experimental conditions (described in Figure S3). Responses of high- (gray line) and low-$Symb_{thermo}$ (black line) corals for (a) holobiont reflectance (dashed lines are the corresponding post-experiment skeletal reflectance), (b) Symbiodinium cell density, (c) chlorophyll a density per Symbiodinium cell, (d) maximal photosynthetic efficiency, (e) effective quantum yield of photosystem II, (f) excitation pressure over photosystem II, (g) non-photochemical quenching, and (h) non-regulated heat dissipation. All error bars are standard error.

Additional file 6. Supporting Text. Supporting methods (1) and supporting references (2).

Additional file 7: Table S1a. Nucleotide sequencing. Primers [79–82] and annealing temperatures used for polymerase chain reaction amplification and nucleotide sequencing.

Additional file 8: Table S1b. Nucleotide sequencing. GenBank accession numbers of genes sequenced in this study.

Additional file 9: Figure S6. Experimental setup. High temperature aquarium encompassing the HT-CL and HT-HL conditions on day 11 of the bleaching experiment (a); black divider separating light arrays, flow baffles, and mounted corals can be seen. Collecting PAM measurements in the control temperature aquarium within the CT-CL condition (CT-HL on the opposite side of black divider) on day 11 of the bleaching experiment (b); positioning of the PAM instrument probes above the coral explants mounted on stone tiles can be seen. Close up of probe holder (custom-machined acrylic block that ensures probes are returned to each explant in the same three-dimensional geometry as previous measurements) supported at a 23° angle by square PVC post (gray) attached to the coral-mounting tile (c); probes are (left to right) temperature, O_2 (data not reported), and PAM fiber optic immobilized in a black PVC tube. Control temperature aquarium encompassing the CT-CL and CT-HL conditions (HT aquarium in background) during acclimation period prior to prescreening and fragmentation (d); photograph taken before installation of the flow baffles and black divider separating light arrays. Hand-held optical fiber attached to a spectrometer to measure R_H (and R_S of cleaned skeleton) with white reflectance standard visible in the background (e). Consent to publish these images has been documented.

Abbreviations

Chl *a*: chlorophyll *a* density; COV: coefficient of variation; CT-CL: control temperature-control light; CT-HL: control temperature-high light; EBS: enhanced backscattering spectroscopy; F': fluorescence yield in actinic light; F_m: maximum fluorescence yield; F_m': maximum fluorescence yield in actinic light; F_v: maximum variable fluorescence yield; F_v/F_m: photochemical efficiency; F_0: minimum fluorescence yield; F_0': minimum fluorescence yield in light-acclimated state; HT-CL: high temperature-control light; HT-HL: high temperature-high light; *I*: light intensity; I_{a}: *Symbiodinium* light-absorption; I_{a1}: skeleton-independent light-absorption; I_{a2}: skeleton-dependent light-absorption; LEBS: low-coherence enhanced backscattering spectroscopy; LMM: linear mixed models; PAM: pulse-amplitude modulation chlorophyll fluorometry; PE: temporal rate of F_v/F_m (photochemical efficiency) reduction; PSII: photosystem II; Q_m: maximum-excitation pressure over photosystem II; R_H: holobiont reflectance; R_S: diffuse reflectance of light from coral skeletons; *T*: temperature; α: amplification of light absorption due to elongation of light paths through the tissue caused by diffuse skeletal reflection of unabsorbed downwelling light; β: accounts for the reentry effect; γ: is the fraction of light that is absorbed by tissue through processes other than I_{a1} or a_2 divided by $(R_S − R_1)$; $\mu_{s,m}'$: microscopic reduced-scattering coefficient or the inverse of the distance a short-path length photon travels before randomization; μ_s' or bulk-μ_s': reduced-scattering coefficient or the inverse of the distance a photon travels before randomization; ρ: *Symbiodinium* cell density; Φ_{PSII}: effective quantum yield of photosystem II; Φ_{NO}: non-regulated heat dissipation; Φ_{NPQ}: non-photochemical quenching.

Authors' contributions

TDS ED JDR VB LAM conceived and designed the experiments. VPS AJR HWG BJK JDR VB measured and analyzed optical properties of corals. TDS ED JH JT LAM performed the experiments. TDS JT performed PAM. HWG EMV performed cell counts and surface area estimates. ED AAG performed HPLC. TDS AG BCB MEW JD VB LAM analyzed the data. MWW KS SL MS GP VB LAM contributed expertise/reagents/materials/animal care/analysis tools. TDS MEW VB LAM prepared the manuscript. All authors read and approved the final manuscript.

Author details

[1] Department of Civil and Environmental Engineering, Northwestern University, 2145 Sheridan Road, Evanston, IL 60208, USA. [2] Department of Zoology, Field Museum of Natural History, 1400 South Lake Shore Drive, Chicago, IL 60605, USA. [3] Department of Biomedical Engineering, Northwestern University, 2145 Sheridan Road, Evanston, IL 60208, USA. [4] Keck Biophysics Facility, Northwestern University, 633 Clark Street, Evanston, IL 60208, USA. [5] Fishes Department, John G. Shedd Aquarium, 1200 South Lake Shore Drive, Chicago, IL 60605, USA. [6] Division of Water Resource Management, Florida Department of Environmental Protection, 2600 Blair Stone Road, Tallahassee 32399, USA.

Acknowledgements

We thank I. Berzins, C. Knapp, M. Moskalick, and L. Nesslar of Shedd Aquarium for helpful discussion, specimens, and engineering; F. Hussain, P. Humecki, J. Juranek, R. Pelkar, and A. Vadlamanu for assisting with experiments; K. Feldheim, P. Sierwald, and S. Ware of the Field Museum for expertise in the molecular and microscopy labs. This research was supported by National Science Foundation (EFRI-1240416 and CBET-1249311), National Institutes of Health (EB 003682), and MacArthur Foundation (to Encyclopedia of Life), and is dedicated to the memory of Corey M. Janczak.

Competing interests

The authors declare that they have no competing interests.

References

1. Wilkinson C. Status of the coral reefs of the world: 2008. Townsville: Global Coral Reef Monitoring Network and Reef and Rainforest Research Centre; 2008.
2. Baker AC, Glynn PW, Riegl B. Climate change and coral reef bleaching: an ecological assessment of long-term impacts, recovery trends and future outlook. Estuar Coast Shelf S. 2008;80:435–71.
3. Hoegh-Guldberg O, Mumby PJ, Hooten AJ, Steneck RS, Greenfield P, Gomez E, et al. Coral reefs under rapid climate change and ocean acidification. Science. 2007;318:1737–42.
4. Lesser MP, Farrell JH. Exposure to solar radiation increases damage to both host tissues and algal symbionts of corals during thermal stress. Coral Reefs. 2004;23:367–77.
5. Marshall PA, Baird AH. Bleaching of corals on the Great Barrier Reef: differential susceptibilities among taxa. Coral Reefs. 2000;19:155–63.
6. Loya Y, Sakai K, Yamazato K, Nakano Y, Sambali H, van Woesik R. Coral bleaching: the winners and the losers. Ecol Lett. 2001;4:122–31.
7. van Woesik R, Sakai K, Ganase A, Loya Y. Revisiting the winners and the losers a decade after coral bleaching. Mar Ecol Prog Ser. 2011;434:67–76.
8. Wooldridge SA. Differential thermal bleaching susceptibilities amongst coral taxa: re-posing the role of the host. Coral Reefs. 2014;33:15–27.
9. Middlebrook R, Hoegh-Guldberg O, Leggat W. The effect of thermal history on the susceptibility of reef-building corals to thermal stress. J Exp Biol. 2008;211:1050–6.
10. Bellantuono AJ, Hoegh-Guldberg O, Rodriguez-Lanetty M. Resistance to thermal stress in corals without changes in symbiont composition. P Roy Soc B Biol Sci. 2012;279:1100–7.
11. Abrego D, Ulstrup KE, Willis BL, van Oppen MJH. Species-specific interactions between algal endosymbionts and coral hosts define their bleaching response to heat and light stress. P Roy Soc B Biol Sci. 2008;275:2273–82.
12. Mieog JC, Olsen JL, Berkelmans R, Bleuler-Martinez SA, Willis BL, van Oppen MJH. The roles and interactions of symbiont, host and environment in defining coral fitness. PLoS One. 2009;4:e6364.
13. Jones AM, Berkelmans R, van Oppen MJH, Mieog JC, Sinclair W. A community change in the algal endosymbionts of a scleractinian coral following a natural bleaching event: field evidence of acclimatization. P Roy Soc B Biol Sci. 2008;275:1359–65.
14. Silverstein RN, Cunning R, Baker AC. Change in algal symbiont communities after bleaching, not prior heat exposure, increases heat tolerance of reef corals. Glob Change Biol. 2015;21:236–49.

15. Enríquez S, Méndez ER, Iglesias-Prieto R. Multiple scattering on coral skeletons enhances light absorption by symbiotic algae. Limnol Oceanogr. 2005;50:1025–32.

16. Kühl M, Cohen Y, Dalsgaard T, Jorgensen BB, Revsbech NP. Microenvironment and photosynthesis of zooxanthellae in scleractinian corals studied with microsensors for O_2, pH and Light. Mar Ecol Prog Ser. 1995;117:159–72.

17. Terán E, Méndez ER, Enríquez S, Iglesias-Prieto R. Multiple light scattering and absorption in reef-building corals. Appl Optics. 2010;49:5032–42.

18. Marcelino LA, Westneat MW, Stoyneva V, Henss J, Rogers JD, Radosevich A, et al. Modulation of light-enhancement to symbiotic algae by light-scattering in corals and evolutionary trends in bleaching. PLoS One. 2013;8:e61492.

19. Wangpraseurt D, Larkum AWD, Franklin J, Szabo M, Ralph PJ, Kühl M. Lateral light transfer ensures efficient resource distribution in symbiont-bearing corals. J Exp Biol. 2014;217:489–98.

20. Ban SS, Graham NAJ, Connolly SR. Evidence for multiple stressor interactions and effects on coral reefs. Glob Change Biol. 2014;20:681–97.

21. Jones RJ, Hoegh-Guldberg O. Diurnal changes in the photochemical efficiency of the symbiotic dinoflagellates (Dinophyceae) of corals: photoprotection, photoinactivation and the relationship to coral bleaching. Plant Cell Environ. 2001;24:89–99.

22. Warner ME, Fitt WK, Schmidt GW. Damage to photosystem II in symbiotic dinoflagellates: a determinant of coral bleaching. P Natl Acad Sci USA. 1999;96:8007–12.

23. Iglesias-Prieto R, Matta JL, Robins WA, Trench RK. Photosynthetic response to elevated temperature in the symbiotic dinoflagellate Symbiodinium microadriaticum in culture. P Natl Acad Sci USA. 1992;89:10302–5.

24. Bhagooli R, Hidaka M. Comparison of stress susceptibility of in hospite and isolated zooxanthellae among five coral species. J Exp Mar Biol Ecol. 2003;291:181–97.

25. Bhagooli R, Hidaka M. Photoinhibition, bleaching susceptibility and mortality in two scleractinian corals, Platygyra ryukyuensis and Stylophora pistillata, in response to thermal and light stresses. Comp Biochem Phys A. 2004;137:547–55.

26. Mumby PJ, Chisholm JRM, Edwards AJ, Andrefouet S, Jaubert J. Cloudy weather may have saved Society Island reef corals during the 1998 ENSO event. Mar Ecol Prog Ser. 2001;222:209–16.

27. Stambler N, Dubinsky Z. Corals as light collectors: an integrating sphere approach. Coral Reefs. 2005;24:1–9.

28. Wangpraseurt D, Larkum AWD, Ralph PJ, Kühl M. Light gradients and optical microniches in coral tissues. Front Microbiol. 2012;3:316.

29. Rodríguez-Roman A, Hernández-Pech X, Thomé PE, Enríquez S, Iglesias-Prieto R. Photosynthesis and light utilization in the Caribbean coral Montastraea faveolata recovering from a bleaching event. Limnol Oceanogr. 2006;51:2702–10.

30. Kaniewska P, Magnusson SH, Anthony KRN, Reef R, Kühl M, Hoegh-Guldberg O. Importance of macro-versus microstructure in modulating light levels inside coral colonies. J Phycol. 2011;47:846–60.

31. Anthony KRN, Hoogenboom MO, Connolly SR. Adaptive variation in coral geometry and the optimization of internal colony light climates. Funct Ecol. 2005;19:17–26.

32. Wangpraseurt D, Polerecky L, Larkum AWD, Ralph PJ, Nielsen DA, Pernice M, Kühl M. The in situ light microenvironment of corals. Limnol Oceanogr. 2014;59:917–26.

33. Salih A, Larkum A, Cox G, Kühl M, Hoegh-Guldberg O. Fluorescent pigments in corals are photoprotective. Nature. 2000;408:850–3.

34. Stolarski J. Three-dimensional micro- and nanostructural characteristics of the scleractinian coral skeleton: a biocalcification proxy. Acta Palaeontol Pol. 2003;48:497–530.

35. Cuif JP, Dauphin Y. The environment recording unit in coral skeletons—a synthesis of structural and chemical evidences for a biochemically driven, stepping-growth process in fibres. Biogeosciences. 2005;2:61–73.

36. Enríquez S, Mendez E, Hoegh-Guldberg O, Iglesias-Prieto R. Morphological dependence of the variation in the light amplification capacity of coral skeleton. In: 11th International Coral Reef Symposium; Ft. Lauderdale: 2008. p. 5–18.

37. Rogers JD, Çapoğlu IR, Backman V. Nonscalar elastic light scattering from continuous random media in the Born approximation. Opt Lett. 2009;34:1891–3.

38. Ralph PJ, Larkum AWD, Kühl M. Photobiology of endolithic microorganisms in living coral skeletons: 1. Pigmentation, spectral reflectance and variable chlorophyll fluorescence analysis of endoliths in the massive corals Cyphastrea serailia, Porites lutea and Goniastrea australensis. Mar Biol. 2007;152:395–404.

39. Krause GH, Weis E. Chlorophyll fluorescence and photosynthesis. The basics. Ann Rev Plant Phys. 1991;42:313–49.

40. Iglesias-Prieto R, Beltrán VH, LaJeunesse TC, Reyes-Bonilla H, Thomé PE. Different algal symbionts explain the vertical distribution of dominant reef corals in the eastern Pacific. P Roy Soc B Biol Sci. 2004;271:1757–63.

41. Warner ME, LaJeunesse TC, Robison JD, Thur RM. The ecological distribution and comparative photobiology of symbiotic dinoflagellates from reef corals in Belize: potential implications for coral bleaching. Limnol Oceanogr. 2006;51:1887–97.

42. Kramer DM, Johnson G, Kiirats O, Edwards GE. New fluorescence parameters for the determination of QA redox state and excitation energy fluxes. Photosynth Res. 2004;79:209–18.

43. Robison JD, Warner ME. Differential impacts of photoacclimation and thermal stress on the photobiology of four different phylotypes of Symbiodinium (Pyrrhophyta). J Phycol. 2006;42:568–79.

44. Hill R, Larkum AWD, Prasil O, Kramer DM, Szabo M, Kumar V, Ralph PJ. Light-induced dissociation of antenna complexes in the symbionts of scleractinian corals correlates with sensitivity to coral bleaching. Coral Reefs. 2012;31:963–75.

45. Warner ME, Lesser MP, Ralph PJ. Chlorophyll fluorescence in reef building corals. In: Suggett DJ, Prášil O, Borowitzka MA, editors. Chlorophyll a fluorescence in aquatic sciences: methods and application, vol 4. New York: Springer; 2010. p. 209–22.

46. Ragni M, Airs RL, Hennige SJ, Suggett DJ, Warner ME, Geider RJ. PSII photoinhibition and photorepair in Symbiodinium (Pyrrhophyta) differs between thermally tolerant and sensitive phylotypes. Mar Ecol Prog Ser. 2010;406:57–70.

47. Warner ME, Fitt WK, Schmidt GW. The effects of elevated temperature on the photosynthetic efficiency of zooxanthellae in hospite from four different species of reef coral: a novel approach. Plant Cell Environ. 1996;19:291–9.

48. Hill R, Frankart C, Ralph PJ. Impact of bleaching conditions on the components of non-photochemical quenching in the zooxanthellae of a coral. J Exp Mar Biol Ecol. 2005;322:83–92.

49. Jimenez IM, Larkum AWD, Ralph PJ, Kühl M. Thermal effects of tissue optics in symbiont-bearing reef-building corals. Limnol Oceanogr. 2012;57:1816–25.

50. Lesser MP. Coral bleaching: causes and mechanisms. In: Coral reefs: an ecosystem in transition. 2011. p. 405–19.

51. Dove S. Scleractinian corals with photoprotective host pigments are hypersensitive to thermal bleaching. Mar Ecol Prog Ser. 2004;272:99–116.

52. Brown BE, Downs CA, Dunne RP, Gibb SW. Preliminary evidence for tissue retraction as a factor in photoprotection of corals incapable of xanthophyll cycling. J Exp Mar Biol Ecol. 2002;277:129–44.

53. Middlebrook R, Anthony KRN, Hoegh-Guldberg O, Dove S. Heating rate and symbiont productivity are key factors determining thermal stress in the reef-building coral Acropora formosa. J Exp Biol. 2010;213:1026–34.

54. Berkelmans R, van Oppen MJH. The role of zooxanthellae in the thermal tolerance of corals: a 'nugget of hope' for coral reefs in an era of climate change. P Roy Soc B Biol Sci. 2006;273:2305–12.

55. Cunning R, Baker AC. Excess algal symbionts increase the susceptibility of reef corals to bleaching. Nature Climate Change. 2013;3:259–62.

56. Hume B, D'Angelo C, Burt J, Baker AC, Riegl B, Wiedenmann J. Corals from the Persian/Arabian Gulf as models for thermotolerant reef-builders: prevalence of clade C3 Symbiodinium, host fluorescence and ex situ temperature tolerance. Mar Pollut Bull. 2013;72:313–22.

57. Fitt WK, Gates RD, Hoegh-Guldberg O, Bythell JC, Jatkar A, Grottoli AG, et al. Response of two species of Indo-Pacific corals, Porites cylindrica and Stylophora pistillata, to short-term thermal stress: the host does matter in determining the tolerance of corals to bleaching. J Exp Mar Biol Ecol. 2009;373:102–10.

58. Howells EJ, Beltran VH, Larsen NW, Bay LK, Willis BL, van Oppen MJH. Coral thermal tolerance shaped by local adaptation of photosymbionts. Nature Climate Change. 2012;2:116–20.

59. Lesser MP, Stat M, Gates RD. The endosymbiotic dinoflagellates (*Symbiodinium* sp.) of corals are parasites and mutualists. Coral Reefs. 2013;32:603–11.

60. Doyle JJ, Doyle JL. A rapid DNA isolation procedure for small quantities of fresh leaf tissue. Phytochem Bull. 1987;19:11–5.

61. Veron JEN. Corals of the World, vol 1, 2, and 3. Townsville: Australian Institute of Marine Science; 2000.

62. Kim YL, Liu Y, Turzhitsky VM, Roy HK, Wali RK, Backman V. Coherent backscattering spectroscopy. Opt Lett. 2004;29:1906–8.

63. Roy H, Turzhitsky V, Kim Y, Goldberg M, Watson P, Rogers J, et al. Association between rectal optical signatures and colonic neoplasia: potential applications for screening. Cancer Res. 2009;69:4476–83.

64. Turzhitsky V, Rogers JD, Mutyal NN, Roy HK, Backman V. Characterization of light transport in scattering media at subdiffusion length scales with low-coherence enhanced backscattering. IEEE J Sel Top Quant. 2010;16:619–26.

65. Radosevich AJ, Mutyal NN, Turzhitsky V, Rogers JD, Yi J, Taflove A, Backman V. Measurement of the spatial backscattering impulse-response at short length scales with polarized enhanced backscattering. Opt Lett. 2011;36:4737–9.

66. Radosevich AJ, Eshein A, Nguyen T-Q, Backman V. Subdiffusion reflectance spectroscopy to measure tissue ultrastructure and microvasculature: model and inverse algorithm. J Biolmed Opt. 2015;20:97002.

67. Hoegh-Guldberg O, Jones RJ. Photoinhibition and photoprotection in symbiotic dinoflagellates from reef-building corals. Mar Ecol Prog Ser. 1999;183:73–86.

68. Rogers JE, Marcovich D. A simple method for the extraction and quantification of photopigments from *Symbiodinium* spp. J Exp Mar Biol Ecol. 2007;353:191–7.

69. Veal CJ, Carmi M, Fine M, Hoegh-Guldberg O. Increasing the accuracy of surface area estimation using single wax dipping of coral fragments. Coral Reefs. 2010;29:893–7.

70. West BT, Welch KB, Galecki AT. Linear mixed models: a practical guide using statistical software. Abingdon-on-Thames: Taylor & Francis; 2006.

71. Kaandorp JA, Sloot PMA, Merks RMH, Bak RPM, Vermeij MJA, Maier C. Morphogenesis of the branching reef coral *Madracis mirabilis*. P Roy Soc B Biol Sci. 2005;272:127–33.

72. Ulstrup KE, Ralph PJ, Larkum AWD, Kühl M. Intra-colonial variability in light acclimation of zooxanthellae in coral tissues of Pocillopora damicornis. Mar Biol. 2006;149:1325–35.

73. Jimenez IM, Kühl M, Larkum AWD, Ralph PJ. Effects of flow and colony morphology on the thermal boundary layer of corals. J Roy Soc Interface. 2011;8:1785–95.

74. Cohen AL, Smith SR, McCartney MS, van Etten J. How brain corals record climate: an integration of skeletal structure, growth and chemistry of *Diploria labyrinthiformis* from Bermuda. Mar Ecol Prog Ser. 2004;271:147–58.

75. Fisher PL, Malme MK, Dove S. The effect of temperature stress on coral–*Symbiodinium* associations containing distinct symbiont types. Coral Reefs. 2012;31:473–85.

76. Sampayo EM, Ridgway T, Bongaerts P, Hoegh-Guldberg O. Bleaching susceptibility and mortality of corals are determined by fine-scale differences in symbiont type. P Natl Acad Sci USA. 2008;105:10444–9.

77. Wang JT, Meng PJ, Chen YY, Chen CA. Determination of the thermal tolerance of *Symbiodinium* using the activation energy for inhibiting photosystem II activity. Zool Stud. 2012;51:137–42.

78. LaJeunesse TC, Smith RT, Finney J, Oxenford H. Outbreak and persistence of opportunistic symbiotic dinoflagellates during the 2005 Caribbean mass coral 'bleaching' event. P Roy Soc B Biol Sci. 2009;276:4139–48.

79. Folmer O, Black M, Hoeh W, Lutz R, Vrijenhoek R. DNA primers for amplification of mitochondrial cytochrome c oxidase subunit I from diverse metazoan invertebrates. Mol Mar Biol Biotech. 1994;3:294–9.

80. Fukami H, Budd AF, Levitan DR, Jara J, Kersanach R, Knowlton N. Geographic differences in species boundaries among members of the *Montastraea annularis* complex based on molecular and morphological markers. Evolution. 2004;58:324–37.

81. McFadden CS, Donahue R, Hadland BK, Weston R. A molecular phylogenetic analysis of reproductive trait evolution in the soft coral genus *Alcyonium*. Evolution. 2001;55:54–67.

82. Santos SR, Taylor DJ, Kinzie RA, Hidaka M, Sakai K. Coffroth MA (2002) Molecular phylogeny of symbiotic dinoflagellates inferred from partial chloroplast large subunit (23S)-rDNA sequences. Mol Phylogenet Evol. 2002;23:97–111.

2

Non-associative versus associative learning by foraging predatory mites

Peter Schausberger[1,2]* ⓘ and Stefan Peneder[2]

Abstract

Background: Learning processes can be broadly categorized into associative and non-associative. Associative learning occurs through the pairing of two previously unrelated stimuli, whereas non-associative learning occurs in response to a single stimulus. How these two principal processes compare in the same learning task and how they contribute to the overall behavioural changes brought about by experience is poorly understood. We tackled this issue by scrutinizing associative and non-associative learning of prey, Western flower thrips *Frankliniella occidentalis*, by the predatory mite, *Neoseiulus californicus*. We compared the behaviour of thrips-experienced and -naïve predators, which, early in life, were exposed to either thrips with feeding (associative learning), thrips without feeding (non-associative learning), thrips traces on the surface (non-associative learning), spider mites with feeding (thrips-naïve) or spider mite traces on the surface (thrips-naïve).

Results: Thrips experience in early life, no matter whether associative or not, resulted in higher predation rates on thrips by adult females. In the no-choice experiment, associative thrips experience increased the predation rate on the first day, but shortened the longevity of food-stressed predators, a cost of learning. In the choice experiment, thrips experience, no matter whether associative or not, increased egg production, an adaptive benefit of learning.

Conclusions: Our study shows that both non-associative and associative learning forms operate in foraging predatory mites, *N. californicus*. The non-rewarded thrips prey experience produced a slightly weaker, but less costly, learning effect than the rewarded experience. We argue that in foraging predatory mites non-associative learning is an inevitable component of associative learning, rather than a separate process.

Keywords: Associative, Foraging, Learning, Predation, Mites, Non-associative, Thrips

Background

Learning, changed behaviour following experience, is ubiquitous in animals, from protozoans to primates [1–3]. At large, the huge variety of learning processes can be categorized into non-associative and associative [2, 4]. Associative learning occurs through the association of two previously unrelated stimuli, and includes reinforcement, whereas non-associative learning occurs in response to a single stimulus, without reinforcement. Distinction between these two principal learning categories is not clear-cut and under debate, e.g. [3–6]. Nonetheless, studies rigorously tackling this issue, by, for example, comparing the relative importance of

non-associative and associative experiences on the learning success in a given task, such as host or prey recognition, are scarce ([7] for parasitoids). Associative learning involves Pavlovian (classical) and operant (instrumental) conditioning ([8, 9] for honeybees; [10] for cockroaches; [11] for *Drosophila*; [12] for review), while non-associative learning involves sensitization, habituation and imprinting ([1–3] for reviews).

Here, we addressed the behavioural aspects of non-associative vs. associative learning in foraging predatory mites, *Neoseiulus californicus*. *N. californicus* is a plant-inhabiting generalist predator feeding on herbivorous mites such as spider mites and rust and gall mites, small insects such as thrips, and plant-derived substances such as pollen [13–15]. *N. californicus* has a ranked food preference. Among the possible food options, spider mites

*Correspondence: peter.schausberger@univie.ac.at
[1] Department of Behavioural Biology, University of Vienna, Vienna, Austria
Full list of author information is available at the end of the article

such as the two-spotted mite *Tetranychus urticae* are the primary prey [13, 16]. Difficult-to-grasp small insects such as larvae of the Western flower thrips *Frankliniella occidentalis* are an alternative, secondary prey, e.g. [17]. *Neoseiulus californicus* has five life stages—egg, larva, protonmyph, deutonymph, adult—and is able to improve its foraging performance by imprinting on a given prey in a sensitive phase early in life, i.e. in the larval and early protonymphal stage [14]. The larvae are, compared to later life stages, little mobile, because of having only six legs, and usually do not feed; the next developmental stage, the protonymph, has eight legs and is the first obligatory feeding stage [18]. The predators are eyeless and use primarily chemo- and mechano-sensory cues to sense their environment, including recognizing suitable prey [19]. For prey, such as thrips, which is difficult to grasp and overwhelm by the fragile small juvenile predators, mere prey contact in early life suffices to establish persistent memory, allowing improving foraging on this prey by the larger adult predators [14]. While food imprinting early in life, a non-associative form of learning [20], produces prey-specific, long-lasting, life stage-crossing effects in foraging *N. californicus* [14], it is unclear how these effects compare to the effects of associative experience made by the predators. Moreover, which prey cues are learned, probably body odours or chemical traces left on the surface, is unknown. These are important issues, from both fundamental and applied perspectives. Studies comparing the operation of different learning processes and their relative contribution to a given learning effect are scarce [7] but inevitable for a thorough mechanistic understanding of learning at the behavioural, perceptual and neuronal levels. Detailed understanding of the learning processes and cues has also relevance to the use of natural enemies, such as *N. californicus*, in biological control, because possibly allowing priming them on a target pest ([21, 22] for parasitoids).

We conducted two experiments, no-choice and choice, to determine which features of the alternative prey Western flower thrips, *F. occidentalis*, are learned by *N. californicus* early in life, and to compare the effects produced by non-associative and associative experience. The prey cues presented to young predatory mites during the learning phase varied in complexity and information content, ranging from (1) prey traces left on the surface, to (2) prey traces left on the surface plus chemical, behavioural and morphological traits on the body of live prey, to (3) prey traces left on the surface plus chemical, behavioural and morphological traits on the body of live prey plus dead prey allowing easy feeding. Treatments (1) and (2) represent non-associative learning paradigms, while treatment (3) represents an associative learning

paradigm, because of involving tasting, feeding and/or satiation rewards.

Results

No-choice experiment

Thrips experience (GEE: *Wald* $x_{1}^{2} = 5.56$, $p = 0.02$) but not type of experience (*Wald* $x_{2}^{2} = 1.67$, $p = 0.43$) and the interaction between thrips and type of experience (*Wald* $x_{1}^{2} = 2.22$, $p = 0.14$) affected the number of thrips killed and sucked out over the 4 days experimental period (Fig. 1). Thrips-experienced predators killed more thrips than thrips-naïve predators. Only on the 1st day, associative thrips learners, i.e. those that had experienced thrips by feeding, killed and sucked out more thrips than non-associative thrips learners and thrips-naïve predators (Bonferroni, $p < 0.05$ for each pairwise comparison; Fig. 1). The number of eggs produced did neither vary with thrips experience (GEE: *Wald* $x_{1}^{2} = 0.47$, $p = 0.49$) nor type of experience (*Wald* $x_{2}^{2} = 0.27$, $p = 0.87$) nor their interaction (*Wald* $x_{1}^{2} = 2.46$, $p = 0.12$) over time (Fig. 2). Type of experience affected predator longevity (GLM: *Wald* $x_{2}^{2} = 19.61$, $p < 0.001$), no matter of thrips experience (*Wald* $x_{1}^{2} = 2.29$, $p = 0.13$) and the interaction between thrips and type of experience (*Wald* $x_{1}^{2} = 0.12$, $p = 0.74$) (Fig. 3). Predators that had only experienced prey traces survived longer than predators with feeding experience (Bonferroni: $p < 0.05$); longevity of predators that had contacted prey was intermediate, but not statistically separable ($p > 0.05$ in pairwise comparisons) from longevity of predators experienced with prey traces and those with feeding experience (Fig. 3).

Choice experiment

Thrips-experienced predators consumed more prey in total (spider mites plus thrips) than thrips-naïve predators (GEE: *Wald* $x_{1}^{2} = 7.69$, $p = 0.006$), no matter of the type of experience (*Wald* $x_{2}^{2} = 0.86$, $p = 0.65$) and the interaction between thrips and type of experience (*Wald* $x_{1}^{2} = 1.87$, $p = 0.17$) (Fig. 4). This was primarily due to thrips-experienced predators consuming more thrips than thrips-naïve predators (GEE: *Wald* $x_{1}^{2} = 5.16$, $p = 0.02$), no matter of the type of experience (*Wald* $x_{2}^{2} = 0.06$, $p = 0.97$) and the interaction between thrips and type of experience (*Wald* $x_{1}^{2} = 0.22$, $p = 0.64$) (Fig. 5). In contrast, predation on spider mites did neither vary with thrips experience (GEE: *Wald* $x_{1}^{2} = 1.14$, p = 0.29) nor type of experience (*Wald* $x_{2}^{2} = 0.72$, p = 0.70) nor their interaction (*Wald* $x_{1}^{2} = 1.89$, p = 0.17) (Fig. 6). Egg production was marginally significantly higher in thrips-experienced than -naïve predators (GEE: *Wald* $x_{1}^{2} = 3.37$, $p = 0.06$), no matter of the type of experience (*Wald* $x_{2}^{2} = 0.61$, $p = 0.74$) and the interaction between thrips and type of experience (*Wald* $x_{1}^{2} = 0.00$, $p = 0.99$) (Fig. 7).

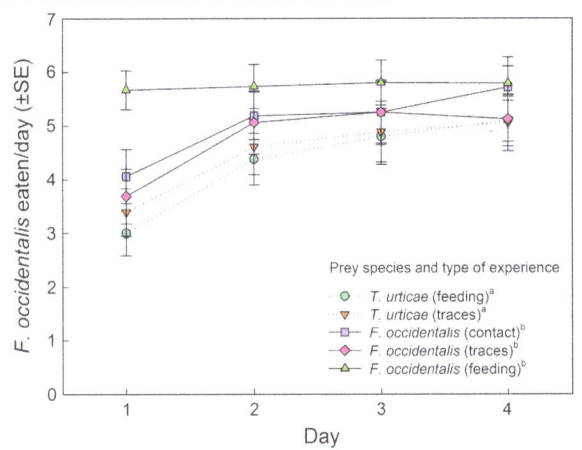

Fig. 1 Predation on first larvae of thrips *F. occidentalis* by gravid thrips-experienced and -naïve (spider mite *T. urticae*-experienced) *N. californicus* females over time, in dependence of the predators' type of experience early in life. Type of experience was either contact with live prey but no feeding (contact), feeding on prey (feeding), or contact with prey traces left on the surface (traces). *Different superscript letters* accompanying prey species and type of experience indicate significant differences (GEE; *P* < 0.05)

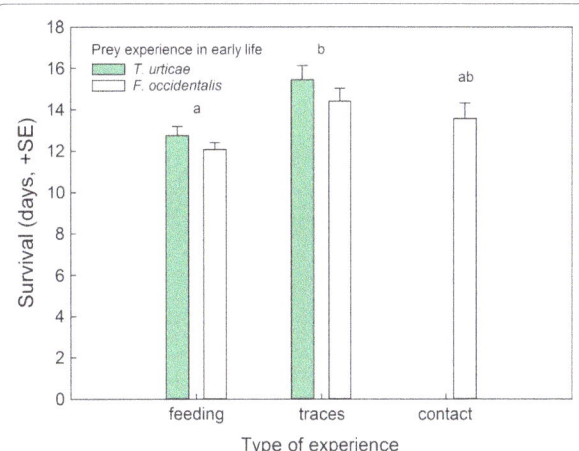

Fig. 3 Survival of gravid thrips-experienced and -naïve (spider mite *T. urticae*-experienced) *N. californicus* females offered first larvae of thrips *F. occidentalis* as prey, in dependence of the predators' type of experience early in life. Type of experience was either contact with live prey but no feeding (contact), feeding on prey (feeding), or contact with prey traces left on the surface (traces). *Different letters* on *top of bars* indicate significant differences among types of experience (Bonferroni following GLM; *P* < 0.05)

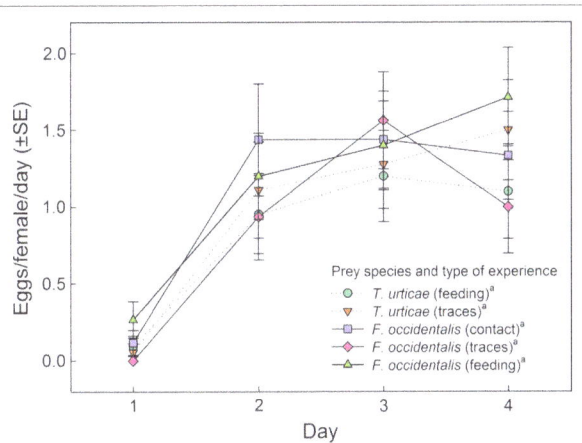

Fig. 2 Oviposition by gravid thrips-experienced and -naïve (spider mite *T. urticae*-experienced) *N. californicus* females offered first larvae of thrips *F. occidentalis* as prey over time, in dependence of the predators' type of experience early in life. Type of experience was either contact with live prey but no feeding (contact), feeding on prey (feeding), or contact with prey traces left on the surface (traces). The *same superscript letter* accompanying prey species and type of experience indicates non-significance (GEE; *P* > 0.05)

Discussion

Our study reveals that both non-associative and associative learning processes operate in foraging predatory mites *Neoseiulus californicus*. Mere contact with the prey *F. occidentalis* or its traces left on the surface was sufficient for learning and establishing long-lasting

(considering the predators' longevity of ~50 days at 25 °C) memory [14]. However, reinforcement of the prey experience made early in life, by pairing external prey cues with a feeding reward (taste and/or satiation), strengthened or intensified the learning effect, as indicated by the initially higher predation rate of predators with thrips feeding experience than those with only thrips contact or traces experience in the no-choice experiment. At the behavioural level, this could mechanistically represent non-associative learning plus the added associative effect, or graded intensities of one and the same learning process (i.e., increasing intensity with increasing cue variety and/or quality), or completely distinct processes. The latter is unlikely because, chronologically, first, orientation on, and recognition of, external prey cues is required, in both non-associative and associative learning, and only then, after recognition and acceptance as suitable prey, in associative learning gustatory cues and satiation come into play, reinforcing learning through the feeding reward. In principle, every associative experience can have, or can build on, components of non-associative learning but this has rarely been assessed ([7] for parasitoids).

At the neuronal level, associative learning may either strengthen or intensify the changes in the same neuronal pathways, as compared to non-associative learning, or establish additional interconnected or separate pathways than non-associative learning. The latter is true for the distinction between short- and long-term memory ([23] for *Drosophila*; [24] for honey bees), which, at the

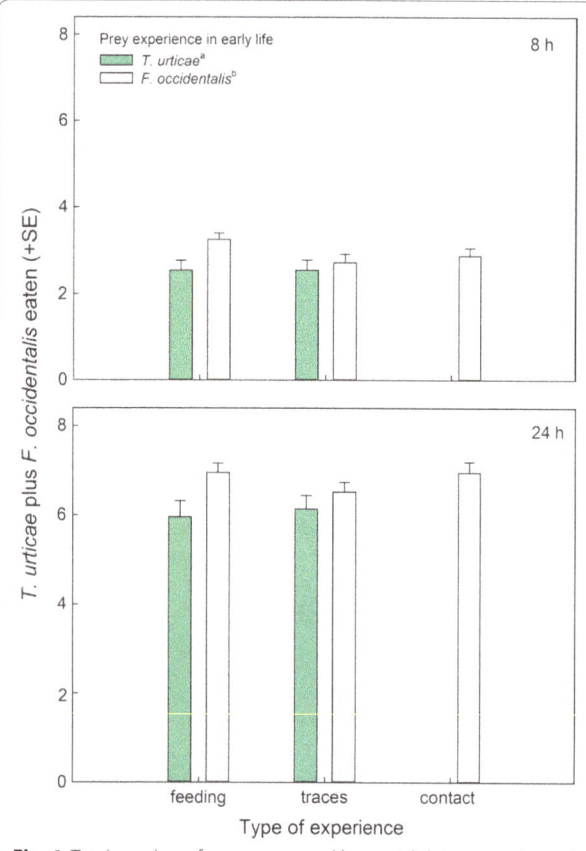

Fig. 4 Total number of prey consumed by gravid thrips-experienced and -naïve (spider mite *T. urticae*-experienced) *N. californicus* females simultaneously offered four spider mite nymphs, *T. urticae*, plus four first larvae of thrips, *F. occidentalis*, after 8 and 24 h, in dependence of the predators' type of experience early in life. Type of experience was either contact with live prey but no feeding (contact), feeding on prey (feeding), or contact with prey traces left on the surface (traces). *Different superscript letters* accompanying prey species experience indicate a significant difference (GEE; P < 0.05)

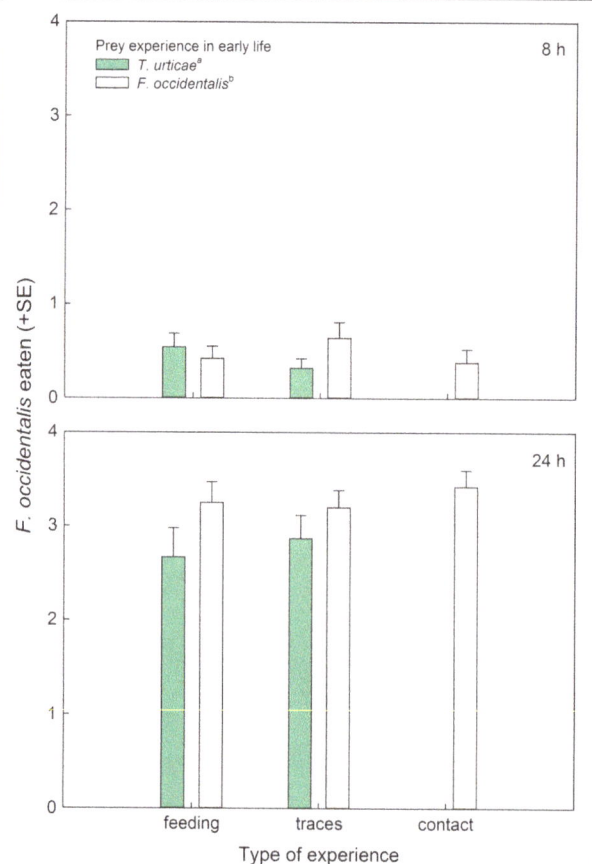

Fig. 5 Number of first larvae of thrips, *F. occidentalis*, consumed by gravid thrips-experienced and -naïve (spider mite *T. urticae*-experienced) *N. californicus* females, simultaneously offered four spider mite nymphs, *T. urticae*, plus four thrips larvae, *F. occidentalis*, after 8 and 24 h, in dependence of the predators' type of experience early in life. Type of experience was either contact with live prey but no feeding (contact), feeding on prey (feeding), or contact with prey traces left on the surface (traces). *Different superscript letters* accompanying prey species experience indicate a significant difference (GEE; P < 0.05)

molecular level, trigger differing protein syntheses ([25, 26] for reviews). While in honey bees, non-associative and associative learning could be reflected in different memory retention times [27, 28], this is not applicable to predatory mites, because also non-associative imprinting may produce long-lasting effects [14]. At the genetic level, short- and long-term memory, and possibly non-associative and associative learning, may involve genetically distinct, functionally different components (e.g. [29] for *Drosophila*). At the epigenetic level, short- and long-term memory may be discernible in methylation of learning-related genes (e.g. [30, 31] for honey bees), which may also be true for epigenetic marks produced by non-associative and associative experiences. For predatory mites, any evidence of the involvement of different genes and/or differing epigenetic regulation of non-associative

and associative learning remains elusive until identification of learning-associated genes.

Apart from feeding experience increasing the initial predation rate on thrips in the no-choice experiment, we did not observe any differences between the types of experience within thrips-experienced predators. However, prey feeding experience early in life, no matter whether thrips or spider mites, had an effect on survival of the experimental animals, that is, it shortened predator longevity. This might represent an operating cost of associative learning, i.e., a trade-off between learning and life history traits [32, 33]. Energy needed to form new, or strengthen existing, neuronal connections and pathways, was traded off against energy used for basic physiological maintenance and processes, no matter whether the

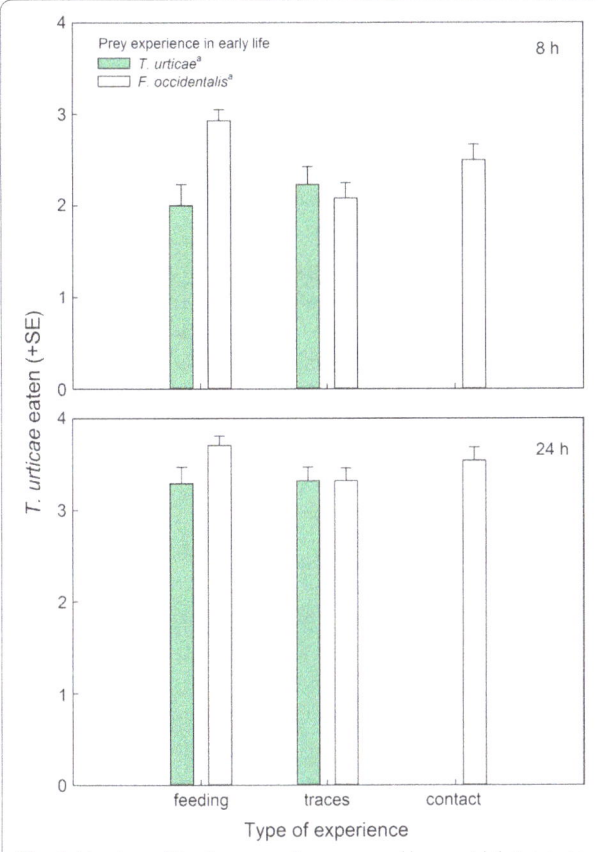

Fig. 6 Number of *T. urticae* nymphs consumed by gravid thrips-experienced and -naïve (spider mite *T. urticae*-experienced) *N. californicus* females, simultaneously offered four spider mite nymphs, *T. urticae*, plus four thrips larvae, *F. occidentalis*, after 8 and 24 h, in dependence of the predators' type of experience early in life. Type of experience was either contact with live prey but no feeding (contact), feeding on prey (feeding), or contact with prey traces left on the surface (traces). The *same superscript letter* accompanying prey species experience indicates non-significance (GEE; *P* > 0.05)

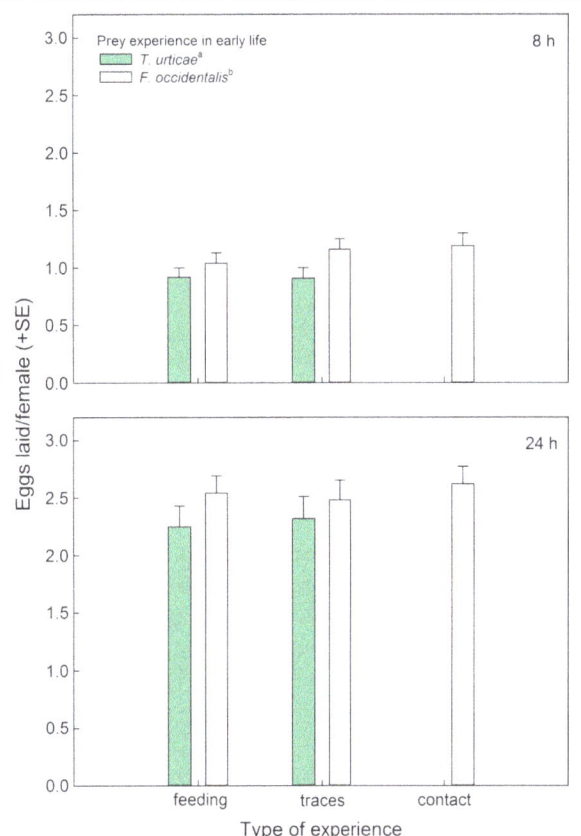

Fig. 7 Number of eggs laid by gravid thrips-experienced and -naïve (spider mite *T. urticae*-experienced) *N. californicus* females, simultaneously offered four spider mite nymphs, *T. urticae*, plus four first larvae of thrips, *F. occidentalis*, within 8 and 24 h, in dependence of the predators' type of experience early in life. Type of experience was either contact with live prey but no feeding (contact), feeding on prey (feeding), or contact with prey traces left on the surface (traces). *Different superscript letters* accompanying prey species experience indicate a marginally significant difference (GEE; *P* = 0.06)

predators then received the prey experienced early in life or a novel prey. Predation rate over time was higher in thrips-experienced than -naïve predators but did not differ among types of thrips-experience (traces vs. contact vs. feeding) in the no choice-experiment. Lacking difference among types of experience within prey species was also evident in the choice experiment. Similar to the no-choice experiment, thrips-experienced predators consumed more thrips and laid more eggs than thrips-naïve predators, no matter of the type of thrips experience (traces vs. contact vs. feeding). Higher egg production of thrips-experienced than thrips-naïve predators points at the adaptive benefits of thrips learning [33]. The choice experiment underlines that the behavioural changes brought about by thrips experience early in life are thrips specific and not the result of unspecific sensitization [14,

33]. If thrips experience in early life would have sensitized the predators, thrips-experienced predators should have fed more on any prey, including spider mites, which was not the case.

At the perceptual level, two or three chemosensory modalities were involved in associative learning, (1) volatile and/or (2) tactile chemoreception and (3) gustation by ingestion, whereas it was just one or two, (1) volatile and/or (2) tactile chemoreception, in non-associative learning. In associative learning, satiation came as an additional internal stimulus into play. Knowledge about the sensory modalities and prey cues playing a role in learning has relevance for the use of *N. californicus,* and other natural enemies, in biological control. Commonly, these predators are mass-reared on other than target prey, possibly compromising their performance against the target pest after release in the crop. Adding chemical extracts or dead corpses from target prey to the mass rearing might enhance the efficacy of the predators, at least in the short term after release. Pertinent proofs of concept are available for parasitoids [21, 22]: parasitoids primed on the target host during rearing performed better against this host in the field than target host-naïve parasitoids.

Conclusions

Learning processes can be broadly categorized into non-associative and associative but how these two processes compare in the same learning task is poorly understood. We tackled this issue by investigating the effects produced by non-associative and associative learning in plant-inhabiting predatory mites *Neoseiulus californicus* in foraging contexts. Adult predatory mite females memorized, after three molting events, prey, Western flower thrips *Frankliniella occidentalis*, experienced in early life. Associative, rewarded experience produced slightly stronger, but physiologically more costly, learning effects than non-associative experience. Both learning processes resulted in persistent memory. We argue that non-associative learning is an inevitable component of associative learning rather than a completely distinct process.

Methods
Predator and prey rearing

Neoseiulus californicus used in experiments derived from a laboratory population founded with specimens obtained from Koppert (NL). The predators were reared in piles of detached leaves of common bean, *Phaseolus vulgaris* L., infested by two-spotted spider mites, *Tetranychus urticae* Koch. Detached leaves were piled up on an artificial arena consisting of an acrylic tile (15 × 15 cm) resting on a water-saturated foam cube kept in a plastic box (20 × 20 × 6 cm) half-filled with water. Moist tissue paper was folded over the edges of the tile to prevent the mites from escaping. To obtain predator eggs used for experiments, gravid females were randomly withdrawn from the population, transferred to detached leaf arenas (called oviposition arenas) and provided with mixed *T. urticae* stages. Oviposition arenas consisted of bean leaves placed upside down on water-saturated foam cubes kept in plastic boxes half-filled with water. Moist tissue paper was wrapped around the stem of the leaf to maintain leaf turgidity, and folded over its edges to prevent mite escaping. Eggs laid by the predator females were collected after 24 h for use in experiments. The predator rearing unit, leaf arenas and experimental units were kept in an environmental chamber at 25 ± 1 °C, 60 ± 5% RH and 16:8 h (light:dark) photoperiod.

Tetranychus urticae nymphs used as prey in experiments were randomly collected from a population reared on whole bean plants, *P. vulgaris*. Western flower thrips, *Frankliniella occidentalis* (Pergande), was reared on detached bean leaves embedded in 1% water agar (Fluka, Vienna) in closed plastic Petri dishes (14.5 cm diameter). The Petri dish lids were perforated but closed by gauze for ventilation. Nescofilm® was used to tightly connect the lids and the bottom parts to prevent thrips escaping. To obtain first larval stages used as prey in experiments, adult female thrips were transferred to separate detached bean leaves and allowed to lay eggs for 24 h. Every 24 h, the females were transferred to a new leaf. First larvae emerged after 4–6 days and were then used in experiments. Petri dishes were kept in an environmental chamber at 25 ± 1 °C, 60 ± 5% RH and 16:8 h (light:dark) photoperiod.

Pre-experimental procedures

Predator eggs (<24 h old), giving rise to the experimental animals, were randomly withdrawn from the predator oviposition arenas and placed inside acrylic cages. Each acrylic cage consisted of a circular cavity (Ø1.5 cm) laser-cut into an acrylic plate, covered on the bottom side by gauze and on the upper side by a removable microscope slide [34]. The cages were checked daily for hatching larvae, which were then singly transferred to new acrylic cages for the learning phase (dubbed learning cages).

Before placing the predator larvae into the learning cages, the cages were prepared according to one of five treatments, three of which generated thrips-experienced predators and two of which generated thrips-naïve predators. For the three treatments used to generate thrips-experienced predators, each cage received three first larvae of *F. occidentalis* for 24 h. Before adding the predatory mite larva to the cage for the 24 h conditioning phase, for treatment 1 (thrips feeding) one prey larva was left alive and two were killed immediately before, for treatment 2 (thrips contact) all three prey larvae were left alive, and for treatment 3 (thrips traces) all three prey larvae were removed so that only their traces, such as metabolic waste products, remained in the cage. For the two treatments 4 and 5, which were used to generate thrips-naïve predators, each cage received three nymphs of *T. urticae* for 24 h. Before adding the predatory mite larva to the cage, for treatment 4 (spider mite feeding) all three prey nymphs were left alive, and for treatment 5 (spider mite traces) all three prey nymphs were removed immediately before so that only their webbing and traces, such as metabolic waste products, remained in the cage. A treatment "spider mite contact" could not be established because the predators inevitably attack, kill and feed on the spider mites upon encounter. Predators that had possibly fed on thrips in the thrips contact group, determined when a dead thrips was found after 24 h, were discarded. Predators of treatment (1) were considered associative thrips learners, predators of (2) and (3) were considered non-associative thrips learners and those of treatments (4) and (5) were thrips-naïve. After the 24 h learning phase, the predator larvae (or freshly moulted protonymphs) were removed and singly transferred into cages containing mixed spider mite stages as prey (replenished as needed) and left there until reaching adulthood, lasting three to four days. Feeding on thrips by the associative learners in treatment (1) and feeding on spider mites in treatment (4) was verified by the coloured content of the digestive tract of the predators.

No-choice experiment

Upon reaching adulthood, the predator females were singly transferred to cages that had been previously loaded with seven live first larvae of *F. occidentalis*; a male, randomly withdrawn from the rearing, was added for mating, and the cages checked for the occurrence and number of killed and sucked out thrips larvae after 24 h. After 24, 48 and 72 h the now gravid predator females were singly transferred to new cages, each containing seven thrips larvae, and the number of killed and sucked out thrips larvae and eggs laid by the predators counted, and removed, the next day. Following the fourth thrips counting, i.e., 96 h after starting the experiment, the predators were left in their cages, without replenishing prey, and their survival checked once per day until natural death. Each of the five treatments was replicated 15–21 times.

Choice experiment

To start the choice experiment, gravid females conditioned and raised according to one of the five treatments described in the pre-experimental procedures, i.e. (1) thrips feeding, (2) thrips contact, (3) thrips traces, (4) spider mite feeding and (5) spider mite traces, were singly placed inside acrylic cages containing four spider mite nymphs plus four first larvae of thrips. The numbers of killed and sucked out spider mites and thrips, and eggs laid by the predators were assessed after 8 and 24 h. Each of the five treatments was replicated 22–26 times.

Statistical analyses

We used IBM SPSS 23 (IBM Corp., USA) for all statistical analyses. The raw data of both experiments, no-choice and choice, are provided in Additional File 1: Table S1. In the no-choice experiment, we used separate generalized estimating equations (GEE; Poisson distribution, log link) to analyse the influence of thrips experience and type of experience (traces, contact, feeding) on the predation

rate on thrips and egg production with thrips prey over the 4 days experimental period (used as auto-correlated inner subject variable). A generalized linear model (GLM; normal distribution, identity link) was used to compare post-experimental survival as affected by thrips and type of experience. In the choice experiment, we used separate generalized estimating equations (GEE) to analyse the influence of thrips experience and type of experience (traces, contact, feeding) on total predation rate (spider mites plus thrips), predation on thrips and predation on spider mites after 8 and 24 h (Poisson distribution, log link; time used as inner subject variable), and eggs laid by the predators within 8 and 24 h (normal distribution, identity link; time used as inner subject variable).

Authors' contributions
PS perceived the study idea and acquired the funding. PS and SP designed the experiments. SP carried out the experiments. PS and SP analysed the data and wrote the manuscript. Both authors read and approved the final manuscript.

Author details
[1] Department of Behavioural Biology, University of Vienna, Vienna, Austria.
[2] Group of Arthropod Ecology and Behavior, Department of Crop Sciences, University of Natural Resources and Life Sciences, Vienna, Austria.

Acknowledgements
We thank Inga C. Christiansen, M. Seiter and D. Çekin for comments on a previous version of the manuscript.

Competing interests
The authors declare that they have no competing interests.

Funding
This work was supported by the Austrian Science Fund (FWF): P 25876-B25.

References
1. Papaj DR, Lewis AC. Insect learning: ecological and evolutionary perspectives. New York: Chapman and Hall; 1999.
2. Goodenough J, McGuire B, Wallace RA. Perspectives on Animal Behaviour. 2nd ed. New York: Wiley; 2001.
3. Pearce JM. Animal learning and cognition, an introduction. 3rd ed. Hove: Psychology Press; 2008.
4. Moore BR. The evolution of learning. Biol Rev. 2004;79:301–35.
5. Van Kampen HS. Filial imprinting and associative learning: similar mechanisms? Neth J Zool. 1993;43:143–54.
6. Gallistel CR, Balsam PD. Time to rethink the neuronal mechanisms of learning and memory. Neurobiol Learn Mem. 2014;108:136–44.
7. Kaiser L, Perez-Maluf R, Sandoz JC, Pham-Delegue MH. Dynamics of odour learning in *Leptopilina boulardi*, a hymenopterous parasitoid. Anim Behav. 2003;66:1077–84.

8.   Menzel R, Müller U. Learning and memory in honeybees: from behavior to neuronal substrates. Annu Rev Neurosci. 1996;19:379–404.

9.   Kisch J, Erber J. Operant conditioning of antennal movements in the honey bee. Behav Brain Res. 1999;99:93–102.

10. Eisenstein EM. Selecting a model system for neurobiological studies of learning and memory. Behav Brain Res. 1997;82:121–32.

11. McGuire SE, Deshazer M, Davis RL. Thirty years of olfactory learning and memory research in Drosophila melanogaster. Prog Neurobiol. 2005;76:328–47.

12. Giurfa M, Devaud J-M, Sandoz JC. Invertebrate learning and memory. Lausanne: Frontiers; 2011.

13. McMurtry JA, Croft BA. Life styles of phytoseiid mites and their roles in biological control. Annu Rev Entomol. 1997;42:291–321.

14. Schausberger P, Walzer A, Hoffmann D, Rahmani H. Food imprinting revisited: early learning in foraging predatory mites. Behaviour. 2010;147:883–97.

15. Nguyen DT, Vangansbeke D, De Clercq P. Performance of four species of phytoseiid mites on artificial and natural diets. Biol Control. 2015;80:56–62.

16. Blackwood JS, Schausberger P, Croft BA. Prey-stage preference in generalist and specialist phytoseiid mites (Acari: Phytoseiidae) when offered Tetranychus urticae (Acari: Tetranychidae) eggs and larvae. Environ Entomol. 2001;30:1103–11.

17. Walzer A, Paulus HF, Schausberger P. Ontogenetic shifts in intraguild predation on thrips by phytoseiid mites: the relevance of body size and diet specialization. Bull Entomol Res. 2004;94:577–84.

18. Schausberger P, Croft BA. Activity, feeding and development among larvae of specialist and generalist phytoseiid mite species (Acari: Phytoseiidae). Environ Entomol. 1999;28:322–9.

19. Sabelis MW, Dicke M. Long-range dispersal and searching behaviour. In: Helle W, Sabelis MW, editors. Spider mites, their biology, natural enemies and control, vol. 1B. Amsterdam: Elsevier; 1985. p. 141–60.

20. Immelmann K. Ecological significance of imprinting and early learning. Annu Rev Ecol Syst. 1975;6:15–37.

21. Hare JD. Priming Aphytis: behavioral modification of host selection by exposure to a synthetic contact kairomone. Entomol Exp Appl. 1996;78:263–9.

22. Hare JD, Morgan DJW. Mass-priming Aphytis: Behavioral improvement of insectary reared biological control agents. Biol Control. 1997;10:207–14.

23. Margulies C, Tully T, Dubneu J. Deconstructing memory in Drosophila. Curr Biol. 2005;15:R700–13.

24. Eisenhardt D. Learning and memory formation in the honeybee (Apis mellifera) and its dependency on the cAMP-protein kinase A pathway. Anim Biol. 2006;56:259–78.

25. Hoedjes KM, Kruidhof HM, Huigens ME, Dicke M, Vet LEM, Smid HM. Natural variation in learning rate and memory dynamics in parasitoid wasps: opportunities for converging ecology and neuroscience. Proc R Soc B. 2011;278:889–97.

26. Eisenhardt D. Molecular mechanisms underlying formation of long-term reward memories and extinction memories in the honeybee (Apis mellifera). Learn Mem. 2014;21:534–42.

27. Hammer M, Menzel R. Learning and memory in the honeybee. J Neurosci. 1995;15:1617–30.

28. Menzel R. Memory dynamics in the honeybee. J Comp Physiol A. 1999;185:323–40.

29. Tully T, Preat T, Boynton SC, Del Vecchio M. Genetic dissection of consolidated memory in Drosophila. Cell. 1994;79:35–47.

30. Biergans SD, Jones JC, Treiber N, Galizia CG, Szyszka P. DNA methylation mediates the discriminatory power of associative long-term memory in honeybees. PLoS ONE. 2012;7:e39349.

31. Biergans SD, Galizia CG, Reinhardt J, Claudianos C. Dnmts and Tet target memory associated genes after appetitive olfactory training in honey bees. Sci Rep. 2015;5:16223.

32. Mery F, Kawecki TJ. An operating cost of learning in Drosophila melanogaster. Anim Behav. 2004;68:589–98.

33. Christiansen IC, Szin S, Schausberger P. Benefit-cost trade-offs of early learning in foraging predatory mites Amblyseius swirskii. Sci Rep. 2016;6:23571.

34. Schausberger P. Inter- and intraspecific predation on immatures by adult females in Euseius finlandicus, Typhlodromus pyri and Kampimodromus aberrans (Acari: Phytoseiidae). Exp Appl Acarol. 1997;21:131–50.

3

Chemical disguise of myrmecophilous cockroaches and its implications for understanding nestmate recognition mechanisms in leaf-cutting ants

Volker Nehring[1,2]* ⓘ, Francesca R. Dani[3,4], Luca Calamai[3,5], Stefano Turillazzi[2,3], Horst Bohn[6], Klaus-Dieter Klass[7] and Patrizia d'Ettorre[1,8]

Abstract

Background: Cockroaches of the genus *Attaphila* regularly occur in leaf-cutting ant colonies. The ants farm a fungus that the cockroaches also appear to feed on. Cockroaches disperse between colonies horizontally (via foraging trails) and vertically (attached to queens on their mating flights). We analysed the chemical strategies used by the cockroaches to integrate into colonies of *Atta colombica* and *Acromyrmex octospinosus*. Analysing cockroaches from nests of two host species further allowed us to test the hypothesis that nestmate recognition is based on an asymmetric mechanism. Specifically, we test the U-present nestmate recognition model, which assumes that detection of undesirable cues (non-nestmate specific substances) leads to strong rejection of the cue-bearers, while absence of desirable cues (nestmate-specific substances) does not necessarily trigger aggression.

Results: We found that nests of *Atta* and *Acromyrmex* contained cockroaches of two different and not yet described *Attaphila* species. The cockroaches share the cuticular chemical substances of their specific host species and copy their host nest's colony-specific cuticular profile. Indeed, the cockroaches are accepted by nestmate but attacked by non-nestmate ant workers. Cockroaches from *Acromyrmex* colonies bear a lower concentration of cuticular substances and are less likely to be attacked by non-nestmate ants than cockroaches from *Atta* colonies.

Conclusions: Nest-specific recognition of *Attaphila* cockroaches by host workers in combination with nest-specific cuticular chemical profiles suggest that the cockroaches mimic their host's recognition labels, either by synthesizing nest-specific substances or by substance transfer from ants. Our finding that the cockroach species with lower concentration of cuticular substances receives less aggression by both host species fully supports the U-present nestmate recognition model. Leaf-cutting ant nestmate recognition is thus asymmetric, responding more strongly to differences than to similarities.

Keywords: *Acromyrmex*, *Atta*, *Attaphila*, Camouflage, Cuticular hydrocarbons, Leaf-cutting ants, Mimicry, Myrmecophily, Nestmate recognition

Background

The coordination of complex societies requires precise communication. It is particularly important to defend the community from overt attacks and subtle forms of theft by outsiders. Potential intruders may be individuals from competing societies, but also specialised social parasites. The latter are a strong threat since they have evolved mechanisms to intrude efficiently.

Social insects use the most ancient modality to discriminate colony members from intruders: olfaction. Each individual bears a cuticular chemical profile, and between-colony variation in the profiles is informative

*Correspondence: volker.nehring@biologie.uni-freiburg.de
[2] Present Address: Department for Ecology and Evolution, Biology I, Freiburg University, Hauptstr. 1, 79104 Freiburg, Germany
Full list of author information is available at the end of the article

about the colony identity, making the profile an ideal 'nestmate recognition cue' [1, 2]. When one individual ("discriminator") encounters another individual whose cues do not match that of the discriminator's colony, the discriminator typically attacks the encountered individual. The nestmate recognition process employed by social insects has led to adaptations by social parasites to copy their host's chemical cues, either by synthesizing the respective substances and/or by acquiring them from their hosts [3, 4].

Nestmate recognition appears to be asymmetric. When individuals from two different colonies A and B meet, they are not necessarily equally aggressive towards each other, even if they are in principle equally motivated to attack intruders [5, 6]. The discrepancy between the cuticular profiles of colonies A and B may be perceived differently by A- and B-individuals. For instance, discriminators from colony A may perceive the odour dissimilarity as larger, and therefore be more likely to treat B-individuals as non-nestmates, than vice versa. This effect is possible because odour differences do not appear to be measured by the ants by a simple equivalent of Euclidean distance. Instead, it has been proposed that discriminators only react aggressively to non-nestmate profiles when these contain substances that are novel to the discriminator ants or when a given substance is more concentrated in the opponent's cuticular profile than in the discriminator's own profile (u-present model, [5, 7]). In the example laid out above, the odour blends of colonies A and B would include the same substances, but colony B would have an additional substance [5]. Such a recognition asymmetry can be caused by a desensitization of olfactory receptor neurons by the constant exposure to colony-specific substances. The neurons will then sensitively react to an increase in the quantity of any substance, but not to a reduction (pre-filter hypothesis [8]). The asymmetry could also be explained by a process of habituation, a form of non-associative learning [5]. In any case, social insects are more sensitive to differences than to similarities.

The "asymmetry hypotheses" also explain why callow workers can easily be transferred between social insect colonies without receiving aggression by non-nestmate workers. Callows bear a very low concentration of cuticular substances and will thus not be detected as intruders by ants with fully developed cuticular profiles [9]. Some social parasites exploit this effect to intrude host colonies by reducing the amount of recognition cues they bear (chemical insignificance, [4]). Lacking a profile can mean that the concentration of cuticular substances is generally low [10–13], or that the substances are not relevant for recognition (e.g. linear alkanes [14, 15]). Studying how social parasites gain entrance to social insect colonies can

thus improve our understanding of the nestmate recognition process in general.

Using behavioural experiments and chemical analyses, we investigated the chemical strategies that myrmecophilous cockroaches of the genus *Attaphila* employ to enter leaf-cutting ant colonies. *Attaphila* are small (ca. 3 mm body length) cockroaches with apterous females and brachypterous males (Fig. 1). Leaf-cutting ant colonies are fruitful targets for social parasites because the ants cultivate a fungus with nutritional hyphae for food that intruders can also feed on. Hitherto, six different *Attaphila* species have been described, which are typically found in *Atta* and *Acromyrmex* colonies [16–21]. The cockroaches disperse between colonies by following the ants' foraging trails and clinging to female *Atta* sexuals that depart for the nuptial flight [22, 23]. We collected *Attaphila* cockroaches from colonies of *Atta colombica* and *Acromyrmex octospinosus* leaf-cutting ants in Panama.

We tested whether the cockroaches bear colony-specific cuticular chemical profiles and whether cockroaches are recognised as intruders when they encounter discriminator ants from non-nestmate colonies. The experiments were designed to examine whether the cockroaches copy their host colony's recognition label or evolved another way of intruding into host colonies. Having access to cockroaches from colonies of two different leaf-cutting ant genera that differ in the concentration of their cuticular chemical profile, we could also test the hypothesis that nestmate recognition is asymmetric.

Fig. 1 An *Attaphila* male on the fungus garden of a leaf-cutting ant colony

Results

General observations

Ant workers from the two species were of similar size (U-Test n = 33, p = 0.81; head width *Atta* n = 18, $\bar{x}$ = 1.69 mm, sd = 0.31 mm; *Acromyrmex* n = 15, $\bar{x}$ = 1.71 mm, sd = 0.37 mm). Similarly, no morphometric measurements, including surface area, differed between *Atta*- and *Acromyrmex*-associated cockroaches (U-Tests, n = 42, p > 0.4 in all cases; head width of *Atta*-associated cockroaches n = 27, $\bar{x}$ = 1.52 mm, sd = 0.21 mm; *Acromyrmex*-associated cockroaches n = 15, $\bar{x}$ = 1.42 mm, sd = 0.35 mm). All cockroach morphometric measurements were highly correlated (typically r > 0.7, except eye distance with body and head length, where r = 0.66 and r = 0.50, respectively) and a principal component analysis yielded only a single principal component (PC) with an eigenvalue larger than one. Not surprisingly, nymphs were smaller than adults (generalized linear model (glm) on the PC, p = 0.04), but the ant species whose colonies the cockroaches were collected from did not affect the PC values (p = 0.55, interaction age x species p = 0.94).

Although *Attaphila* individuals from *Atta* and *Acromyrmex* colonies did not differ in size, closer inspection of their morphology revealed that they were in fact two different and so far not described species (HB, unpublished data). This was confirmed by sequences of the genes 12S, 16S, 18S, 28S, and H3 (Marie Djernæs and KK, unpublished data). The two species appear to be highly host-specific. All 24 *Atta*-associated cockroach individuals were categorized by morphology to belong to *Attaphila* sp. *A*, while all 11 *Attaphila* sp. *B* were associated with *Acromyrmex* colonies.

During the behavioural observations, several *Attaphila* individuals were observed to manipulate fungus fragments with their mouthparts, which suggests that they feed on the fungus and may thus negatively affect their host's fitness, at least when occurring in large numbers.

Behavioural experiment

The aggression an *Acromyrmex*-associated cockroach received depended on the workers it encountered (glmm p < 0.001; Fig. 2); nestmate workers were not aggressive, while allospecific (i.e. *Atta*) workers were most aggressive. The aggression of conspecific (*Acromyrmex*) non-nestmates was intermediate and differed from that of nestmates (p = 0.036) and allospecifics (p = 0.005). The pattern was similar for *Atta*-associated cockroaches (interaction between cockroach and worker origin p > 0.99), but these received overall more aggression than *Acromyrmex*-associated cockroaches (factor cockroach origin p < 0.05; Fig. 2). In total, more cockroaches survived for 48 h in nestmate

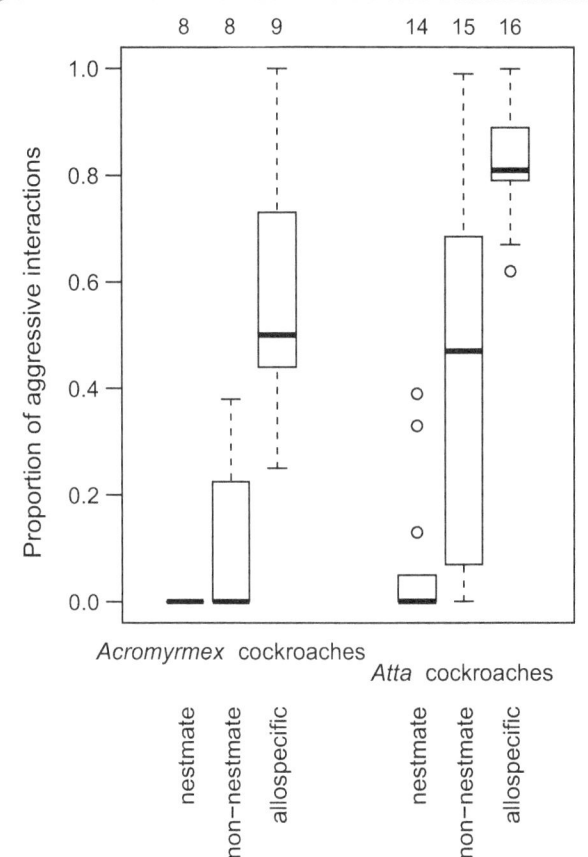

Fig. 2 Aggression received by *Attaphila* cockroaches. Cockroaches associated with *Acromyrmex octospinosus* and *Atta colombica* colonies received aggression from non-nestmate, but hardly from nestmate workers. Allospecific workers were always more aggressive than conspecific non-nestmate workers, and cockroaches from *Atta* colonies received more aggression than those that were associated with *Acromyrmex* colonies. The *boxplots* indicate median (*horizontal mark*), interquartile range (*box*), data range, and outliers. The sample size is specified above the *boxes*

subcolonies (17 out of 22) than in non-nestmate and allospecific subcolonies (9/23 non-nestmate and 10/25 allospecific subcolonies, Pearson's χ^2 = 8.58, df = 2, p = 0.014;). Whether cockroaches would die could be predicted from the proportion of aggressive encounters, with a predicted 34 % of the cockroaches dying when there was no aggression, and 74 % dying when all encounters were aggressive (glm with binomial errors, n = 70, p = 0.015). Dead cockroaches were typically transported into the subcolony's trash pile. Mortality among non-attacked cockroaches was likely due to the aggression test arenas not being optimal for long-term housing (cockroaches did not find suitable hideouts and were constantly fleeing from ants, and may have suffered from desiccation).

Chemical analysis

We found qualitative differences between samples collected from nests of the two different ant species, but no qualitative differences between cockroaches and workers collected from nests of the same species (Fig. 3; Table 1). In total, 46 GC-peaks of *Acromyrmex* workers and *Acromyrmex*-associated cockroaches, and 44 peaks from *Atta* workers and *Atta*-associated cockroaches

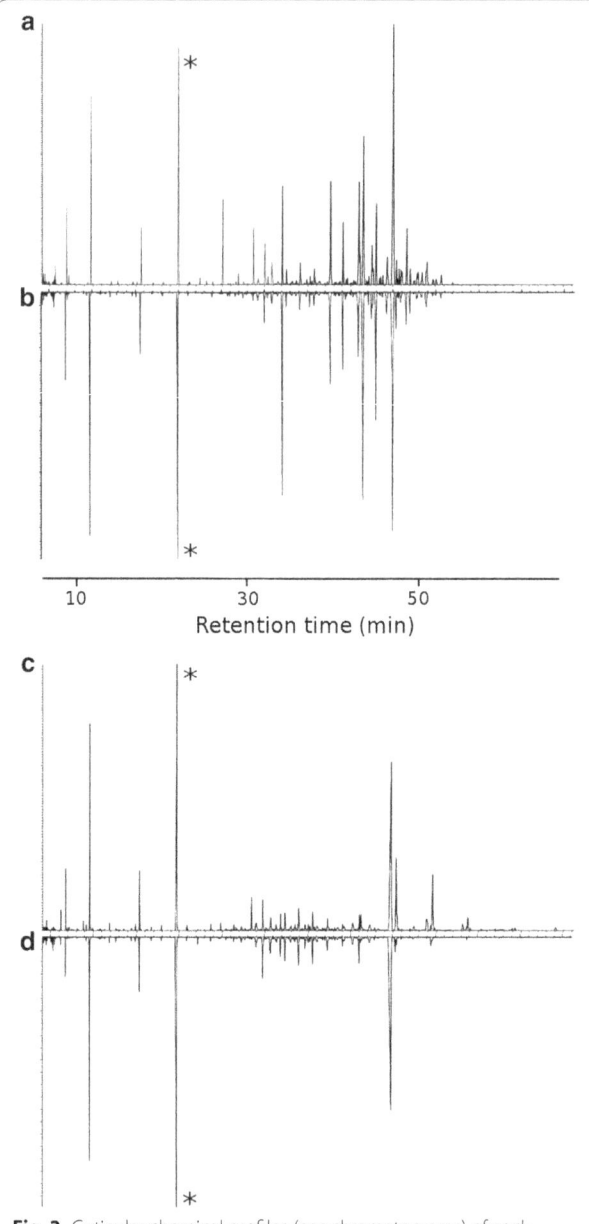

Fig. 3 Cuticular chemical profiles (gas chromatograms) of cockroaches and leaf-cutting ant workers. **a** *Atta* worker; **b** *Atta*-associated cockroach; **c** *Acromyrmex* worker; **d** *Acromyrmex*-associated cockroach. The profiles of the cockroaches are more similar to those of their host workers than among each other. All profiles include the peak of an internal standard (*)

were used for further analysis (Table 1). The chemical profiles of *Acromyrmex* workers and *Acromyrmex*-associated cockroaches consisted largely of a row of unsaturated hydrocarbons that were absent from *Atta* colonies (C29:1, C31:1, C31:2, C31:1, C37:2; Table 1). Specific to samples from *Atta*-colonies were trimethyls C31 and C34, which were very abundant in these samples. Samples from colonies of both species contained large quantities of docoseneamide and small amounts of octadecenamide. Amides have been previously found in nests of leaf-cutting ants by Richard et al. [24], who suggested that they may be produced by the symbiotic fungus. Such substances have otherwise rarely been described from social insects, apart from the Dufour gland and inside of the body of *Polistes* paper wasps [25]. These unusual substances could in theory be contaminants from plasticware, but are also known to be used in arthropod communication [26, 27].

While there were no qualitative differences, short-chained *n*-alkanes (*n*-C20 up to *n*-C22) were more abundant in the *Acromyrmex*-associated cockroaches than in *Acromyrmex* ants. In contrast, longer-chain *n*-alkanes (mainly *n*-C27) were more abundant in *Acromyrmex* workers (Table 1). Overall, however, the relative amount of linear alkanes did not differ between workers and cockroaches (ANOVA p = 0.087, and neither did we find differences for methylated, unsaturated, or non-hydrocarbon substances (ANOVA p > 0.35 in all cases; Table 2).

The difference between *Acromyrmex* workers and *Acromyrmex*-associated cockroaches was also evident in a multivariate analysis (Fig. 4a; Wilks MANOVA using the first four principal components (PCs), $n_W = 15$, $n_C = 12$, $\lambda = 0.21$, p < 0.001), and profiles also differed between colonies ($\lambda = 0.26$, p < 0.001). We did not find a difference between the profiles of cockroach adults (only n = 3) and nymphs (n = 9, $\lambda = 0.89$, p = 0.53). The difference between *Acromyrmex* workers and *Acromyrmex*-associated cockroaches was evident in the second, third, and forth PC, where the linear alkanes described above had high loadings (Additional file 1). In a discriminant analysis, 22 out of 27 *Acromyrmex* samples ($n_C = 12$, $n_W = 15$) were correctly classified according to their colony (81 %), which is significantly more than would be expected by chance (Fig. 4c; p < 0.001, as compared to a median of 48 % , 95 %-quantile of 63 %, and maximum of 74 % in a permutation test with random groups); all five misclassified samples were cockroaches.

Atta workers and *Atta*-associated cockroaches did not differ in the relative abundance of branched or unsaturated alkanes and non-hydrocarbons (ANOVA p > 0.09 in all cases; Table 2); however, workers had higher relative amounts of linear alkanes (p = 0.02), which was

Table 1 The cuticular substances of *Acromyrmex octospinosus* and *Atta colombica* workers and *Attaphila* cockroaches

No.	Substance	Rt	Type	Acromyrmex colonies				Atta colonies			
				Worker		Cockroach		Worker		Cockroach	
				$\bar{x}$	sd	$\bar{x}$	sd	$\bar{x}$	sd	$\bar{x}$	sd
1	Unidentified	5.3	Other	0.34	0.23	0.70	0.73	0.32	0.30	0.77	1.11
2	C12:OH	6.7	Other	2.42	1.46	2.86	2.00	2.48	2.10	5.21	2.23
3	n-C18	11.2	Linear	0.64	1.36	1.14	1.20	0.06	0.07	0.27	0.60
4	C16-OH	12.6	Other	0.21	0.21	0.30	0.28	–	–	–	–
5	n-C20	14.9	Linear	0.06	0.05	0.83	0.96	–	–	–	–
6	C18-OH	16.6	Other	0.38	0.22	0.61	0.26	0.27	0.40	0.35	0.27
7	n-C22	19.0	Linear	0.10	0.07	0.64	0.70	0.03	0.02	2.19	8.89
8	n-C23	20.9	Linear	0.31	0.24	0.29	0.29	0.26	0.12	0.15	0.19
9	Octadecenamide	22.1	Other	0.70	0.53	0.46	0.49	0.44	0.61	0.32	0.42
10	n-C24	22.8	Linear	0.24	0.17	0.77	0.79	0.26	0.16	0.26	0.50
11	n-C25	24.8	Linear	1.19	0.79	0.55	0.53	9.07	2.84	0.36	0.36
12	n-C26	26.6	Linear	0.52	0.22	0.78	0.83	0.86	0.26	0.41	0.50
13	7-, 8-, 9-, 10-, 11-, 12-, 13-MeC26	27.2	Methyl	1.01	0.55	1.06	0.48	0.61	0.27	0.99	0.36
14	n-C27	28.4	Linear	4.55	2.40	0.59	0.44	4.83	1.11	0.53	0.33
15	7-, 9-, 11-, 13-MeC27	28.9	Methyl	1.94	1.14	2.33	1.53	0.83	0.57	1.71	0.92
16	Docosenamide	29.7	Other	13.97	14.65	15.57	11.07	8.37	5.93	8.86	8.16
17	n-C28	30.0	Linear	0.55	0.39	0.67	0.68	0.68	0.33	1.51	3.83
18	9-, 10-, 11-, 12-, 13-, 14-MeC28	30.6	Methyl	2.87	4.27	3.47	6.17	1.07	0.71	1.95	1.21
19	C29:2a	30.7	Unsat	0.45	0.68	0.19	0.19	–	–	–	–
20	C29:2b	30.8	Unsat	0.47	0.62	0.27	0.24	–	–	–	–
21	C29:1	31.4	Unsat	10.90	12.01	2.60	2.31	–	–	–	–
22	n-C29	31.8	Linear	3.06	1.92	5.61	1.82	9.93	3.83	12.19	6.00
23	9-, 11-, 13-MeC29	32.2	Methyl	3.88	2.53	4.64	2.21	1.41	0.88	3.05	2.38
24	3-MeC29 + methylated alkanes	32.9	Methyl	1.32	1.22	0.95	0.69	3.09	3.95	2.18	2.99
25	10-, 11-, 12-, 13-, 14-, 15-MeC30	33.8	Methyl	3.60	2.87	2.35	1.30	2.19	1.47	3.91	2.61
26	C31:2	34.1	Unsat	5.88	9.33	5.94	3.61	–	–	–	–
27	C31:1	34.6	Unsat	3.71	3.81	5.65	2.50	–	–	–	–
28	N-C31 and 2Me-C30	34.9	Linear	1.31	0.87	4.06	2.88	0.81	0.49	1.49	0.81
29	9-, 11-, 13-, 15-Me-C31	35.4	Methyl	4.26	2.44	5.18	1.80	1.44	1.04	2.52	1.61
30	7,11-diMe-C31	36.0	Methyl	–	–	–	–	0.55	0.81	0.84	0.86
31	10-, 11-, 12-, 13-, 14-, 15-, 16-MeC32	37.2	Methyl	1.62	1.86	2.19	0.97	–	–	–	–
32	3,7,11-triMe-C31	37.3	Methyl	–	–	–	–	11.91	3.17	7.95	3.44
33	C33:2	37.4	Unsat	4.02	4.49	6.83	3.37	0.22	0.09	0.22	0.16
34	8,12-diMe-C32	37.7	Methyle	0.84	0.90	1.33	1.31	–	–	–	–
35	C33:2	37.7	Unsat	0.84	0.90	1.33	1.31	0.98	2.86	0.69	2.37
36	6,10-diMe-C32	37.9	Methyl	–	–	–	–	0.15	0.13	0.33	0.25
37	n-C33 + methylated alkane	38.2	Linear	0.71	0.52	0.93	0.43	0.27	0.22	0.84	0.61
38	9-, 11-, 13-, 15-MeC33	38.8	Methyl	2.04	1.73	2.80	1.21	–	–	–	–
39	9,13-diMe-C33	39.5	Methyl	0.59	0.69	0.14	0.11	0.35	0.68	0.18	0.11
40	3Me-C33	40.0	Methyl	1.49	1.43	1.27	1.51	0.37	0.31	1.28	1.77
41	C35:1 + C35:2	41.0	Unsat	3.33	4.31	5.31	4.43	–	–	–	–
42	Methyle alkane mixture	42.0	Methyl	–	–	–	–	4.74	1.63	4.69	1.52
43	4,8,12-triMe-C34	42.6	Methyl	–	–	–	–	12.60	4.34	14.60	5.91
44	13-, 15-, 17-MeC35	42.6	Methyl	0.64	0.78	1.11	0.40	–	–	–	–
45	x-MeC36:1	42.9	Unsat	–	–	–	–	0.40	0.47	0.35	0.75
46	x-MeC36:1	43.1	Unsat	–	–	–	–	0.31	0.25	0.91	3.06
47	x-MeC36:1	43.3	Unsat	–	–	–	–	0.46	0.28	0.32	0.21

Table 1 continued

No.	Substance	Rt	Type	Acromyrmex colonies				Atta colonies			
				Worker		Cockroach		Worker		Cockroach	
				$\bar{x}$	sd	$\bar{x}$	sd	$\bar{x}$	sd	$\bar{x}$	sd
48	5,9- and 5,11-diMe-C35	43.9	Methyl	–	–	–	–	3.75	1.21	3.32	1.09
49	3,7,11-triMe-C35	45.0	Methyl	–	–	–	–	5.05	6.62	3.81	2.00
50	C37:2 + C37:1	45.0	Unsat	5.59	6.78	3.50	2.07	–	–	–	–
51	x-MeC37:1	45.3	Unsat	–	–	–	–	0.55	0.72	0.65	0.83
52	4,8,12-triMe-C36	46.6	Methyl	–	–	–	–	2.61	0.79	2.65	0.77
53	x-MeC38:1	47.0	Unsat	0.75	0.42	0.64	0.54	1.07	0.72	0.93	0.70
54	x-MeC40:1	48.5	Unsat	–	–	–	–	2.16	1.23	2.47	1.22
55	x-MeC40:1	49.2	Unsat	4.44	5.83	1.96	1.25	0.66	0.30	0.59	0.27
56	x-MeC41:1	50.1	Unsat	–	–	–	–	1.13	0.43	0.99	0.35
57	x-MeC41:1	51.4	Unsat	–	–	–	–	0.37	0.47	0.22	0.11
58	x-MeC41:2	53.8	Unsat	0.72	1.03	0.21	0.20	–	–	–	–
59	13-, 15-Me-C41	55.1	Methyl	0.38	0.47	0.13	0.10	–	–	–	–
60	Alkyl ester 1	58.8	Other	0.55	0.70	0.20	0.29	–	–	–	–
61	x-MeC44:1	59.2	Unsat	0.53	0.50	0.30	0.19	–	–	–	–
62	Alkyl ester 2	63.9	Other	0.90	0.83	0.15	0.09	–	–	–	–

Cockroach samples are separated according to the ant species they were collected with. Shown are mean and standard deviations of the relative abundances as well as the retention time (Rt) and substance class (Type). Some substances were not found in one of the species (–)

Table 2 Percentage of different substance classes from the total cuticular chemical profile of ants and cockroaches

Substance class	Acromyrmex cockroaches	Acromyrmex workers	Atta cockroaches	Atta workers
Linear Alkanes	16.9 ± 6.4	13.2 ± 5.3	20.2 ± 8.5	27.1 ± 6
Branched Alkanes	29.6 ± 9.6	26.8 ± 12.8	56.7 ± 12.4	53.1 ± 6.7
(Branched) Alka(di)enes	34.3 ± 10.6	41.7 ± 19.1	8.3 ± 4.4	8.3 ± 3.2
Non-Hydrocarbons	20.1 ± 10.6	18.2 ± 14.7	14.7 ± 7.7	11.6 ± 6.3

Mean ± standard deviation

mainly caused by differences in n-C25 and n-C27 (Table 1). In the multivariate analysis, the cuticular profiles of samples collected from Atta colonies also differed between cockroaches and workers (Fig. 4b; Wilks MANOVA, $n_W = 19$, $n_C = 20$, $\lambda = 0.14$, $p < 0.001$) and between colonies ($\lambda = 0.03$, $p < 0.001$) in a MANOVA on the first eight principal components. There may be a difference between cockroach life stages (6 adults and 14 nymphs; $\lambda = 0.61$, $p = 0.07$). Inspecting the contribution of the eight PCs and the loadings of all substances on each PC separately revealed that the difference between workers and cockroaches was mostly evident in the two first PCs (see Fig. 4b; Additional file 1), but no single substances were responsible for this difference as loadings were evenly distributed on these PCs. The discriminant analysis correctly identified the colony origin of 34 out of 39 samples ($n_C = 20$, $n_W = 19$) from Atta colonies (87 %; in a permutation test with random grouping, a median of 38 %, 95 %-quantile of 49 %, and maximum of 59 % of the

samples were classified correctly; $p < 0.001$; Fig. 4d). The algorithm predicted the colony for three workers and two cockroaches incorrectly.

The concentration of cuticular substances on workers from Acromyrmex colonies was lower than on those from Atta colonies (U-Test, $n = 32$, $p < 0.01$; Fig. 5), and there was a similar trend in the cockroaches ($n = 36$, $p = 0.067$).

Discussion

Attaphila cockroaches were mostly attacked by non-nestmate but rarely by nestmate workers, suggesting that the cockroaches bear colony-specific recognition cues that can be detected by the ants. Our chemical analyses confirmed this: the cockroaches did not only share a species-specific cuticular chemical profile with their host ants, but also a colony-specific label, which is often found to be the case in social parasites and myrmecophiles [4, 28–30]. The cockroaches could blend into their host

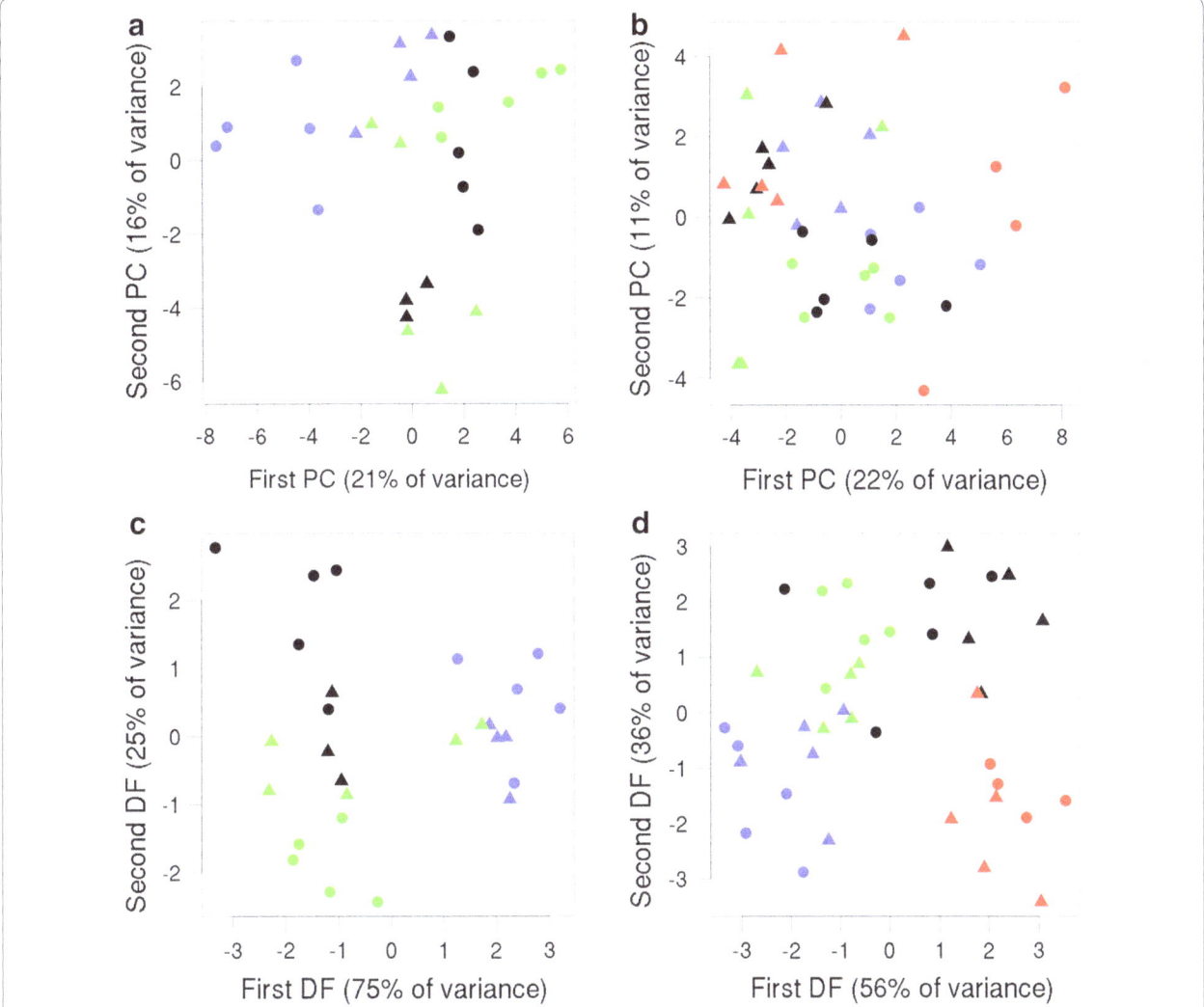

Fig. 4 Multivariate representations of cockroach and worker cuticular chemical profiles. The first two principle components (PCs, *panels* **a**, **b**) and discriminant functions (DFs, *panels* **c**, **d**) for chemical profiles of workers and cockroaches from *Acromyrmex* (**a**, **c**) and *Atta* (**b**, **d**) colonies. The discriminant analyses were set up to discriminate between colonies. *Triangles* represent cockroaches and *circles* workers; colours code for the different colonies

colonies either by acquiring a colony-specific chemical profile through substance transfer (camouflage [4]) or by synthesizing relevant chemicals (mimicry [31]). Currently we cannot exclude either possibility. As the cockroaches from *Atta* and *Acromyrmex* appear to be two specialised species, mimicry would be a reasonable explanation. Alternatively, both cockroach species could bear a similar "totipotent" profile that can be camouflaged into resembling any ant species by substance transfer. To discern these two hypotheses, cockroaches would need to be "crossfostered" for a while in colonies of another host species (e.g. *Atta*-associated cockroaches crossfostered in *Acromyrmex* colonies) or be kept in isolation before the cuticular chemical profiles are sampled.

Hypotheses regarding the other chemical mechanism for disguise, chemical insignificance, are comparably hard to test in the *Attaphila*-attine system. Ant and cockroach morphology differ vastly, so that it is difficult to calculate an estimate of surface area that is reliably comparable between ants and cockroaches. We can thus not test whether the cockroach cuticular chemical substances are less concentrated than those on the ants, which would be one way to achieve insignificance. The other way would be a relative reduction of recognition-relevant substances within the chemical profile, with no need for systematic variation in the total substance concentration. *Acromyrmex insinuator*, an ant social parasite of *Acromyrmex echinatior*, reduced the overall

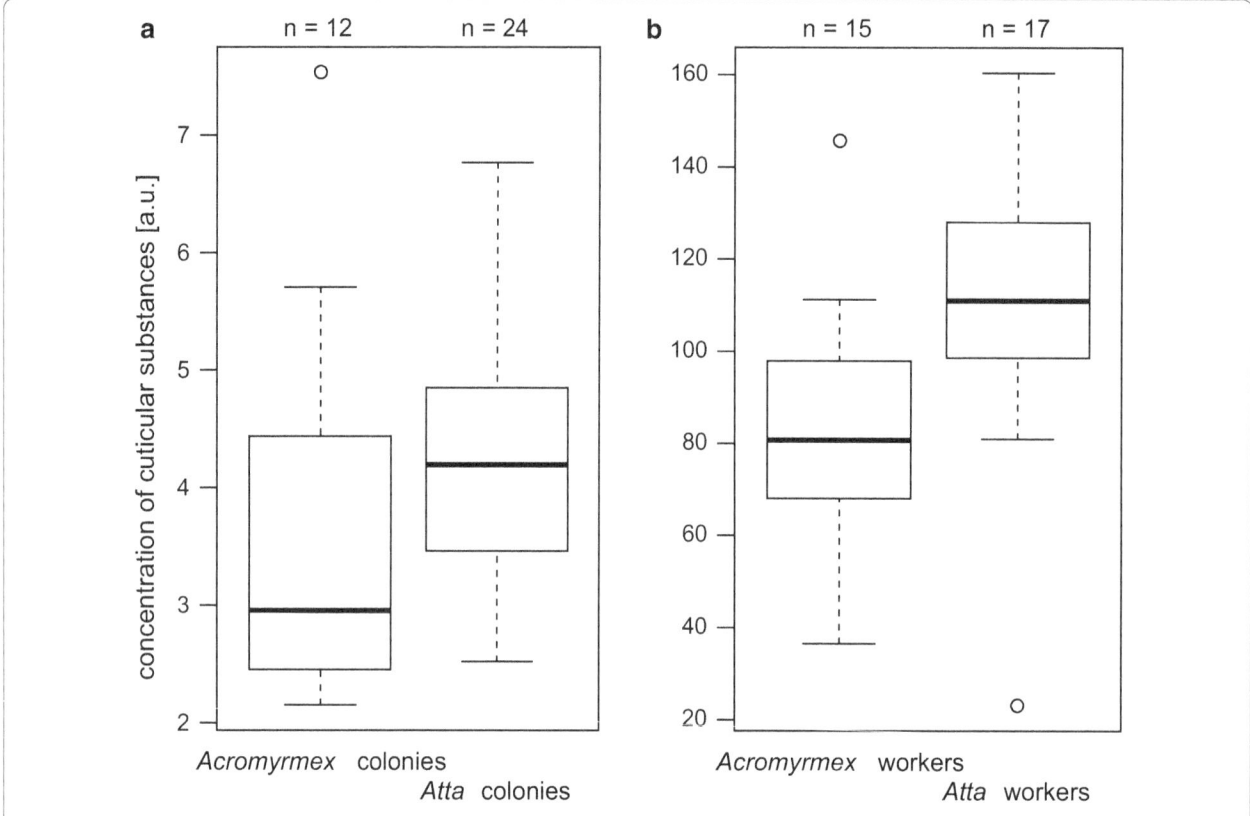

Fig. 5 The concentration of all cuticular substances for *Attaphila* cockroaches (**a**) and for ant workers (**b**). The units are arbitrary since the concentrations per surface area could not be calculated exactly. The measurements only serve to compare among ants and cockroaches, respectively. Values are not comparable between ants and cockroaches. *Boxplots* indicate median (*horizontal mark*), interquartile range (*box*), data range, and outliers; sample size is specified above the boxes

concentration of the cuticular hydrocarbons and in addition produces relatively large quantities of linear alkanes as compared to its host [29, 32]. Linear alkanes are often not used for nestmate recognition, thus their overproduction may be a way to achieve chemical insignificance [7, 15]. Indeed, the chemical profiles differed between cockroaches and ant workers. However, we did not find any general pattern in the relative abundance of different substance classes. One exception is that *Atta* workers bear more *n*-alkanes than *Atta*-associated cockroaches, so that there is currently no evidence for chemical insignificance in the cockroaches. That said, leaf-cutting ants may be a special case with regard to the substances used for nestmate recognition. Non-hydrocarbons, in particular volatile substances, appear to be involved in leaf-cutting ant nestmate recognition [33, 34]. The overall picture is not entirely clear since the substances in question vary between species and non-volatile substances and hydrocarbons such as those we analysed in this paper also play a role (VN, unpublished data, [35–37]). In any case, our analysis may have missed some very volatile substances from the ants' glands that could potentially also affect

nestmate recognition, and a formal analysis of the substances used in different leaf-cutting ant species would be in order before drawing any final conclusions regarding chemical insignificance in *Attaphila* cockroaches.

Chemical strategies may not be the only possibility to facilitate social parasitism. During the behavioural experiments, the cockroaches seemed to actively avoid inspection by ant workers. Cockroaches also often hid in crevices in the fungus. This behaviour exposed only their round and smooth backs (Fig. 1; cf. [16]), making it difficult for ants to grab or bite the cockroaches. Successful attacks by the ants seem unlikely unless a cockroaches becomes exposed, for example when it is flipped onto its back.

We found that the concentration of chemical substances on workers and cockroaches collected from *Atta* colonies was higher than that on workers and cockroaches from *Acromyrmex* colonies. The difference allows for a correlative test of the hypothesis that nestmate recognition is asymmetric, i.e. sensitive to increases but ignorant to a reduction of profile concentration [5, 8]. All else being equal, individuals with lower concentrations of recognition cues should be less likely to be attacked by discriminators [13]. Even if relative abundance of substances in the recognition labels differed between the discriminator and a non-nestmate, the discriminator is expected to not perceive these differences if the substances in question are below a hypothetical physiological detection threshold (low concentration, [38]). Therefore, theory predicts that cockroaches from *Acromyrmex* colonies should receive less aggression than those from *Atta* colonies. Indeed, we observed this pattern in our aggression experiments with cockroaches from nests of the two leaf-cutting ant species. Our results fully support the U-present model [5] and are compatible with the pre-filter hypothesis [8]. Further experiments, including experimentally manipulated cuticular profiles and electrophysiology, specifically tailored to test the models, are called for to refine our understanding of ant nestmate recognition.

Conclusions

Attaphila cockroaches found in colonies of *Atta colombica* and *Acromyrmex octospinosus* bear colony-specific recognition cues. Thus, the cockroaches are accepted by nestmate but attacked by non-nestmate ant workers. A comparative analysis of chemical and behavioural data supports that nestmate recognition is based on an asymmetric mechanism, i.e. it is more sensitive to an increase than to a reduction of chemical recognition cues.

Methods
Animals
We collected fragments of six mature *Atta colombica* ant colonies and four entire *Acromyrmex octospinosus* ant colonies that contained individuals of *Attaphila* cockroach species in Gamboa, Panama. We set up the colonies in plastic bowls (20–30 cm diameter), with Fluon-covered sides, in a laboratory close to the field site. The fungus gardens of the ant colonies were covered by smaller plastic bowls (10–20 cm diameter) to keep the appropriate humidity. The colonies were housed under natural temperature and light conditions and were fed with *Lagestroemia speciosa* leaves and mango fruits. We used cockroaches and ant workers in behavioural experiments and also analysed their cuticular chemicals.

Behavioural experiment
We set up subcolonies with 200 mg of fungus in Petri dishes (6 cm diameter), the minor workers that were naturally present in it, four medium workers with a head width of 1–2 mm, and one larger worker (2–2.5 mm head width). The lid was closed during all experiments, and half of the dish's floor was covered with moist filter paper to prevent desiccation. We let the ants acclimate for 1 h,

and then introduced a single cockroach through a small hole in the lid. We observed the behaviour of the ant workers towards the cockroach for 5 min and counted the aggressive (threat, bite) and non-aggressive (antennation, indifference) interactions.

We used 45 cockroaches from the six *Atta colombica* "donor" colonies and 25 cockroaches from the four *Acromyrmex octospinosus* donor colonies for the aggression tests. Each donor colony was paired with a non-nestmate colony of the same ant species and an allospecific colony (*A. octospinosus* colonies for *Atta*-associated cockroaches and vice versa) from which subcolonies were prepared. Nestmate subcolonies contained workers and fungus from the donor colony itself. For the aggression tests, roughly equal numbers of cockroaches were introduced into nestmate, conspecific non-nestmate, and allospecific subcolonies. All subcolonies and cockroaches were used in one aggression test only, and the experimenter was blind to the origin of the cockroaches. After the behavioural observations, we kept the subcolonies intact for 48 h and regularly checked whether the cockroaches were still alive.

For each aggression test, we calculated the proportion of encounters that were aggressive and analysed them in a generalized linear mixed model with binomial errors. We used the ant species the cockroach was associated with (*Atta* vs *Acromyrmex*) and the origin of the ant workers relative to the cockroach (nestmate, non-nestmate, or allospecific colony) as fixed factors, and the identity of the colony the cockroach was collected from as a random factor. We tested for the significance of the fixed factors and their interaction using log likelihood and Akaike information criteron [39].

Chemical analysis
We extracted the cuticular chemical profiles of 3–5 cockroaches and 4–5 medium sized workers from each of three *Acromyrmex octospinosus* and four *Atta colombica* colonies by immersing each freeze-killed individual in 200 μl of *n*-pentane for 5 min. We did not extract individuals from all ten colonies used in the behavioural experiment because some of these did not contain sufficient numbers of cockroaches. The pentane was evaporated and then 15 μl of *n*-heptane were added, which contained 5 ng/μl of nonadecanoic acid as internal standard. Three microlitres of the sample were then injected into an Agilent 7890A gas chromatograph (ZB-5 column 30 m × 0.25 mm, 0.1 μm thickness) coupled to a 5975C mass spectrometer. The column temperature was initially held at 70 °C for 1 min, then increased by 30 °C/min to 150 °C, then by 4 °C/min to 270 °C, at 2 °C/min to 310 °C, and finally at 4 °C/min to 320 °C, where it was held for 10 min. Inlet and transfer line were set to 300 °C.

We integrated the areas under peaks that contributed at least an average of 0.1 % to the total chemical profile of all samples of an ant species or cockroaches from colonies of an ant species. We transformed the peak areas of all substances but the internal standard according to Aitchison [40] and submitted the transformed areas to two separate principal component analyses [41], one for all samples collected in *Acromyrmex* colonies, and the other for all samples from *Atta* colonies. We used the principal components (PCs) judged informative by the broken stick method [42] for a MANOVA with the factors species (cockroach vs. ant worker), cockroach life stage (adult or nymph, based on presence/absence of tegmina and the morphology of the terminal sternites), and colony identity. We also exploratorily checked which principal components contributed to any effects found in the MANOVA and inspected the factor loadings for conspicuous substance variation. To investigate whether the cuticular profiles of cockroaches were as colony-specific as those of ant workers, we conducted a discriminant analysis with leaving-one-out crossvalidation using colony identity as the only grouping variable, not differentiating between ants and cockroaches [43]. We estimated p-values for the discriminant analysis in a permutation test with 1000 randomly drawn groups [44].

We calculated the relative amounts of different substance classes (linear alkanes, branched alkanes, unsaturated hydrocarbons, other substances) and compared them (log-transformed) between cockroaches and ant workers and among colonies using ANOVA. In samples from *Atta* colonies, two linear alkanes co-eluted with methyl-alkanes (Table 1). These two substances could not be separated for the analysis. We thus conducted the analysis twice, once attributing the peaks as linear alkanes and once as methyl-alkanes. The results obtained were not qualitatively affected and in this manuscript we report the results for classifying the peaks as linear alkanes.

We estimated the variation in cuticular substance amounts across individuals from the internal standard's area relative to that of all other substances. Since the substance amount per individual is likely to depend on the individual's size and surface area, we also took morphometric measures of the samples used in the chemical analysis. After extracting the cuticular substances, we transferred the samples individually into vials with ethanol for storage until we could take the morphometric measurements. Four cockroach and two worker samples lack morphometric measurements since labels on ethanol samples were lost during transport; these individuals were omitted from the analysis of substance concentrations. We measured head width for ant workers and body and head length and width, as well as minimum eye distance, for cockroaches. We approximated cockroach surface area treating the cockroach as a prolate spheroid with

$$A = 2\pi a^2 \times \left(1 + \frac{c}{a \times e} \times \arcsin(e)\right)$$

where a and c are body width and length and $e = \sqrt{\frac{1-a^2}{c^2}}$. As we cannot estimate worker body size by head width alone, we used squared head width as a proxy since surface scales quadratically with diameter and related measures (however, using linear or squared head width made no difference in the effects observed). We then tested whether the substance concentration, i.e. amount per surface area, differed between samples from *Acromyrmex* and *Atta* colonies. Note that the calculation would in principle yield the concentration of cuticular substances in ng per mm^2 surface area for cockroaches or per mm^2 squared head width for ants. However, the use of only a single non-hydrocarbon standard and the simplified calculation of the surface area only allow for rough estimates, so that we will only refer to "arbitrary units" to avoid the impression of an accurate measurement of concentration.

Additional files

Additional file 1. The cuticular substances of *Acromyrmex octospinosus* and *Atta colombica* workers and of *Attaphila* cockroaches collected from colonies of the two leaf-cutting ant species, with the relative abundances (mean and SD), retention times (rt), and principal component loadings (PC). Some substances were not found in one of the species (-). We also list the substance class as used for comparative statistics. "x-Me" refers to methylated alkenes with undetermined branch position.

Additional file 2. The raw chemical data (total ion count) per sample for *Acromyrmex* colonies. Peaks are numbered according to Table 1 and additional file 1; sample names include information about colony number and whether the sample is a cockroach ("roach") or a worker ("mw").

Additional file 3. The raw chemical data (total ion count) per sample for *Atta* colonies. Peaks are numbered according to Table 1 and additional file 1; sample names include information about colony identity (first word) and whether the sample is a cockroach ("roach") or a worker ("mw").

Additional file 4. The data collected in the behavioural experiment. Included are raw numbers of the occurrences of different behaviours (ignorance, investigation, grooming, threat, biting) per replicate and information on whether (1) or not (0) a cockroach died within 48 h in the aggression arena.

Authors' contributions
VN carried out the experiments and analysed the data. FRD, LC and ST participated in analysing the chemical samples and identifying chemical substances. HB and KDK classified cockroach species and life stages. VN and PdE conceived of the study and wrote the manuscript. All authors read and approved the final manuscript.

Author details
[1] Centre for Social Evolution, University of Copenhagen, Copenhagen, Denmark. [2] Present Address: Department for Ecology and Evolution, Biology I, Freiburg University, Hauptstr. 1, 79104 Freiburg, Germany. [3] Centro di Servizi di Spettrometria di Massa (CISM), University of Florence, Florence, Italy. [4] Dipartimento di Biologia, University of Florence, Florence, Italy. [5] Dipartimento di

Scienza del Suolo e Nutrizione della Pianta, University of Florence, Florence, Italy. [6] Zoologische Staatssammlung München, Munich, Germany. [7] Senckenberg Naturhistorische Sammlungen Dresden, Dresden, Germany. [8] Laboratoire d'Ethologie Expérimentale et Comparée (LEEC), Université Paris 13, Sorbonne Paris Cité, Villetaneuse, France.

Acknowledgements
We thank The Smithsonian Tropical Research Institute for providing facilities in Gamboa that allowed us to collect ant colonies and conduct the behavioural assays. We also want to thank all members of the Centre for Social Evolution for creating a pleasant working environment.

Competing interests
The authors declare that they have no competing interests.

Funding
This study was supported by The Danish National Research Foundation, a Marie Curie Excellence Grant (CODICES, EXT-CT-2004-014202), and the German Academic Exchange Service (DAAD).

References
1. d'Ettorre P, Lenoir A. Nestmate recognition. In: Lach L, Parr C, Abbott K, editor. Ant ecology. Oxford: Oxford University Press; 2010. pp. 194–208.
2. Sturgis SJ, Gordon DM. Nestmate recognition in ants (Hymenoptera: Formicidae): a review. Myrmecol News. 2012;16:101–10.
3. Nash DR, Boomsma JJ. Communication between hosts and social parasites. In: d'Ettorre P, Hughes DP, editors. Sociobiology of communication. Oxford: Oxford University Press; 2008. pp. 55–79.
4. Lenoir A. d'Ettorre P, Errard C, Hefetz A: Chemical ecology and social parasitism in ants. Annu Rev Entomol. 2001;46:573–99.
5. Guerrieri FJ, Nehring V, Jørgensen CG, Nielsen J, Galizia CG, d'Ettorre P. Ants recognize foes and not friends. Proc R Soc B. 2009;276:2461–8.
6. Couvillon MJ, Caple JP, Endsor SL, Kärcher MH, Russell TE, Storey DE, Ratnieks FLW. Nest-mate recognition template of guard honeybees (Apis mellifera) is modified by wax comb transfer. Biol Lett. 2007;3:228–30.
7. van Zweden JS, d'Ettorre P. Nestmate recognition in social insects and the role of hydrocarbons. In: Blomquist GJ, Bagneres AG, editors. Insect hydrocarbons: biology, biochemistry, and chemical ecology. Cambridge: Cambridge University Press; 2010. p. 222–43.
8. Ozaki M, Hefetz A. Neural mechanisms and information processing in recognition systems. Insects. 2014;5:722–41.
9. Breed MD, Perry S, Bjostad LB. Testing the blank slate hypothesis: why honey bee colonies accept young bees. Insectes Soc. 2004;51:12–6.
10. Jeral JM, Breed MD, Hibbard BE. Thief ants have reduced quantities of cuticular compounds in a ponerine ant. Ectatomma ruidum. Physiol Entomol. 1997;22:207–11.
11. Kroiss J, Schmitt T, Strohm E. Low level of cuticular hydrocarbons in a parasitoid of a solitary digger wasp and its potential for concealment. Entomol Sci. 2009;12:9–16.
12. Lorenzi MC. The result of an arms race: the chemical strategies of Polistes social parasites. Ann Zool Fennici. 2006;43:550–63.
13. Cini A, Gioli L, Cervo R. A quantitative threshold for nest-mate recognition in a paper social wasp. Biol Lett. 2009;5:459–61.
14. Martin SJ, Takahashi J-I, Ono M, Drijfhout FP. Is the social parasite Vespa dybowskii using chemical transparency to get her eggs accepted? J Insect Physiol. 2008;54:700–7.
15. Kilner RM, Langmore NE. Cuckoos versus hosts in insects and birds: adaptations, counter-adaptations and outcomes. Biol Rev Camb Philos Soc. 2011;86:836–52.
16. Wheeler WM. A new myrmecophile from the mushroom gardens of the Texan Leaf-Cutting Ant. Am Nat. 1900;107:731.
17. Bolívar I. Un nuevo ortóptero mirmecófilo Attaphila Bergi. Com Mus Nac Buenos Aires. 1901;1:331–6.
18. Bolívar I. Les blattes myrmécophiles. Mitt Schweiz ent Ges. 1905;11:134–41.
19. Gurney AB. Studies in certain genera of American Blattidae (Orthoptera). Proc Ent Soc Wash. 1937;39:101–12.
20. Brossut R. Étude morphologique de la blatte myrmécophile Attaphila fungicola Wheeler. Insectes Soc. 1976;23:167–74.
21. Wheeler WM. The social insects, their origin and evolution. London: Kegan Paul, Trench, Trubner & Co., LTD.; 1928.
22. Moser JC. Inquiline roach responds to trail-marking substance of leaf-cutting ants. Science. 1048;1964:143.
23. Moser JC. Mating activities of Atta texana (Hymenoptera, Formicidae). Insectes Soc. 1967;14:295–312.
24. Richard F-J, Poulsen M, Drijfhout FP, Jones GR, Boomsma JJ. Specificity in chemical profiles of workers, brood and mutualistic fungi in Atta, Acromyrmex, and Sericomyrmex fungus-growing ants. J Chem Ecol. 2007;33:2281–92.
25. Dani FR. Cuticular lipids as semiochemicals in paper wasps and other social insects. In: Ann Zool Fennici, Vol 43. Helsinki: Suomen Biologian Seura Vanamo. 1964; 2006:500–514.
26. McDonald GR, Hudson AL, Dunn SMJ, You H, Baker GB, Whittal RM, Martin JW, Jha A, Edmondson DE, Holt A. Bioactive contaminants leach from disposable laboratory plasticware. Science. 2008;322:917.
27. Zhang D, Terschak JA, Harley MA, Lin J, Hardege JD. Simultaneously hermaphroditic shrimp use lipophilic cuticular hydrocarbons as contact sex pheromones. PLoS One 2011; 6.
28. Martin SJ, Bayfield J. Is the bee louse Braula coeca (Diptera) using chemical camouflage to survive within honeybee colonies? Chemoecology. 2014;24:165–9.
29. Nehring V, Dani FR, Turillazzi S, Boomsma JJ, D'Ettorre P. Integration strategies of a leaf-cutting ant social parasite. Anim Behav. 2015;108:55–65.
30. Guillem RM, Drijfhout F, Martin SJ. Chemical deception among ant social parasites. Curr Zool. 2014;60:62–75.
31. Dettner K, Liepert C. Chemical mimicry and camouflage. Annu Rev Entomol. 1994;39:129–54.
32. Lambardi D, Dani FR, Turillazzi S, Boomsma JJ. Chemical mimicry in an incipient leaf-cutting ant social parasite. Behav Ecol Sociobiol. 2007;61:843–51.
33. Hernández JV, Goitía W, Osio A, Cabrera A, Lopez H, Sainz C, Jaffé K. Leaf-cutter ant species (Hymenoptera: Atta) differ in the types of cues used to differentiate between self and others. Anim Behav. 2006;71:945–52.
34. Hernández JV, Lopez H, Jaffé K. Nestmate recognition signals of the leaf-cutting ant Atta laevigata. J Insect Physiol. 2002;48:287–95.
35. Larsen J, Fouks B, Bos N, d'Ettorre P, Nehring V. Variation in nestmate recognition ability among polymorphic leaf-cutting ant workers. J Insect Physiol. 2014;70:59–66.
36. Sainz-Borgo C, Leal B, Cabrera A, Hernández JV. Mandibular and postpharyngeal gland secretions of Acromyrmex landolti (Hymenoptera: Formicidae) as chemical cues for nestmate recognition. Rev Biol Trop. 2013;61:1261–73.
37. Valadares L, Nascimento D, Nascimento FS. Foliar substrate affects cuticular hydrocarbon profiles and intraspecific aggression in the leafcutter ant Atta sexdens. Insects. 2015;6:141–51.
38. Nehring V, Wyatt TD, d'Ettorre P. Noise in chemical communication. In: Brumm H, editor. Animal communication and noise. New York: Springer; 2013. p. 373–405.
39. Bates D, Maechler M, Bolker BM: lme4: Linear mixed-effects models using S4 classes. 2011.
40. Aitchison J. The statistical analysis of compositional data. New Jersey: Blackburn Press; 1986.
41. R Development Core Team. R: a language and environment for statistical computing. Vienna: R Foundation for Statistical Computing; 2006.
42. Oksanen J, Blanchet FG, Kindt R, Legendre P, O'Hara RB, Simpson GL, Solymos P, Stevens MHH, Wagner H. vegan: Community Ecology Package. 2011.
43. Venables WN, Ripley BD. Modern applied statistics with S. Fourth. New York: Springer; 2002.
44. Nehring V, Evison SEF, Santorelli LA, D'Ettorre P, Hughes WOH. Kin-informative recognition cues in ants. Proc R Soc B. 2011;278:1942–8.

4

High genetic diversity in the offshore island populations of the tephritid fruit fly *Bactrocera dorsalis*

Chunyan Yi[†], Chunyan Zheng[†], Ling Zeng[*] and Yijuan Xu[*]

Abstract

Background: Geographic isolation is an important factor that limit species dispersal and thereby affects genetic diversity. Because islands are often small and surrounded by a natural water barrier to dispersal, they generally form discrete isolated habitats. Therefore, islands may play a key role in the distribution of the genetic diversity of insects, including flies.

Results: To characterize the genetic structure of island populations of *Bactrocera dorsalis*, we analyzed a dataset containing both microsatellite and mtDNA loci of *B. dorsalis* samples collected from six offshore islands in Southern China. The microsatellite data revealed a high level of genetic diversity among these six island populations based on observed heterozygosity (Ho), expected heterozygosity (H_E), *Nei's* standard genetic distance (*D*), genetic identity (*I*) and the percentage of polymorphic loci (PIC). These island populations had low F_{ST} values ($F_{ST} = 0.04161$), and only 4.16 % of the total genetic variation in the species was found on these islands, as determined by an analysis of molecular variance. Based on the mtDNA COI data, high nucleotide diversity (0.9655) and haplotype diversity (0.00680) were observed in all six island populations. *F*-statistics showed that the six island populations exhibited low or medium levels of genetic differentiation among some island populations. To investigate the population differentiation between the sampled locations, a factorial correspondence analysis and both the unweighted pair-group method with arithmetic mean and Bayesian clustering methods were used to analyze the microsatellite data. The results showed that Hebao Island, Weizhou Island and Dong'ao Island were grouped together in one clade. Another clade consisted of Shangchuan Island and Naozhou Island, and a final, separate clade contained only the Wailingding Island population. Phylogenetic analysis of the mtDNA COI sequences revealed that the populations on each of these six islands were closely related to different populations on mainland China.

Conclusions: Our study suggests that these island populations have high genetic diversity, experience frequent gene flow and exhibit low or medium levels of genetic differentiation among some island populations. Therefore, the geographic isolation of the six islands does not appear to be a major dispersal barrier to *B. dorsalis*. Such knowledge is helpful for a better understanding of evolutionary processes of the species of island populations.

Keywords: Oriental fruit fly, Population genetic structure, Mitochondrial DNA, Microsatellite, Genetic diversity, Island isolation

*Correspondence: zengling@scau.edu.cn; xuyijuan@scau.edu.cn;
xuyijuan@yahoo.com
†Chunyan Yi and Chunyan Zheng contributed equally to this work
Laboratory of Insect Ecology, Department of Entomology, South China
Agricultural University, Guangzhou 510640, China

Background

Genetic diversity is a critical component of biodiversity and affects the survival and evolution of species [1, 2]. Geographic isolation is an important factor that limits species dispersal and affects the genetic diversity of species [3, 4]. By definition, islands are smaller than continents, surrounded by water, and therefore form discrete habitats isolated from other terrestrial habitats [5, 6]. Generally, the water surrounding islands acts as a geographic barrier to dispersal that limits gene flow both between island populations and between island and mainland populations. Consequently, the genetic diversity of insular species tends to be more complex due to multiple factors that include the natural dispersal distance of flight-capable species and dispersal that is mediated by human activity. In such special cases, island populations have lower genetic diversity than do mainland populations due to founder effects/bottlenecks and continued isolation from the mainland [3, 7–10]. To maintain a population's fitness and adapt to an island environment, a population may lose genetic diversity [11]. Previous bottlenecks or continued isolation from the mainland can also decrease genetic diversity. Moreover, the loss of genetic diversity in island populations can often be caused by founder effects, breeding rates, and dispersal ability, among other factors. However, the island populations of some species, i.e., mammals and birds, have higher levels of genetic diversity than do mainland populations [8, 11, 12].

Bactrocera dorsalis (Hendel) (Diptera: Tephritidae), the oriental fruit fly, is a quarantine pest that is found worldwide, is highly fecund and highly adaptable. This fly species causes serious economic losses to fruit production in tropical and subtropical areas [13–16]. The original source of *B. dorsalis* populations has been reported to be from tropical or subtropical Asia [17]. To date, *B. dorsalis* has spread over Asia and to many regions around the Pacific, including Hawaii [18]. *Bactrocera dorsalis* was first recorded in Taiwan in 1911 but has since spread to many other provinces in China [19, 20]. Wan et al. [20] examined the genetic diversity and genetic differentiation of *B. dorsalis* in the Chongqing region in China. Their results indicated that the height of these mountains in this region was insufficient to prevent the long-distance dispersal of *B. dorsalis* and suggested that there would be a high frequency of gene flow among fly populations. However, Li et al. [16] demonstrated that geographic isolation from mountains and canyons distributed across China, Vietnam and Thailand slowed the dispersal of *B. dorsalis* and has resulted in genetic differentiation between these regions. To date, no studies have explored the genetic diversity of *B. dorsalis* on islands. In this study, we hypothesized that six island populations

of *B. dorsalis* in South China would have lower levels of genetic diversity due to their isolation from the mainland. Furthermore, the isolation of islands may also result in high levels of genetic differentiation between different island populations. In this study, we generated and analyzed a dataset of both microsatellite and mtDNA loci from *B. dorsalis* samples collected from six offshore island populations in South China to estimate the genetic divergence and dispersal ability of these flies. Our results may also suggest possible strategies for the control of this species based on their dispersal patterns.

Methods

Sample collection

Bactrocera dorsalis were sampled from six offshore islands in South China (Fig. 1), and the longitude and latitude information of all locations were recorded. No permissions were required to collect these samples from the field. The map for sample distribution was generated by DIVA-GIS version 7.1. Shangchuan Island, Hebao Island, Dong'ao Island and Wailingding Island are located in Guangdong Province, and Naozhou Island and Weizhou Island are located in Guangxi Province. A number of fruits, including mango, banana and papaya, are cultivated on the islands. However, there is an exceptionally low population density of *B. dorsalis* on the islands. Male *B. dorsalis* individuals were captured using methyl eugenol-baited traps on each island. After collection, the samples were preserved in 90 % ethanol prior to DNA extraction.

PCR amplification

DNA from individual specimens was extracted using the TIANamp Genomic DNA Kit (Qiagen, Hilden, Germany). In total, DNA was extracted from 20 flies from each island. Eight polymorphic microsatellite loci (MS4, MS6, MS12A, 4.3A, 6.8A, 4.6A, Ccmic32 and Bo-D48) were analyzed, all of which were previously developed [14, 21–23] (see Additional file 1: Table S1). Primers were synthesized by Sangon Biotech Co. Ltd (Shanghai, China). PCR reactions were performed in 25 μL volumes containing 2 μL of DNA extract (40 ng/μL), 2.5 μL of 10× PCR buffer (containing Mg^{2+}), 2 μL of a dNTP mixture (10 mmol/L), 1.5 μL of each primer (10 mmol/L), 0.2 μL of Taq DNA polymerase (5 U/μL) and 15.3 μL of ddH_2O. Cycling conditions for amplification were as follows: 3 min at 94 °C followed by 35 cycles of 30 s at 94 °C, 30 s at 48 °C, 45 s at 72 °C, and an extension of 5 min at 72 °C.

The PCR-amplified products were separated using a DNF-900 High Sensitivity Large Fragment Analysis Kit in a fragment analyzer (Advanced Analytical Technologies, Inc., USA). Alleles were scored using PROSize 2.0 (Advanced Analytical Technologies, Inc., USA).

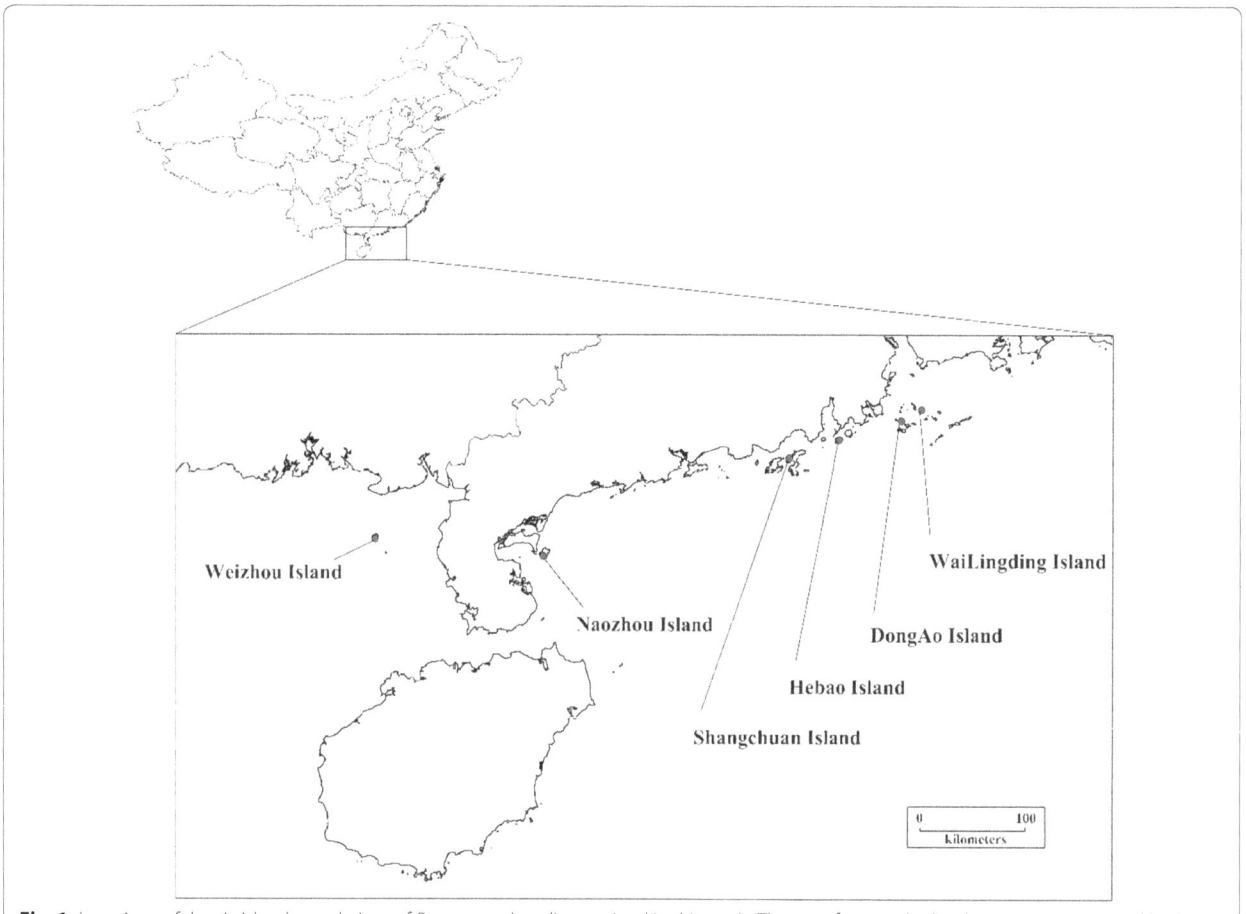

Fig. 1 Locations of the six island populations of *Bactrocera dorsalis* examined in this study. The map for sample distribution was generated by DIVA-GIS version 7.1

A 759-bp fragment of the cytochrome oxidase subunit I (COI) gene was amplified using primers (COI-F: CAACATTTATTTTGATTTTTTGG; COI-R: TCCATT GCACTAATCTGCCATATTA; synthesized by Sangon Biotech Co., Ltd (Shanghai, China)) and following the method described by Tang et al. [24, 25]. The PCR reaction volumes were the same as the microsatellite DNA amplification volumes, except the annealing temperature duration was 45 s.

Microsatellite data
Deviations from Hardy–Weinberg proportions and linkage disequilibrium were calculated using Genepop version 5.0 (http://www.wbiomed.curtin.edu.au/Genepop/Genepop_op1.html) [26]. We estimated null allele frequencies for each locus using the expectation maximization (EM) algorithm method in the FREENA software (http://www.montepllier.inra.fr/URLB) [27]. Population genetic parameters included the observed number of alleles (NA), the effective number of alleles (N_E), Shannon's information index (I), observed heterozygosity

(H_o), expected heterozygosity (H_E), the percentage of polymorphic loci (P), Nei's genetic diversity (Nei's), gene flow (N_{em}), Nei's original measures of genetic identity (I) and genetic distance (D) for paired populations. These parameters were then analyzed using Popgene Version 1.3.1 [28]. The population genetic structure was analyzed with STRUCTURE v.5.0 [29], which uses a Bayesian Markov Chain Monte Carlo (MCMC) method to analyze the numbers of K genetic clusters. We ran K values from 1 to 6 using an admixture model and an allele frequencies correlated model. The first 100,000 repetitions were discarded as burn-in, and 10,000 MCMC repetitions were then run. Multiple runs are used to check consistency between runs. Then, the results after running STRUCTURE were then uploaded to the Structure Harvester website (http://www.taylor0.biology.ucla.edu/structureHarvester/#). Structure Harvester was used to calculate six iterations per K-value and obtained the best K-value using the ΔK method of Evanno. [30]. *Nei's* genetic distance (D) for the six island populations was reconstructed for the population phylogenetic tree using the unweighted pair-group method with arithmetic mean (UPGMA) in the tools for population genetic analyses (TFPGA) program [31]. We used a multidimensional factorial correspondence analysis (FCA) within the GENETIX program to identify clusters of individuals with similar genotypes based on allele frequencies and genotype [32].

Isolation by distance was measured using the TFPGA software. Matrix correlations using pairwise genetic distance versus geographic distance were estimated and significance was determined using the Mantel test.

Analysis of molecular variance (AMOVA) was performed with ARLEQUIN ver. 3.5 to partition the genetic variance within and among populations based on the numbers of different alleles (F_{ST}) and the sum of squared size differences [33].

Mitochondrial data

Published COI sequences of *B. dorsalis* from mainland China populations and mainland populations from other countries (Cambodia, Mandalay, Laos and America) were downloaded from NCBI [16, 34] to characterize the genetic relationships between these mainland populations and the studied six island populations in South China. We cut and joined the amplification sequences using one pair of primers with DNAStar and manually corrected any obvious errors. Next, we aligned all sequences using MEGA6 software [35]. A population phylogenetic tree was reconstructed using the UPGMA method in MEGA6 based on genetic distances. Haplotype diversity (*Hd*), the average number of differences (*k*), and nucleotide diversity (*π*) were calculated for each population using DnaSP 5.0 [20, 36].

The fixation index (pairwise F_{ST}) and gene flow (N_{em}) were estimated using ARLEQUIN ver. 3.5 [33]. Generally, gene exchange leading to low genetic differentiation between populations occurs when $N_{em} > 4$ [37]. The significance level was assessed using 1000 permutations. The index was interpreted as follows: a low degree of genetic differentiation ($0 \leq F_{ST} < 0.05$); a medium degree of genetic differentiation ($0.05 \leq F_{ST} < 0.15$); a high degree of genetic differentiation ($0.15 \leq F_{ST} \leq 0.25$); and a very high degree of genetic differentiation ($F_{ST} > 0.25$) [38, 39].

Results
Microsatellite analysis
Hardy–Weinberg and genetic diversity
All loci and samples significantly deviated from Hardy–Weinberg proportions (P < 0.01). The average values of F_{IS} for each locus in each population ranged from 0.0365

to 0.8353, which indicated that there were deficiencies in heterozygotes. Over all populations and loci, the locus MS12A for the DAD population had the highest null allele frequency ($N_a = 0.41419$). The null allele frequency of the other loci and populations ranged from 0 to 0.37780. The presence of null alleles was the primary contributor to the deficiencies in heterozygotes and the departure from the Hardy–Weinberg equilibrium. There was no significant linkage disequilibrium between pairs of loci for all populations (P > 0.05). Therefore, the eight loci were inherited independently.

In total, 107 alleles were observed for eight microsatellite loci among the six island populations, and the number of alleles per locus ranged from 7 (Locus 4.3A) to 19 (Locus 6.8A), with an average of 13.375 (Additional file 1: Table S2). The expected heterozygosity (H_E) of each locus ranged from 0.5397 to 0.8806, and the observed heterozygosity (H_O) ranged from 0.1008 to 0.6441. The observed number of alleles per island population ranged from 5.75 (Dong'ao Island) to 8 (Wailingding Island) (Table 1), with an average of 7.3214. The expected heterozygosity per population ranged from 0.7532 (Wailingding Island) to 0.2442 (Hebao Island). The results showed that the six island populations all exhibited high levels of genetic diversity. The Hebao Island population exhibited a lower level of genetic diversity than did the other five island populations. Nei's genetic diversity index for the Hebao Island population was 0.1769, which was far lower than the other five island populations (Table 1).

Population genetic structure
We analyzed the genetic structure of the different *B. dorsalis* populations using the Bayesian clustering analysis method from the STRUCTURE software. The results showed that the best possible ΔK was 3 (Fig. 2). This finding suggested that the six island populations were divided into three clusters based on the allele frequencies of the geographic populations. Hebao Island, Weizhou Island and Dong'ao Island were clustered into one branch (highlighted in blue). Shangchuan Island and Naozhou Island were grouped into another branch (highlighted in red). The last branch consisted only of the Wailingding Island population (highlighted in green) (Fig. 3).

We used TFPGA software to construct an UPGMA population phylogenetic tree, and the results showed that the six island populations could be divided into three groups (Fig. 4). Hebao Island, Weizhou Island, and Dong'ao Island were classified into one group; Shangchuan Island and Naozhou Island were classified into another group; and Wailingding Island was classified into the final group. The results from the phylogenetic tree

Table 1 Indices of genetic diversity of *Bactrocera dorsalis* populations from six islands based on microsatellite data

Population	N_A	N_E	I	H_O	H_E	Nei's	Np	P (%)
Dong'ao Island	5.7500 ± 2.1876	3.7076 ± 1.8766	1.3450 ± 0.4904	0.2970 ± 0.2371	0.6740 ± 0.1850	0.6571 ± 0.1804	8	100
Wailingding Island	8.0000 ± 2.3905	4.4046 ± 1.8152	1.6184 ± 0.4118	0.4313 ± 0.2359	0.7532 ± 0.1131	0.7344 ± 0.1103	8	100
Hebao Island	7.1250 ± 2.1671	3.4046 ± 1.6079	1.3886 ± 0.4488	0.3355 ± 0.2442	0.2442 ± 0.1769	0.1769 ± 0.1725	8	100
Shangchuan Island	7.6250 ± 2.8754	3.8322 ± 1.7547	1.4875 ± 0.5302	0.3500 ± 0.3218	0.6697 ± 0.1983	0.6869 ± 0.1933	8	100
Weizhou Island	6.8750 ± 2.7999	3.4946 ± 1.5458	1.4143 ± 0.4487	0.2812 ± 0.2103	0.6796 ± 0.1494	0.6627 ± 0.1457	8	100
Naozhou Island	7.8750 ± 2.7999	4.4092 ± 1.9838	1.6249 ± 0.4329	0.4016 ± 0.2906	0.7546 ± 0.1072	0.7356 ± 0.1046	8	100
Mean	7.3214 ± 0.8192	3.8755 ± 0.4386	1.4798 ± 0.1192	0.3494 ± 0.0583	0.6292 ± 0.1926	0.6089 ± 0.2143	8	100

N_A Observed number of alleles; N_E Effective number of alleles [42]; I Shannon's information index [43]; Nei's Nei's gene diversity; H_O observed heterozygosity; H_E expected heterozygosity; Np number of polymorphic loci; P percentage of polymorphic loci

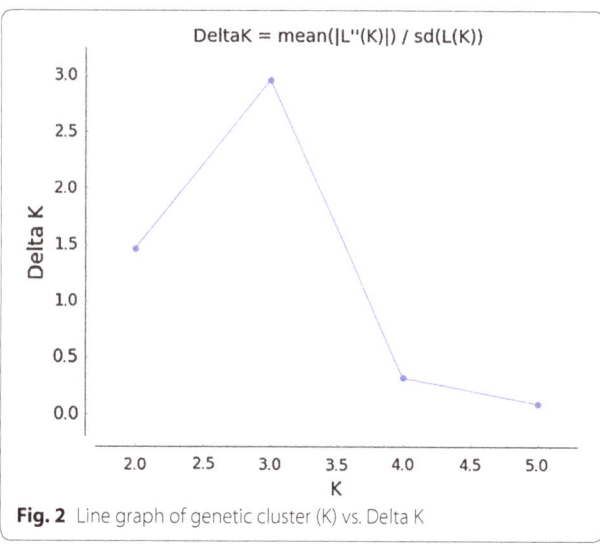

Fig. 2 Line graph of genetic cluster (K) vs. Delta K

using UPGMA based on microsatellite data were consistent with those obtained from the STRUCTURE analysis.

We analyzed the genetic divergence among populations based on the individual genotypes and constructed the three-dimensional FCA shown in Fig. 5. According to this graph, the results of the FCA were the same as the results from STRUCTURE and the phylogenetic tree.

Hierarchical *F*-statistics were estimated for all populations as a single group and for all populations partitioned into three groups based on the results of the Bayesian clustering analysis. AMOVA results indicated that genetic variation primarily contributed to variation among individuals within populations and to variation within individuals (Table 2). The genetic variation that was divided into one group and three groups was 48.6 % ($F_{IT} = 0.52765$, P < 0.001) and 47.9 % ($F_{IS} = 0.50631$, P < 0.001) of the total variation, respectively. In addition, within individual variation accounted for 47.23 % ($F_{IT} = 0.52765$, P < 0.01) and 46.75902 % ($F_{IT} = 0.53241$, P < 0.01) of the total variation for the one group and three groups analyses, respectively. Most of the total genetic variation was explained by the variation among individuals within populations.

As shown in Table 3, *Nei's* standard genetic distance (*D*) (below the diagonal) among six island populations ranged from 0.0853 to 0.4021 and *Nei's* genetic identity (I) (above the diagonal) varied from 0.6689 to 0.9182. The population genetic identity (I) between the Wailingding island population and the Weizhou island population was 0.6689, which was the minimum among all paired

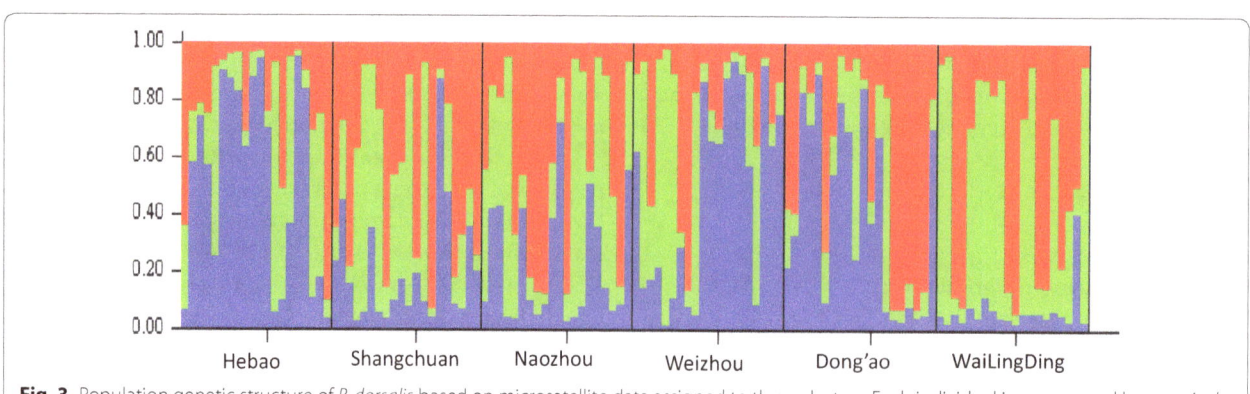

Fig. 3 Population genetic structure of *B. dorsalis* based on microsatellite data assigned to three clusters. Each individual is represented by a *vertical bar*

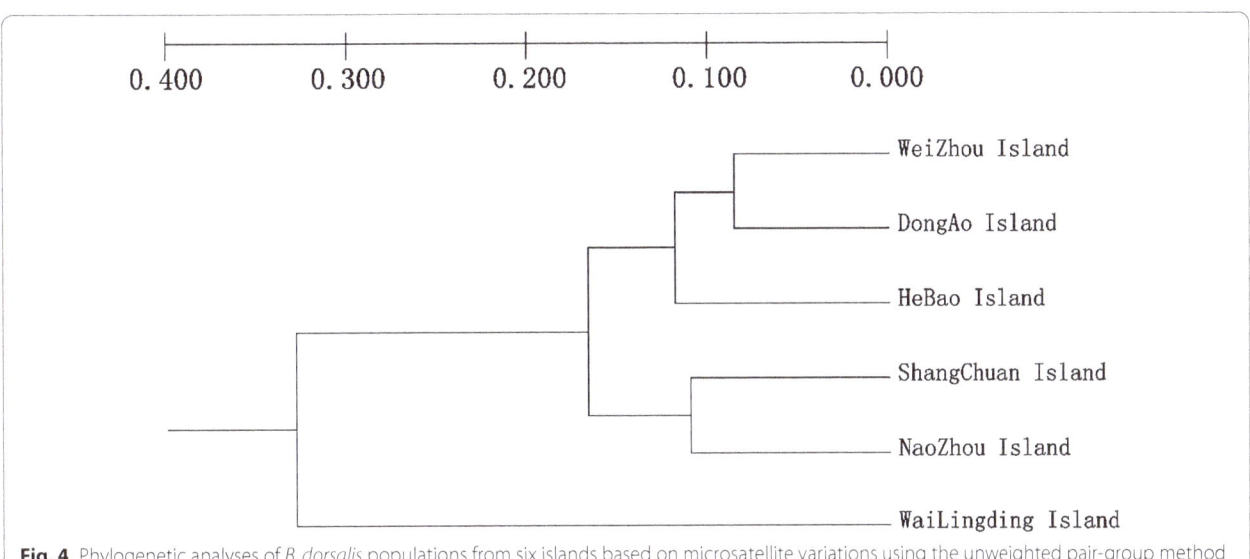

Fig. 4 Phylogenetic analyses of *B. dorsalis* populations from six islands based on microsatellite variations using the unweighted pair-group method with arithmetic mean (UPGMA)

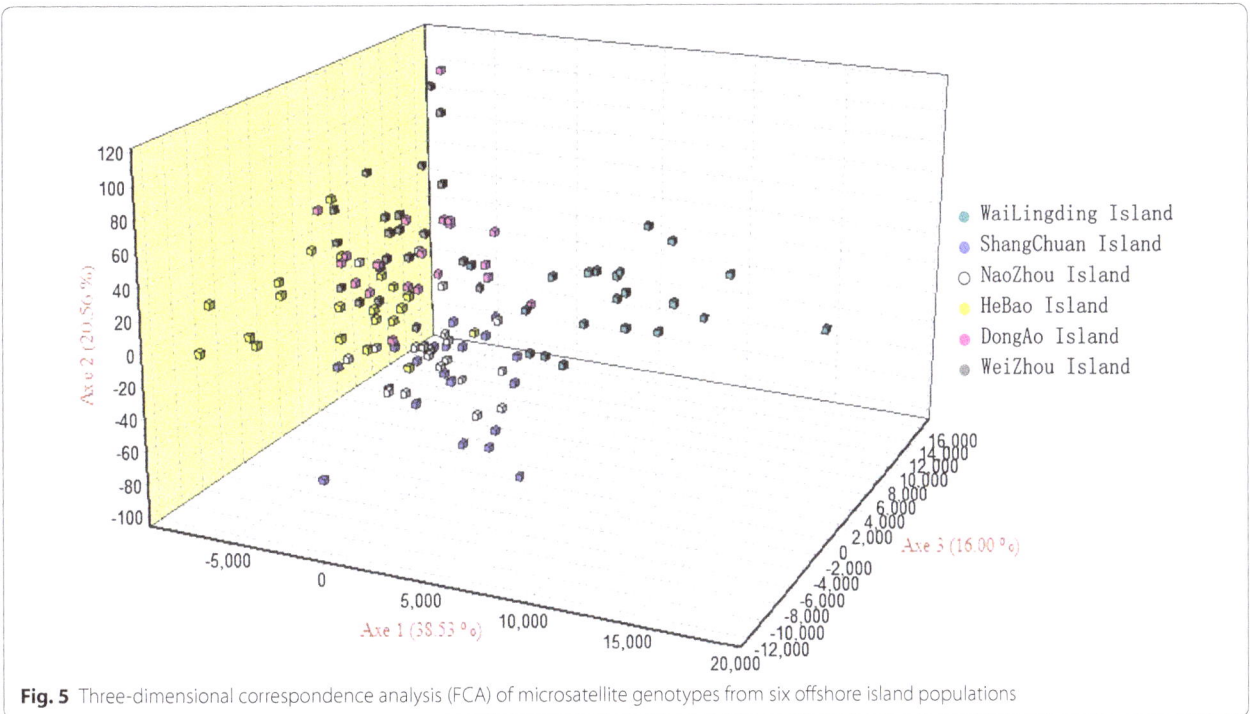

Fig. 5 Three-dimensional correspondence analysis (FCA) of microsatellite genotypes from six offshore island populations

island populations. Accordingly, there was high genetic distance, which was 0.4021 between these paired island populations (Table 3).

The relationship between genetic distance and geographic distance was determined among island ranges. The result showed that there was no relationship between genetic distance and geographic distance (Fig. 6; $R^2 = 0.0027$, P = 0.855).

Mitochondrial analysis

We obtained 120 sequences from COI with a length of 759 bp. In total, 66 different mitochondrial haplotypes were detected in the six island populations. All populations on the six islands had high levels of nucleotide (0.0068) and haplotype diversity (0.9665). Haplotype diversity (*Hd*) for each population ranged from 0.889 to 0.995. The average number of differences (*k*) ranged from

Table 2 AMOVA results based on microsatellite genotypes

Source of variation	d.f.	Variance components	Percentage of variation	Fixation indices
Among populations	5	0.12261 Va	4.16	$F_{ST} = 0.04161$**
Among individuals within populations	114	1.43202 Vb	48.6	$F_{IS} = 0.50715$**
Within individuals	120	1.39167 Vc	47.23	$F_{IT} = 0.52765$**
Among groups	2	0.13294 Va	4.46153	FCT = 0.04462**
Among populations within groups	2	0.02457 Vb	0.82472	FSC = 0.00863
Among individuals within populations	114	1.42888 Vc	47.95	FIS = 0.50631**
Within individuals	120	1.39325 Vd	46.75902	FIT = 0.53241**

Table 3 Nei's original measures of genetic identity (above diagonal) and genetic distance (below diagonal)

Population	Hebao	Shangchuan	Naozhou	Weizhou	Dong'ao	Wailingding
Hebao	****	0.8095	0.8433	0.8897	0.8883	0.6752
Shangchuan	0.2113	****	0.8968	0.8238	0.8631	0.7691
Naozhou	0.1704	0.1090	****	0.8889	0.8539	0.7570
Weizhou	0.1169	0.1938	0.1177	****	0.9182	0.6689
Dong'ao	0.1184	0.1473	0.1580	0.0853	****	0.7410
Wailingding	0.3928	0.2625	0.2784	0.4021	0.2998	****

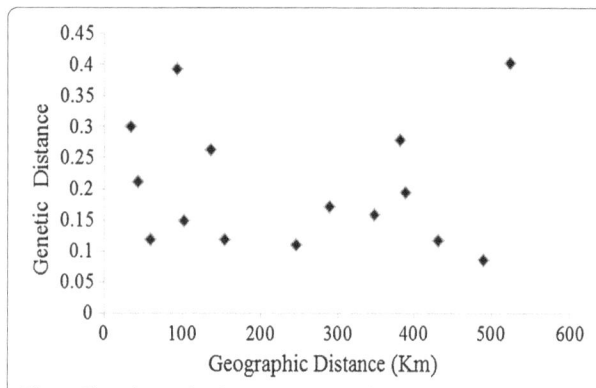

Fig. 6 The relationship between genetic distance and geographic distance of island populations

3.895 (Wailingding Island) to 6.553 (Naozhou Island). Nucleotide diversity (π) for each population ranged from 0.513 % (Wailingding Island) to 0.863 % (Naozhou Island) (Table 4).

There was genetic divergence between the six island populations, as estimated by population pairwise F_{ST} significance tests. The F_{ST} values between paired groups ranged from −0.03194 to 0.05413. The Dong'ao Island population had a low degree of differentiation with the Wailingding Island, Hebao Island and Shangchuan Island populations ($0.01024 \leq F_{ST} \leq 0.01895$). However, Dong'ao differentiated from the Weizhou and Naozhou Island populations ($0.05065 \leq F_{ST} \leq 0.05413$, P < 0.05)

(Table 5). Hebao Island had no differentiation from the Wailingding, Shangchuan and Naozhou Island populations ($−0.00179 \leq F_{ST} < 0$) (Table 5). There were low and medium levels of differentiation among some paired island populations. The value of gene flow (N_{em}) between each pair of populations was over four, which suggested that there was a full exchange of genes between the six island populations. The isolation of the islands did not prevent gene flow between the populations. These results were consistent with the microsatellite data.

The topology of the B. dorsalis population maximum likelihood tree suggested that there were no independent groups in the island populations. In the phylogenetic tree, these six island populations of Southern China were clustered in one branch with different mainland populations of China (Fig. 7). We also found that the Shangchuan and Naozhou island populations were more closely related to populations from Cambodia and Myanmar than to populations from Laos and the United States (Figs. 5, 7).

Discussion
Null alleles
Null alleles, geographic factors and likely founder effects among populations caused heterozygote deficiencies and emerged as significant departures from the Hardy–Weinberg equilibrium [40]. In this study, the frequency of null alleles in an isolated microsatellite of *Bactrocera dorsalis* was high and exceeded 0.4 only among locus 6.4. We checked the heterozygotes and found significant

Table 4 Haplotype diversity (Hd), average number of differences (k) and nucleotide diversity (π) for each population

Population	N	Haplotype (frequency)	Hd	π	k
Dong'ao Island	20	H1(1), H2(1), H3(5), H4(2), H5(5), H6(1), H7(1), H8(1), H9(1), H10(1), H11(1)	0.889 ± 0.049	0.00633	4.805
Wailingding Island	20	H3(3), H5(1), H10(1), H13(1), H19(4), H23(2), H57(1), H105(1), H142(1), H143(1), H144(1), H145(1), H146(1), H147(1)	0.947 ± 0.034	0.00513	3.895
Hebao Island	20	H3(2), H5(4), H8(1), H13(1), H19(1), H25(1), H74(1), H77(1), H102(1), H103(1), H104(1), H105(1), H106(1), H107(1), H108(1), H109(1)	0.963 ± 0.033	0.0062	4.705
Shangchuan Island	20	H3(2), H5(2), H8(1), H11(1), H13(1), H19(2), H37(1), H77(2), H105(1), H133(1), H134(1), H135(1), H136(1),H137(1), H138(1), H139(1)	0.979 ± 0.021	0.00659	5.005
Weizhou Island	20	H3(2), H5(2), H44(1), H62(1), H77(1), H107(1), H129(1), H140(1), H142(1), H148(1), H149(1), H150(1), H151(1), H152(1), H153(2), H154(1), H155(1)	0.984 ± 0.020	0.00689	5.232
Naozhou Island	20	H5(2), H8(1), H25(1), H52(1), H77(1), H119(1), H120(1), H121(1), H122(1), H123(1), H124(1), H125(1), H126(1), H127(1), H128(1), H129(1), H130(1), H131(1), H132(1)	0.995 ± 0.018	0.00863	6.553
Total	120	66	0.9665 ± 0.008	0.00680	5.112

N sample size; Hd haplotype diversity; π nucleotide diversity; k average number of differences

Table 5 F_{ST} values (below diagonal) and N_{em} (above diagonal) between the six island populations of *Bactrocera dorsalis*

Population	Dong'ao	Wailingding	Hebao	Shangchuan	Weizhou	Naozhou
Dong'ao	****	48.3333	39.8018	25.8889	8.7377	9.3713
Wailingding	0.01024	****	inf	inf	42.6287	14.8611
Hebao	0.01241	−0.01058	****	inf	80.6048	inf
Shangchuan	0.01895	0	−0.03194	****	178.3258	inf
Weizhou	0.05413*	0.01159	0.00616	0.0028	****	134.2085
Naozhou	0.05065*	0.03255*	−0.00179	−0.00089	0.00371	****

Data with asterisks indicate significant difference (P < 0.05)

heterozygote deficiencies, except for locus MS12A, in several island populations.

Genetic diversity

Population genetic diversity is a product of evolutionary change over several generations [41]. Thus, population genetic diversity reflects the ability of populations to adapt to local environments [42]. The percentage of polymorphism, the level of heterozygosity, the number of alleles, nucleotide diversity and haplotype diversity were used to estimate population genetic diversity. Our research showed that populations of the oriental fruit fly from islands in South China had high levels of genetic diversity and were closely related to populations from mainland China based on COI data analysis. Using the mtDNA COI from five Yunnan province populations of *B. dorsalis*, Shi et al. [43] showed that the average haplotype diversity and nucleotide diversity of the five populations were 0.9786 and 0.9038 %, respectively. Thus, the five Yunnan populations had higher levels of genetic diversity than did the island populations examined in this study. Analysis of the genetic diversity of *B. dorsalis* from China, Laos, and Thailand with microsatellite markers showed that the average *Nei's* genetic diversity and Shannon's information index were 0.6464 and 0.7870,

respectively. The average percentage of polymorphic loci in all populations was 94.45 %. This finding suggests that mainland South China has high levels of population genetic diversity [44].

However, many studies have shown that the gene flow of mammals, birds and other species from islands is blocked or reduced due to isolation, which results in lower levels of genetic diversity of island populations compared to mainland populations [8, 14, 45]. Isolated islands have discrete boundaries that are generally thought to reduce migration between populations on islands [3]. Long-term bottlenecks, founder effects, genetic drift and inbreeding may all reduce genetic diversity [46, 47]. Mitochondrial and microsatellite data have been previously used to estimate the genetic structure of *Apis mellifera* populations from the Canary Islands [7]. The results showed that there was a lower level of genetic variation based on the average number of alleles and heterozygosity in populations on the Canary Islands than mainland populations in Iberia and Morocco [48]. Boessenkool et al. reported that natural populations of robins on the Breaksea and Nukuwaiata Islands have lower levels of genetic diversity than larger mainland populations. In addition, some alleles have been lost, which is a result of bottlenecks and isolation from the mainland

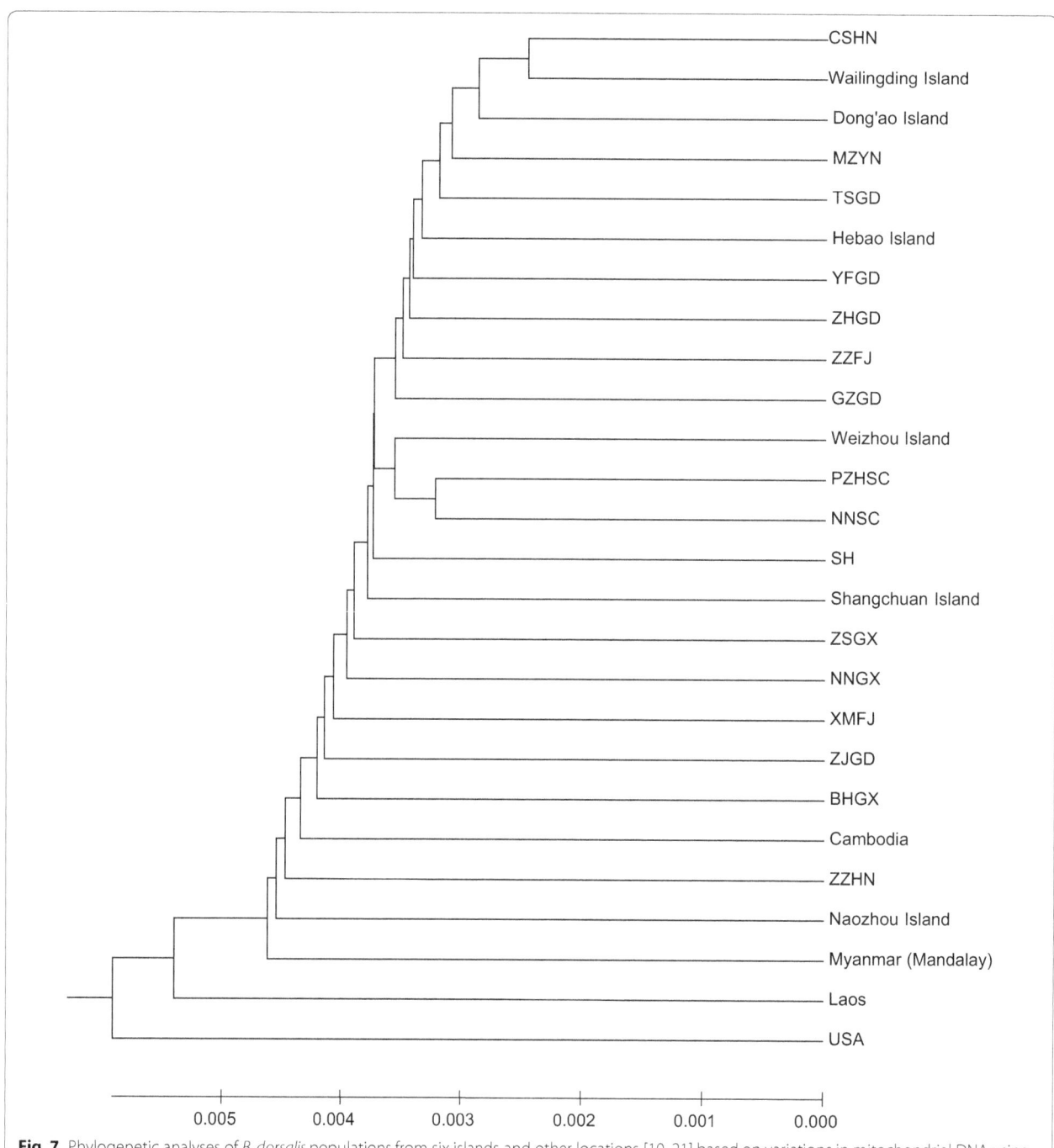

Fig. 7 Phylogenetic analyses of *B. dorsalis* populations from six islands and other locations [10, 21] based on variations in mitochondrial DNA using the unweighted pair-group method with arithmetic mean (UPGMA). CSHN (Changsha, Hunan), MZYN (Mengzi, Yunnan), TSGD (Taishan, Guangdong), YFGD (Yunfu, Guangdong), ZHGD (Zhuhai, Guangdong), ZZFJ (Zhangzhou, Fujian), GZGD (Guangzhou, Guangdong), PZHSC (Panzhihua, Sichuan), NNSC (Ningnan, Sichuan), SH (Shanghai), ZSGX (Zhongshan, Guangxi), NNGX (Nanning, Guangxi), XMFJ (Xiamen, Fujian), ZJGD (Zhanjiang, Guangdong), BHGX (Beihai, Guangxi), and ZZHN (Zhengzhou, Henan) were all from China

[49]. Jensen et al. [3] reported that the genetic diversity of house sparrows along a coastal latitudinal gradient from middle to Northern Norway tended to decrease in island populations compared to mainland populations. However, in our study, the results showed that the genetic

diversity of *B. dorsalis* from island populations was high. The geographic isolation imposed by the sea may not hinder gene exchange. There are many reasons for high genetic diversity among island populations, including the dispersal ability of this species and both environmental and anthropogenic activities that cause frequent contact with mainland populations. The six islands have been isolated for a long period of time, and most of the islands originated from the accumulation of volcanic matter. As the islands developed, commercial activities on the islands increased, which has incidentally promoted the spread of species throughout the islands. *B. dorsalis* can use most commercial plants and fruits as host plants, and the flies can disperse between the mainland and islands by relying on the wind. According to the phylogenetic tree analysis, there were two main routes of invasion from inland China. One route was from Southeast China, and the other route was from Southwest China [50]. However, whether *B. dorsalis* regularly makes migratory flights from the nearby mainland to these islands is unclear.

Genetic divergence

For COI data, the value of N_{em} among populations all exceeded 4 ($N_{em} > 4$), which showed that among populations there was a high levels of gene flow and a low and medium level of differentiation among some paired island populations. These results were consistent with the results from our microsatellite data that shows a low *Nei's* standard genetic distance (D) and a high genetic identity (I). Pairwise F_{ST} values indicated that there was a medium degree of population differentiation between Weizhou Island and Naozhou Island. The microsatellite data indicated that population genetic variation was mostly partitioned among populations. We divided the six island populations into three groups, which were supported by the FCA. Our results were also similar to the levels of population differentiation between mainland China populations. Wan et al. [20] analyzed the genetic diversity of *B. dorsalis* from six populations in Qiongqing using eight microsatellite loci. The results showed that the Qiongqing populations had low levels of population differentiation. Yao et al. [51] also examined the genetic relationships among populations from Fujian, Hainan, Guangdong, Yunnan and Sichuan provinces. Except for the Fujian population, all populations had low levels of population differentiation because of the proximity between provinces. Using mtDNA from 25 populations on the China mainland, flies from mainland China had lower levels of genetic divergence than those from Thailand, Japanese and American populations [50]. Li et al. [16] reported that the average Nei's standard genetic distance was 0.8049 and 0.9397, for South China and Southeast Asia populations, respectively. Li et al. [16] also reported that F_{ST} was 0.25 between South China and Southeast Asia populations, which suggested that these populations had genetic divergence due to

geographic isolation. Our study suggested that island *B. dorsalis* populations had high levels of genetic diversity and a low or medium level of differentiation among some paired island populations and the genetic distance of pairing populations had no correlation with geographic isolation ($R^2 = 0.0027$, P = 0.855). There are several possible explanations for this pattern. First, Tephritid fruit flies are capable fliers that can fly more than 25 km under windless conditions [52, 53]. The distance between each island and the mainland ranged from 8.2 to 38.4 km. The distance between islands ranged from 32.4 to 519.7 km. Hebao Island, Wailingding Island and Dong'ao Island are located in Zhuhai province and are closer in proximity than are Weizhou Island and Naozhou Island. Thus, Hebao Island populations were barely differentiated from the Wailingding Island, Shangchuan Island and Naozhou Island populations. However, there was a medium degree of genetic differentiation between Dong'ao Island and Weizhou Island populations and between Dong'ao Island and Naozhou Island populations. *Bactrocera dorsalis* can migrate to and from islands on its own or with the aid of typhoons [52]. The isolation of islands by water limits gene flow. However, in our study, the islands did not appear to restrict gene flow: the population genetic diversity and differentiation on the islands were similar to values observed for mainland populations. Second, based on the results of our study, commercial activities between the island and the mainland have affected the gene pool and diversity of species. These six islands all have suitable host plants for *B. dorsalis*, such as banana, pawpaw and others. Many tropical fruits, such as guava, carambola, and mango, are imported and exported from the mainland to the islands. *Bactrocera dorsalis* can lay eggs on these fruits, and the infested fruits can subsequently be transported to other areas. Therefore, gene flow is likely frequent between islands. The low levels of genetic diversity are also likely caused by the short distance between the islands and the mainland.

Conclusions

This study showed that island isolation may not significantly influence the genetic diversity of tephritid fruit flies. We have found that offshore island populations of *B. dorsalis* have relatively high levels of genetic diversity, whereas populations from offshore islands exhibited low genetic differentiation.

Authors' contributions
YJX conceived and designed the experiments. CYY performed the experiments. YJX, LZ, CYY and CYZ analyzed the data and wrote the manuscript. All authors read and approved the final manuscript.

Acknowledgements

We wish to thank G. Q. Wang, C. Cui, and Y. Q. Huang from South China Agricultural University for their help in collecting the material used in this study.

Competing interests

The authors declare that they have no competing interests.

Funding

This study was supported by the Science Foundation for Excellent Youth Scholars of Guangdong Province (No. Yq2013031). The funders played no role in the design of the study, data collection and analysis or the decision to publish and prepare the manuscript.

References

1. Booy G, Hendriks R, Smulders M, Groenendael JV, Vosman B. Genetic diversity and the survival of populations. Plant Biol. 2000;2(4):379–95.
2. Ellstrand NC, Elam DR. Population genetic consequences of small population size: implications for plant conservation. Annu Rev Ecol Syst. 1993;24:217–42.
3. Jensen H, Moe R, Hagen IJ, Holand AM, Kekkonen J, Tufto J, Sæther BE. Genetic variation and structure of house sparrow populations: is there an island effect? Mol Ecol. 2013;22(7):1792–805.
4. Petit RJ, Excoffier L. Gene flow and species delimitation. Trends Ecol Evol. 2009;24(7):386–93.
5. Gillespie RG, Roderick GK. Arthropods on islands: colonization, speciation, and conservation. Annu Rev Entomol. 2002;47(1):595–632.
6. Martinkova N, Barnett R, Cucchi T, Struchen R, Pascal M, Pascal M, Fischer MC, Higham T, Brace S, Ho SY, et al. Divergent evolutionary processes associated with colonization of offshore islands. Mol Ecol. 2013;22(20):5205–20.
7. De la Rúa P, Galián J, Serrano J, Moritz R. Genetic structure and distinctness of Apis mellifera L. populations from the Canary Islands. Mol Ecol. 2001;10(7):1733–42.
8. Frankham R. Do island populations have less genetic variation than mainland populations? Heredity. 1997;78(3):311–27.
9. Miller HC, Allendorf F, Daugherty CH. Genetic diversity and differentiation at MHC genes in island populations of tuatara (Sphenodon spp.). Mol Ecol. 2010;19(18):3894–908.
10. Wolf AT, Harrison SP. Effects of habitat size and patch isolation on reproductive success of the serpentine morning glory. Conserv Biol. 2001;15(1):111–21.
11. Boessenkool S, Taylor SS, Tepolt CK, Komdeur J, Jamieson IG. Large mainland populations of South Island robins retain greater genetic diversity than offshore island refuges. Conserv Genet. 2006;8(3):705–14.
12. Brekke P, Bennett PM, Santure AW, Ewen JG. High genetic diversity in the remnant island population of hihi and the genetic consequences of re-introduction. Mol Ecol. 2011;20(1):29–45.
13. Clarke A, Armstrong K, Carmichael A, Milne J, Raghu S, Roderick G, Yeates D. Invasive phytophagous pests arising through a recent tropical evolutionary radiation: the Bactrocera dorsalis complex of fruit flies. Annu Rev Entomol. 2005;50:293–319.
14. Dai S, Lin C, Chang C. Polymorphic microsatellite DNA markers from the oriental fruit fly Bactrocera dorsalis (Hendel). Mol Ecol Notes. 2004;4(4):629–31.
15. Huang S, Han R. Advance in the research on the quarantine pest Bactrocera dorsalis. Chin Bull Entomol. 2005;5:000.
16. Li Y, Wu Y, Chen H, Wu J, Li Z. Population structure and colonization of Bactrocera dorsalis (Diptera: Tephritidae) in China, inferred from mtDNA COI sequences. J Appl Entomol. 2012;136(4):241–51.
17. Aketarawong N, Bonizzoni M, Thanaphum S, Gomulski L, Gasperi G, Malacrida A, Gugliemino C. Inferences on the population structure and colonization process of the invasive oriental fruit fly, Bactrocera dorsalis (Hendel). Mol Ecol. 2007;16(17):3522–32.
18. Drew RA, Hancock DL. The Bactrocera dorsalis complex of fruit flies (Diptera: Tephritidae: Dacinae) in Asia. Bull Entomol Res Suppl Ser. 1994;2:1–68.
19. Shi W, Kerdelhué C, Ye H. Genetic structure and inferences on potential source areas for Bactrocera dorsalis (Hendel) based on mitochondrial and microsatellite markers. PLoS One. 2012;7(5):e37083.
20. Wan X, Liu Y, Zhang B. Invasion history of the oriental fruit fly, Bactrocera dorsalis, in the Pacific-Asia region: two main invasion routes. PLoS One. 2012;7(5):e36176.
21. Augustinos A, Stratikopoulos E, Zacharopoulou A, Mathiopoulos K. Polymorphic microsatellite markers in the olive fly, Bactrocera oleae. Mol Ecol Notes. 2002;2(3):278–80.
22. Baliraine F, Bonizzoni M, Osir E, Lux S, Mulaa F, Zheng L, Gomulski L, Gasperi G, Malacrida A. Comparative analysis of microsatellite loci in four fruit fly species of the genus Ceratitis (Diptera: Tephritidae). Bull Entomol Res. 2003;93(01):1–10.
23. Wang Y, Yu H, Raphael K, Gilchrist A. Genetic delineation of sibling species of the pest fruit fly Bactocera (Diptera: Tephritidae) using microsatellites. Bull Entomol Res. 2003;93(04):351–60.
24. Mun J, Bohonak AJ, Roderick GK. Population structure of the pumpkin fruit fly Bactrocera depressa (Tephritidae) in Korea and Japan: pliocene allopatry or recent invasion? Mol Ecol. 2003;12(11):2941–51.
25. Tang K. Molecular identification of Bactrocera (Diptera: Tephritidae) larvae by using microsatellite markers and population genetic structure of Bactrocera dorsalis in China base on COI data Master. Guilin: Guangxi Normal University; 2006.
26. Raymond M, Rousset F. GENEPOP (version 1.2): population genetics software for exact tests and ecumenicism. J Hered. 1995;86(3):248–9.
27. Chapuis M-P, Estoup A. Microsatellite null alleles and estimation of population differentiation. Mol Biol Evol. 2007;24(3):621–31.
28. Yeh F, Yang R, Boyle T. POPGENE version 1.31 quick user guide. Canada: University of Alberta and Center for International Forestry Research; 1999.
29. Vorsino AE, Wieczorek AM, Wright MG, Messing RH. Genetic analysis of an introduced biological control agent reveals temporal and geographic change, with little evidence of a host mediated shift. Biol Control. 2014;77:41–50.
30. Evanno G, Regnaut S, Goudet J. Detecting the number of clusters of individuals using the software STRUCTURE: a simulation study. Mol Ecol. 2005;14(8):2611–20.
31. Miller MP. Tools for population genetic analyses (TFPGA) 1.3: a windows program for the analysis of allozyme and molecular population genetic data. 1997;4:157. (Computer software distributed by author).
32. Belkhir K, Borsa P, Chikhi L, Raufaste N, Bonhomme F. GENETIX 4.05, logiciel sous Windows TM pour la génétique des populations. Laboratoire génome, populations, interactions, CNRS UMR. 1996;5000:1996–2004.
33. Excoffier L, Lischer HEL. Arlequin suite ver 3.5: a new series of programs to perform population genetics analyses under Linux and Windows. Mol Ecol Res. 2010;10(3):564–7.
34. Wu G, Li Y, Li Z, Liu L. The genetic relation of the population of Bactrocera dorsalis (Diptera: Tephritidae) in Shanghai District based on mtDNA CO I gene. J China Agric Univ. 2012;2:94–101.
35. Tamura K, Stecher G, Peterson D, Filipski A, Kumar S. MEGA6: molecular evolutionary genetics analysis version 6.0. Mol Biol Evol. 2013;30:2725–9.
36. Librado P, Rozas J. DnaSP v5: a software for comprehensive analysis of DNA polymorphism data. Bioinformatics. 2009;25(11):1451–2.
37. Wright S. Variability within and among natural populations. Evolution and the genetics of populations: a treatise in four volumes. Vol 4. Chicago: University of Chicago Press; 1978.
38. Chakraborty R, Danker-Hopfe H. 7 Analysis of population structure: a comparative study of different estimators of wright's fixation indices. Handbook Stat. 1991;8:203–54.
39. Rousset F. Genetic differentiation and estimation of gene flow from F-statistics under isolation by distance. Genetics. 1997;145(4):1219–28.

40. Wen YFUK, Han WJ, Ueno S, Xie WD, Xu GB, Tsumura Y. Null alleles in microsatellite markers. Biodivers Sci. 2013;21(1):117–26.

41. Rieseberg LH, Soltis D. Phylogenetic consequences of cytoplasmic gene flow in plants: evolutionary trends in Plants; 1991.

42. Pogson GH, Mesa KA, Boutilier RG. Genetic population structure and gene flow in the Atlantic cod Gadus morhua: a comparison of allozyme and nuclear RFLP loci. Genetics. 1995;139(1):375–85.

43. Shi W, Ye H. Genetic differentiation in five geographic populations of the oriental fruit fly, Bactrocera dorsalis (Hendel) (Diptera: Tephritidae) in Yunnan Province. Acta Entomol Sin. 2003;47(3):384–8.

44. Li W, Yang L, Tang K, Zeng L, Liang G. Microsatellite polymorphism of Bactrocera dorsalis (Hendel) populations in China. Acta Entomol Sin. 2007;50(12):1255–62.

45. Francisco-Ortega J, Santos-Guerra A, Kim S-C, Crawford DJ. Plant genetic diversity in the Canary Islands: a conservation perspective. Am J Bot. 2000;87(7):909–19.

46. Green EC, Tremetsberger K, Jiménez A, Gómez-González S, Stuessy TF, Baeza CM, López PG. Genetic diversity of pioneer populations: the case of Nassauvia argentea (Asteraceae: Mutisieae) on Volcán Lonquimay, Chile. Plant Syst Evol. 2012;298(1):109–19.

47. Nei M, Maruyama T, Chakraborty R. The bottleneck effect and genetic variability in populations. Evolution. 1975;29(1):1–10.

48. Franck P, Garnery L, Solignac M, Cornuet J-M. The origin of west European subspecies of honeybees (Apis mellifera): new insights from microsatellite and mitochondrial data. Evolution. 1998;52(4):1119–34.

49. Boessenkool S, Taylor SS, Tepolt CK, Komdeur J, Jamieson IG. Large mainland populations of South Island robins retain greater genetic diversity than offshore island refuges. Conserv Genet. 2007;8(3):705–14.

50. Wang M, Xu L, Zhang R, Zhang G, Yu D. Population genetic differentiation analysis of the oriental fruit fly, Bactrocera dorsalis (Diptera: Tephritidae) based on mtDNA COI gene. Acta Entomol Sin. 2014;57(12):1424–38.

51. Yao TSHJ, Li HJ, Ran C, Liu HQ, Lei HD, Xiao T. Microsatellite markers for genetic polymorphism in five geographic population of the Bactrocera dorsalis (Hendel). Chin Agric Sci Bull. 2010;26(8):234–6.

52. Fletcher B. The biology of dacine fruit flies. Annu Rev Entomol. 1987;32:115–44.

53. Koyama J, Kakinohana H, Miyatake T. Eradication of the melon fly, Bactrocera cucurbitae, in Japan: importance of behavior, ecology, genetics, and evolution. Ann Rev Entomol. 2004;49(1):331–49.

5

A century of morphological variation in Cyprinidae fishes

Stephen J. Jacquemin[1]* and Mark Pyron[2]

Abstract

Background: Aquatic habitats have been altered over the past century due to a variety of anthropogenic influences. Ecomorphology is an area of aquatic ecology that can both directly and indirectly assess the effects of habitat alterations on organisms. However, few studies have explored long term trends in morphological variation. Long term changes in morphology can potentially impact niche and ultimately contribute to organismal success and the ecosystem. Therefore, in this study we assessed long term morphological variation with body size, sex, time, and hydrology using museum collections of five species of Cyprinidae (Minnows) from lentic and lotic systems over the past 100 years to gain insight into long term patterns in morphology.

Results: Variation in Cyprinidae morphology tended to relate to: body size—indicating strong allometric growth patterns with robustness of larger individuals; sex—indicating a level of fecundity selection for deeper bodies in females compared with males; and year—indirectly suggesting responses to habitat changes over the past century. In lotic ecosystems, Cyprinidae morphology tended to be more fusiform in conjunction with lower mean annual discharge or higher variation in discharge. In lentic ecosystems, change in morphology was observed but no historic habitat variables were available to discern potential mechanisms. Interestingly, not all species responded in the same magnitude or directionality.

Conclusions: Long term changes in morphological variation provide a link to exploring functional relationships between taxa and their environment and have implications for understanding ecosystem attributes, community assembly patterns, and conservation.

Keywords: Geometric morphometrics, North American fish morphology, Long term morphology, Global change, Long term hydrological variation, Habitat alteration, Contemporary evolution

Background

Freshwater ecosystems have become increasingly altered over the past several centuries through a variety of anthropogenic changes to watersheds that have resulted in degraded physical habitats, declines in water quality, and disruption of flow regimes on both local and regional scales [1–3]. Understanding the potential effects of widespread ecosystem alterations on biota and their implications have become primary conservation issues. Central to understanding the impact of changes to habitats over the past century are species-environment relationships [4], particularly how organismal morphology may play a part in the functional role of an individual in an ecosystem. These larger scale relationships between morphology and environment can improve predictability and understanding of potential assemblage and ecosystem level attributes [5–7]. However, few studies have used long term collections or datasets to assess what variables may relate to variation in morphology over a long term period. Therefore, the purpose of this study was to apply an exploratory technique incorporating a suite of tests to identify potential morphological covariates in fishes over the past century.

Freshwater fishes display a range of morphological adaptations across a wide variety of physiological states and environmental conditions as a result of genetic

*Correspondence: stephen.jacquemin@wright.edu
[1] Department of Biological Sciences, Dwyer Hall, Wright State University-Lake Campus, Celina, OH 45822, USA
Full list of author information is available at the end of the article

divergence and/or phenotypic plasticity [5, 8, 9]. Physiological influences are frequently identified as morphological relationships with body size (allometry), diet (feeding performance), and sex (sexual selection) while plasticity tends to respond to local environmental variation related to niche patterns of resource utilization, behavior, and/or habitat use [5, 8–11]. Typically, as fishes move through their ontogeny there are allometric trends in shape [12, 13]. These allometric trajectories can follow different patterns in individual species but are often shown to relate to body depth and robustness of an individual [12, 14]. These changes with ontogeny can be further explored in light of sexual dimorphism as individuals reach sexual maturity, whereby females tend to take on a deeper body shape compared with males. In addition to body size and sex, diet and resource utilization affect morphology, whereby particular dietary items can induce morphological change within or among populations [8]. Lastly, all of these factors can be complicated by environmental variation with phenotypic plasticity. For example, Langerhans et al. [15] identified increasingly fusiform body shapes of Neotropical fishes in main channel habitats compared to lower flow lagoon habitats coinciding with variation in flow conditions. Similarly, Haas et al. [16] found that deeper-bodied *Cyprinella* (Cyprinidae) are indicative of reservoir compared to riverine habitats. More recently, Dugas et al. [17] studied body streamlining (fusiform) with river velocity in *Cyprinella whipplei* males and did not find positive correlations but did suggest that deeper body shape may provide increased maneuverability during male–male interactions during combat and that this may be of greater importance than increased streamlining. Regardless, while these field documented effects of local habitats influence on morphology are varied they have been reproduced and manipulated in laboratory experiments [18]. Yet, while these documented trends in physiology and plasticity exist in contemporary studies, it is impossible to quantify everything that drives body shape trends. Further challenging is dealing with questions of long term organismal changes [19] where a paucity of biological specimens and/or environmental information impedes long term morphological studies.

Natural history museum collections may provide the best opportunity to observe and test for some of these long term changes in morphological traits of taxa concurrent with habitat variation. Biological and environmental data are most readily available in the form of historical museum collections of preserved fish specimens (potentially from 1880 to present [20, 21]) and associated specimen lot description tags. For example, a recent study utilizing museum collection lots and data identified rapid change in the morphology of a stream fish in the decades that followed construction of a series of reservoirs

in the United States [19]. And while these long term collections that are available do not encompass all taxa from all locales, these resources can help facilitate ecological hypotheses across a large temporal scale of taxa, assemblages, and ecosystems [22].

Specific to long term morphology of fishes, morphological variation that is correlated with physiological parameters such as body size or sex can be directly assessed from measurement or dissection of preserved specimens and analyzed by collection year. These types of studies become more accessible compared with those invoking plasticity or immigration trends with environmental change as local habitat conditions at the time of these long term collections may not be as readily apparent with variation in detailed collection records. One interesting solution to sparse habitat records is to use time as an indirect proxy for habitat change and follow up with historical climate change data of precipitation, temperature, landscape development trends, or with direct records such as flow regime (daily hydrology data available for many streams from 1900 to present—http://waterdata.usgs.gov/).

Global climate change models were particularly helpful in facilitating our understanding of ecosystem responses to changes in precipitation, temperature, and human landscape impacts [23]. Of particular interest are changes in flow regime, as flow regime provides an overarching influence on aquatic ecosystems [1], either directly in streams or indirectly for lakes and can serve as a proxy for habitat [24]. The effects of flow regime changes on freshwater fishes have been documented in numerous taxa and communities (reviewed by Poff and Zimmerman [25]) and provide inferences into population responses to environmental alterations [16, 26]. Alterations to flow regimes have been linked to physical habitat homogenization and degradation [27], changes in aquatic assemblages [28], reductions in ecological community resilience [29], changes in assemblage functional attributes [30, 31], changes to structure of assemblage level morphological diversity [32], and long term hydrology effects on contemporary assemblages [33]. However, fewer studies have linked long term changes in freshwater fish taxa with variation in environment [34].

Despite numerous studies that have established relationships between morphology and suites of abiotic and biotic factors, the patterns that emerge from long term tests for these relationships are virtually unknown. Therefore, the objective of this study was to describe shape variation for five species in the family Cyprinidae (Minnows) relative to body size, sex, time, and environment (e.g. hydrology where available) over 100 years using long term collections from both lotic and lentic sites. We predicted that long term collections would exhibit morphological signal as a result of both physiological and environmental influences. Ultimately, ecomorphological perspectives may improve our understanding of shape variability while providing potential links with functional performance in changing ecosystems.

Methods

Fish and morphology

Museum collections of five species of Cyprinidae from the Illinois Natural History Survey (INHS) were selected based on long term repeated collections of approximately ten individuals representing both males and females from each collection effort. Species found to meet this criteria were Striped Shiner (*Luxilus chrysocephalus*), Redfin Shiner (*Lythrurus umbratilis*), Emerald Shiner (*Notropis atherinoides*), Sand Shiner (*Notropis stramineus*), and Bluntnose Minnow (*Pimephales notatus*) (see Appendix for INHS catalog numbers). Collection dates ranged from 1899 to 2006. All collections were in Illinois, USA (Fig. 1; Table 2) and included long term biological monitoring sites on the Sangamon River (Champaign/McLean Co.), Mackinaw River (Tazewell/Woodford Co.), Little Vermillion River (Vermillion Co.), Salt Creek (Logan/DeWitt Co.), Illinois River (LaSalle/Grundy Co.), and Lake Michigan (Cook Co.). Individuals were photographed using a tripod-mounted digital Nikon D70 camera (Nikkor AF-S DX Macro Zoom Lens) against a grid display for scale reference. Images were digitized by one person (SJJ) using a series of 11 predefined landmarks [14] (see Fig. 2 for landmark placement and description) within the software

tpsDig [35]. To account for bent specimens in collection jars, four additional lateral landmarks, not included in the shape analyses, were added along the midline of each individual to digitally unbend specimens using the 'unbend function' within tpsUtil [36]. Only adult individuals were photographed and digitized. Centroid size was calculated to infer body size of digitized individuals [37]. Sex was determined by gonad inspection. Only collections preserved for a minimum of five years were used to avoid preservation bias in shape that can occur during initial formalin fixation and long term ethanol storage using standard museum preservation protocols [38].

Habitat of rivers and lakes

To characterize flow habitat of lotic sites, daily discharge data from United States Geological Survey (USGS) stream gauging stations (http://waterdata.usgs.gov/) were used to describe site flow conditions (Table 1). Gauging stations were selected from the same streams (less than 15 km from site) as biological collections. Daily discharge data were analyzed using two approaches. First, annual mean and variation (coefficient of variation) were calculated from daily discharge data spanning a full calendar year prior to collection to specifically capture the flow conditions leading up to each individual collection effort. Second, overarching site level trends using all available years (including both collection and non-collection years) were analyzed for larger scale emergent patterns with year using the Indicators of Hydrologic Alteration (IHA) [39] flow regime methodology (calculated using IHA software ver. 7.1, The Nature Conservancy). For this broad scale hydrology analysis, all available years were used to provide a hydrology description of the entire site depicting the dominant hydrological trends through time that each lotic fish population had experienced over time. IHA calculates 32 hydrological variables in five categories including magnitude, frequency of occurrence, duration of flow condition, timing, and flashiness [1] and provides a measure of change in these individual variables with time using regression analysis. These IHA variables were identified as ecologically relevant, particularly with regard to feeding, reproduction, maturity, timing of life cycle periods, and stress [39]. In summary, magnitude is a measure of wet volume and can serve as an indication of high versus low flows. Frequency of occurrence is a measure of how episodic extremes in flow magnitude (high or low) occur. Duration of flow condition is a measure of the longevity of different flow events. Timing is a measure of when in a Julian water year extreme conditions occur. The flashiness category is a measure of the rate of change attributed with variation in flow conditions. Departures from the natural flow regime were linked to anthropogenic modifications of rivers, either

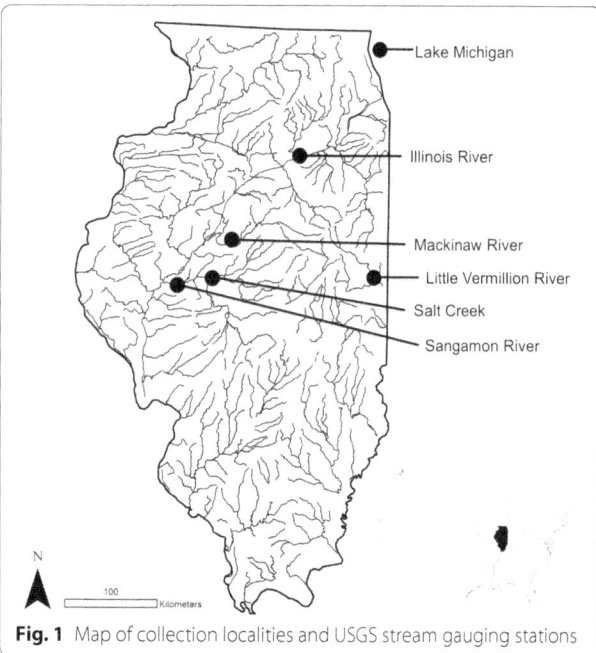

Fig. 1 Map of collection localities and USGS stream gauging stations

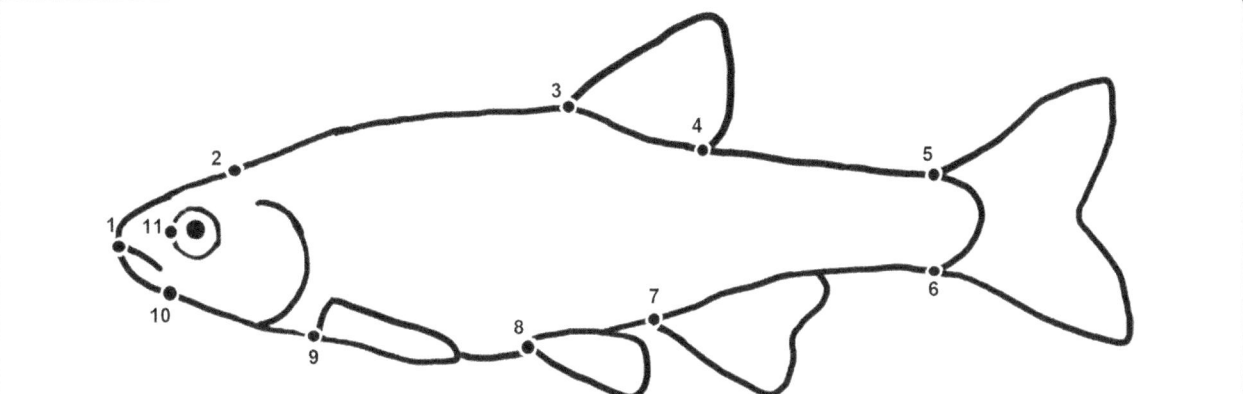

Fig. 2 Location of morphology landmarks. Landmarks correspond to anterior snout (*1*), superior margin of head (*2*), anterior dorsal origin (*3*), posterior dorsal origin (*4*), superior posterior caudal peduncle (*5*), inferior posterior caudal peduncle (*6*), anterior anal fin origin (*7*), anterior pelvic fin origin (*8*), superior pectoral fin origin (*9*), ventral opercular isthmus (*10*), and anterior medial edge of orbital socket (*11*)

Table 1 List of USGS gauging stations by drainage basin and include years

Site code	Drainage basin	Site	Hydrology years
USGS 05582000	Illinois River	Salt Creek at Greenview	1940–2003
USGS 05583000	Illinois River	Sangamon River at Oakford	1910–2005
USGS 05543500	Illinois River	Illinois River at Marseilles	1919–2003
USGS 05568000	Illinois River	Mackinaw River at Green Valley	1921–2000
USGS 03339000	Wabash River	Little Vermillion River at Danville	1914–2002

directly or indirectly [1, 39]. Since hydrological information was only available for lotic sites, and no direct habitat information was available for lentic sites, only time was able to be used as an indirect approximation of habitat change [40]. Time also served as a covariate in lotic system analyses. Although not mechanistic in nature, time is indirectly linked to changes in water quality and habitat in the Midwestern United States over the past century [40].

Modeling morphology

General procrustes analysis (GPA; least squares method) was used to superimpose digitized individuals to a reference or consensus shape by removing effects of scaling, rotation, and translation, allowing comparisons among individuals [36]. Relative Warp Analysis (RWA) was performed in tpsRelw [37] and used to assess total morphological variation among individuals. RWA is essentially a morphometric principal components analysis of relative landmark positions (using variance/covariance matrix) and produces a series of orthogonal axes ranked by the amount of shape variation explained [41]. All shape axes were retained for analyses in this study. Two sets of RWA analyses were performed on individuals, including a family level analysis which incorporated all individuals from all species to test for macro level interspecific effects across species as well as a series of separate RWAs for each species at each site to test for intraspecific variation.

Table 2 List of taxa with collection site, collection year, and number of collections per site

Species	Collection site	Collection years	Number of collections	Mean sample (±SE)
Striped Shiner *Luxilus chrysocephalus*	Sangamon River	1901–2001	9	n = 15 (2.4)
Redfin Shiner *Lythrurus umbratilis*	Little Vermillion River	1899–2002	9	n = 14 (2.1)
Emerald Shiner *Notropis atherinoides*	Illinois River	1897–2003	8	n = 13 (1.2)
	Lake Michigan	1900–1999	7	n = 15 (3.6)
Sand Shiner *Notropis stramineus*	Sangamon River	1901–2005	9	n = 17 (3.4)
	Mackinaw River	1901–2000	11	n = 14 (2.4)
Bluntnose Minnow *Pimephales notatus*	Salt Creek	1900–2003	9	n = 10 (1.0)
	Lake Michigan	1900–1998	6	n = 13 (4.5)

Shape axes were treated as dependent variables in multivariate analysis of covariance (MANCOVA) models to test for relative influence of the following factors: species, site, body size, hydrology, sex, year, and all biologically relevant interactions (interpreted through heterogeneous slopes; [42]). Given the exploratory nature of the study, the following terms and interactions were included in the family combined model: species, site, body size, sex, year, daily discharge mean, daily discharge variation, body size * species, sex * species, year * species, daily discharge mean * species, and daily discharge variation * species while the following terms and interactions were included in the species specific models: body size, sex, year, daily discharge mean, daily discharge variation, body size * sex, body size * daily discharge mean, body size * daily discharge variation, daily discharge mean * year, daily discharge variation * year, daily discharge mean * sex, daily discharge variation * sex, and sex * year. Initial models revealing insignificant interaction terms were rerun without interaction terms. An unbalanced MANCOVA design was used to accommodate some limited cases of missing data of certain columns, such as where hydrological records were missing [43]. We felt inclusion of all samples and time periods were relevant as they contained information particular to specific trends (e.g. testing for temporal trends in cases where hydrological information was missing). Prior to a given terms inclusion in the family or species level MANCOVA models, issues with multicollinearity were assessed using variance inflation factor (VIF) statistics with a cutoff of 2.5 [44]. In instances where a variable exceeded this VIF threshold, that variable was dropped from the model and the model was rerun. Wilk's lambda was used to calculate F values and Pillai's partial eta^2 was used to infer effect size by variance explained. Partial eta^2 reflects the proportion of variation explained by a given independent variable while parsing out variation from other independent variables. As such, partial eta^2 values are not necessarily able to be summed to 1.0 but can provide a degree of variation explained, particularly relative to other variables included in the model [45]. In addition to relative warp analyses with MANCOVA models, total morphological disparity was calculated to compare species. Disparity indices can be useful in discerning relative degrees of variability between species or groups and may contribute to our understanding of specific variation over long sampling time periods or provide general evolutionary information to larger order taxonomic groups [46, 47]. Disparity was calculated from Procrustes coordinates using a randomized residual permutation procedure using 1000 permutations [48]. All models were implemented in the statistical language R (version 2.14.1, R Development Core Team, 2011) using the heplots [45], fmsb [43], geomorph [48], and base stats

packages. Thin plate spline deformation grids [49] (produced in tpsRegr [50]) were used to visualize shape variation indicated in MANCOVA models. Alpha was 0.05 for all tests of significance.

Results

Museum collections of Cyprinidae were from long term (100+ year) collections at five lotic sites (Sangamon River, Salt Creek, Little Vermillion River, Mackinaw River, and Illinois River), all with adjacent USGS daily discharge monitoring sites, and one lentic site (Lake Michigan). Specimen collection dates ranged from 1899 to 2006 and spanned 98 to 106 collection years (mean number of collection events per site = 8.5) across the sites (Table 2; Appendix). Number of individuals per collection event included in the analyses ranged from 9 to 34 (mean number of individuals = 13.9). All sites included the available range of body sizes of sexually mature individuals of both males and females. United States Geological Survey (USGS) hydrological archives of daily discharge ranged from 63 to 96 consecutive years at the five lotic sites, with the earliest data collected in 1910 (Table 1). Indicators of Hydrologic Alteration analyses revealed significant alterations for all sites across the five primary IHA categories, including frequency and duration of pulses, magnitude and duration of annual extremes, magnitude of monthly conditions, rate and frequency of water condition changes, and timing of annual extreme conditions (Appendix). Mean number of hydrological alterations (of 32 possible) at sites was 11.4 (range of 1–23 alteration types) with varying degrees of magnitude, indicating that sites were variable in numbers of alterations and the extent these alterations modified the flow regime (Appendix). Overall, Cyprinidae morphology was related to a combination of variables and interaction terms, including body size, sex, hydrology (lotic sites), and time but was not uniformly consistent across species (Tables 3, 4).

Cyprinidae

The combined analysis at the family level included 908 individuals from 68 lots representing both lotic and lentic sites spanning 1897–2003. Total morphological variation was captured along 18 relative warp axes which explained 100 % of shape variation. However, shape was primarily driven by the first six axes which accounted for 80 % of this variation. The primary morphological gradient extracted from the family level MANCOVA model (aside from the species grouping variable) was body size, which explained 45 % of the total morphological variation. In addition to body size, sex (11 %) and year (8 %) were also recovered as important contributors to morphological variation. In lotic populations, both discharge

Table 3 Results from family level MANCOVA model assessing relationships between morphological variables and body size, hydrology, sex, year, and interactions for five species

Family	Effect	Variance explained	F	d.f.	P
Cyprinidae	Species	0.70	2.8	72, 2004	<0.001
	Body size	0.45	22.7	18, 498	<0.001
	Sex	0.11	3.6	18, 498	<0.001
	Year	0.08	2.4	18, 498	<0.001
	Discharge mean	0.11	3.2	18, 498	<0.001
	Discharge CV	0.08	2.3	18, 498	<0.001
	Body size * species	0.10	2.8	72, 2004	<0.001
	Sex * species	0.06	1.7	72, 2004	<0.001
	Year * species	0.11	3.4	72, 2004	<0.001
	Discharge mean * species	0.08	2.5	72, 2004	<0.001
	Discharge CV * species	0.09	2.6	72, 2004	<0.001

See Table 4 for taxon specific analyses and Figs. 3, 4, 5, 6, 7 for visualizations of shape changes in individual species

mean (11 %) and variation (8 %) were recovered across all taxa. Testing for differences in trajectories of these relationships between species through the species interaction term uncovered significant variation in all cases indicating that there are species level differences in body shape responses to size, sex, year, and hydrology (Table 3). These differences in overall trajectory or slope between species renders an overall visualization less useful than discerning species specific trends as presented in subsequent results with species level visualizations. However, relative to overall morphological variation, disparity analysis indicated significantly different levels of variation across species (P < 0.05) and that the highest amount of morphological variation was found in Striped Shiner followed by Redfin Shiner, Emerald Shiner, Bluntnose Minnow, and Sand Shiner, respectively. Interesting, both species with lake populations, Emerald Shiner and Bluntnose Minnow, exhibited appreciably more variation in the lentic populations compared with their lotic counterparts while the only species with multiple lotic samples (Sand Shiner) were not significantly different from one another.

Striped Shiner

Striped Shiner collections were from a single lotic site, the Sangamon River, and included 131 individuals from nine lots spanning 1901–2001. Total morphological variation was captured along 18 relative warp axes which explained 100 % of shape variation. This variation was primarily driven by the first 8 relative warp axes which

accounted for 87 % of the shape variation. The primary morphological gradient extracted from the Sangamon River MANCOVA model was body size, which explained 63 % of the total morphological variation. Smaller individuals tended to have narrower body, caudal peduncle areas, and fin bases compared to larger individuals. Secondarily, year (49 %) was identified as an important predictor of morphology as individuals from more recent collections tended to exhibit more fusiform morphologies than the deeper bodied individuals from past collections. Hydrological analyses of the Sangamon River site provide a potential causal mechanism that undergirds this year variable. IHA analyses showed significant alterations to 9 (of 32) different types of hydrological variables between 1910 and 2005 (Appendix). Regression slopes for IHA variables with time indicated that the largest hydrologic alteration types occurred in magnitude and duration of annual extremes. Although it was not possible to explicitly incorporate IHA variables into the morphological model, a more direct test of hydrology, variation in annual discharge (29 %) and mean annual discharge (24 %), also supported a relationship between shape and hydrological variation. Striped Shiner morphologies with downturned heads, shorter and more robust caudal peduncle areas, and generally deeper bodies coincided with higher annual discharge and higher variation in annual discharge. A significant interaction between both year and annual discharge mean, as well as variation in discharge, indicated that something outside of these hydrological parameters is also likely influencing morphological variation independent of time. This variation could be attributed to some other hydrological attribute, such as one measured by the IHA, or reflect an unmeasured habitat variable that has also changed with time. Additionally, interactions recovered between body size and hydrology metrics (daily mean and variation in daily mean) indicated that while there was an overall effect of hydrology on morphology, this relationship was not consistent across all sizes of individuals. Lastly, interactions between sex and year indicated that males and females responded to the time variable differently, despite the lack of overall significance between parent terms. No interaction between sex and hydrology was recovered. See Table 4 and Fig. 3 for model configuration and deformation grids depicting shape change.

Redfin Shiner

Redfin Shiner collections were from a single lotic site, the Little Vermillion River, and included 104 individuals from nine lots spanning 1899–2002. Total morphological variation was captured along 18 relative warp axes which explained 100 % of shape variation. This variation was primarily driven by the first seven relative

Table 4 Results from MANCOVA models assessing relationships between morphological variables and body size, hydrology, sex, year, and interactions for five species

Species	Site	Effect	Variance explained	F	d.f.	P
Striped Shiner	Sangamon River	Body size	0.63	24.1	8, 113	<0.001
Luxilus chrysocephalus	(1901–2001)	Discharge mean	0.24	4.6	8, 113	<0.001
		Discharge CV	0.29	5.9	8, 113	<0.001
		Sex	0.11	1.6	8, 113	0.14
		Year	0.49	13.7	8, 113	<0.001
		Body size * discharge mean	0.16	2.7	8, 113	0.01
		Body size * discharge CV	0.15	2.4	8, 113	0.02
		Discharge mean * year	0.22	3.9	8, 113	<0.001
		Discharge CV * year	0.35	7.7	8, 113	<0.001
		Sex * year	0.27	5.4	8, 113	<0.001
Redfin Shiner	Little Vermillion River	Body size	0.18	2.3	7, 73	0.04
Lythrurus umbratilis	(1899–2002)	Discharge mean	0.14	1.7	7, 73	0.11
		Discharge CV	0.08	0.85	7, 73	0.55
		Sex	0.28	3.9	7, 73	<0.001
		Year	0.21	2.8	7, 73	0.01
		Body size * discharge mean	0.23	3.1	7, 73	<0.001
		Discharge mean * year	0.26	3.6	7, 73	<0.001
Emerald Shiner	Illinois River	Body size	0.35	4.3	8, 71	<0.001
Notropis atherinoides	(1897–2003)	Discharge mean	0.23	2.7	8, 71	0.02
		Discharge CV	0.69	19.5	8, 71	<0.001
		Sex	0.32	4.2	8, 71	<0.001
		Year	*	*	*	*
	Lake Michigan	Body size	0.51	14.8	6, 85	<0.001
	(1900–1999)	Sex	0.27	5.3	6, 85	<0.001
		Year	0.41	9.2	6, 85	<0.001
		Body size * sex	0.19	3.2	6, 85	<0.01
		Sex * year	0.17	2.9	6, 85	<0.01
Sand Shiner	Sangamon River	Body size	0.39	8.6	8, 108	<0.001
Notropis stramineus	(1901–2005)	Discharge mean	0.41	9.3	8, 108	< 0.001
		Discharge CV	0.42	9.7	8, 108	<0.001
		Sex	0.24	4.2	8, 108	<0.001
		Year	0.11	1.6	8, 108	0.14
	Mackinaw River	Body size	0.3	3.5	8, 64	<0.001
	(1901–2000)	Discharge mean	0.19	1.9	8, 64	0.05
		Discharge CV	0.08	0.69	8, 64	0.71
		Sex	0.11	0.95	8, 64	0.48
		Year	*	*	*	*
Bluntnose Minnow	Salt Creek	Body size	0.61	11.1	7, 49	<0.001
Pimephales notatus	(1900–2003)	Discharge mean	0.24	2.2	7, 49	0.05
		Discharge CV	0.06	0.4	7, 49	0.87
		Sex	0.42	5.1	7, 49	<0.001
		Year	0.16	1.3	7, 49	0.26
		Body size * sex	0.38	4.4	7, 49	<0.001
	Lake Michigan	Body size	0.64	17.5	7, 68	<0.001
	(1900–1998)	Sex	0.31	4.4	7, 68	<0.001
		Year	0.18	2.1	7, 68	0.05

See Figs. 3, 4, 5, 6, 7 for visualizations of shape changes

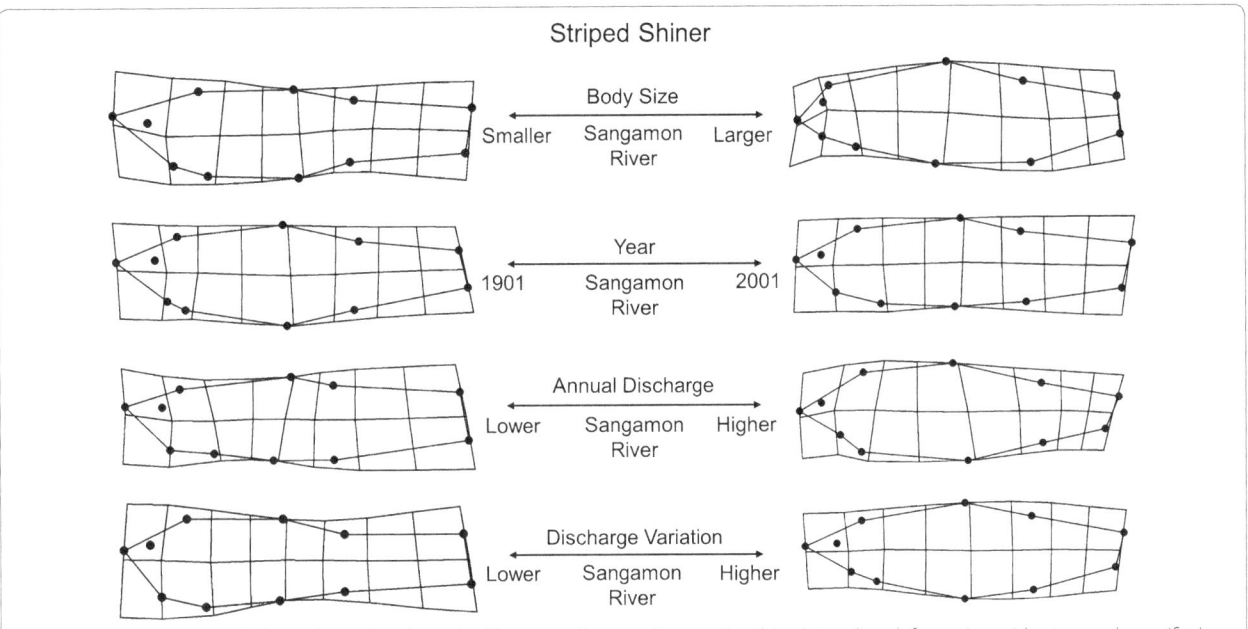

Fig. 3 Striped Shiner morphological variation along significant predictor gradients using thin plate spline deformation grid extremes (magnified ×3 for visualization)

warp axes which accounted for 81 % of the shape variation. The primary morphological gradient extracted from the MANCOVA model was sex, which explained 28 % of the total morphological variation. Females were found to exhibit deeper abdomens than males, however, this was not found to relate with any interactions with body size, hydrology, or year indicating consistent differences among body sizes, hydrology variables, sex differences, and year. Secondarily, year (28 %) and body size (18 %) were both significant main factors in morphological variation. More recent collections tended to exhibit comparatively smaller head shapes, a more arched dorsal surface, increasingly posterior pelvic and dorsal fin placements, and deeper caudal peduncles than older collections. Related to time, IHA analyses of the Little Vermillion River site showed significant hydrologic alterations in 19 (of 32) different types of IHA variables between 1915 and 2002 (Appendix). Regression slopes for IHA variables with time indicated that the largest alterations occurred in the magnitude and duration of annual extremes. Relative to body size, larger fish were found to be deeper bodied with shorter head areas compared with smaller bodied individuals. Although neither mean daily discharge or variation in daily discharge were significant, the model did recover interaction between mean daily discharge and time as well as body size, related to 26 % and 23 % of the variation, respectively. See Table 4 and Fig. 4 for model configuration and deformation grids depicting shape change.

Emerald Shiner

Emerald Shiner collections were from a single lotic site, the Illinois River, and one lentic site, Lake Michigan, and included 198 individuals from 15 lots spanning 1897–2003. Total morphological variation for Illinois River and Lake Michigan individuals was captured along 18 relative warp axes which encompassed 100 % of the variation in each analysis, but was primarily driven by 8 and 6 relative warp axes that accounted for 86 and 84 % of shape variation Illinois River and Lake Michigan populations, respectively.

The primary morphological gradients extracted from the Illinois River model were hydrological, including variation in annual discharge (69 %) and mean annual discharge (23 %). More fusiform individuals with downward turned and extended caudal peduncle areas tended to occur during periods of lower mean annual discharge and higher variation in this discharge. In addition to hydrology, shape of Illinois River individuals also correlated with variation in body size (35 %) and sex (32 %). Larger individuals tended to exhibit deeper bodies, more robust caudal peduncles, and narrower heads compared with smaller individuals. In addition, females also tended to exhibit deeper bodies with more robust caudal peduncles compared with males. Issues with multicollinearity (VIF > 2) necessitated the exclusion of year from the Illinois River model. No interaction terms were recovered in this model, indicating a lack of sex-specific allometry or hydrological covariates. Although time was excluded

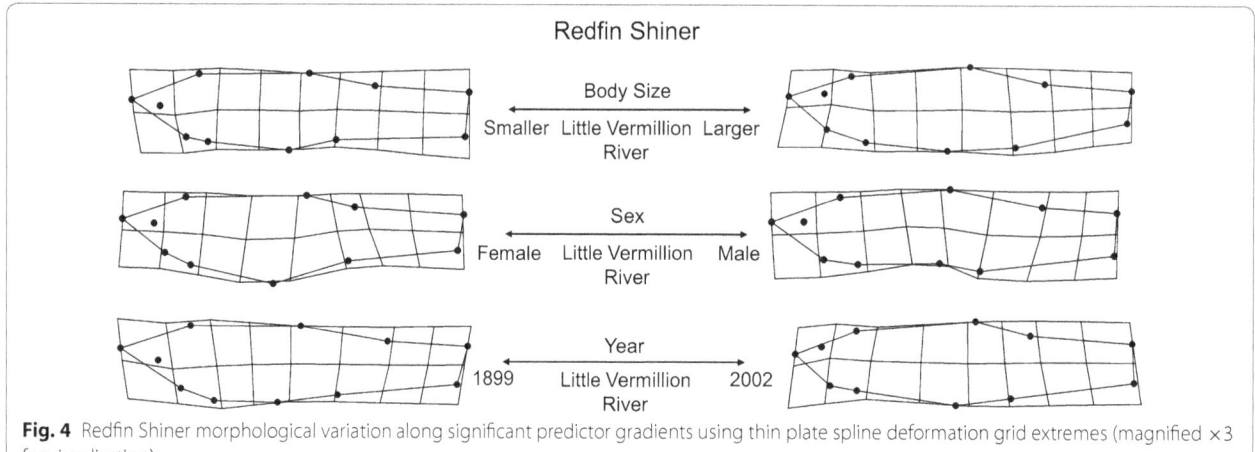

Fig. 4 Redfin Shiner morphological variation along significant predictor gradients using thin plate spline deformation grid extremes (magnified ×3 for visualization)

from the model due to issues with multicollinearity, hydrological analyses of the Illinois River site revealed significant hydrologic alterations to 23 (of 32) IHA variables between 1920 and 2003 (Appendix). Regression slopes for IHA variables with time indicated that the largest alterations occurred with magnitude and duration of annual extremes.

The primary morphological gradients extracted from the Lake Michigan model were body size (51 %), year (41 %), and sex (27 %). Similar to collections from the Illinois River, larger individuals and females tended to be deeper bodied with more upturned heads and comparatively robust caudal peduncle areas. Interestingly, a significant interaction between body size and sex (19 %), not recovered in the Illinois River model, was recovered in the Lake Michigan model, indicating a degree of sex-specific allometry present in Lake Michigan collections. Lastly, year was explained by more recent collections of increasingly fusiform shaped individuals with upturned and narrower head and caudal shapes compared with older collections. This relationship between body shape and time was compounded by an interaction with sex, indicating a difference in degree of response to time between males and females. See Table 4 and Fig. 5 for model configuration and deformation grids depicting shape change.

Sand Shiner

Sand Shiner collections were from two lotic sites, the Sangamon River and Mackinaw River, and included 306 individuals from 20 lots spanning 1901–2005. Total morphological variation explained for Sangamon River and Mackinaw River individuals was captured along 18 relative warp axes that explained 100 % of the total variation in each population. This variation was primarily driven by eight axes in each case which accounted for

84 and 83 % of the total morphological variation among Sangamon and Mackinaw River individuals, respectively. Among the primary morphological gradients extracted from both the Sangamon River and Mackinaw River MANCOVA models was body size, which explained 39 and 30 % of the total morphological variation, respectively. Both sites exhibited similar allometric shape trends related to increased dorsal fin base, anterior placement of pectoral fin, as well as deepening of the body and head concurrent with increased body size. Sangamon River populations were further described by high levels of sex-specific (24 % variance explained) differences, primarily related to more distended abdomens in females compared with males. Lack of an interaction between body size and sex retained in the final model indicated homogeneity of slopes between males and females in the size range included in this analysis, despite differences in overall body shape. This high degree of sex difference in body shape was different from Mackinaw River populations where sex was related to only 11 % of variation and was not recovered as statistically significant in the final model. Hydrological variables provided explanation for Sangamon River and Mackinaw River morphologies as higher levels of mean discharge for both the Sangamon (41 %) and Mackinaw (19 %) Rivers were positively related to deeper bodied and more robust individuals. Higher levels of variation in mean daily discharge tended to relate (42 %) to increasingly fusiform shapes with higher arched dorsal surfaces in Sangamon River populations only. Lastly, time was not recovered as a significant predictor of morphology in either model as issues with multicollinearity (VIF > 2) warranted exclusion of year from the Mackinaw River model and non-significant model terms precluded interpretation in the Sangamon River model. However, hydrological analyses of the Mackinaw River site elicited significant hydrologic

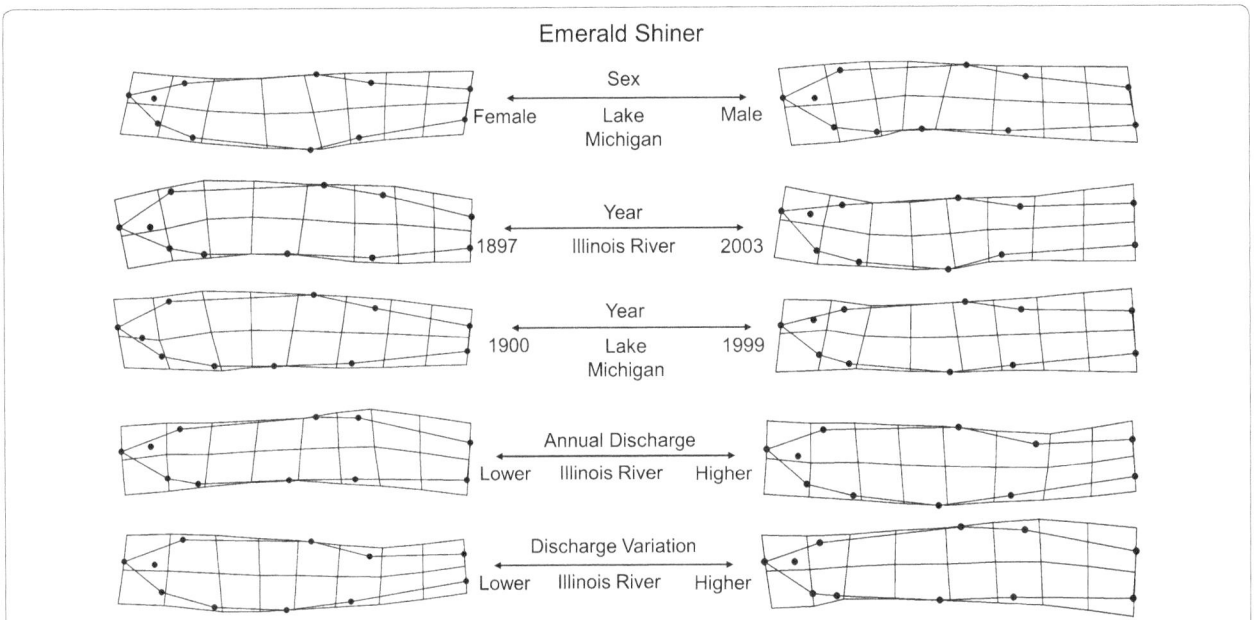

Fig. 5 Emerald Shiner morphological variation along significant predictor gradients using thin plate spline deformation grid extremes (magnified ×3 for visualization)

alterations to 1 (32) type of IHA variable between 1922 and 2000 (Appendix). Regression slopes for IHA variables with time indicated a reduction in high pulse numbers in more recent Mackinaw River years. Similarly, hydrological analyses of the Sangamon River showed significant alterations to 9 (of 32) of the measured IHA type variables between 1910 and 2005 (Appendix). Regression slopes for IHA variables with time indicated that the largest alterations in the Sangamon River occurred in magnitude and duration of annual extremes. No relationships were recovered between sex and hydrology, sex and body size, or sex and time at either site, indicating similar slopes. See Table 4 and Fig. 6 for model configuration and deformation grids depicting shape change.

Bluntnose Minnow

Bluntnose Minnow collections were from one lotic site, Salt Creek, and one lentic site, Lake Michigan, and included 169 individuals from 15 lots spanning 1900–2003. Total morphological variation explained for Salt Creek and Lake Michigan individuals was captured along 18 relative warp axes that explained 100 % of the total variation in each population. This variation was primarily driven by seven axes in each case which accounted for 83 and 87 % of the total morphological variation among Salt Creek and Lake Michigan individuals, respectively. The primary morphological gradient extracted from both Salt Creek and Lake Michigan MANCOVA models was body size, which explained 61 and 64 % of the

total morphological variation, respectively. Both sites exhibited similar shape trends with increasing body size, including an increasingly downward facing snout, deeper body, and more robust caudal region. In addition to body size, both sites exhibited significant variation attributable to sex (42 and 31 %, respectively). Females from both sites differed from males by deeper abdomens and less overall arching of the dorsal surface. Collections from Salt Creek also demonstrated strong signal (38 %) from the body size and sex interaction term, indicating a higher degree of sex-specific allometric slope differences compared with Lake Michigan individuals. Relative to time, collections from Lake Michigan (18 %) indicated that individuals from collections in recent years had longer and narrower caudal regions, more terminal snouts (compared with superior snouts in past years), more fusiform overall body shapes, and a reduction in dorsal fin base length than past years. Time did not have a significant effect on morphology of individuals from Salt Creek. However, hydrological analyses of Salt Creek site during this time identified significant alterations in 6 (of 32) types of IHA variables between 1942 and 2005 (Appendix). Regression slopes for IHA variables with time indicated that the largest hydrologic alterations occurred in magnitude and duration of annual extreme flow events. Salt Creek hydrology variables supported a relationship between morphology and daily discharge mean (24 %) as higher values coincided with increasingly fusiform individuals compared with more robust individuals typical of

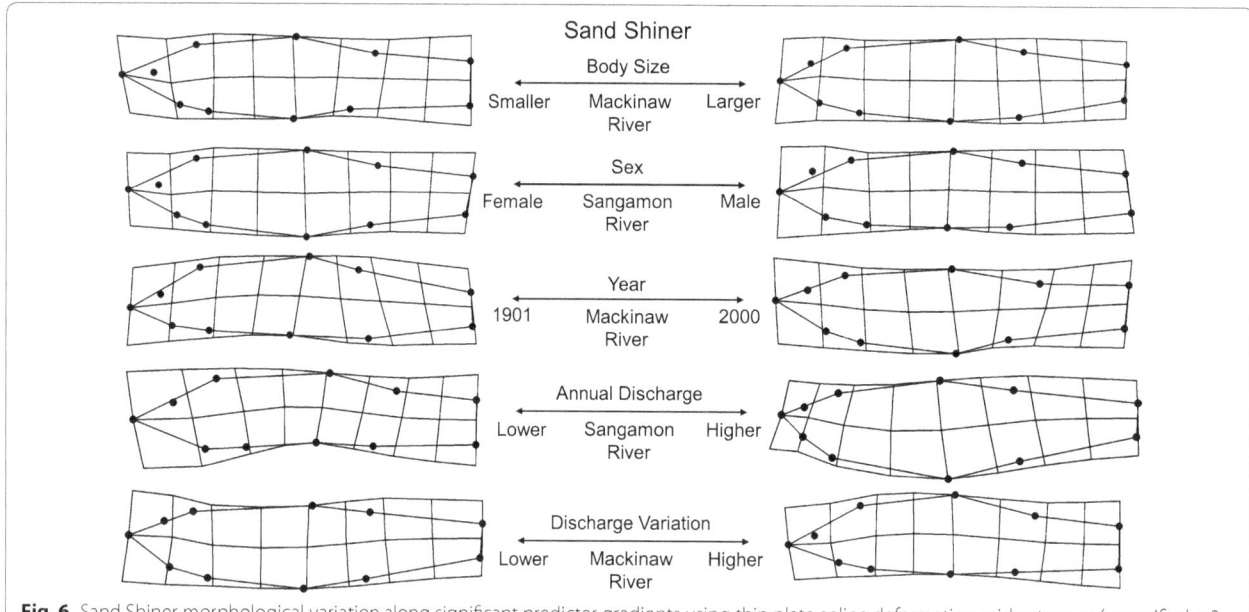

Fig. 6 Sand Shiner morphological variation along significant predictor gradients using thin plate spline deformation grid extremes (magnified ×3 for visualization)

lower daily discharge mean values. Salt Creek Bluntnose Minnow morphological variation over these years did not demonstrate a significant effect from time. Similarly, no interaction terms were recovered between sex and hydrology, body size and hydrology, or sex and year. See Table 4 and Fig. 7 for model configuration and deformation grids depicting shape change.

Discussion

Morphological models identified significant shape relationships with body size, sex, hydrology, time, and a series of interactions in this exploratory study of fish shape variation over the past 100 years. Several trends emerged that indicated similarities and differences in the relative contributions of each variable. However, no universal pattern or formula described Cyprinidae shape over the past century consistently across species. Overall, body size trends tended to be similar among taxa as larger individuals tended to be deeper bodied, exhibit narrower head shapes, show arching of the dorsal surface, and have compressed caudal peduncle regions relative to smaller individuals. In addition, sex-specific differences among species were present in the species measured and were generally driven by females which exhibited deeper or distended abdominal regions compared with males. In lotic sites only, responses to hydrological variables such as mean annual discharge or variation in discharge were present in most species but directionality of morphological responses was not consistent among species as some tended to be more fusiform with higher

discharge or variation in discharge, while others demonstrated increased robustness. Lastly, time was recovered in almost every model as an important contributor to morphological variation, however directionality also varied between species. This variability associated with time was not surprising, as the time variable is without a true mechanism measured over this century time span, and may reflect habitat variation at local and/or regional scales. Interactions among variables were far less consistent but did provide important controls for detecting differences in slopes between independent variables [42]. At the family level, the species level interaction term uncovered differences in slopes by species while at the intraspecific level interaction terms uncovered variation in allometry, sexual dimorphism, responses to flow, and time.

Body size and allometry

Morphological variation along body size gradients was expected as resource utilization in fishes typically changes during development into adulthood consistent with ontogenic relationships to niche [12, 13]. These aspects of body size and morphological relationships are frequently linked to prey selection, competition, microhabitat choice, and predator–prey dynamics [8, 51]. Results of this study supported a general change in morphology across species consistent with body size; however, directly linking this variation to local environments was not possible. An exception to this was information gleaned from the interaction terms with body size, as

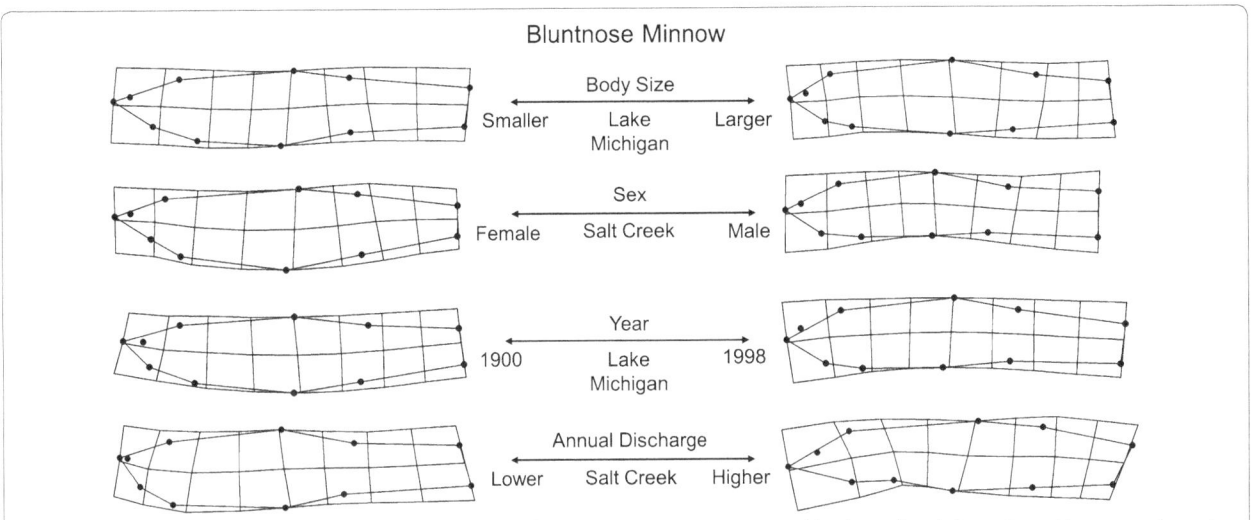

Fig. 7 Bluntnose Minnow morphological variation along significant predictor gradients using thin plate spline deformation grid extremes (magnified ×3 for visualization)

several species showed relationships between size and sex as well as size and hydrology, suggesting that the allometric scaling relationship is contingent on a variety of factors that may affect survival. Interestingly, these interactions were not consistent across taxa or sites. Moreover, the effect of body size was not consistent in strength of relationships across taxa. Consequently, this might suggest different levels of selection for the body shape and body size relationship across Cyprinidae.

Sexual dimorphism

Morphological variation by sex was predicted as an effect of functional selection [52, 53] related to fecundity constraints due to ovary size relative to testes [54–56]. However, sexual shape dimorphism is not ubiquitous in Cyprinidae taxa [57]. Our results support this mixed result as sex-specific shape differences were not consistent across taxa or sites as evidenced by varying effects of sex between species and sites, as well as results failing to meet initial expectations of heterogeneity of body size slopes with sex. We suggest that this could be a result of either a real signal (or lack thereof) of sex-specific shape differences that are taxon-specific within Cyprinidae, or an artifact of the body size ranges available and used in this study. Interestingly, shape variation between males and females also seemed to fluctuate among sites, particularly for comparison of the lake to rivers, where lotic samples reflected a higher degree of sexual shape dimorphism in both species tested (Bluntnose Minnow and Emerald Shiner with lotic and lentic sites) than either of their lentic system counterparts. In addition, there were no interactions between sex and hydrology, where one might expect male morphologies to show a stronger

response to flow than females given reproductive constraints on body shape [57]. Our results did not support our a priori hypotheses [52, 53] regarding constraints on female morphology with selection for increased fecundity, as both male and female fishes responded similarly to varying hydrology. It seems plausible that a factor beyond often cited fecundity constraints may be responsible for the varying signal. For example, subtle differences in shape between sexes may translate to variation in niche space (e.g. diet, resource utilization, etc.) and potential release of competition. More research in the field and lab is needed into the variability within and between species and between sexes that explore potential hypotheses relating to niche, hydrology, fecundity, and morphological variability.

Hydrological influence

Changes in morphology with hydrology indicate the presence of environmental influences on morphology. And while hydrology is by no means the only potential environmental corollary of morphology, it represents a strong potential environmental mechanism to explain changes over time. This variation in morphology could be rooted in selective variation or occur as a result of phenotypic plasticity, however, these two potential mechanisms are not necessarily mutually exclusive. It is unlikely that field data without the addition of any population genetic component could be used to distinguish between the two. However, phenotypic plasticity likely contributes to the trends elicited in this study as several of the fishes exhibited exceptionally strong relationships with both mean annual discharge and with variation of discharge. Unfortunately, the relative contribution of plasticity in response to local environmental conditions or population genetic divergence to morphological variation is unknown for most taxa [11]. Determining whether plasticity, directional selection towards a specific morphology, or some combination thereof results in morphological covariation with the environment can further distinguish potential adaptations. However, this requires genetic information that is difficult to obtain following formalin fixation and long term ethanol museum storage [58].

This study complements contemporary hydrology and morphology studies, which typically only address hypotheses involving current environmental measurements and current fish samples by adding a long term component. The predominant trend among short term intraspecific morphology—hydrology studies is that individuals with deeper-bodied morphologies tend to occur in lower flow velocity reaches of lotic or in lentic ecosystems, whereas individuals with more fusiform morphologies tend to occur in high flow conditions [14–16]. We found that not all Cyprinidae taxa exhibited a significant signal with hydrology variation. Moreover, not all species demonstrated the same response to high levels of mean annual flow or variation of annual flows. High flow velocities are expected to select for morphology that is a fusiform shape, and the opposite with low flows because of the resistance—performance relationship resulting from hydrological drag in high vs low velocity waters that can accompany flow volumes [59]. One caveat that may explain the differences identified here is that high discharge does not necessarily translate to velocity, and in fact may have an opposite effect, as higher volumes can overflow into floodplains, resulting in reaches with relatively low velocity. Furthermore, although not a direct objective of this study, visualization of consensus shapes between lentic and lotic sites suggested that individuals from lentic sites tended to be deeper-bodied than lotic individuals. However, future studies should directly quantify this with additional Cyprinidae species to test whether the pattern emerges across all species or is taxon specific since this study was based more on site level replicates through time and less on spatial replicates [26].

Long term habitat change
In this study, while hydrology served as a direct mechanism of morphological change, time was used an indirect metric of change and can function as a notice of some unmeasured abiotic or biotic alteration. In several instances, time was excluded because it was essentially synonymous with variation in included hydrology variables. In these instances, hydrology is likely the primary determinant of environmentally driven shape change. However, in other instances where time was not related to mean annual discharge or discharge variation, it

provides a unique perspective in the MANCOVA models independent of the hydrological factors directly assessed. However, hydrological variation can be assessed using a variety of metrics or approaches and the IHA analyses elicited several other hydrological parameters that also fluctuated with time. Other variables, in addition to hydrological variation, that would improve understanding of Cyprinidae morphology include disentangling the relative contributions from microhabitat, individual movement and dispersal potential, community structure, and phylogenetic history [15, 60–66]. Incorporation of these additional details into Cyprinidae morphology studies may provide a better understanding of morphological evolution—and by extension improve understanding of large scale ecosystem, community, and conservation issues.

Systematic, ecosystem, and community implications
Temporal variation in morphology can result from immigration, selection, or other causes [67]. Therefore the presence of temporal morphological variation can prevent or disrupt long term ecological conclusions based off of a single sample from a single site. This study utilized multiple samples from numerous sites for several species and could provide a useful template or model for understanding long term changes in the environment and resulting morphologies of fishes. The variation in morphology documented herein can contribute to our understanding of variation in large scale ecosystem processes in light of ecomorphology [7, 68–70]. Ecosystem attributes such as primary production, nutrient cycling, and stability are linked to functional richness and diversity [71]. Therefore, we may better understand how these changes can influence large scale processes by providing clear mechanisms to what drives form. For example, in a recent Cyprinidae study, Wanink and Witte [72] demonstrated morphological differentiation that coincided with a pelagic to benthic niche shift as a result of an ecological disturbance. Similarly, Eklöv and Svanbäck [9] identified a predator- and habitat-induced shift in morphology that coincided with resource specialization. However, despite few studies that documented linkages between morphological variation and ecosystem attributes, more direct comparisons or experiments are necessary. This study describes morphological variation that coincides with several metrics that relate to organismal function. As morphology changed with time as documented in the model one can postulate potential shifts in the ecosystem pertaining to energy, habitat, or other members of the community. However, studies that directly address the hypothesis of ecosystem covariation with intraspecific morphological variation may be confounded by multiple interspecific population responses contributing to an

ecosystem process. In addition, hydrological alterations directly influence ecosystem processes and are present in the majority of lotic and lentic systems worldwide [25]. Thus, the ability to manipulate a single population within an ecosystem and measure an indirect effect of morphological change on ecosystem properties may be possible only in laboratories or mesocosm experiments.

Morphological variation may also impact or influence community structure. Gatz [5] described resource utilization within an assemblage of freshwater stream fishes on the basis of their associated morphologies. He discerned similar morphological patterns among disparate assemblages indicating non-random distribution of resource niche space. Gatz's [5] conclusion of non-random morphological assembly was supported by Winston [6] for Cyprinid shape, where a lack of morphological overlap was attributed to interspecific competition. This non-random distribution in morphological space within an assemblage could potentially result in large overall impacts following a shift in one or more taxa. Environmental alterations that induce a change in morphology and thus functional roles could instigate increased or relaxed competition, directly or indirectly.

Conclusions

Discerning linkages between morphology and hydrology has conservation applications for understanding the capacity of species to respond to changing ecosystems. The results presented herein provide a template for species that have persisted in a variety of sites over the past century. Given that these sites are disturbed and altered (as evidenced by the IHA analyses), future work should attempt to repeat this study with different taxa from less altered sites to test whether these species are successful solely because they are plastic in their morphologies. Morphological variation may also aid in understanding the invasion potential for ecosystems [72] through competition overlap and resilience. One prediction of the ecosystem resilience model is that competition among taxa of similar shape [6] can prevent invasive taxa from becoming established [73]. Linking variation in morphology via disparity indices with assemblage shifts could provide one avenue to address these sorts of questions by establishing a metric based on morphology that could be used to look at ecosystem level resilience. Morphological variation within communities is linked to the zoogeographical history of a region [32] and can be applied at a taxon specific level to better understand historic and current environmental conditions and biogeographical patterns. Understanding individual response mechanisms may increase understanding of selective pressures that influenced the evolutionary history of the Cyprinidae.

Abbreviations
INHS: Illinois Natural History Survey; USGS: United States Geological Survey; IHA: indicators of hydrologic alteration; GPA: generalized procrustes analyses; RWA: relative warp analysis; MANCOVA: multivariate analysis of covariance; VIF: variance inflation factor.

Authors' contributions
SJJ performed all analyses and writing and MP contributed to writing. Both authors read and approved the final manuscript.

Author details
[1] Department of Biological Sciences, Dwyer Hall, Wright State University-Lake Campus, Celina, OH 45822, USA. [2] Department of Biology, Aquatic Biology and Fisheries Center, Ball State University, Muncie, IN 47306, USA.

Acknowledgements
We are grateful to the Illinois Natural History Survey for specimen loans and to Emily Brink (Schritter) for assistance in the laboratory.

Competing interests
The authors declare that they have no competing interests.

Funding
No funding was received to complete this study.

Appendix
See Tables 5, 6.

Table 5 Lot numbers for preserved specimens from the Illinois Natural History Survey (INHS)

Species	INHS	River	Year	Illinois county
Luxilus chrysocephalus	8513	Sangamon River	1965	Piatt
Luxilus chrysocephalus	16093	Sangamon River	1965	Macon
Luxilus chrysocephalus	21385	Sangamon River	1968	McLean
Luxilus chrysocephalus	43181	Sangamon River	1989	Champaign
Luxilus chrysocephalus	84335	Sangamon River	1901	Champaign
Luxilus chrysocephalus	84370	Sangamon River	1901	Champaign
Luxilus chrysocephalus	84385	Sangamon River	1901	Champaign
Luxilus chrysocephalus	91871	Sangamon River	2001	Champaign
Luxilus chrysocephalus	96038	Sangamon River	1988	Champaign
Luxilus chrysocephalus	96063	Sangamon River	1987	Champaign
Luxilus chrysocephalus	98050	Sangamon River	1988	Champaign
Lythrurus umbratilis	11950	Little Vermilion	1962	Vermilion
Lythrurus umbratilis	32769	Little Vermilion	1994	Vermilion
Lythrurus umbratilis	37727	Little Vermilion	1996	Vermilion
Lythrurus umbratilis	41540	Little Vermilion	1997	Vermilion
Lythrurus umbratilis	42177	Little Vermilion	1997	Vermilion
Lythrurus umbratilis	43139	Little Vermilion	1997	Vermilion
Lythrurus umbratilis	65032	Little Vermilion	1989	Vermilion
Lythrurus umbratilis	86476	Little Vermilion	1899	Vermilion
Lythrurus umbratilis	94588	Little Vermilion	2002	Vermilion
Notropis atherinoides	632	Lake Michigan	1964	Cook
Notropis atherinoides	25287	Illinois River	1957	Grundy
Notropis atherinoides	25460	Illinois River	1962	LaSalle
Notropis atherinoides	25480	Illinois River	1963	LaSalle
Notropis atherinoides	26591	Lake Michigan	1975	Cook
Notropis atherinoides	27303	Lake Michigan	1979	Lake
Notropis atherinoides	49876	Illinois River	1997	Grundy
Notropis atherinoides	51058	Illinois River	1902	LaSalle
Notropis atherinoides	51173	Lake Michigan	1998	Cook
Notropis atherinoides	52188	Illinois River	1999	LaSalle
Notropis atherinoides	53920	Lake Michigan	1999	Cook
Notropis atherinoides	84565	Lake Michigan	1900	Cook
Notropis atherinoides	85583	Illinois River	1897	Mason
Notropis atherinoides	98018	Illinois River	2003	LaSalle
Notropis stramineus	8022	Sangamon River	1959	Champaign
Notropis stramineus	10902	Mackinaw River	1958	Woodford
Notropis stramineus	14879	Mackinaw River	1963	Tazewell
Notropis stramineus	15063	Mackinaw River	1966	Tazewell
Notropis stramineus	16134	Sangamon River	1968	Macon
Notropis stramineus	21240	Sangamon River	1965	McLean
Notropis stramineus	33080	Mackinaw River	1994	Woodford
Notropis stramineus	40803	Mackinaw River	1995	Tazewell
Notropis stramineus	41475	Mackinaw River	1997	Tazewell
Notropis stramineus	42050	Mackinaw River	1996	Tazewell
Notropis stramineus	42188	Sangamon River	1997	Champaign
Notropis stramineus	52941	Mackinaw River	1999	Woodford
Notropis stramineus	57216	Sangamon River	2000	Macon
Notropis stramineus	84387	Sangamon River	1901	Champaign

Table 5 continued

Species	INHS	River	Year	Illinois county
Notropis stramineus	85832	Mackinaw River	1901	McLean
Notropis stramineus	93187	Mackinaw River	2000	McLean
Notropis stramineus	93223	Mackinaw River	2000	McLean
Notropis stramineus	96042	Sangamon River	1988	Champaign
Notropis stramineus	99815	Sangamon River	2003	Champaign
Notropis stramineus	101741	Sangamon River	2005	Macon
Pimephales notatus	617	Lake Michigan	1964	Cook
Pimephales notatus	13714	Salt Creek	1963	DeWitt
Pimephales notatus	13731	Salt Creek	1967	DeWitt
Pimephales notatus	18681	Salt Creek	1971	Logan
Pimephales notatus	42450	Salt Creek	1997	DeWitt
Pimephales notatus	42834	Salt Creek	1997	Logan
Pimephales notatus	56877	Lake Michigan	1980	Cook
Pimephales notatus	84567	Lake Michigan	1900	Cook
Pimephales notatus	85438	Salt Creek	1900	Logan
Pimephales notatus	85460	Salt Creek	1901	Logan
Pimephales notatus	89383	Salt Creek	2000	Mason
Pimephales notatus	96955	Lake Michigan	1999	Cook
Pimephales notatus	96960	Lake Michigan	1998	Cook
Pimephales notatus	97146	Lake Michigan	1987	Cook
Pimephales notatus	97588	Salt Creek	2003	Mason

Table 6 Results from IHA analyses with time

Variable	IHA category	Illinois River Slope	R²	Mackinaw River Slope	R²	Salt Creek Slope	R²	Sangamon River Slope	R²	Vermillion River Slope	R²
High pulse count	Frequency and duration of pulses	*0.03*	0.05	*−0.07*	0.18	−0.02	0.02	*0.03*	0.07	*0.05*	0.1
High pulse duration		−0.02	0.01	0.15	0.06	*0.14*	0.09	0.07	0.04	−0.01	0.01
Low pulse count		*0.16*	0.19	0	0	0.01	0	0	0	0.01	0.01
Low pulse duration		*0.04*	0.07	−0.05	0.01	−0.1	0.01	−0.62	0.08	−0.19	0.09
1-day maximum	Magnitude and duration of annual extremes	*295.2*	0.15	−15.08	0	21.41	0	129.6	0.04	*81.05*	0.06
1-day minimum		−61.55	0.51	−0.1	0.02	*0.84*	0.12	*1.74*	0.07	*0.44*	0.18
30-day maximum		*58.76*	0.05	2.91	0	21.19	0.03	58.36	0.06	*17.57*	0.07
30-day minimum		−65.06	0.49	−0.4	0.05	*1.31*	0.1	0.48	0	*0.75*	0.13
3-day maximum		*241.7*	0.14	−1.38	0	21.84	0	122.8	0.04	*65.77*	0.07
3-day minimum		−63.79	0.55	−0.12	0.02	0.80	0.11	*1.62*	0.06	*0.46*	0.18
7-day maximum		*168.40*	0.12	−0.08	0	18.83	0	*101.8*	0.05	*37.48*	0.06
7-day minimum		−64.05	0.55	−0.12	0.02	*0.99*	0.12	1.51	0.04	*0.48*	0.17
90-day maximum		6.55	0.00	1.4	0	13.7	0.04	*33.14*	0.05	*10.19*	0.08
90-day minimum		−51.73	0.29	−0.09	0	1.73	0.01	4.17	0.02	*3.02*	0.06
Baseflow index		0	0.56	0	0.01	0	0.03	0	0.02	*0*	0.06
April	Magnitude of monthly conditions	−38.09	0.02	−4.48	0.02	2.82	0	−1.15	0	5.77	0.03
August		*−41.95*	0.17	1.29	0.03	6.62	0.03	−9.85	0.01	2.21	0.02
December		−42.04	0.06	−1.51	0.01	9.13	0.03	*31.28*	0.08	*6.32*	0.06
February		*−39.67*	0.05	−3.34	0.02	3.78	0	14.22	0.01	3.02	0.01
January		*−61.71*	0.11	−1.32	0	0.36	0	4.07	0	−5.45	0.02
July		−11.87	0.01	2.99	0.03	4.49	0.01	−1.19	0	3.07	0.02
June		3.09	0.00	4.72	0.07	10.42	0.03	13.86	0.01	4.24	0.03
March		−18.53	0.01	1.18	0	19.43	0.06	27.55	0.04	8.99	0.07
May		−25.76	0.02	7.83	0.04	17.66	0.04	29.63	0.03	4.31	0.01
November		*−51.93*	0.10	−0.78	0	2.28	0	10.67	0.02	*6.26*	0.05
October		*−52.70*	0.13	−1.31	0.01	4.14	0.01	−0.31	0	3.48	0.05
September		*−38.86*	0.07	0.41	0	5.15	0.04	−12.2	0.02	1.49	0.01
Fall rate	Rate and frequency of water condition changes	−3.67	0.18	0.07	0	−0.11	0	−0.32	0.01	*−0.37*	0.08
Number of reversals		*0.47*	0.11	0.12	0.03	0.12	0.03	*0.52*	0.41	*0.36*	0.17
Rise rate		*1.98*	0.05	−0.54	0.02	−0.13	0	−0.15	0	0.47	0.02
Date of maximum	Timing of annual extreme conditions	0.47	0.02	0.4	0.02	*1.07*	0.07	−0.06	0	0.19	0
Date of minimum		*0.92*	0.05	−0.02	0	−0.1	0	0.43	0.02	−0.08	0

See Tables 1 and 2 for site descriptions. Significant (P < 0.05) slope values are in italic

References

1. Poff NL, Allan JD, Bain MB, Karr JR, Prestegaard KL, Richter BD, Sparks RE, Stromberg JC. The natural flow regime. Bioscience. 1997;47:769–84.
2. Richter BD, Braun DP, Mendelson MA, Master LL. Threats to imperiled freshwater fauna. Conserv Biol. 1997;11:1081–93.
3. Xenopoulos MA, Lodge DM. Going with the flow: using species-discharge relationships to forecast losses in fish biodiversity. Ecology. 2006;87:1907–14.
4. Ross ST, Matthews WJ, Echelle AA. Persistence of stream fish assemblages: effects of environmental change. Am Nat. 1985;126:24–40.
5. Gatz AJ. Community organization in fishes as indicated by morphological features. Ecology. 1979;60:711–8.
6. Winston MR. Co-occurrence of morphologically similar species of stream fishes. Am Nat. 1995;145:527–45.
7. Tilman D, Knops J, Wedin D, Reich P, Ritchie M, Siemann E. The influence of functional diversity and composition on ecosystem processes. Science. 1997;277:1300–2.
8. Wainwright PC, Reilly SM. Ecological morphology: integral organismal biology. Chicago: University of Chicago Press; 1994.
9. Eklöv P, Svanbäck R. Predation risk influences adaptive morphological variation in fish populations. Am Nat. 2006;167:440–52.
10. Van Valen L. Morphological variation and width of ecological niche. Am Nat. 1965;99:377–90.
11. Langerhans RB. Predictability of phenotypic differentiation across flow regimes in fishes. Integr Comp Biol. 2008;158:699–708.

12. Hood CS, Heins DC. Ontogeny and allometry of body shape in the black-tail shiner, *Cyprinella venusta*. Copeia. 2000;2000:270–5.
13. Svanbäck R, Eklöv P. Effects of habitat and food resources on morphology and ontogenetic growth trajectories in perch. Oecologia. 2002;131:61–70.
14. Jacquemin SJ, Martin E, Pyron M. Morphology of bluntnose minnow *Pimephales notatus* (Cyprinidae) covaries with habitat in a central Indiana watershed. Am Midl Nat. 2013;169:137–46.
15. Langerhans RB, Layman CA, Langerhans AK, Dewitt TJ. Habitat-associated morphological divergence in two Neotropical fish species. Biol J Linn Soc. 2003;80:689–98.
16. Haas TC, Blum MJ, Heins DC. Morphological responses of a stream fish to water impoundment. Biol Lett. 2010;6:803–6.
17. Dugas MB, Franssen NR, Bastille MO, Martin RA. Morphological correlates of river velocity and reproductive development in an ornamented stream fish. Evol Biol Fish. 2016;30:21–33.
18. Pakkasmaa S, Piironen J. Water velocity shapes juvenile salmonids. Evol Ecol. 2001;14:721–30.
19. Cureton JC II, Broughton RE. Rapid morphological divergence of a stream fish in response to changes in water flow. Biol Lett. 2014;10:20140352.
20. Forbes SA, Richardson RE. The Fishes of Illinois. Urbana: Illinois State Laboratory of Natural History; 1908.
21. Smith PW. The Fishes of Illinois. Champaign: University of Illinois Press; 1979.
22. Pyke GH, Ehrlich PR. Biological collections and ecological/environmental research: a review, some observations and a look to the future. Biol Rev. 2010;85:247–66.
23. Ficke AD, Myrick CA, Hansen LJ. Potential impacts of global climate change on freshwater fisheries. Rev Fish Biol Fish. 2007;17:581–613.
24. Harris NM, Gurnell AM, Hannah DM, Petts GE. Classification of river regimes: a context for hydroecology. Hydrol Process. 2000;14:2831–48.
25. Poff NL, Zimmerman KH. Ecological responses to altered flow regimes: a literature review to inform the science and management of environmental flows. Freshw Biol. 2010;55:194–205.
26. Haas TC, Heins DC, Blum MJ. Predictors of body shape among populations of a stream fish (*Cyprinella venusta*, Cypriniformes: Cyprinidae). Biol J Linn Soc. 2015;115:842–58.
27. Dewson ZS, James ABW, Death RG. A review of the consequences of decreased flow for instream habitat and macroinvertebrates. J N Am Benthol Soc. 2007;26:401–15.
28. Taylor CM, Millican DS, Roberts ME, Slack WT. Long-term change to fish assemblages and the flow regime in a southeastern US river system after extensive aquatic ecosystem fragmentation. Ecography. 2008;31:1–11.
29. Johnson PTJ, Olden JD, Vander Zanden MJ. Dam invaders: impoundments facilitate biological invasions into freshwaters. Front Ecol Environ. 2008;6:357–63.
30. Poff NL, Allan JD. Functional organization of stream fish assemblages in relation to hydrologic variability. Ecology. 1995;76:606–27.
31. Pyron M, Lauer TE. Hydrological variation and fish assemblage structure in the middle Wabash River. Hydrobiologia. 2004;525:203–13.
32. Hoagstrom CW, Berry CR. Morphological diversity among fishes in a Great Plains river drainage. Hydrobiologia. 2008;596:367–86.
33. Webb M, Reid M, Thoms M. The influence of hydrology and physical habitat character on fish assemblages at different temporal scales. River Syst. 2011;19:283–99.
34. Koel TM, Sparks RE. Historical patterns of river stage and fish communities as criteria for operations of dams on the Illinois River. River Res Appl. 2002;18:3–19.
35. Rohlf FJ. tpsDig—thin plate spline digitizer, version 2.11. Stony Brook: State University of New York at Stony Brook; 2001.
36. Rohlf FJ. tpsUtil—thin plate spline utility program, version 1.46. Stony Brook: State University of New York at Stony Brook; 2010.
37. Zelditch ML, Swiderski DL, Sheets HD. Geometric morphometrics for biologists: a primer. Cambridge: Elsevier Academic Press; 2004.
38. Jawad LA. The effect of formalin, alcohol and freezing on some body proportions of *Alepes djeddaba* (Pisces: Carangidae) collected from the Red Sea coast of Yemen. Rev Biol Mar Oceanogr. 2003;38:77–80.
39. Richter BD, Baumgartner JV, Powell J, Braun DP. A method for assessing hydrologic alteration within ecosystems. Conserv Biol. 1996;10:1163–74.
40. Chapra SC, Dove A, Warren GJ. Long-term trends of Great Lakes major ion chemistry. J Great Lakes Res. 2012;38:550–60.
41. Rohlf FJ. In: Marcus LF, Bello E, Garcia-Valdecasa AG, editors. Contributions to morphometrics. Madrid: Museo Nacional de Ciencias Naturales; 1993. p. 131–59.
42. Sheets HD, Zelditch ML. Studying ontogenetic trajectories using resampling methods and landmark data. Hystrix. 2013;24:67–73.
43. Scheiner SM. In: Scheiner SM, Gurevitch J, editors. Design and analysis of ecological experiments 2nd edition. Oxford: Oxford University Press; 2001. p. 99–115.
44. Nakazawa M. fmsb: functions for medical statistics book with some demographic data. R package version 0.5.2. 2015. http://CRAN.R-project.org/package=fmsb. Accessed Mar 2016.
45. Fox J, Friendly M, Monette G. Heplots: visualizing tests in multivariate linear models. R package version 1.0-0. 2012. http://CRAN.R-project.org/package=heplots. Accessed Mar 2016.
46. Collar DC, Near TJ, Wainwright PC. Comparative analysis of morphological diversity: does disparity accumulate at the same rate in two lineages of centrarchid fishes? Evolution. 2005;59:1783–94.
47. Friedman M. Explosive morphological diversification of spiny-finned teleost fishes in the aftermath of the end-cretaceous extinction. Proc R Soc B. 2010. doi:10.1098/rspb.2009.2177.
48. Adams D, Collyer M, Kaliontzopoulou A, Sherratt E. Geomorph: geometric morphometric analyses of 2D/3D landmark data. R package version 3.0.1. 2016. http://CRAN.R-project.org/package=geomorph. Accessed July 2016.
49. Querino RB, Moraes RCB, Zucchi RA. Relative warp analysis to study morphological variation in the genital capsule of *Trichogramma pretiosum* Riley (Hymenoptera: Trichogrammatidae). Neotrop Entomol. 2002;31:217–24.
50. Rohlf FJ. tpsRegr—thin plate splines regression analysis, version 1.43. Stony Brook: State University of New York at Stony Brook; 2016.
51. Rincón PA, Bastir M, Grossman GD. Form and performance: body shape and pre-capture success in four drift-feeding minnows. Oecologia. 2007;152:345–55.
52. Hedrick AV, Temeles EJ. The evolution of sexual dimorphism in animals: hypotheses and tests. Trends Ecol Evol. 1989;4:136–8.
53. Parker GA. The evolution of sexual size dimorphism in fish. J Fish Biol. 1992;41(Supplement B):1–20.
54. Ostrand KG, Wilde GR, Strauss RE, Young RR. Sexual dimorphism in plains minnow, *Hybognathus placitus*. Copeia. 2001;2001:563–5.
55. Skelton CE. New dace of the genus *Phoxinus* (Cyprinidae: Cypriniformes) from the Tennessee River Drainage, Tennessee. Copeia. 2001;2001:118–28.
56. Pyron M, Fincel M, Dang M. Sexual size dimorphism and ecomorphology of spotfin shiner (*Cyprinella spiloptera*) from the Wabash River watershed. J Freshw Ecol. 2007;22:687–96.
57. Douglas ME. Analysis of sexual dimorphism in an endangered cyprinid fish (*Gila cypha* Miller) using video image technology. Copeia. 1993;1993:334–43.
58. Chakraborty A, Sakai M, Iwatsuki Y. Museum fish specimens and molecular taxonomy: a comparative study on DNA extraction protocols and preservation techniques. J Appl Ichthyol. 2006;22:160–6.
59. Vogel S. Life in moving fluids: the physical biology of flow. 2nd ed. Princeton: Princeton University Press; 1994.
60. Hart JS. Geographical variation of some physiological and morphological characters in certain freshwater fish. Univ Toronto Stud Biol Ser 60. 1952;72:1–79.
61. Schmitt RJ, Holbrook SJ. Ontogeny of prey selection by black surfperch *Embiotoca jacksoni* (Pisces: Embiotocidae): the roles of fish morphology, foraging behavior, and patch selection. Mar Ecol. 1984;18:225–39.
62. Meyer A. Phenotypic plasticity and heterochrony in *Cichlasoma managuense* (Pisces, Chichlidae) and their implications for speciation in cichlid fishes. Evolution. 1987;41:1357–69.
63. Laurent P, Perry SF. Environmental effects on gill morphology. Physiol Zool. 1991;64:4–25.
64. Schluter D. Experimental evidence that competition promotes divergence in adaptive radiation. Science. 1994;266:798–801.
65. Hegrenes S. Diet-induced phenotypic plasticity of feeding morphology in the orangespotted sunfish, *Lepomis humilis*. Ecol Freshw Fish. 2001;10:35–42.
66. Langerhans RB, Makowicz AM. Shared and unique features of morphological differentiation between predator regimes in *Gambusia caymanensis*. J Evol Biol. 2009;22:2231–42.

67. Lessios HA, Weinberg JR, Starczak VR. Temporal variation in populations of the marine isopod *Excirolana*: how stable are gene frequencies and morphology? Evolution. 1994;48:549–63.

68. Hjelm J, Weerd GH, Sibbing FA. Functional link between foraging performance, functional morphology, and diet shift in roach (*Rutilus rutilus*). Can J Fish Aquat Sci. 2003;60:700–9.

69. Ruehl CB, DeWitt TJ. Trophic plasticity and foraging performance in red drum, *Sciaenops ocellatus* (Linnaeus). J Exp Mar Biol Ecol. 2007;349:284–94.

70. Ellers J. In: Verhoef HA, Morin PJ, editors. Community ecology: processes, models, and applications. Oxford: Oxford University Press; 2010. p. 151–62.

71. Loreau M, Naeem S, Inchausti P, Bengtsson J, Grime JP, Hector A, Hooper DU, Huston MA, Raffaelli D, Schmid B, Tilman D, Wardle DA. Biodiversity and ecosystem functioning: current knowledge and future challenges. Science. 2001;294:804–8.

72. Wanink JH, Witte F. Rapid morphological changes following niche shift in the zooplanktivorous cyprinid *Rastrineobola argentea* from Lake Victoria. Neth J Zool. 2000;50:365–72.

73. Mason NWH, Mouillot D, Lee WG, Wilson JB. Functional richness, functional evenness and functional divergence: the primary components of functional diversity. Oikos. 2005;111:112–8.

6

Seasonal cues induce phenotypic plasticity of *Drosophila suzukii* to enhance winter survival

Peter W. Shearer[1], Jessica D. West[2], Vaughn M. Walton[3], Preston H. Brown[1], Nicolas Svetec[4] and Joanna C. Chiu[2*]

Abstract

Background: As global climate change and exponential human population growth intensifies pressure on agricultural systems, the need to effectively manage invasive insect pests is becoming increasingly important to global food security. *Drosophila suzukii* is an invasive pest that drastically expanded its global range in a very short time since 2008, spreading to most areas in North America and many countries in Europe and South America. Preliminary ecological modeling predicted a more restricted distribution and, for this reason, the invasion of *D. suzukii* to northern temperate regions is especially unexpected. Investigating *D. suzukii* phenology and seasonal adaptations can lead to a better understanding of the mechanisms through which insects express phenotypic plasticity, which likely enables invasive species to successfully colonize a wide range of environments.

Results: We describe seasonal phenotypic plasticity in field populations of *D. suzukii*. Specifically, we observed a trend of higher proportions of flies with the winter morph phenotype, characterized by darker pigmentation and longer wing length, as summer progresses to winter. A laboratory-simulated winter photoperiod and temperature (12:12 L:D and 10 °C) were sufficient to induce the winter morph phenotype in *D. suzukii*. This winter morph is associated with increased survival at 1 °C when compared to the summer morph, thus explaining the ability of *D. suzukii* to survive cold winters. We then used RNA sequencing to identify gene expression differences underlying seasonal differences in *D. suzukii* physiology. Winter morph gene expression is consistent with known mechanisms of cold-hardening such as adjustments to ion transport and up-regulation of carbohydrate metabolism. In addition, transcripts involved in oogenesis and DNA replication were down-regulated in the winter morph, providing the first molecular evidence of a reproductive diapause in *D. suzukii*.

Conclusions: To date, *D. suzukii* cold resistance studies suggest that this species cannot overwinter in northern locations, e.g. Canada, even though they are established pests in these regions. Combining physiological investigations with RNA sequencing, we present potential mechanisms by which *D. suzukii* can overwinter in these regions. This work may contribute to more accurate population models that incorporate seasonal variation in physiological parameters, leading to development of better management strategies.

Keywords: *Drosophila suzukii*, Phenotypic plasticity, Cold tolerance, Diapause, High-throughput sequencing, Transcriptome

Background

It is estimated that insects account for 18 % of global crop production losses [1]. An increase in average global temperature will likely intensify the damage caused by insect pests, as higher average temperature is predicted to increase insect populations through greater overwintering survival, higher reproductive rates, and an increased number of generations [2, 3]. In particular, invasive species may have an advantage over indigenous species in such conditions [4–6]. Therefore, it is

*Correspondence: jcchiu@ucdavis.edu
[2] Department of Entomology and Nematology, University of California, Davis, CA 95616, USA
Full list of author information is available at the end of the article

imperative to understand how invasive species can successfully invade and compromise ecosystems, sometimes in a very short timeframe. It has been hypothesized that high levels of phenotypic plasticity play an important role in the success of invasive species in changing conditions, such as those caused by global climate change [6, 7].

One such invasive species, *Drosophila suzukii* Matsumura (Diptera: Drosophilidae) was first discovered in the continental USA (Watsonville, CA) in 2008 [8] and has rapidly spread to become an established pest of fruit crops all over the world, including North and South America and much of Europe, [9–13]. Commonly known as the Spotted Wing Drosophila, this vinegar fly has an enlarged, serrated ovipositor, allowing adult females to penetrate the skin of soft-skinned, ripening fruit and lay eggs inside, where the larvae feed and destroy the fruit [11]. *Drosophila suzukii* most commonly infests cherries, blackberries, raspberries, and strawberries, but has also been found to oviposit into grapes, plums, peaches, and other fruits [11]. In the U.S. alone, *D. suzukii* invasions have caused significant crop losses, and costs directly related to management practices are estimated to vary between $129 and 172 million (6–8 % of farmgate value) annually [14].

Drosophila suzukii has a wide climatic presence, causing economic losses of affected fruits in areas ranging from mild subtropical production regions to severe continental climates [15]. In Asia, where this species is native, *D. suzukii* are preferentially found at higher altitudes and higher latitudes when compared to other closely related species [16]. Previous studies conducted on *D. suzukii* cold tolerance predict that this species will likely not survive extended periods of cold such as those found in production regions in Canada, Eastern Oregon, Washington, and Michigan [17]. Despite these predictions, *D. suzukii* is now an established pest in those regions [12, 15, 18], and in fact has proven successful in a wide range of environments ranging from Southern California to British Columbia, Canada [13], raising the question of how this species can adapt to the harsh climates in more northern locations.

Insects exhibit a wide variety of strategies to increase cold tolerance and overwinter. There are two main classes of cold-hardening: (1) seasonal cold-hardening, which is induced over a timescale of days to weeks, and (2) rapid cold-hardening, which can occur in minutes or hours, and is induced by a sudden drop in temperature like a cold snap [19]. Both seasonal and rapid cold-hardening mechanisms include adjustments to ion transport and membrane restructuring to increase membrane fluidity at low temperatures. The synthesis of cryoprotectants, typically polyols such as glycerol, sorbitol, or inositol, is an important mechanism in seasonal cold-hardening, but

it is unclear whether it is associated with rapid cold-hardening. Up-regulation of antifreeze proteins and ice nucleating agents are mechanisms that increase cold tolerance in seasonally cold-hardened insects, but are not associated with rapid cold-hardening. Inhibition of apoptotic cell death, MAP kinase signaling, and calcium signaling have all been found to be important mechanisms of the rapid cold-hardening response, but have not been found to occur in seasonal cold-hardening [19].

Seasonal cold-hardening is often an essential component of winter diapause, a process characterized by developmental arrest, decreased metabolic activity, and a general state of dormancy [19–21]. Reproductive diapause is an adaptation that allows an organism to temporarily cease reproduction in order to conserve resources to survive unfavorable conditions, continuing reproduction when more favorable conditions arise [22]. In fact, evolution of diapause has been linked to enabling range expansion in other invasive species [23, 24]. Diapause is found to occur in many *Drosophila* species, including *D. melanogaster*, yet can vary significantly among clinally distributed natural populations [25–28]. In *D. suzukii*, high rates of reproductively immature adult females have been observed in the months leading up to winter in Hokkaido, Japan, suggesting that a reproductive diapause may occur in this species [29]. Additionally, *D. suzukii* has a relatively low nucleotide substitution rate when compared to other Drosophilids [16]. This is consistent with presence of a reproductive diapause in this species, as a low substitution rate may be caused by fewer generations per year. Diapause incidence has also been assessed via ovary dissection (Anna K. Wallingford, Jana C. Lee, Gregory M. Loeb, personal communications). Wallingford et al. found that at a photoperiod of 12:12 L:D and 10 °C, there were almost no reproductively mature females in laboratory conditions, and no reproductively mature females in December at field collection sites in Oregon and New York.

In addition to undergoing reproductive diapause, Drosophilids are known to exhibit multiple strategies to survive suboptimal cold temperatures and low humidity. These strategies include accumulation of cryoprotectants such as maltose, trehalose, proline, and myo-inositol [30–34], altered composition of membrane phospholipids [30, 35], and increased expression of stress-induced genes such as heat shock proteins [36–38]. Darker cuticle pigmentation has been hypothesized to be involved in thermoregulation of ectotherms in cold environments, resulting in increased ultraviolet absorption and increased ability to warm up [39, 40]. However, increased melanization has also been implicated in immunity and increased desiccation resistance [41–43]. A larger body size in colder environments may also be advantageous in colder temperatures, as it may be involved in thermoregulation [44–46] or cost-benefits of altered membrane fluidity [45], but the relationship between body size and temperature is not well understood [47, 48]. Latitudinal clines of adult body size are known to exist for many ectotherms, including several *Drosophila* species [49, 50], with smaller flies observed in warmer places, but this phenomenon is not well conserved in insects and other ectotherms [47, 51]. In addition to the inverse relationship of temperature and body size observed in natural environments, most ectotherms, including *D. melanogaster*, grow to be smaller sizes when raised in warmer temperatures in the laboratory [45, 47, 48].

Phenotypic plasticity is a phenomenon by which one genotype can lead to multiple phenotypes in different environmental conditions [52]. Phenotypic plasticity often occurs in response to seasonal changes in order for the insect to display traits that best suit seasonal conditions, producing a seasonal morph [22]. In some cases, seasonal morphs are tightly linked to diapause [22]. A recent study reported on *D. suzukii* seasonal morphs [53], in which they found that *D. suzukii* winter morphs are able to survive lower temperatures than *D. suzukii* summer morphs, helping to explain the wide climatic presence of this invasive pest. In this study, we characterized seasonal phenotypic plasticity in *D. suzukii* in both field-collected populations and flies placed under simulated seasonal conditions in the laboratory. We quantified phenotypic characters and measured differences in low-temperature survival rates between the summer and winter morphs of *D. suzukii*. Finally, we used RNA sequencing to examine transcriptomic differences that underlie the observed phenotypic divergence between the adult stages of the two seasonal morphs. Knowledge of *D. suzukii* survival strategies in the winter is essential and of great economic importance, as it will direct the development of more effective management strategies.

Results

D. suzukii exhibit seasonal variations in phenological traits in the field

Female *D. suzukii* collected seasonally from traps located in Hood River, OR, USA displayed large variation in individual body size throughout the 2011 and 2012 samples (Fig. 1a). To determine the possible factors underlying seasonal variation in body size, we monitored two abiotic factors: temperature of development and day-length. As proxy for developmental temperature, we used the average temperature for the 12-day period preceding the collection date. To ascertain whether there is a significant relationship between these factors and wing length, we then performed, for each sex, linear regressions of wing length (i.e. a proxy of body size) over temperature and

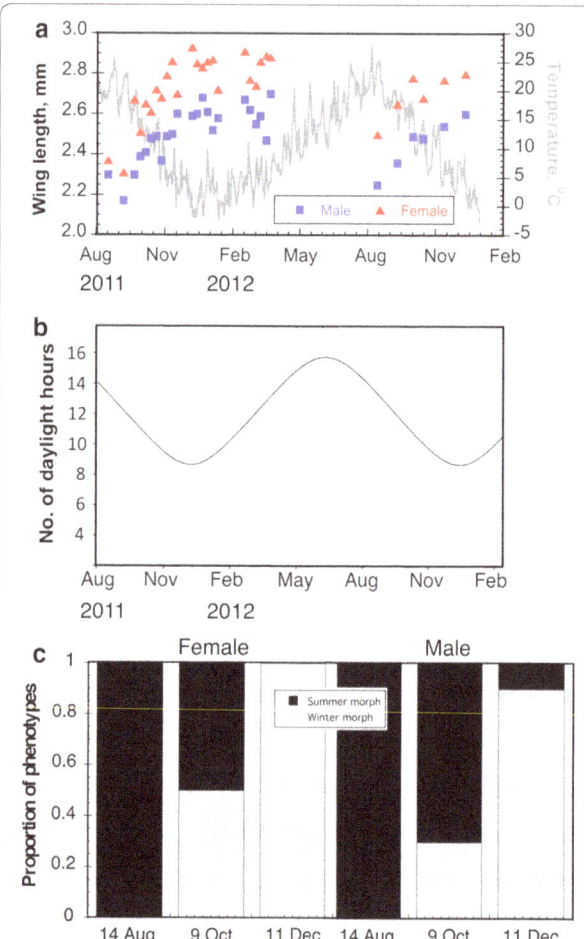

Fig. 1 *Drosophila suzukii* exhibit phenotypic plasticity in size and melanization in the field. **a** Seasonal changes in wing length of female *D. suzukii* collected from the field (Hood River, OR, USA) in apple cider vinegar baited traps. Both wing length (mm) (*left axis*) and temperature (°C) (*right axis*) are represented on the Y-axes, and chronological time (between August 2011 to February 2013) is represented on the X-axis. Male wing length is plotted in *blue* and female wing length is plotted in *red*. Temperature is plotted in *gray*. Mean daily air temperatures are plotted from the Hood River, Oregon AgriMet Weather station (HOXO) [73]. **b** Seasonal change in photoperiod from August 2011 to February 2013. The day-length (in hours) is shown on the *Y-axis* and date is shown on the *X-axis*. Day-length values were *plotted* for Hood River, Oregon [74]. **c** Percent summer morph (lighter pigmentation) and winter morph (*darker pigmentation*) of female (*left*) and male (*right*) *D. suzukii* from August to December in Hood River, OR during 2011. Summer and winter morphs are represented by *black* and *gray* respectively

day-length. The 12-day mean temperature explained 68 % ($R^2 = 0.68$) of the observed variation in wing length for females where it explained 77 % ($R^2 = 0.68$) of the observed variation for male wing size (both linear regressions: $P < 0.0001$). The mean wing length was also negatively correlated to day-length (Fig. 1b) Day-length explained 40 % ($R^2 = 0.40$) of the observed variation in female wing length, and 47 % ($R^2 = 0.4658$) of the observed variation in males' wing length (both linear regressions: $P < 0.0001$). Moreover, the seasonal composition of both male and female *D. suzukii* winter morph increased from levels of 0 % of both sexes to 100 and 95 % respectively when examining dates starting on 14 August to 11 December, 2011 (Fig. 1c).

Simulated summer and winter conditions in the laboratory induce differences in phenological traits

To investigate the impact of environmental factors in influencing seasonal variations in *D. suzukii* morphology and physiology, we tested effects of simulated summer and winter laboratory conditions by varying photoperiod and temperature. Differences in adult *D. suzukii* body color and wing length reared at 16:8 Light:Dark (L:D) in hours and 20 °C as compared to 12:12 L:D and 10 °C were visually apparent (Fig. 2), and closely resemble summer and winter morphs that we observed in the field. Abdominal melanization ratings were significantly higher for fourth abdominal segments of female flies subjected to 12:12 L:D and 10 °C compared to female flies that were reared at 16:8 L:D and 20 °C (t = −20.6; df = 16; $P < 0.0001$) (Table 1). Similarly, abdominal melanization ratings were significantly higher for the third abdominal segments for male flies that were reared in 12:12 L:D and 10 °C when compared to male flies that were housed in 16:8 L:D and 20 °C (t = −13.5; df = 27; $P < 0.0001$) (Table 1).

We then conducted a series of experiments to examine intergeneration effects of photoperiod and temperature on wing length. We first examined the effect

Fig. 2 Phenotypic variation of laboratory-reared *D. suzukii* expressed by different photoperiod and temperature regimes. Summer morph adults are reared at 20 °C and 16:8 L:D photoperiod (*top panels*); winter morph adults are reared at 10 °C and 12:12 L:D photoperiod (*bottom panels*)

Table 1 Average melanization rating of dorsal abdominal bands of female and male *Drosophila suzukii* seasonal morphs

Seasonal morph and sex[b]	Melanization rating[a]				
	Abdominal segment				
	1st	2nd	3rd	4th	5th
Summer female	1.1	2.0	1.9	1.8[*,c]	4.6
Winter female	2.8	2.9	3.5	5.0	5.0
Summer male	1.4	2.1	2.6[*]	5.0	5.0
Winter male	3.5	3.6	4.9	5.0	5.0

n = 13 for winter male and female; n = 17 for summer male and female

[a] Melanization rating based on visual rating of the thickness of the transverse dark line of each dorsal abdominal segment (Additional file 1: Figure S1): *1* thin dark line, *5* completely dark

[b] Summer morphs were reared at 16:8 L:D and 20 °C; winter morphs were reared at 12:12 L:D and 10 °C

[c] Means followed by an asterisk are significantly different within a sex (t test, $P \leq 0.05$)

of photoperiod alone on wing length (Table 2). Holding temperature constant at 20 °C, we either kept parent flies (F0), which were reared in 16:8 L:D photoperiod, in the same photoperiod (16:8 L:D) or transferred the adult parent flies (F0) to 12:12 L:D and examine the resulting offsprings (F1). Not surprisingly, there was no difference in wing length for female offsprings when parents were maintained in a photoperiod of 16:8 L:D as compared to their female parents (Table 2). However, when the parents were transferred to 12:12 L:D, their female offsprings displayed significantly increased wing length compared to female offsprings with parents reared under 16:8 L:D ($F = 37.7$; df = 2, 32; $P < 0.0001$) (Table 2).

We next examined the effect of both photoperiod and temperature on wing length (Table 3). In a photoperiod of 16:8 L:D, if parents (F0) were transferred from 20 to 10 °C, their female offsprings had significantly increased wing length when compared to their parents, which were originally raised in 20 °C ($F = 215.9$; df = 2, 33; $P < 0.0001$) (Table 3). However, compounding the change

Table 2 Effect of photoperiod on female *Drosophila suzukii* wing length over two generations

Environmental parameters				Wing length (mm)[a]
Parental (F0)		Offspring (F1)		
Day-length (L:D)	Temp (°C)	Day-length (L:D)	Temp (°C)	
16:8	20			2.2 b
		16:8	20	2.2 b
		12:12	20	2.5 a

[a] Mean values of wing length followed by the same letter are not significantly different, ANOVA, Tukey. n = 5 for each mean

Table 3 Effect of photoperiod and temperature on female *Drosophila suzukii* wing length over two generations

Environmental parameters				Wing length (mm)[a]
Parental (F0)		Offspring (F1)		
Day-length (L:D)	Temp (°C)	Day-length (L:D)	Temp (°C)	
16:8	20			2.4 b
		16:8	10	2.9 a
		12:12	10	2.9 a

[a] Mean values of wing length followed by the same letter are not significantly different, ANOVA, Tukey. n = 5 for each mean

in temperature (from 20 to 10 °C) with a shortening of day-length (16–12 h) did not further affect wing length of female offsprings (Table 3).

We further investigated the effect of shortening day-length and decrease in temperature on female *D. suzukii* wing length and melanization (Table 4). When parent flies (F0), which were originally reared in 16:8 L:D and 20 °C, were transferred to a photoperiod of 12:12 L:D either at 10 or 20 °C, the average wing length of their offsprings were significantly longer as compared to their parents. Offsprings produced under 12:12 L:D and 10 °C had the largest wings ($F = 134.31$; df = 2, 38; $P < 0.0001$) (Table 4). The melanization rating of the fourth abdominal segment was greater for female offsprings produced at 10 °C and 12:12 L:D than at 20 °C and 12:12 L:D, while the width of the bands of their female parents, which were originally produced at 16:8 L:D and 20 °C, were intermediate compared with band width of their offspring ($F = 48.38$; df = 2, 42; $P < 0.0001$ with regard to melanization ratings).

Our results point to a complex interaction between photoperiod and temperature in affecting wing length and abdominal melanization. Although either shorter day-length or lower temperature can independently induce increase in wing length (Tables 2, 3), transition from long to short day-length did not appear to provide added positive effect on increased wing length if it is accompanied by a decrease in temperature (Table 3). Interestingly, transition from summer-like (20 °C) to winter-like (10 °C) temperature showed an additive effect if accompanied by a decrease in day-length (Table 4). Unlike wing length, which increases in response to changes that signal winter (short day-length and lower temperature), abdominal melanization appeared to be differentially regulated by these two cues that signal winter: shorter day-length decreased the melanization while lower temperature greatly increased abdominal melanization (Table 4).

Survival of summer and winter morphs at different temperatures

To examine whether a transition to winter morphs provided a survival advantage in winter conditions, specifically low temperature, we subjected summer and winter morphs of *D. suzukii* to various temperature conditions (1, 5, 10, 20 and 28 °C) and measured their survival rates (Fig. 3). Paired t tests performed on estimated LT_{50} values (days) for each sex at each temperature revealed that adult female winter morph *D. suzukii* lived significantly longer than adult female summer morph ($LT_{50} = 115$ vs. $LT_{50} = 28$ d, respectively) at 1 °C ($t = -6.36$; df = 7; $P = 0.0004$) (Figs. 3, 4). No other statistical differences in LT_{50} survival between female morphs were observed at the four other temperature regimes ($t = -0.46$ to 2.35; df = 7; $P = 0.07–0.95$). Male winter morph *D. suzukii* had a higher LT_{50} value than male summer morph *D. suzukii* ($LT_{50} = 93$ vs. $LT_{50} = 11$ d, respectively) at 1 °C ($t = -9.37$; df = 7; $P < 0.0001$) (Figs. 3, 4). Conversely, male summer morph *D. suzukii* survived longer at 28 °C than male winter morph ($LT_{50} = 8$ vs. $LT_{50} = 3$ d, respectively) ($t = 2.72$; df = 7; $P = 0.03$). No other statistical differences in survival time between male morphs were observed at the other three temperature regimes ($t = -1.74$ to 0.02; df = 7; $P = 0.13–0.98$).

Table 4 Effect of shortening day-length and decrease in temperature on female *Drosophila suzukii* wing length and abdominal melanization over two generations

Environmental parameters				Wing length (mm)[a]	Melanization rating of segment 4[a]
Parental (F0)		Offspring (F1)			
Day-length (L:D)	Temp (°C)	Day-length (L:D)	Temp (°C)		
16:8	20			2.1 c	3.1 b
		12:12	10	2.9 a	4.8 a
		12:12	20	2.5 b	2.2 c

[a] Mean values within a column followed by the same letter are not significantly different, ANOVA: Wing length: F = 134.31; df = 2, 38; P < 0.0001. Melanization rating: F = 48.38; df = 2, 42; P < 0.0001. Tukey. n = 5 for each mean

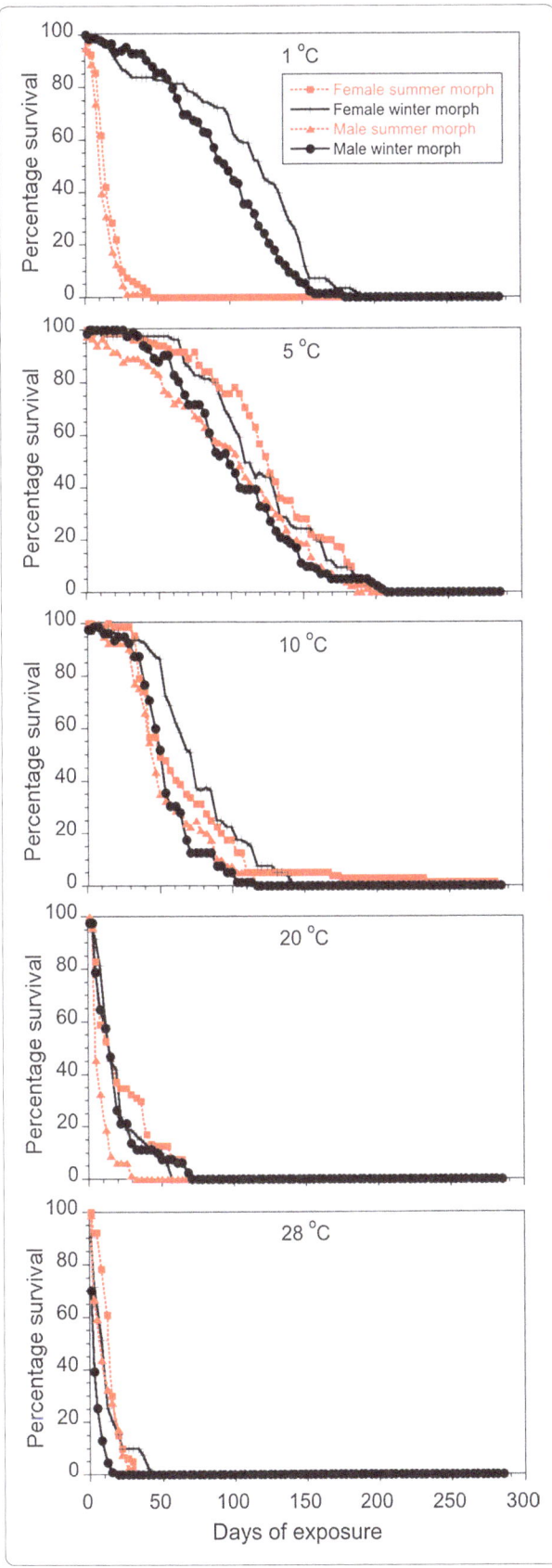

Fig. 3 Mortality curves for summer and winter morphs of *D. suzukii* at five controlled temperatures. Summer and winter morphs of *D. suzukii* (male and female adults) were maintained at 1, 5, 10, 20, and 28 °C, and their survival were assessed

We also tested for sex-specific differences in survival at 1 °C. There was a significant difference in the LT_{50} values for male (mean = 93.1 d; SEM = 9.8) and female (mean = 112.9 d; SEM = 9.8) winter morphs held at 1 °C (t = 2.114; df = 7; $P = 0.036$). In this instance, the LT_{50} for females was approximately 20 days longer than for males.

Gene expression differences in summer and winter morph
To determine global gene expression differences that result in the morphological and physiological differences between summer and winter morphs of *D. suzukii*, we performed differential expression analysis using RNA sequencing between summer and winter morphs. Examination of gene expression in heads and bodies separately revealed a higher number of genes that are differentially expressed (up- or down-regulated) in bodies relative to heads [q value (FDR-adjusted p value) <0.05] (Figs. 5, 6a), even though the head and body transcriptomes contain similar number of genes that could be mapped to the reference genome (Additional file 2: Table S1). A scatter plot of FPKM values clearly illustrates that there are more differentially expressed genes (DEGs) between the two morphs in the body (Fig. 5b; Pearson's correlation coefficient r = 0.6396) than in heads (Fig. 5a; Pearson's correlation coefficient r = 0.9101).

Moreover, the extent (fold change) of differential expression appeared much higher in bodies (Fig. 6b). It is possible that since the bodies contain some of the most metabolically active organs and tissues, e.g. fat body and muscles, many of the highly differentially expressed genes (DEGs) could be involved in the regulation of cellular metabolism, and possibly altered in winter morphs to enable winter survival. To systematically identify enriched categories of genes and molecular pathways that are differentially regulated between the summer and winter morphs, we performed Gene Ontology (GO) enrichment analysis using two independent methods, BiNGO 3.0.3 [54] and DAVID [55], which provided us with similar results. The output for BiNGO is presented in Figs. 7, 8, and the results from DAVID is presented in Additional files 3, 4, 5 and 6: Tables S2, S3, S4, and S5.

Up-regulated genes in bodies of winter morphs
The most significantly enriched terms in the winter bodies were those involved in glycolysis, the tricarboxylic acid (TCA) cycle, the electron transport chain, and ATP

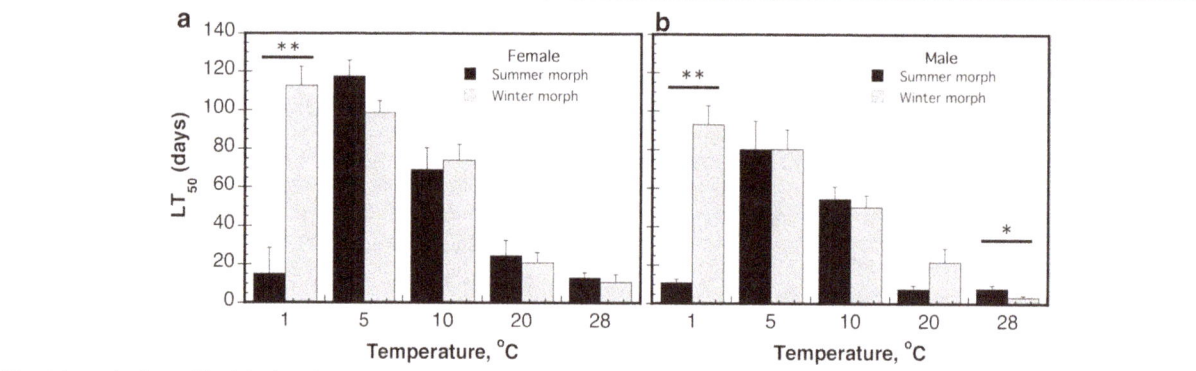

Fig. 4 Length of time (days) for female and male morphs to reach 50 % mortality (LT_{50}) at various constant temperatures. LT_{50} was calculated for the summer and winter morphs of (**a**) female *D. suzukii*, and (**b**) male *D. suzukii* when maintained at 1, 5, 10, 20, and 28 °C. Paired t tests were performed on estimated LT_{50} values (days) for each sex between summer and winter morph at each temperature. Significant differences were indicated by asterisks. (* indicates $P < 0.05$ and ** indicates $P < 0.01$)

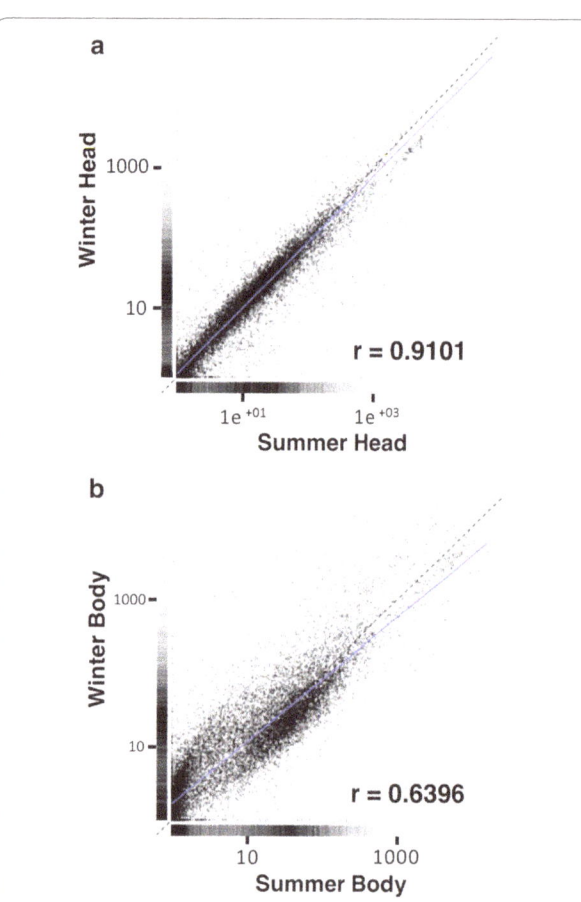

Fig. 5 Correlation of gene expression between summer and winter morphs of *D. suzukii*. The FPKM values for all transcripts were plotted for summer and winter morphs by averaging across the biological replicates. Pairwise FPKM comparisons were generated using the csScatter() function from CummeRbund in (**a**) heads and (**b**) bodies. Pearson's correlation coefficient were calculated using the cor() function in R. *Dotted line* represents r = 1. *Solid line* represents deduced "r" value as calculated using the data

synthase (Fig. 7a and Additional file 3: Table S2). Based on the DAVID output, GO Terms involved with oxidative phosphorylation and the electron transport chain had the highest enrichment score, followed by glucose metabolism, TCA cycle, and terms involved with glycogen metabolism. Other notable enriched categories of genes that are up-regulated are involved in morphogenesis, development, and pigmentation, which is consistent with the enlarged, more melanized winter morph phenotype. Finally, genes involved in circadian rhythm and rhythmic behavior (e.g. *period, shaggy, timeless*) were also up-regulated in the bodies. There has been previous evidence suggesting that these genes are up-regulated in diapausing insects [25]. Although not listed as one of the enriched GO category, some of the most highly up-regulated genes in the winter bodies are genes involved in chitin biosynthesis and metabolism (Additional file 7: Table S6). The genes *CG14301, zye, kkv, Cpr76Bd, verm, Cpr47Ec, Cpr49Ae, obst-B,* and *Gasp* are all involved in chitin binding, structure, or metabolism and have a $\log_2$(fold change) greater than 5.0 in the winter bodies.

Down-regulated genes in bodies of winter morphs

The most significantly depleted terms in winter bodies were associated with the chromosome, chromatin organization, mitotic cell cycle, DNA replication, and DNA repair (Additional file 4: Table S3 and Fig. 7b). The enrichment score for most of these categories are very high, with many genes within these GO categories being down-regulated simultaneously. In addition, terms associated with the chorion, eggshell formation, oogenesis, and female meiosis were all enriched in down-regulated genes in the winter bodies. These results suggest a high likelihood that these female winter morphs are overwintering in reproductive diapause.

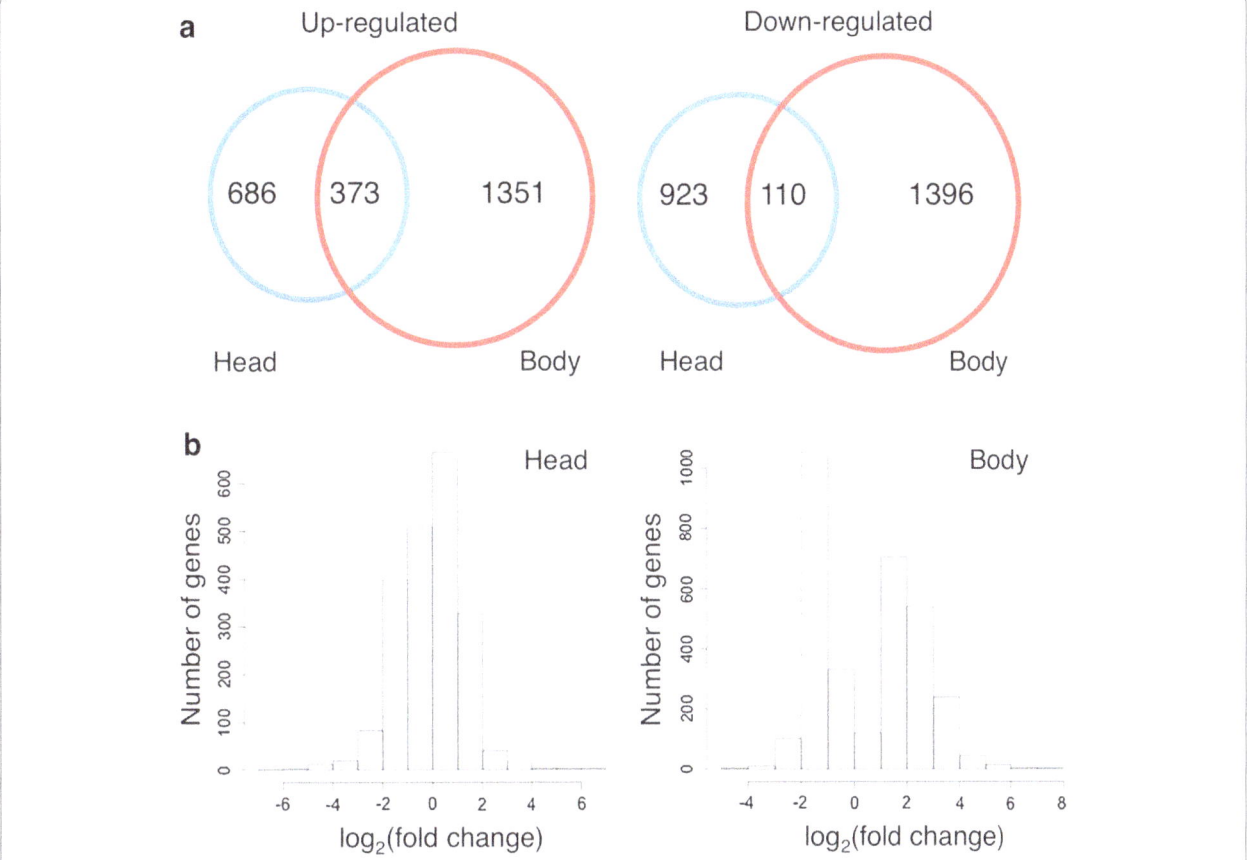

Fig. 6 Summary of differentially expressed genes between summer and winter morphs of *D. suzukii*. **a** Venn diagram showing the number of up- and down-regulated genes in heads and bodies of winter morphs relative to summer morphs. **b** Histograms showing the distribution of fold changes of differentially expressed genes in summer and winter morphs. *Fold changes* represent the ratio of expression levels of winter to summer morphs. Genes are binned into groups based on log$_2$ (fold change). Up- and down-regulated genes have positive and negative log$_2$ (fold change) values respectively. *Left* and *right panels* show the histograms for head and body transcriptome data respectively. Histograms were generated using R. Cutoff q value (FDR-adjusted p value) < 0.05

Up-regulated genes in heads of winter morphs

Based on DAVID output, the most significantly enriched GO terms were associated with immunoglobulin, plasma membrane, transmembrane, neuron development, ion transport, and muscle development (Additional file 5: Table S4). This is consistent with the BiNGO output, in which GO terms involved in ion channel activity, transporter activity, and multicellular organismal development are most enriched (Fig. 8a). In addition to being up-regulated in the body, genes involved in circadian rhythm were also up-regulated in the heads of winter morphs.

Down-regulated genes in heads of winter morphs

Some of the most significantly depleted terms in winter heads were those associated with the ribosome (high enrichment score), dehydrogenase, and organic acid biosynthesis (Additional file 6: Table S5). BiNGO also shows genes involved in carbohydrate metabolism, protein metabolic process, lipid catabolism, and translation being significantly down-regulated (Fig. 8b).

Discussion

D. suzukii can tolerate colder climates by transitioning into a winter morph

We first characterized seasonal morphs of *D. suzukii* in the field, and observed that wing length increased as winter approached and average temperature decreased. The proportion of flies with the winter morph phenotype also increased as the seasons progressed (Fig. 1). We then determined the abiotic factors needed to induce the winter morph phenotype that is observed in field-collected flies in the month leading up to winter. A laboratory-simulated winter-like photoperiod (12:12 L:D) and temperature (10 °C) was sufficient to induce higher levels of melanization and larger wing size. These changes in physiological characters are associated with an increase in cold hardiness as measured by longevity at 1 °C. While

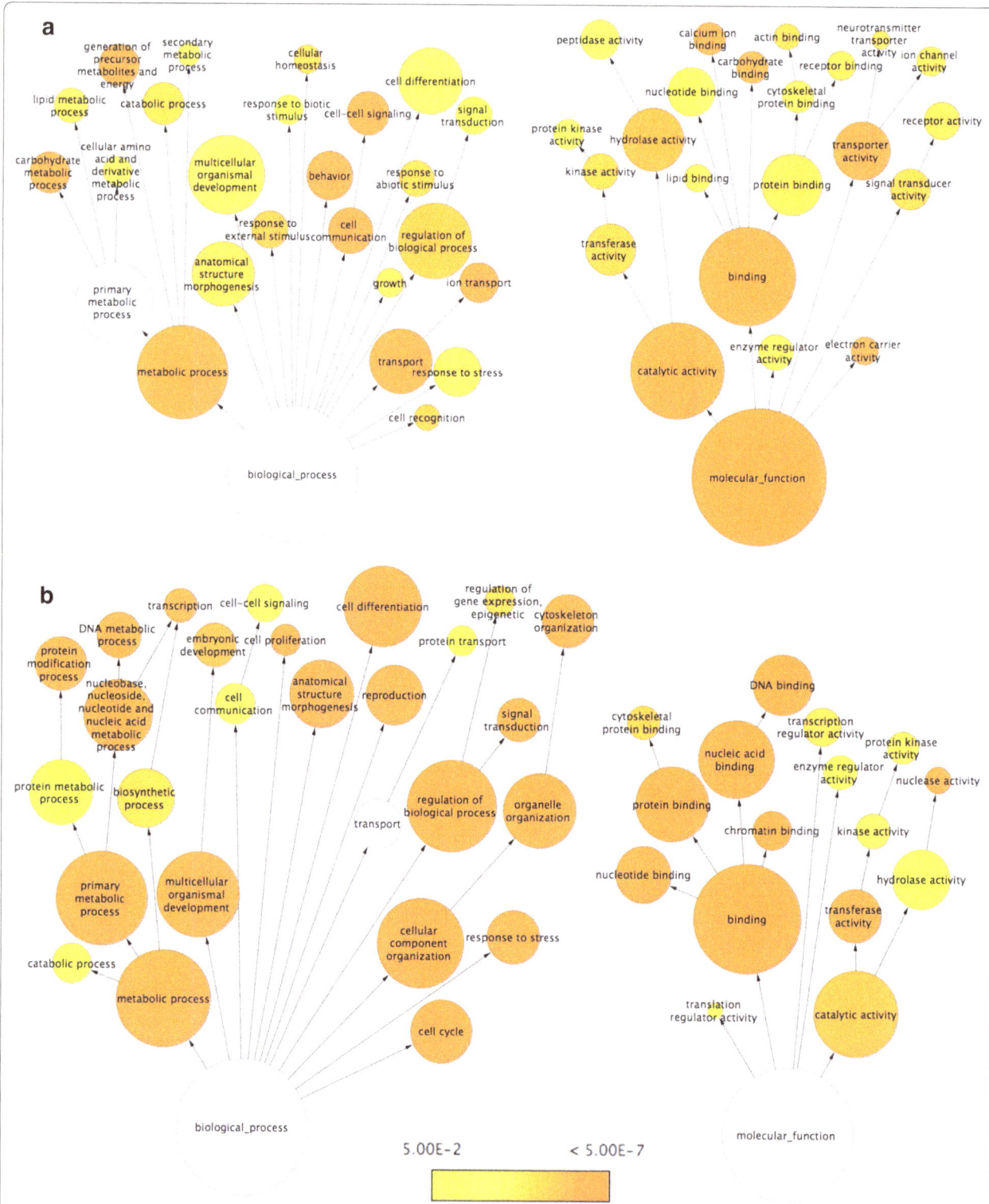

Fig. 7 Cytoscape BiNGO visualization of overrepresented Gene Ontology (GO) categories in differentially expressed genes between summer and winter bodies of *D. suzukii* in the context of the GO hierarchy. Enriched GO terms that are (**a**) up-regulated and (**b**) down-regulated in winter bodies relative to summer bodies are classified by biological process (*left*) and molecular function (*right*). The size of each *circle* represents the number of genes that are included in each GO term and the *color of the circle* indicates the enrichment p value for the labeled GO term. As indicated in the enrichment scale, *orange* represents the highest enrichment and *yellow* represents the minimum enrichment above the cutoff (FDR corrected = 0.05). *White circles* represent nodes that are not enriched; they are shown in the figure to illustrate the GO term hierarchy and are only present if their "leaf nodes" are enriched. The hierarchical layout in Cytoscape was used to arrange the networks with manual adjustment of the nodes to allow for visualization of the text labels

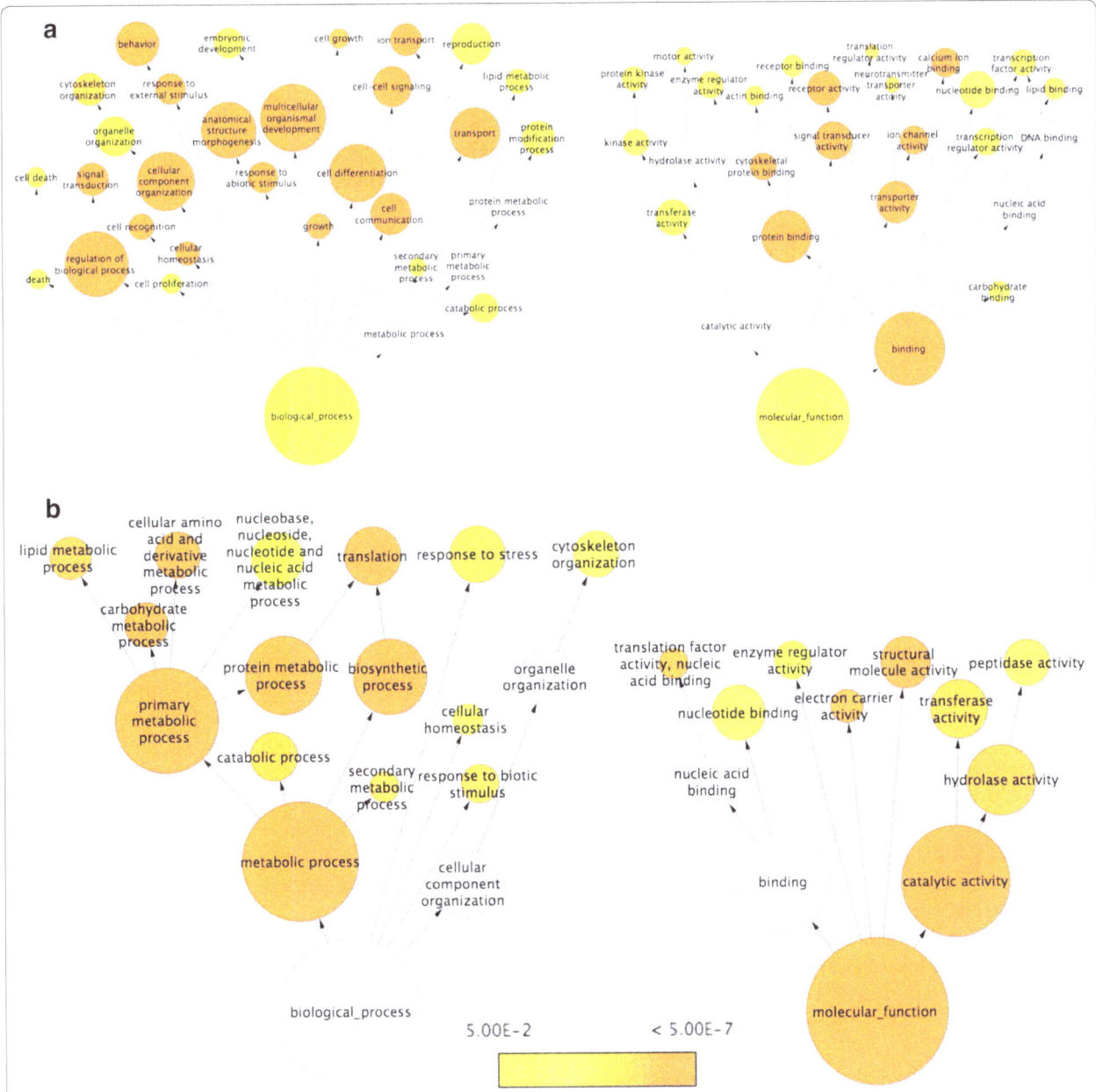

Fig. 8 Cytoscape BiNGO visualization of overrepresented Gene Ontology (GO) categories in differentially expressed genes between summer and winter heads of *D. suzukii* in the context of the GO hierarchy. Enriched GO terms that are (**a**) up-regulated and (**b**) down-regulated in winter heads relative to summer heads are classified by biological process (*left*) and molecular function (*right*). The *size of each circle* represents the number of genes that are included in each GO term and the *color of the circle* indicates the enrichment p value for the labeled GO term. As indicated in the enrichment scale, *orange* represents the highest enrichment and *yellow* represents the minimum enrichment above the cutoff (FDR corrected = 0.05). *White circles* represent nodes that are not enriched; they are shown in the figure to illustrate the GO term hierarchy and are only present if their "leaf nodes" are enriched. The hierarchical layout in Cytoscape was used to arrange the networks with manual adjustment of the nodes to allow for visualization of the text labels

larger and darker forms of *D. suzukii* have been documented to occur in the fall [56], this is the first report characterizing intergeneration transition of *D. suzukii* seasonal morphs in field-collected populations resulting from environmental cues.

An increase in body size, using wing length as a proxy, in *D. suzukii* may be advantageous in colder environments, as it may aid in thermoregulation [44]. A larger body size may also allow for increased storage of sugars and fats, as overwintering insects often have an enlarged fat body [57]. However, a larger wing size may also be involved in increased dispersion capabilities. Interestingly, photoperiod and temperature appear to affect melanization in opposite ways. We found that shortening the photoperiod to a "winter" photoperiod, while keeping the temperature constant at 20 °C actually led to a decrease in melanization. This may be due to UV radiation protection; with a longer photoperiod, the flies are exposed to more light and therefore may need more melanin to protect from UV radiation [58]. Indeed, melanization of flies in sub-Saharan Africa most strongly correlate with UV radiation intensity when compared with other environmental factors, implicating a role for melanin in UV photoprotection [59]. Rearing the flies at a lower temperature of 10 °C significantly increases their melanization rating compared to flies reared at 20 °C. This suggests that melanization may play a role in cold tolerance in addition to its potential role in UV protection. An increase in melanization at low temperatures may increase UV absorption, increasing the ability to warm up. Further experiments are necessary to precisely identify the role of melanization in overwintering capability.

Past studies conducted on *D. suzukii* cold tolerance suggest that they have relatively low levels of cold tolerance [17, 18]. However, low levels of cold tolerance may represent a cost to improved plasticity [60]. In the experiments by Dalton et al. [17] and Jakobs et al. [18], flies were reared under summer conditions (25 °C) and then subjected to rapid or long-term cold-hardening. These conditions did not allow for developmental or intergeneration cold-hardening to occur, which is the focus of this current study. One recent study found that *D. suzukii* raised in simulated winter conditions has increased survival when briefly exposed to subzero temperatures, but this study did not investigate survival rates at prolonged periods of low temperature [53]. In our study, *D. suzukii* flies displaying the winter morph phenotype have low-temperature survival rates higher than those reported previously [17, 18]. In Dalton et al. [17], the LT$_{50}$ period at 1 °C was 3 days within intra-generational acclimated adult flies, as opposed to the current study where the same level of mortality was reached at 115 days. These previous studies suggest that *D. suzukii* cannot survive winter conditions, but results presented in our study demonstrate that intergenerational and/or developmental plasticity in cold-tolerance may explain the presence of *D. suzukii* in cold northern locations. Our results support previous studies suggesting that *D. suzukii* overwinter in the adult stage [17, 29, 56, 61], and show that *D. suzukii* is more likely to successfully survive extended periods of cold in an adapted physiological state. Additionally, we show that male and female *D. suzukii* flies

display differences in survival rates as measured by LT$_{50}$. Stephens et al. [53] did not find any differences between sexes in supercooling points or lower lethal temperatures. Given our data, males may be more susceptible to cold, while females may preferentially survive the winter to reproduce in the spring. This is in agreement with the observation from Ometto et al. [16], suggesting a male bottleneck in *D. suzukii* population.

One aspect of current population modeling for *D. suzukii* that is significantly lacking is overwintering parameters. Our findings may be incorporated as seasonal parameters (e.g. [61]) in order to more accurately predict population levels and refine current management decisions. It is clear from phenological and physiological studies on *D. suzukii* [17, 61, 62] that winter is the bottleneck period for *D. suzukii* survival, as is the case for most insects. Knowledge of *D. suzukii* overwintering strategies and mechanisms is therefore of major importance when conducting risk assessment for the crop season immediately following the winter period.

It is known that *D. suzukii* is established in regions where temperatures frequently fall below freezing [13]. Although we did not test winter morph survival at subzero temperatures, Stephens et al. [53] predict that 50 % of *D. suzukii* adult summer and winter morphs die when exposed to approximately −10.01 and −15.3 °C, respectively. These basic physiological findings are not the only factors contributing to winter survival of *D. suzukii*. Behavior, suitable winter refuge sites, and suitable food sources will likely contribute to increased winter survival [63]. In addition, changes in humidity, which was not tested in our experiments but can be investigated in future experiments, may also be an environmental factor that trigger the transitions between summer to winter morphs and/or affect survival of seasonal morphs. [64]. Nevertheless, our findings greatly contribute to the understanding of *D. suzukii* overwintering mechanisms.

Gene expression differences between summer and winter morphs of *D. suzukii* suggest altered metabolism and reproductive diapause in winter morphs

To begin to identify the molecular pathways that underlie the morphological and physiological differences observed in summer and winter morphs of *D. suzukii*, we performed global gene expression analysis using RNA sequencing. Among the many categories of DEGs, the biological processes that appear to be significantly altered in winter morphs of *D. suzukii* relative to the summer morphs include cellular metabolism, protein synthesis and translation, cell cycle and DNA replication, and chitin and cuticular protein synthesis.

Cellular respiration and metabolism

We found that genes involved in cellular respiration (i.e. glycolysis, TCA cycle, and electron transport chain) and glycogen metabolism were the most significantly enriched groups in genes that were up-regulated in winter morphs. This is somewhat surprising, as suppressed metabolism is a hallmark of diapausing insects [22]. However, up-regulation of glycolysis has previously been associated with an increase in polyol synthesis to yield increased level of cryoprotectants, and higher rates of anaerobic respiration in response to hypoxic conditions [37, 65]. These studies also found a decrease in TCA cycle enzymes, supporting the idea that anaerobic respiration dominates in some diapausing insects. However, many of these studies investigated insects that diapause in the pupal stage, which are more likely to overwinter in hypoxic conditions such as in soil. Since *D. suzukii* likely overwinter as adults, they may not be exposed to hypoxic conditions and therefore may be able to maintain a high rate of aerobic respiration as long as food resources are available. Indeed, in *D. melanogaster*, adult diapausing females had increased levels of glycolytic transcripts but also some increase in the TCA cycle transcripts [66].

Up-regulation of genes involved in cellular respiration may reflect the need for increased glycogen and fat stores. Genes involved with glycogen metabolism were highly enriched in bodies of winter morphs. Interestingly, both glycogen phosphorylase, which catalyzes glycogen breakdown, and glycogen synthase, which catalyzes glycogen synthesis, are both significantly up-regulated. The fact that both catabolic and anabolic enzymes are up-regulated may suggest a high rate of glycogen turnover. Futile cycling is a process by which two opposing pathways are active simultaneously where the only net effects are to hydrolyze ATP and to produce heat. Glycogen futile cycling has been observed in bacteria [67] and futile cycling has been found to produce significant amounts of heat in bumblebees [68]. High metabolic rates may be advantageous for ectotherms at low temperatures, as it may produce heat to raise body temperature. Alternatively, metabolic genes may also need to be up-regulated to compensate for lowered enzyme efficiency at lower temperatures. Since we are sequencing mRNA transcripts from entire bodies, the large transcript abundance of genes involved in cellular respiration may also arise from differences in tissue size. A highly enlarged fat body, as often seen in diapausing insects [57], may explain the enrichment of metabolic terms. For these reasons, physiological metabolic rate (as measured by heat rate, oxygen consumption, and carbon dioxide production) should be measured to further investigate overall metabolic differences between the summer and winter morphs.

Although our study focused on photoperiod and temperature as environmental cues to induce *D. suzukii* winter morph phenotype, other cues, such as food availability are likely important in regulating metabolism. Because our study did not address food scarcity issues that likely occur in the wild, further research is needed to determine metabolic gene expression differences in conditions that are more ecologically relevant.

Cell cycle, DNA replication, and protein synthesis

Genes involved in DNA replication, female meiosis, and egg production were highly down-regulated in the bodies of winter morphs, suggesting that they may be in reproductive diapause. Our gene expression data is consistent with studies in diapausing vs. nondiapausing *D. melanogaster* [66], showing down-regulation of similar gene classes in diapausing females. Suppression of DNA replication, growth, and decreased metabolic activity are hallmarks of diapause [21]. Increased metabolic rates, discussed earlier, may allow for accumulation of glycogen and/or lipid reserves that is associated with diapause.

Additionally, protein translation and ribosome biogenesis appeared to be substantially down-regulated in winter morphs of *D. suzukii*. This may be an adaptive mechanism that allows an insect to allocate energy and metabolites to more important processes such as increasing cold tolerance [69].

Chitin and cuticular protein synthesis

Among the over 1500 genes that were up-regulated in bodies of winter morphs, genes involved in chitin metabolism were the most highly up-regulated transcripts in terms of fold change. Increased chitin synthesis has been implicated in desiccation resistance [70]. These cuticular proteins may also be involved in repairing damage caused by desiccation [71]. An increase in cuticular lipids has been associated with decreased water loss and an increase in freeze-tolerance [72]. Finally, the increased expression of these genes may also be necessary for the large size of the winter morphs of *D. suzukii*. The fact that this class of genes are among the most up-regulated genes further warrant future investigation of humidity as a factor to trigger transition between summer to winter morphs and a variable that regulates survival of the seasonal morphs.

Conclusions

In this study, we examined seasonal variations in morphology and physiology in *D. suzukii* and investigated the role of phenotypic plasticity in facilitating its rapidly expanding range. We investigated the role of temperature and photoperiod in the induction of seasonal phenotypic plasticity in this invasive species. Our study is a first step to better understand mechanisms employed by *D. suzukii* to survive harsh winter conditions and successfully expand its global range. Future work is necessary to determine the complex interaction of photoperiod, temperature, and other environmental cues, and the mechanisms by which they affect physiological responses and adaptation. In addition, experiments will be necessary to determine the developmental stage at which the cues need to be received and the mechanisms that enable transition between different phenotypic morphs.

Our gene expression analysis identified candidates that are important for the observed differences between the seasonal morphs and set the stage for functional characterization. In particular, many unannotated genes are highly expressed in the winter morph and have no known function. These genes may be involved in cold tolerance and/or diapause and could be of broad interest to the investigation of organismal physiology and adaptation. Moreover, transcript abundance is only one piece of the puzzle. Post-transcriptional and post-translational regulation of molecular pathways can certainly play additional roles in modulating the overall biochemical makeup of the organisms in response to environmental cues. Metabolomic profiling of *D. suzukii* summer and winter morphs can therefore provide further insight into biochemical mechanisms of increased cold tolerance and desiccation resistance.

Drosophila suzukii is emerging as a powerful model for ecological genetics due to its close phylogenetic relationship with the model organism *D. melanogaster*, its recently sequenced genome, and the expanding worldwide population monitoring and sampling network because of its economic importance as an agricultural pest. *Drosophila suzukii* is ideal for bridging the gap between laboratory model organisms for which molecular tools are readily available and ecological models used to study adaptations in natural environments.

Methods

Observation of phenology traits of *D. suzukii* in the field

Adult *D. suzukii* were captured in traps placed in the field (Hood River, OR, USA GPS coordinates 45°41′12.39″ N 121°32′53.27″ W) and were then measured for wing length and abdominal melanization (see below). Traps were constructed from clear 946 ml plastic food containers (Solo Cup Co., Lake Forest, IL). Each trap had 10–0.5 cm holes in the sides near the top. Traps were baited with 150 ml of clear apple cider vinegar and then capped. Approximately 1 ml of unscented dish soap (Ultra Pure Clear, Colgate-Palmolive Co., New York, NY) was added per liter of vinegar to break the surface tension. The vinegar attractant was replaced weekly. When present, adult *D. suzukii* were removed from the traps,

sexed and placed in vials containing 70 % ethanol separated by collection date. Mean daily air temperatures were plotted from the Hood River, Oregon AgriMet Weather station (HOXO) [73]. Day-length values were plotted from the Astronomical Applications Department for the U.S. Naval Observatory for Hood River, Oregon [74].

Measurement of abdominal melanization and wing length

Abdominal melanization was quantitatively rated using a visual scoring system under a stereomicroscope as described in [43, 72]. The thickness of each melanized band along the dorso-ventral line of abdominal segments 1–5 was estimated and scored on a scale from 0 (no melanization) to five (complete melanization) (Additional file 1: Figure S1). Melanization scores for the five melanized abdominal bands were compared between morphs for each sex.

The length of wings of field-collected *D. suzukii* as well as adults reared from day-length and temperature-controlled studies were measured as a proxy for body size as described in [75–77]. Wing size measurements were conducted on two segments along vein IV [75], [77, 78] of the left wing (Additional file 1: Figure S1). The first segment (L1) was measured from the base of the fourth longitudinal vein to the posterior cross vein. The second segment (L2) was measured from the posterior cross vein to the distal extreme of the fourth longitudinal vein. Wings were first dissected and then slide mounted in order to take digital photographs of the wings using a Leica camera (Leica DFC480, Buffalo Grove, Il) mounted on a binary microscope (Leica MZ12A, Buffalo Grove, Il). Images were then imported into imaging software (ImagePro Plus, MediaCybernetics, Rockville, MD) where length measurements were obtained.

Drosophila suzukii strains and culture conditions

A *D. suzukii* stock colony was started from 200 individuals sourced from field collections during September 2012. All individuals were reared from wild blackberries, Rosaceae: *Rubus discolor*, collected in Hood River, OR USA (GPS coordinates 45°41′12.39″ N 121°32′53.27″ W). Permission for fly and blackberry collections as well as access to collection site was not required. Adult *D. suzukii* that emerged from the berries were placed in Bugdorm (299 × 299 × 299 cm, Model 1452, Bioquip, Rancho Dominguez, CA) rearing cages. These cages were modified by gluing clear plastic film (Flex-O-Glass, Warp Bros., Chicago, IL) over the screened walls to maintain humidity and prevent cross contamination. *D. suzukii* laboratory colonies were subsequently maintained at 23 ± 1 °C and a photoperiod of 16:8 Light:Dark (L:D) in hours.

Within the laboratory, *D. suzukii* was reared using a commercial *Drosophila* diet (Formula 4–24 Instant drosophila medium, Carolina Biological, Burlington, NC). Disposable polystyrene petri dishes (100 × 15 mm, VWR International, Radnor, PA) were filled with 100 ml of *Drosophila* diet and 100 ml of distilled water. A yeast paste (~1.5 ml) was made by mixing 15 g of yeast (Red Star, Lesaffre Yeast Corp., Milwaukee, WI) with 20 ml of water until a creamy consistency was achieved. The yeast paste was then applied as a thin strip to the top of the diet. Six petri dishes were added to each Bugdorm cage and left in the cage for 1 week to allow *D. suzukii* to oviposit. Each cage was also provided with three water containers with wicks and one 45 % sucrose solution (w/v) container with a wick. Petri dishes were replaced weekly and water and sugar water containers were refilled. Petri dishes removed from the cages were placed into clean Bugdorm cages to rear out future generations of *D. suzukii*. Once the next generation of *D. suzukii* adults began emerging, one petri dish with diet and yeast, water and sugar water were added to the cage. One week after the first *D. suzukii* adults emerged, petri dishes and additional pupae were removed from the cage in order to standardize the fly age to 1–7 day-old individuals. Six new petri dishes with diet and yeast were then added to the cage to begin rearing the next generation of *D. suzukii*. This process was repeated to maintain colonies for experiments.

Drosophila suzukii from the stock colony were reared in biological incubators (Model: l-36-LLVL, Percival, Perry, IA) under two regimes, 20 °C and 16:8 L:D photoperiod ("summer conditions") and 10 °C and 12:12 L:D photoperiod ("winter conditions") with relative humidity set to 70 %. The "summer" and "winter" photoperiod and temperature were chosen to reflect conditions in Hood River, OR in June and around October, when summer and winter morphs were observed respectively (Fig. 1). Temperature lower than 10 °C were not used due to the difficulty in rearing enough flies for experiments, and given the fact that winter morph phenotypes can be obtained with the simulated winter conditions we used. Adult *D. suzukii* that were maintained in simulated summer conditions (20 °C and 16:8 L:D) were transferred to specific test conditions upon emergence from pupae to allow for mating and seeding of the next generation of flies, and the morphology of the progenies were assessed. Rearing *D. suzukii* populations under the respective environmental conditions was conducted using clear polystyrene rearing vials (wide *Drosophila* vials, Genesee Scientific, San Diego, CA) capped with cellulose acetate plugs (Flugs, Genesee Scientific, San Diego, CA). Approximately 15 ml of water and 15 ml of *Drosophila* diet were added to each vial. After the water was completely absorbed by the

diet, ~0.2 ml of yeast paste (see above) is added on top of the diet. Twenty-five adult male and female *D. suzukii* (F0 individuals) from the stock cultures each were added to vials and these adults were then left inside the vials under each of the four environmental conditions for 7 days in order to allow oviposition. The offspring from these individuals (F1 individuals) were transferred regularly to new vials containing a similar water/yeast mixture as described above. Various measurements of physiological traits were taken from parent flies (F0) and subsequent adult offspring.

Examining the survival of summer and winter morphs at different temperatures

As *D. suzukii* adults emerged each week from rearing containers in the laboratory under either 20 °C and 16:8 L:D photoperiod or 10 °C and 12:12 L:D photoperiods, they were placed within vials containing optimized artificial rearing media based on their rearing environments. Vials contained five adult *D. suzukii* separated by sex and morph phenotype. These vials were then placed in one of five temperature and photoperiod regimes; 1 or 5 °C, both with 12:12 L:D photoperiod or 10, 20 and 28 °C with 16:8 L:D photoperiod. The survivorship of both male and female *D. suzukii* was determined within these vials by counting surviving individuals at 3–4 day intervals for a total observation period of 0–141 days. Eight replications, consisting of 1–3 vials of five 1–7 days-old flies of either sex for each of the five temperature regimes were used for this study. Data was averaged within a replicate if more than one vial of five flies was used. The length of time (d) to reach 50 % mortality (LT_{50}) was estimated for each sex × morph × temperature replicate and then averaged to generate LT_{50} values for each combination of temperature, sex and morph phenotype.

Statistical analyses

Differences in melanization ratings between seasonal morph phenotypes for various abdominal segments were analyzed with t tests for each sex. Wing length measurements for the two morphs were compared with two-way ANOVAs (GLM [79]). Paired t tests performed on estimated LT_{50} values (days) for each sex and morph at each temperature (ProcMeans [79]). A 1-tail paired t test was conducted to compare the LT_{50} of male and female winter morphs held at 1 °C. Wing length measurements were transformed using the square root before ANOVA to stabilize variances [80].

To identify the relationship between body size and two seasonally varying environmental factors, we performed linear regression analysis of wing length (i.e. a good proxy for body size) over temperature and day-length. Wing length data were generated by averaging individual wing length for each sex and collection date. We used the average daily temperature over the 12 days preceding the test date as a proxy for the temperature experienced by the developing fly. Temperature and day-length data were obtained from the Hood River, Oregon AgriMet Weather station (HOXO) [73] and the Astronomical Applications Department for the U.S. Naval Observatory [74].

RNA extraction, transcriptome library preparation, and high-throughput sequencing

Drosophila suzukii used for transcriptome analysis were from the stock colony established from flies collected in Hood River, OR, U.S.A. Summer and winter morphs of *D. suzukii* were reared in simulated summer conditions: a photoperiod of 16:8 L:D and 20 °C and simulated winter condition: a photoperiod of 12:12 L:D and 10 °C respectively and were collected at age 4–5 days. Only females were included to control for sex-specific differences. Each biological replicate contained pools of 15 heads or bodies. Adult females were flash frozen on dry ice 4 h after lights on time and stored in 1.7 mL tubes (Denville, Holliston, MA) at −80 °C. Heads and bodies were separated on dry ice using metal sieves with 425 and 710 μm opening (Newark Wire Cloth Company, Clifton, NJ). To extract total RNA, tissues were first homogenized in 150 μL Tri Reagent (Sigma, St. Louis, MO) on ice using a motorized pestle (Kimble Chase, Vineland, NJ), and 350 μL Tri Reagent was subsequently added to bring the total volume to 500 μL. 100 μL chloroform was added, and tubes were inverted approximately 10 times. Samples were incubated at room temperature for 10 min and then centrifuged for 15 min at 13,000 RPM at 4 °C. Samples were placed on ice and 250 μL of the upper aqueous layer was transferred to a new 1.5 mL DNA LoBind tubes (Eppendorf, Hauppauge, NY). RNA was precipitated with 250 μL isopropanol (Sigma, St. Louis, MO) by incubating at −20 °C overnight. Samples were then centrifuged for 15 min at 13,000 RPM at 4 °C. The supernatant was removed and 800 μL 70 % ethanol was added to wash the RNA pellet. Samples were centrifuged for 5 min at 13,000 RPM at 4 °C and the ethanol was removed. The pellet was left to dry at room temperatures for 20 min. Head and body RNA samples were resuspended in 25 and 50 μL 1X TURBO DNase buffer, respectively. Each sample was treated with 1 μL TURBO DNase (Life Technologies, Carlsbad, CA). Samples were quantified using NanoDrop1000 (Thermo Scientific, Waltham, MA) and their quality assessed using the Experion Bioanalyzer (Bio-Rad, Hercules, CA). All RNA samples had an RNA Quality Indicator (RQI) >7.0. RNA sequencing libraries with insert size of approximately 150 bp were prepared using 1 μg total RNA with the Illumina TruSeq RNA Sample

Preparation kit according to manufacturer instructions (Illumina, San Diego, CA). Libraries were submitted to BGI Americas (Sacramento, CA, USA) where library size and quality was assessed using an Agilent 2100 Bioanalyzer (Agilent Technologies, Santa Clara, CA). Samples were quantified using quantitative PCR, pooled, and sequenced on the Illumina HiSeq 2000 using paired-end 100 bp sequencing. Eleven RNA sequencing libraries were prepared in total: three biological replicates each for summer heads, summer bodies, and winter heads, and two biological replicates for winter bodies.

Transcriptome assembly, identification of differentially expressed genes (DEGs), and Gene Ontology (GO) enrichment analysis

We generated a total of 309 million paired-end reads from 11 libraries. We performed biological triplicates for all treatments, except for bodies of winter morphs, for which biological duplicates were used. Raw reads were mapped to the *D. suzukii* reference genome [81] using Bowtie-based Tophat v2.0.12 [82], resulting in an average of 20.5 million mapped reads per replicate (Additional file 2: Table S1). Pearson correlations of expression levels in FPKM between biological replicates were computed in R (Additional file 8: Table S7). Cufflinks v2.2.1 was used to identify differential expressed (DE) genes (q value (FDR-adjusted p value) < 0.05) [82], and CummeRbund [83] was used to visualize the results (Additional File 7: Table S6; and Additional file 9: Table S8). Up- and down-regulated genes were extracted from the list of DEGs, and analyzed for Gene Ontology (GO) enrichment using the BiNGO 3.0.3 [54] plug-in in Cytoscape. Hypergeometric test with Benjamini and Hochberg False Discovery Rate correction for multiple testing was used to access overrepresentation of generic GO slim terms for each condition. GO annotation of *D. melanogaster* orthologs was obtained from FlyBase FB2015_02 release. BiNGO visualization of overrepresented biological process and molecular function GO categories for body and head are shown in Figs. 7, 8 respectively. Independently, we also used the Functional Annotation Clustering tool in DAVID [55] to perform enrichment analysis and clustering, results of which are presented in Additional files 3, 4, 5, 6: Tables S2, S3, S4, S5.

Availability of supporting data
The RNA sequencing data sets supporting the results of this article are available in the National Center for Biotechnology Information (NCBI) repository. [Bio-Project PRJNA294845 http://www.ncbi.nlm.nih.gov/bio-project/ PRJNA294845/, and Sequence Read Archives SRS1057327 (summer bodies) http://www.ncbi.nlm.nih.gov/ sra/?term = SRS1057327/; SRS1057275 (summer

heads) http://www.ncbi.nlm.nih.gov/ sra/?term = SRS1057275/; SRS1057328 (winter bodies) http://www. ncbi.nlm.nih.gov/ sra/?term = SRS1057328/; SRS1057296 (winter heads) http://www.ncbi.nlm.nih.gov/ sra/?term = SRS1057296/]. NCBI accession for individual replicates are also provided in Additional file 2: Table S1.

Additional files

Additional file 1: Figure S1. Measurements of abdominal melanization and wing length in adult *D. suzukii*. (a) The thickness of the dark abdominal bands was used to differentiate the lighter colored summer morph (left) from darker winter morph (right) of *D. suzukii*. (b) Locations (L1 and L2) where wing measurements were taken from the excised left wings of *D. suzukii*.

Additional file 2: Table S1. Number of mapped reads and NCBI Genbank Accession number for each biological replicate of all four treatments reported by Tophat. S = Summer; W = Winter; H = Head; B = Body. Three biological replicates were performed for each treatment, except for fly bodies of winter *D. suzukii* morphs (WB), which has two replicates.

Additional file 3: Table S2. Functional Annotation Clustering of genes that are significantly up-regulated in the winter fly bodies relative to summer bodies as determined by DAVID.

Additional file 4: Table S3. Functional Annotation Clustering of genes that are significantly down-regulated in the winter fly bodies relative to summer bodies as determined by DAVID.

Additional file 5: Table S4. Functional Annotation Clustering of genes that are significantly up-regulated in the winter fly heads relative to summer heads as determined by DAVID.

Additional file 6: Table S5. Functional Annotation Clustering of genes that are significantly down-regulated in the winter fly heads relative to summer heads as determined by DAVID.

Additional file 7: Table S6. Table of differentially expressed genes in bodies of winter morphs of *D. suzukii* relative to those of summer morphs. Fold change represents the ratio of expression levels of winter to summer morphs.

Additional file 8: Table S7. FPKM correlation matrix for (A) head and (B) body RNA-seq replicates. Three replicates were performed for each condition, except for winter-body (WB), which has two replicates. (S = Summer; W = Winter; H = Head; B = Body).

Additional file 9: Table S8. Table of differentially expressed genes in heads of winter morph of *D. suzukii* relative to those of summer morphs. Fold change represents the ratio of expression levels of winter to summer morphs.

Abbreviations
DEG: differentially expressed genes; FPKM: fragments per kilobase of transcript per million mapped reads; FDR: false discovery rate; GLM: general linear model; GO: gene ontology; L:D: light:dark; LT_{50}: lethal temperature at which 50 % of the individuals perish.

Authors' contributions
Conceived and designed the experiments: PWS, VMW, JDW, and JCC. Performed physiological measurements and experiments and analyze data: PWS, VMW, and PHB. Performed RNA extractions and library preparations: JDW and NS. Performed RNA sequencing data analysis and bioinformatics: JDW and JCC. Wrote the paper: PWS, JDW, VMW, JCC. All authors read and approved the final manuscript.

Author details
[1] Mid-Columbia Agricultural Research and Extension Center, Oregon State University, 3005 Experiment Station Drive, Hood River, OR 97331, USA. [2] Department of Entomology and Nematology, University of California, Davis, CA 95616, USA. [3] Department of Horticulture, Oregon State University, Corvallis, OR 97331, USA. [4] Department of Evolution and Ecology, University of California, Davis, CA 95616, USA.

Acknowledgements
We thank Tuck Contreas for noticing the seasonal morph phenotypes while monitoring *D. suzukii* traps, Kala M. Gonsler, Sarah Hieber, Joana R. Kemper, Matthew T. Riek and Kelley E. Schwaner for measuring fly phenotype parameters, and Kelly A. Hamby, Frank G. Zalom, Ernest K. Lee for fruitful discussions. We thank Christine A. Tabuloc for her assistance in fly dissections. This work was supported by funds received from the USDA Specialty Crops Research Initiative Grant (award number 2010-51181-21167) awarded to PWS and VMW, and the Clarence and Estelle Albaugh Endowment and NSF IOS-1456297 to JCC. JDW is a participant of the BUSP Program at UC Davis, which is supported by NIH-IMSD GM56765 and HHMI Grant 52005892, and a participant of BSHARP program, supported by NIGMS-MARC-U-STAR GM083894.

Competing interests
The authors declare that they have no competing interests.

References
1. Oerke EC. Crop losses to pests. J Agric Sci. 2006;144:31–43.
2. Hatfield J, Takle G, Grotjahn R, Holden P, Izaurralde RC, Mader T, et al. Ch. 6: Agriculture. In: Melillo JM, Richmond TC, Yohe GW, editors. Climate change impacts in the United States: the third national climate assessment. U.S. global change research program. 2014. p. 150–174. doi:10.7930/J02Z13FR.
3. Porter JH, Parry ML, Carter TR. The potential effects of climatic change on agricultural insect pests. Agric For Meteorol. 1991;57:221–40.
4. Diez JM, D'Antonio CM, Dukes JS, Grosholz ED, Olden JD, Sorte CJB, et al. Will extreme climatic events facilitate biological invasions? Front Ecol Environ. 2012;10:249–57.
5. Bellard C, Thuiller W, Leroy B, Genovesi P, Bakkenes M, Courchamp F. Will climate change promote future invasions? Glob Change Biol. 2013;19:3740–8.
6. Chown SL, Slabber S, McGeoch MA, Janion C, Leinaas HP. Phenotypic plasticity mediates climate change responses among invasive and indigenous arthropods. Proc Biol Sci. 2007;274:2531–7.
7. Bock DG, Caseys C, Cousens RD, Hahn MA, Heredia SM, Hubner S, et al. What we still don't know about invasion genetics. Mol Ecol. 2015;24:2277–97.
8. Steck GJ, Dixon W, Dean D. Spotted wing drosophila, *Drosophila suzukii* (Matsumura) (Diptera: Drosophiladae), a new pest to North America. Pest Alerts. 2009; DACS-P-01674. http://www.freshfromflorida.com/pi/pest-alerts/pdf/drosophila-suzukii.pdf.
9. Cini A, Ioriatti C, Anfora G. A review of the invasion of *Drosophila suzukii* in Europe and a draft research agenda for integrated pest management. Bull Insectol. 2012;65:149–60.
10. Goodhue RE, Bolda M, Farnsworth D, Williams JC, Zalom FG. Spotted wing drosophila infestation of California strawberries and raspberries: economic analysis of potential revenue losses and control costs. Pest Manag Sci. 2011;67:1396–402.
11. Walsh DB, Bolda MP, Goodhue RE, Dreves AJ, Lee J, Bruck DJ, et al. *Drosophila suzukii* (Diptera: Drosophilidae): invasive pest of ripening soft fruit expanding its geographic range and damage potential. J Integ Pest Mngmt. 2011;2:1–7.
12. Burrack HJ, Smith JP, Pfeiffer DG, Koeher G, Laforest J. Using volunteer-based networks to track *Drosophila suzukii* (Diptera: Drosophilidae) an invasive pest of fruit crops. J Integ Pest Mngmt. 2012;4:1–5.
13. Asplen MK, Anfora G, Biondi A, Choi D, Chu D, Daane KM, et al. Invasion biology of spotted wing drosophila (*Drosophila suzukii*): a global perspective and future priorities. J Pest Sci. 2015;88:469–94.
14. Bolda MP, Goodhue RE, Zalom FG. Spotted wing drosophila: potential economic impact of a newly established pest. Agric Resour Econ Update. 2010;13:5–8.

15. Isaacs R, Hahn N, Tritten B, Garcia C. Spotted wing drosophila: a new invasive pest of Michigan fruit crops. East Lansing: Michigan State University Extension; 2010. p. E3140.

16. Ometto L, Cestaro A, Ramasamy S, Grassi A, Revadi S, Siozios S, et al. Linking genomics and ecology to unveil the complex evolution of an invasive Drosophila pest. Genome Biol Evol. 2013;5:745–57.

17. Dalton DT, Walton VM, Shearer PW, Walsh DB, Caprile J, Isaacs R. Laboratory survival of Drosophila suzukii under simulated winter conditions of the Pacific Northwest and seasonal field trapping in five primary regions of small and stone fruit production in the United States. Pest Manag Sci. 2011;67:1368–74.

18. Jakobs R, Gariepy TD, Sinclair BJ. Adult plasticity of cold tolerance in a cool-temperate population of Drosophila suzukii. J Insect Physiol. 2015;79:1–9.

19. Teets NM, Denlinger DL. Physiological mechanisms of seasonal and rapid cold-hardening in insects. Physiol Entomol. 2013;38:105–16.

20. Koštál V, Simunkova P, Kobelkova A, Shimada K. Cell cycle arrest as a hallmark of insect diapause: changes in gene transcription during diapause induction in the drosophilid fly, Chymomyza costata. Insect Biochem Mol Biol. 2009;39:875–83.

21. Hahn DA, Denlinger DL. Energetics of insect diapause. Annu Rev Entomol. 2011;56:103–21.

22. Nylin S. Induction of diapause and seasonal morphs in butterflies and other insects: knowns, unknowns and the challenge of integration. Physiol Entomol. 2013;38:96–104.

23. Bean DW, Dalin P, Dudley TL. Evolution of critical day length for diapause induction enables range expansion of Diorhabda carinulata, a biological control agent against tamarisk (Tamarix spp.). Evol Appl. 2012;5:511–23.

24. Urbanski J, Mogi M, O'Donnell D, DeCotiis M, Toma T, Armbruster P. Rapid adaptive evolution of photoperiodic response during invasion and range expansion across a climatic gradient. Am Nat. 2012;179:490–500.

25. Salminen TS, Vesala L, Laiho A, Merisalo M, Hoikkala A, Kankare M. Seasonal gene expression kinetics between diapauses phases in Drosophila virilis group species and overwintering differences between diapausing and non-diapausing females. Sci Rep. 2015;5:11197.

26. Schmidt PS, Matzkin LM, Ippolito M, Eanes WF. Geographic variation in diapause incidence, life history traits and climatic adapation in Drosophila melanogaster. Evolution. 2005;59:1721–32.

27. Schmidt PS, Paaby AB. Reproductive diapause and life-history clines in North American populations of Drosophila melanogaster. Evolution. 2008;62:1204–15.

28. Kimura MT. Cold and heat tolerance of drosophilid flies with reference to their latitudinal distribution. Oecologia. 2004;140:442–9.

29. Mitsui H, Beppu K, Kimura MT. Seasonal life cycles and resource uses of flower- and fruit-feeding drosophilid flies (Diptera: Drosophilidae) in central Japan. Ent Sci. 2010;13:60–7.

30. Koštál V, Korbelová J, Rozsypal J, Zahradníčková H, Cimlová J, Tomčala A, et al. Long-term cold acclimation extends survival time at 0 °C and modifies the metabolomic profiles of the larvae of the fruit fly Drosophila melanogaster. PLoS One. 2011;6:e25025.

31. Vesala L, Salminen TS, Koštál V, Zahradníčková H, Hoikkala A. Myo-inositol as a main metabolite in overwintering flies: seasonal metabolomic profiles and cold stress tolerance in a northern drosophilid fly. J Exp Biol. 2012;215:2891–7.

32. Hariharan R, Hoffman JM, Thomas AS, Soltow QA, Jones DP, Promislow DE. Invariance and plasticity in the Drosophila melanogaster metabolomics network in response to temperature. BMC Syst Biol. 2014;8:139.

33. Pedersen KS, Kristensen TN, Loeschcke V, Petersen BO, Duus JO, Nielsen NC, et al. Metabolic signatures of inbreeding at benign and stressful temperatures in Drosophila melanogaster. Genetics. 2008;180:1233–43.

34. Overgaard J, Malmendal A, Sorensen JG, Bundy JG, Loeschcke V, Nielson NC, Holmstrup M. Metabolomic profiling of rapid cold hardening and cold shock in Drosophila melanogaster. J Insect Physiol. 2007;53:1218–32.

35. Ohtsu TM, Kimura T, Katagiri C. How Drosophila species acquire cold tolerance. Eur J Biochem. 1998;252:608–11.

36. Burton V, Mitchell HK, Young P, Petersen NS. Heat shock protection against cold stress of Drosophila melanogaster. Mol Cell Bio. 1988;8:3550–2.

37. Ragland GJ, Denlinger DL, Hahn DA. Mechanisms of suspended animation are revealed by transcript profiling of diapause in the flesh fly. PNAS. 2010;107:14909–14.

38. Vesala L, Salminen TS, Laiho A, Hoikkala A, Kankare M. Cold tolerance and cold-induced modulation of gene expression in two Drosophila virilis group species with different distributions. Insect Mol Biol. 2012;21:107–18.

39. Harris RM, McQuillan P, Hughes L. A test of the thermal melanism hypothesis in the wingless grasshopper Phaulacridium vittatum. J Insect Sci. 2013;13:51.

40. Trullas SC, van Wyk JH, Spotila JR. Thermal melanism in ectotherms. J Therm Bio. 2007;32:235–45.

41. Kutch IC, Sevgill H, Wittman T, Fedorka KM. Thermoregulatory strategy may shape immune investment in Drosophila melanogaster. J Exp Biol. 2014;217:3664–9.

42. Fedorka KM, Lee V, Winterhalter WE. Thermal environment shapes cuticle melanism and melanin-based immunity in the ground cricket Allonemobius socius. Evol Ecol. 2013;27:521–31.

43. Parkash R, Singh S, Ramniwas S. Seasonal changes in humidity level in the tropics impact body color polymorphism and desiccation resistance in Drosophila jambulina—Evidence for melanism-desiccation hypothesis. J Insect Physiol. 2009;55:358–68.

44. Nielsen ME, Papaj DR. Effects of developmental change in body size on ectotherm body temperature and behavioral thermoregulation: caterpillars in a heat-stressed environment. Oecologia. 2015;77:171–9.

45. Czarnoleski M, Cooper BS, Kierat J, Angilletta MJ. Flies developed small bodies and small cells in warm and in thermally fluctuating environments. J Exp Biol. 2013;216:2896–901.

46. Ghosh SM, Testa ND, Shingleton AW. Temperature-size rule is mediated by thermal plasticity of critical size in Drosophila melanogaster. Proc R Soc B. 2013;280:1–8.

47. Angilletta MJ Jr, Dunham AE. The temperature-size rule in ectotherms: simple evolutionary explanations may not be general. Am Nat. 2003;162:332–42.

48. Partridge L, Barrie B, Fowler K, French V. Evolution and development of body size and cell size in Drosophila melanogaster in response to temperature. Evolution. 1994;48:1269–76.

49. Karan D, Morin JP, Moreteau B, David JR. Body size and developmental temperature in Drosophila melanogaster: analysis of body weight reaction norm. J Therm Biol. 1998;23:301–9.

50. Atkinson D, Sibly RM. Why are organisms usually bigger in colder environments? Making sense of a life history puzzle. Trends Ecol Evol. 1997;12:235–9.

51. Shelomi M. Where are we now? Bergmann's rule sensu lato in insects. Am Nat. 2012;180:511–9.

52. Moczek AP. Phenotypic plasticity and diversity in insects. Philos Trans R Soc Lond B Biol Sci. 2010;365(1540):593–603.

53. Stephens AR, Asplen MK, Hutchison WD, Venette RC. Cold hardiness of winter-acclimated Drosophila suzukii (Diptera: Drosophilidae) adults. Environ Entomol. 2015;44(6):1619–26.

54. Maere S, Heymans K, Kuiper M. BiNGO: a Cytoscape plugin to assess overrepresentation of gene ontology categories in biological networks. Bioinformatics. 2005;21:3448–9.

55. Huang DW, Sherman BT, Lempicki RA. Systematic and integrative analysis of large gene lists using DAVID bioinformatics resources. Nat Protoc. 2009;4:44–57.

56. Kanzawa T. Researctrh into the fruit-fly Drosophila suzukii Matsumura (preliminary report). Kofu: Agricultural Experiment Station; 1935. p. 42.

57. Hodek I. Adult diapause in Coleoptera. Psyche. 2012;. doi:10.1155/2012/249081.

58. Wang Z, Liu R, Wang A, Du L, Deng X. Phototoxic effect of UVR on wild type, ebony and yellow mutants of Drosophila melanogaster: life span, fertility, courtship and biochemical aspects. Sci China Ser C Life Sci Chin Acad Sci. 2008;51:885–93.

59. Bastide H, Yassin A, Johanning EJ, Pool JE. Pigmentation in Drosophila melanogaster reaches its maximum in Ethiopia and correlates most strongly with ultra-violet radiation in sub-Saharan Africa. BMC Evol Biol. 2014;14:179.

60. Stillman JH. Acclimation capacity underlies susceptibility to climate change. Science. 2003;301(5629):64.

61. Wiman NG, Walton VM, Dalton DT, Anfora G, Burrack HJ, Chiu JC, et al. Integrating temperature-dependent life table data into a matrix projection model for Drosophila suzukii population estimation. PLoS One. 2014;9:e106909.

62. Tochen S, Dalton DT, Wiman NG, Hamm C, Shearer PW, Walton VM. Temperature-related development and population parameters for *Drosophila suzukii* (Diptera: Drosophilidae) on cherry and blueberry. Environ Entomol. 2014;43:501–10.

63. Zerulla FN, Schmidt S, Streitberger M, Zebitz CPW, Zelger R. On the overwintering ability of *Drosophila suzukii* in South Tyrol. J Berry Res. 2015;5(1):41–8.

64. Nyamukondiwa C, Terblanche JS. Within-generation variation of critical thermal limits in adult Mediterranean and Natal fruit flies *Ceratitis capitata* and *Ceratitis rosa*: thermal history affects short-term responses to temperature. Physiol Entomol. 2010;35(3):255–64.

65. Bao B, Xu W. Identification of gene expression changes associated with the initiation of diapause in the brain of the cotton bollworm, *Helicoverpa armigera*. BMC Genom. 2011;12:224.

66. Baker DA, Russell S. Gene expression during *Drosophila melanogaster* egg development before and after reproductive diapause. BMC Genom. 2009;10:242.

67. Gaudet G, Forano E, Dauphin G, Delort A. Futile cycling in *Fibrobacter succinogenes* as shown by in situ ^{1}H-NMR and ^{13}C-NMR investigation. Eur J Biochem. 1992;207:155–62.

68. Staples JF, Koen EL, Laverty TM. 'Futile Cycle' enzymes in the flight muscles of North American bumblebees. J Exp Bio. 2004;207:749–54.

69. Boggs CL. Understanding insect life histories and senescence through a resource allocation lens. Func Ecol. 2009;23(1):27–37.

70. Rezende GL, Martins AJ, Gentile C, Farnesi LC, Pelajo-Machado M, Peixoto AA, et al. Embryonic desiccation resistance in *Aedes aegypti*: presumptive role of the chitinized serosal cuticle. BMC Dev Biol. 2008;8:182.

71. Clark MS, Thorne MAS, Purac J, Burns G, Hillyard G, Popovic ZD, et al. Surviving the cold: molecular analyses of insect cryoprotective dehydration in the Arctic springtail *Megaphorura arctica* (Tullberg). BMC Genom. 2009;10:328.

72. Parkash R, Kalra B, Sharma V. Changes in cuticular lipids, water loss and desiccation resistance in a tropical Drosophilid: analysis of variation between and within populations. Fly. 2008;2:189–97.

73. AgriMet cooperative agricultural weather network. http://www.usbr.gov/pn/agrimet/webarcread.html. Accessed 8 Sept 2015.

74. Astronomical application department of the USA. Naval observatory. http://www.aa.usno.navy.mil/data/docs/Dur_OneYear.php. Accessed 8 Feb 2016.

75. Robertson FW, Reeve ECR. Studies in quantitative inheritance. I. The effects of selection for wing and thorax length in *Drosophila melanogaster*. J Genet. 1952;50:416–48.

76. Sokoloff A. Morphological variation in natural and experimental populations of *Drosophila pseudoobscura* and *Drosophila persimilis*. Evolution. 1966;20:49–71.

77. Gilchrist GW, Huey RB, Serra L. Rapid evolution of wing size clines in *Drosophila subobscura*. Genetica. 2001;112–113:273–86.

78. Pegueroles G, Papaceit M, Quintana A, Guillén A, Prevosti A, Serra L. An experimental study of evolution in progress: clines for quantitative traits in colonizing and Palearctic populations of Drosophila. Evol Ecol. 1995;9:453–65.

79. SAS Institute. SAS user's guide. Version 9.4. Cary: SAS Institute; 2014.

80. Zar JH. Biostatistical Analysis. Englewood Cliffs: Prentice Hall; 1984.

81. Chiu JC, Jiang X, Zhao L, Hamm CA, Cridland JM, Saelao P, et al. Genome of *Drosophila suzukii*, the spotted wing Drosophila. G3 (Bethesda). 2013;3:2257–71.

82. Trapnell C, Roberts A, Goff L, Pertea G, Kim D, Kelley DR, et al. Differential gene and transcript expression analysis of RNA-seq experiments with TopHat and Cufflinks. Nat Protoc. 2012;7:562–78.

83. Goff L, Trapnell C, Kelley D. cummeRbund: analysis, exploration, manipulation, and visualization of Cufflinks high-throughput sequencing data. R package version 2.10.0. 2013.

7

Predator cues reduce intraspecific trait variability in a marine dinoflagellate

Sylke Wohlrab[1]* ⓘ, Erik Selander[2] and U. John[1,3]

Abstract

Background: Phenotypic plasticity is commonplace and enables an organism to respond to variations in the environment. Plastic responses often modify a suite of traits and can be triggered by both abiotic and biotic changes. Here we analysed the plastic response towards a grazer of two genotypes of the marine dinoflagellate *Alexandrium fundyense,* evaluated the similarity of this response and discuss potential strain-specific trade-offs. We compared the expression of the known inducible defensive traits paralytic shellfish toxin content, and chain length. The effectiveness of the induced defense was assessed by monitoring grazing rates in both strains.

Results: Our results show that the grazer cues diminish phenotypic variability in a population by driving the phenotype towards a common defended morphotype. We further showed that the expression of the *sxt*A gene that initiates the paralytic shellfish toxin biosynthesis pathway does not correlate with an observed increase in the paralytic shellfish toxin analogue saxitoxin, and that toxin induction differs in its physiological characteristics in both strains.

Conclusion: Induced defense response in *Alexandrium* thus can directly affect further species interactions by reducing phenotypic variation and can result in genotype-dependent ecological trade-offs.

Keywords: *Alexandrium*, Saxitoxin gene expression, Grazer induced defense, Intraspecific trait variation, Predator–prey interaction

Background

Organisms can change their traits and thus their phenotype in response to variations in the environment. However, the ability of various genotypes to produce a common, environmentally cued phenotype may differ in terms of trade-offs associated with its expression caused by genetic variation between individuals [1]. In addition, the induction of one phenotype may indirectly influence the expression of linked traits and expose those traits to selection [2].

Phytoplankton cell size is a master trait that impacts growth, metabolism, and access to resources and therefore shapes the ecological niches of phytoplankton [3]. Cell size and related morphological traits like shape, and coloniality are extremely plastic traits that depend on many environmental variables such as light levels and nutrient concentrations [3]. Response to grazing pressure and related grazer resistance is significantly correlated with morphological traits as most grazers are confined to limited ranges of prey sizes [4]. The grazer's ability to detect, capture, and handle the prey sets the lower limit of size whereas morphological e.g. constrains on handling larger prey set the upper limit [4]. For example, colony forming and size plasticity to escape grazing pressure is common in both marine (e.g. *Phaeocystis globosa*) and freshwater (e.g. *Scenedesmus*) phytoplankton [5]. The marine diatom *Skeletonema marinoi* adjust its chain length in the presence of copepod grazers towards smaller chain sizes but chain length was maintained in cultures exposed to microzooplankton grazers [6]. The bacterium *Flectobacillus* sp. induces filament formation when grazed by the microflagellates as single suspended cells were highly vulnerable to grazing, whereas filamentous cells were resistant to grazing [7].

Here we analysed the plastic response of two genotypes of the marine dinoflagellate *Alexandrium fundyense* in

*Correspondence: Sylke.Wohlrab@awi.de
[1] Department of Ecological Chemistry, Alfred Wegener Institute, Helmholtz Centre for Polar and Marine Research, 27570 Bremerhaven, Germany
Full list of author information is available at the end of the article

response to grazers or grazer cues. The genotypes were different in basal level of paralytic shellfish toxin production, amount of cells in chains, and ability to produce lytic extracellular substances, traits that are considered defensive. *Alexandrium* spp. are marine dinoflagellates and constitute together with diatoms and haptophytes the dominant groups of marine phytoplankton that accounts for approximately half of the global annual net primary production [8]. More than 80% of the marine phytoplankton production is consumed by herbivores, thus grazing is the most important loss factor for phytoplankton [9–11]. Grazing consequently exerts a strong selective pressure on phytoplankton and structures phytoplankton communities [12]. *Alexandrium* combines colony size plasticity and reduced swimming speed to avoid encounters with copepod grazers [13]. In addition, *Alexandrium* produces paralytic shellfish toxins (PSTs) in response to chemical cues from certain copepods [14–19]. Genotypes within a single natural *Alexandrium* population showed phenotypic differences concerning the production of PSTs and further uncharacterized allelochemical compounds [20, 21]. Such an intraspecific trait variation can alter predator–prey dynamics and can lead to altered gene frequencies within a population as a result of strong grazing pressure. Induced defense as response towards grazing yet can maintain genotypic diversity, however it may also bear different trade-offs associated with the inducible defense response for each genotype [1, 22].

Our aim was to investigate genotype dependent plastic responses in *Alexandrium* in order to characterize the effect of grazer cues on trait variability. The two genotypes investigated here differed under standard culture condition in the amount of PSTs produced, the number of cells in colonies, the ability to produce lytic compounds and showed genetic variation in terms of gene expression patterns [23]. We monitored the known inducible traits toxin content and chain length and investigated grazing rates of the copepods. The enhanced PSP-toxin production as response towards copepods and their cues seems to be an induction rather than an accumulations as e.g. caused by phosphorous limitation, We therefore analysed a potential direct correlation between the expression of the *sxtA* gene that initiates the saxitoxin biosynthesis pathway in *Alexandrium* and enhanced toxin production.

Methods

Alexandrium fundyense strains and zooplankton collection

The two clonal strains of *A. fundyense* Balech 1985 emended Anderson [24]; (before named as *Alexandrium tamarense*) were isolated from the North Sea coast east off Scotland [21] and grown in K-medium [25] in a temperature and light controlled room (salinity ~33, 18 °C,

14 h:10 h light–dark cycles, ~150 fmol m^{-2} s^{-1}). The two strains, hereafter named Alex2 and Alex5 (both belong to the same population of the North American clade/ribotype group 1) are both producers of PSTs. Alex2 is further characterized by the presence of allelochemically active, unknown lytic compound(s) [20] whereas strain Alex5 lacks this ability.

Female *Centropages typicus* copepods were collected with vertical work package 2 (WP2) net hauls (200 μm mesh size) from ~20 m depth to the surface in the Gullmars fjord on the Swedish west coast. Only adult females were used in the experiments to minimize the variability between treatments. The copepods were maintained in the laboratory in filtered seawater (0.2 μm; salinity ~33) and fed *Rhodomonas baltica* (from the University of Gothenburg Marine Culture Collection, GUMACC) until the start of the experiments.

Direct grazing experiments

For both *A. fundyense* strains 18 bottles with 500 mL (~5 × 10^6 cells L^{-1}) cultures in K/10 medium [25] was prepared. Half of the bottles received 10 *C. typicus* females per bottle, and the remaining bottles were kept as copepod free controls. All bottles were incubated in a temperature and light controlled room (salinity ~33, 18 °C, 14 h:10 h light–dark cycles, ~150 fmol m^{-2} s^{-1}). Three replicates of the grazed and control treatments were harvested after 12, 48, and 72 h (see "sampling procedure" section below). Clearance rates were calculated using the equations of Frost [26].

Cage experiments

The experiments with fed or starved copepods kept in cages were conducted in 500 mL glass bottles with cages made out of 50 mL polypropylene tubes with a 10 μm plankton mesh at the bottom and incubation conditions as described above. The plankton mesh constrained the organisms to their compartment (flask or cage) while allowing waterborne-cues to move between the compartments. For each *A. fundyense* strain, 9 bottles were set up consisting of three replicates of copepod-free controls, fed *C. typicus*, and starved *C. typicus*. Each bottle received 450 mL of *A. fundyense* culture with ~4 × 10^6 cells L^{-1} in K/10 medium [25]. Cages were deployed into the flasks and filled with 30 mL of the same culture for the control treatments and the fed treatments. The flasks for the starved treatment received 30 mL of K/10 medium instead. Each cage of the fed and starved treatments received 10 *C. typicus* individuals i.e. the same amount as in the direct grazing experiment. The cages were gently moved up and down 5 times a day in order to promote the exchange of chemical cues between the compartments. The experiments were terminated after 48 h.

Sampling procedure

Bottles were sealed and carefully turned over 7 times to ensure equal mixing of the culture. The direct grazing experiment samples were pre-filtered through a submerged 64 μm nylon mesh to remove copepods and copepods eggs. 60–70 mL were sub-sampled for enumeration and sizing of cells and chains using a coulter counter (Micromeritics, Norcross, USA) mounted with a 100 μm orifice tube and continuous stirring. A known volume of the subsamples was then suction filtered onto glass-fibre filters (Whatman) and stored at −20 °C until toxin extraction and analysis.

The remaining culture was concentrated on a 10 μm nylon mesh to collect the *A. fundyense* cells. The 10 μm mesh was rinsed into a 50 mL centrifuge tube with sterile filtered seawater and immediately centrifuged at 4 °C for 5 min for RNA samples. The supernatant was discharged and the pellet immediately mixed with 1 mL 60 °C hot TriReagent (Sigma-Aldrich, Steinheim, Germany) and transferred to a cryovial with acid washed glass beads. The cryovial was vortexed for 10 s and submerged into liquid nitrogen. The samples were stored at −80 °C until RNA isolation.

PST analysis

Paralytic shellfish toxins samples (PST) were freeze-dried and extracted in 1 mL 0.05 M acetic acid (aq) through three consecutive freeze–thaw cycles. Extracts were filtered (GF/F) and stored frozen in glass vials. Samples were hydrolyzed with 0.1 M HCl at 100 °C for 10 min to transform C-toxins into their corresponding carbamates. We used a modified version of the method described by Asp and co-workers [27] to analyze all carbamate PSTs in a single run [16]. HPLC analysis was carried out on a Hitachi-7000 system equipped with a Genesis C8 column, (Vymac, 4 μm, 150*3 mm). A gradient between elution with 2 mM L^{-1} sodiumheptanesulfonate in 10 mM L^{-1} ammonium-phosphate buffer (pH 7.1) and 2 mM L^{-1} sodiumheptanesulfonate in 30 mM L^{-1} ammonium phosphate buffer (pH 7.1): acetonitrile (96:4) was used to separate PSTs. After the separation, toxins were oxidized with 7 mM periodic acid in 50 mM sodium phosphate buffer (pH 9.0) in a PEEK capillary (10 m, 80 °C). The oxidation was terminated with 0.5 M acetic acid before fluorescent detection at $\lambda_{ex} = 330$ nm, $\lambda_{em} = 390$ nm. Toxin standards were obtained from the certified reference materials program, National Research Council, Halifax, Canada.

Total RNA isolation

TriReagent fixed and frozen cells were lysed using a Bio101 FastPrep instrument (Thermo Savant Illkirch, France) at maximum speed (6.5 m s^{-1}) for 2 × 45 s. Lysed cells were cooled on ice, 200 μL chloroform was added and vortexed for 20 s. The samples were transferred to a phase lock tube after 5 min incubation at room temperature (Eppendorf, Hamburg; Germany) and incubated for another 5 min followed by centrifugation for 15 min at 13,000×g and 4 °C. The upper aqueous phase was transferred to a new tube and mixed with the same volume isopropanol, 1/10 volume of 3 M Na-acetate (pH 5.5; Ambion by Life Technologies, Carlsbad, California, USA) and 2 μL linear polyacrylamide (Ambion). Total RNA was precipitated for 90 min at −20 °C and collected by centrifugation for 20 min and 13,000×g at 4 °C. The obtained pellet was washed twice, first with 1 mL 70% EtOH followed by 1 mL EtOH absolute; the RNA pellet was dried for 1 min at 37 °C and resolved in 30 μL RNase free water (Qiagen, Hilden). RNA quality check was performed using a NanoDrop ND-1000 spectrometer (PeqLab, Erlangen, Germany) for purity and the RNA Nano Chip Assay with the 2100 Bioanalyzer device (Agilent Technologies, Santa Clara, California, USA) was just to examine the integrity of the extracted RNA. Potential DNA contaminations of RNA samples were tested by amplification of the 28 s rDNA gene. We considered our samples to be free of DNA as the respective gene could be amplified in positive controls only (containing both, RNA and added DNA).

Expression analysis of the putative *sxt*A gene via RT-qPCR

The quantitative expression of the putative *sxt*A gene in Alex2 was evaluated via RT-qPCR using the method described in Freitag et al. [28]. For this RT-qPCR analysis, we only considered treatments were a significant increase in saxitoxin on the total PST profile compared to control cultures was observed (Alex2, direct grazing experiment, 48 and 72 h). Prior to cDNA synthesis, 500 ng of total RNA of the samples of interest were spiked with 10 ng of MA mRNA and 1 ng of NSP mRNA that served as an external Ref. [28]. The cDNA was synthesized with the Omniscript RT kit (Qiagen, Hilden, Germany) according to the manufacturer's instructions with anchored oligo (dT) 20 primers. The synthesized cDNAs were diluted 1:5 and 2 μL were used to analyze the expression of the putative *sxt*A gene, the MA gene and the NSP gene with a SYBRgreen assay according to the manufactures protocol (Applied Biosystems/Life Technologies, Carlsbad, California, USA). Primers for the putative *sxt*A gene fragment had the sequences: 5′GCGAGACCGACGAGAAGTTC′3 and 5′AGCCGCTTGCGCTGAAG′3; primer sequences for the MA and NSP gene are given by Freitag et al. [28]. The qPCR reaction were carried out in a StepOnePlus™ Real-Time PCR System-device (Applied Biosystem by Life Technologies, Carlsbad, California, USA) with the following cycling parameters: 10 min initial denaturation at

95 °C, 40 cycles of 95 °C for 15 s and 60 °C for 1 min. A melting curve analysis was performed at the end of the reactions to verify the formation of a single PCR product in each reaction. In addition, standard curves with 6 points and a serial dilution factor of 1:10 starting with 10 ng PCR product from each gene were also analyzed with the SYBRgreen assay using the same primers and parameters. PCR products for the standard curves for the *sxt*A gene were obtained from the prepared cDNAs and the following primers: 5'CCGCCATATGTGCTT GTTTG'3 and 5'AGCTCCCTGTACACCTCTGC'3. Expression ratios were calculated using the equations of Pfaffl [29].

Results

Size distribution

The observed size distribution of the two strains of *A. fundyense* shifted in the presence of copepods towards a smaller equivalent spherical diameter (Fig. 1; Additional file 1). This shift was mainly caused by a different proportion of the culture present as single cells, two, or four cell chains as observed by microscopic examination. The shift is present in both the direct grazing experiments and the waterborne-cue experiments, with caged copepods, showing that the observed responses were triggered by the waterborne-cues from the copepods. The response was, however, strongest for Alex2 exposed to direct grazing copepods (Fig. 1A). Here, the average diameter significantly decreased by 10% after 48 h and by 16% after 72 h in grazer exposed cultured (two-way ANOVA $p < 0.05$, TukeyHSD $p < 0.05$). This decrease corresponds to approximately a doubling in the amount of singles cells compared to controls. The mean diameter of Alex2 cells exposed to waterborne-cues of caged fed and starved copepods significantly decreased by ~7% (one-way ANOVA $p < 0.05$, TukeyHSD $p < 0.05$) (Fig. 1B; Additional file 1).

Alex5 was generally less prone to form chains (Fig. 1C, D). Yet, exposure to both direct grazing and grazer cues

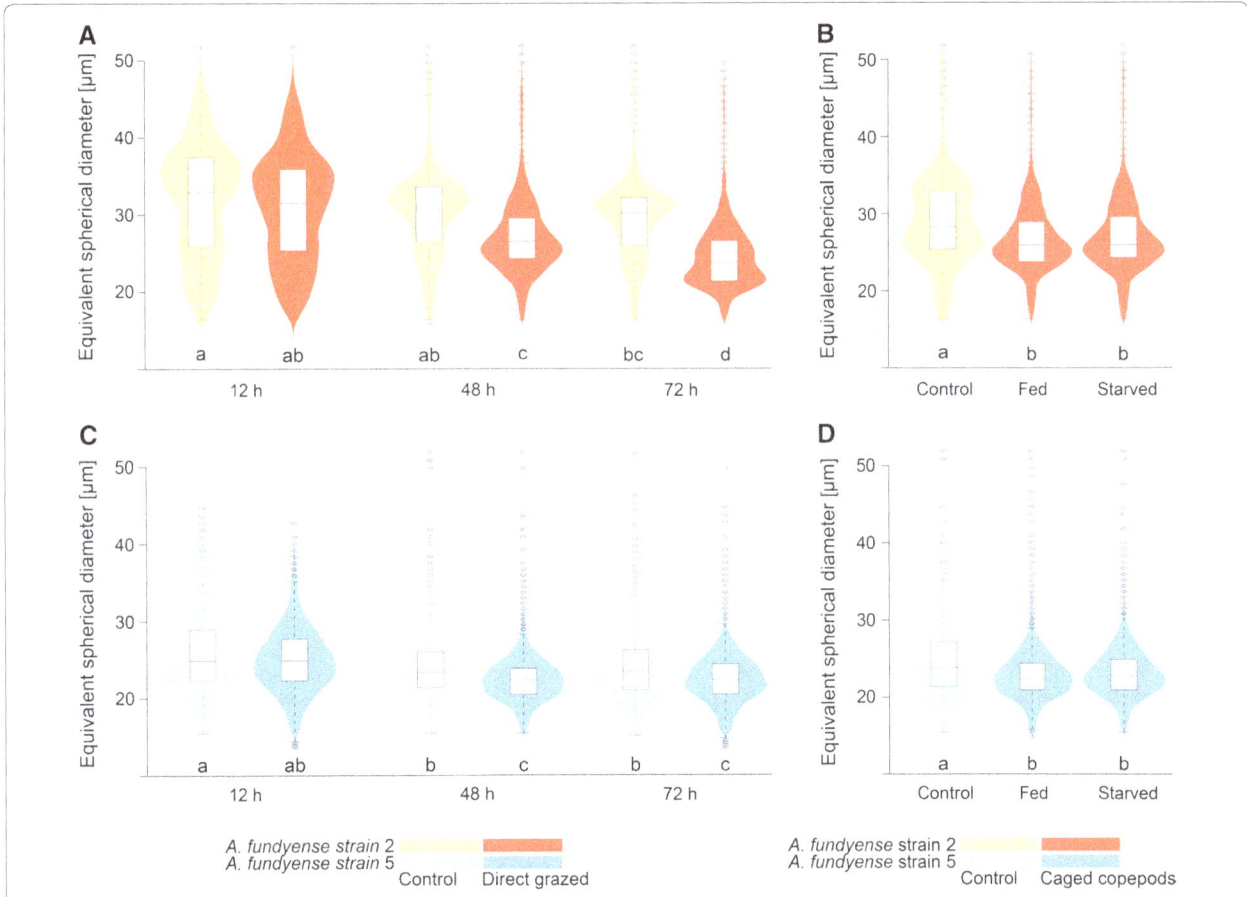

Fig. 1 Cell size distribution of the *Alexandrium fundyense* strains. Cell size distributions of Alex2 (**A**, **B**) and Alex5 (**C**, **D**) after exposure to direct grazing *Centropages typicus* (**A**, **C**) or after exposure to waterborne-cues from caged fed and caged starved *C. typicus* (**B**, **D**). Treatments with waterborne-cues were terminated after 48 h. The diameter of the measured particles (equivalent spherical diameter in μm) is plotted as kernel density estimation with the width corresponding to the relative occurrence of a particle in a respective size class. *Box plots* show the 25th, 50th and 75th percentile; the ends of the whiskers mark the 95% intervals

resulted in a significantly reduced average size (mean diameter) also for this strain (~4% after 48 h and ~6% after 72 h, two-way ANOVA $p < 0.05$, TukeyHSD $p < 0.05$) (Additional file 1).

Different sampling times led to observable variations of cell size distributions in both strains (Fig. 1). Cells sampled at the time point 12 h have been collected just before the onset of the dark cycle that triggers cell division, and therefore also contain several cells that have accumulated enough biovolume to divide. Cells sampled at time points 48 and 72 h yet have been sampled after the dark cycle in the morning after cell division, thus showing a less scattered distribution with smaller overall cell sizes.

In the treatment with waterborne cues from caged fed and caged starved copepods, the cell size means significantly decreased by ~6.5% (one-way ANOVA $p < 0.05$, TukeyHSD $p < 0.05$) (Fig. 1D; Additonal file 1) showing that the response does not require any physical contact between copepods and algal cells.

PST-contents and clearance rates
Exposure of both *Alexandrium* strains to direct grazing *C. typicus* individuals resulted in significantly increased PST contents per biovolume in Alex2 and Alex5 (two-way ANOVA $p < 0.05$, TukeyHSD $p < 0.05$, Fig. 2A, B).

The treatment with waterborne-cues from caged fed and starved copepods however showed significant differences in the mean PST content for Alex2 only (one-way ANOVA $p < 0.05$, Fig. 2B), where PST content increased compared to the control (TukeyHSD $p < 0.05$).

Clearance rates on Alex2 significantly decreased from 1.7 mL per female per hour to 0.4 mL per female per hour (one-way ANOVA $p < 0.05$, Fig. 2A) during the direct grazing experiment. Clearance of copepods grazing on Alex5 cultures, however, remained constant around 0.4 mL per female per hour (Fig. 2C).

PST profiles and differential expression of the *sxt*A gene
The increase in the total PST content in Alex2 cultures after exposure to copepods in the direct grazing treatment

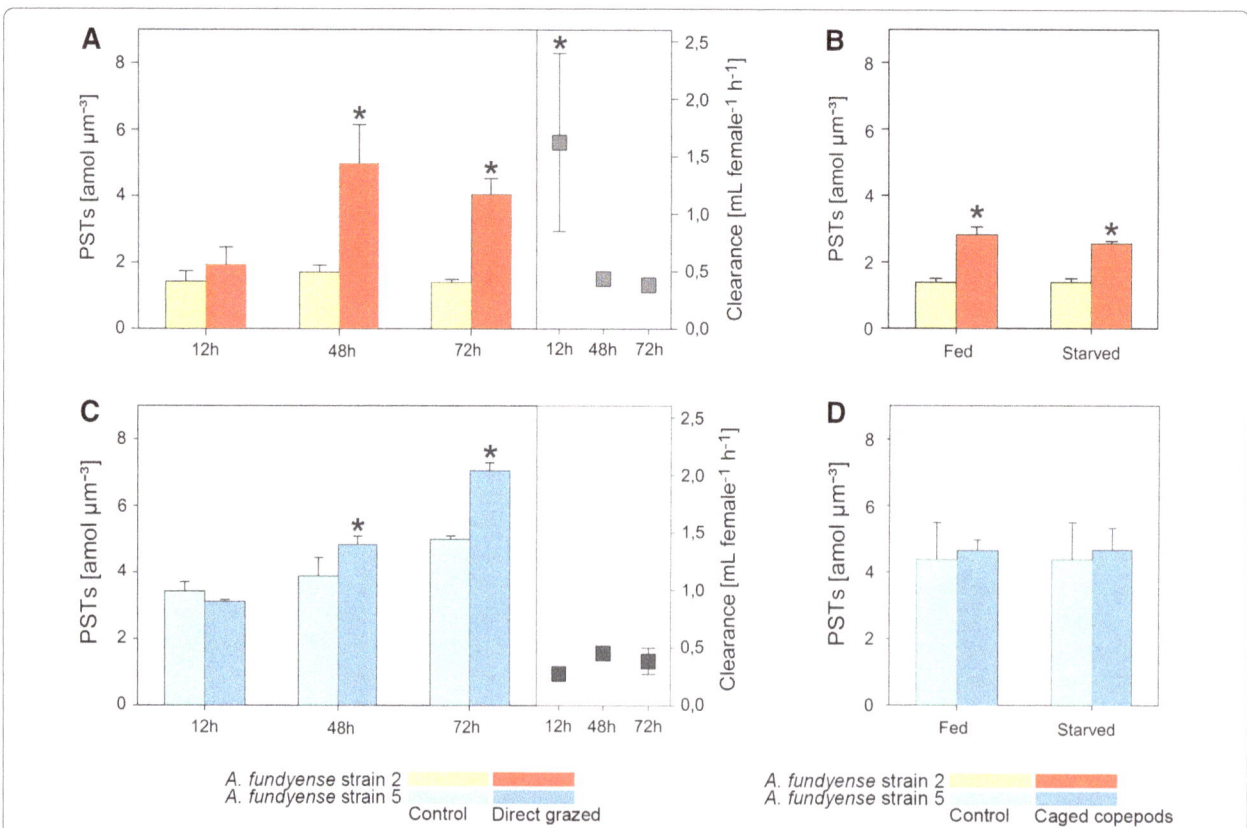

Fig. 2 PST contents of the *Alexandrium fundyense* strains and clearance rates for *Centropages typicus*. PST contents of Alex2 (**A**, **B**) and Alex5 (**C**, **D**). PST contents after exposure to direct grazing *C. typicus* individuals and respective clearance rates are given in **A** and **C**. PST contents after exposure to waterborne-cues from caged fed and caged starved *C. typicus* are given in **B** and **D**. Treatments with waterborne-cues were terminated after 48 h. *Bars* marked with an *asterisk* show significant differences compared to either the control at this time point (for PSTs contents) or between the treatments at different time points (for clearance rates) (ANOVA $p < 0.05$)

and after exposure to caged fed and starved copepods was accompanied by a shift in the PST profile (Fig. 3). Yet, this phenomenon was not observed for Alex5. In Alex2, the amount of saxitoxin in the grazed treatment increased significantly from a mean of ~26% of total PSTs to 48% at 48 h and further to 52% at 72 h (ANOVA $p < 0.05$, TukeyHSD $p < 0.05$). The observed increase in the amount of saxitoxin in Alex2 did not result in a significant higher expression of the sxtA gene neither after 48 h nor at 72 h (Fig. 4).

Discussion

The phenotypic differences show that Alex5 expresses the 'defended' phenotype, with smaller units (cells and/or cell chains) with higher toxin content already before exposure to grazers or grazer cues (Figs. 1, 2). Contrarily, Alex2 exhibited a less defended phenotype, with more cells in chains and lower toxin content in the absence of grazers or grazer cues. This was also accompanied by significant differences in the clearance rate of the copepods on the two genotypes. The clearance rates during the initial 12 h were fourfold lower for Alex5 compared to Alex2 (Fig. 2A, C). However, the grazing rates on Alex2 decrease to the same level after the induction of the defended phenotype

(48 h, Figs. 1, 2). The phenotype expressed by Alex2 prior to defense induction is less vulnerable to unicellular heterotrophic grazers [20] and the allelochemicals produced by this strain did not seem to protect them against grazing copepods. Hence, the consequence of grazer exposure for Alex2 is a phenotypic shift towards less vulnerability towards copepod grazing. The splitting of cell chains in Alex2 may cause ecological costs under natural conditions i.e. a reduction of the ability to perform vertical migrations to retrieve nutrients at depth [13] and to resist unicellular grazers and competitors due to a reduced amount of cells in chains accompanied by a decreased concentration of allelochemicals surrounding the cells of this strain [30, 31]. These ecological costs may not only be strain-specific as they can indirectly influence the survival of conspecifics protected by the allelochemical cloud of this strain [23, 32]. Yet, the ability of A. fundyense to induce defense at different levels (morphological and physiological, as well as behavioural (see also Selander et al. [13]), offers flexibility in the applied defense strategy and allows to lower the predation risk at feasible costs depending on trade-offs that are themselves dependent on the environment and genotype [33, 34].

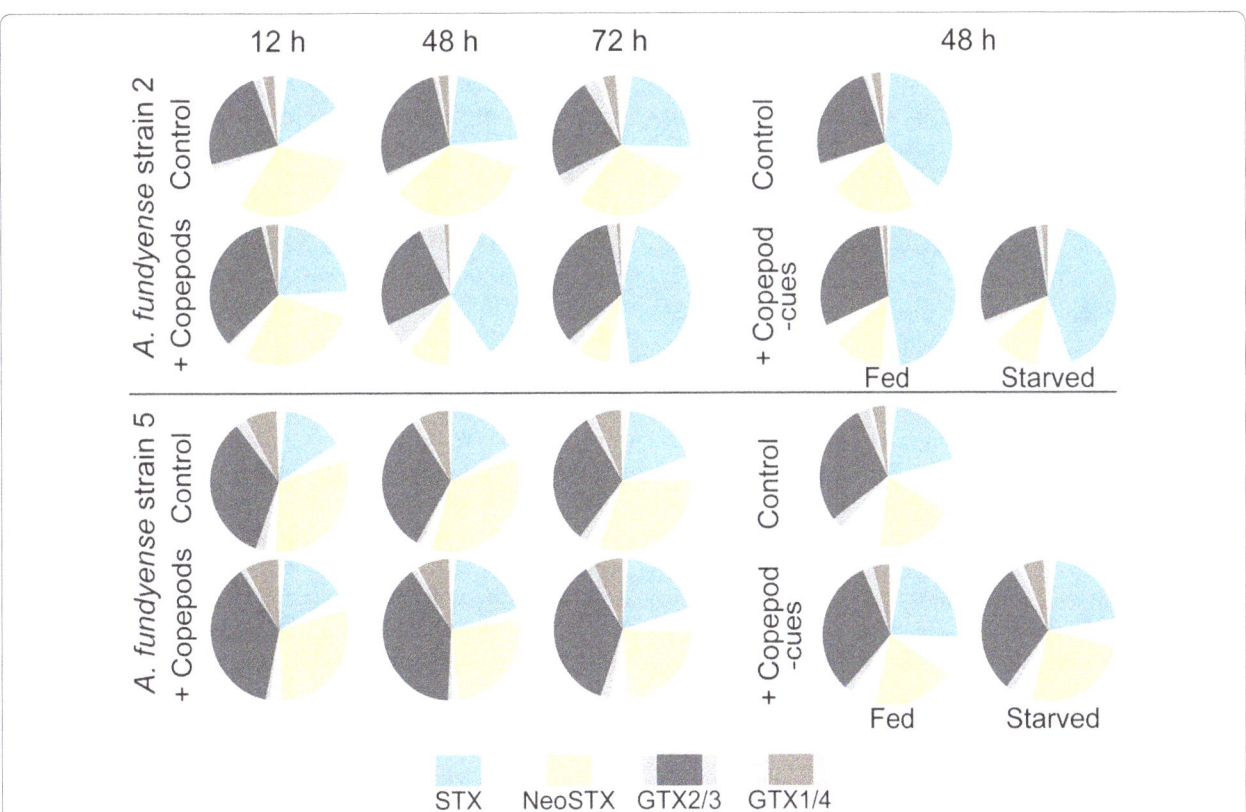

Fig. 3 PST profiles of Alex2 and Alex5. PST profiles of Alex2 after exposure to direct grazing Centropages typicus, waterborne-cues of C. typicus and the respective controls are shown in the *upper panel*. PST profiles of Alex5 after exposure to direct grazing C. typicus, waterborne-cues of C. typicus and the respective controls are shown in the *lower panel*. *Shaded colors* indicate confidence intervals

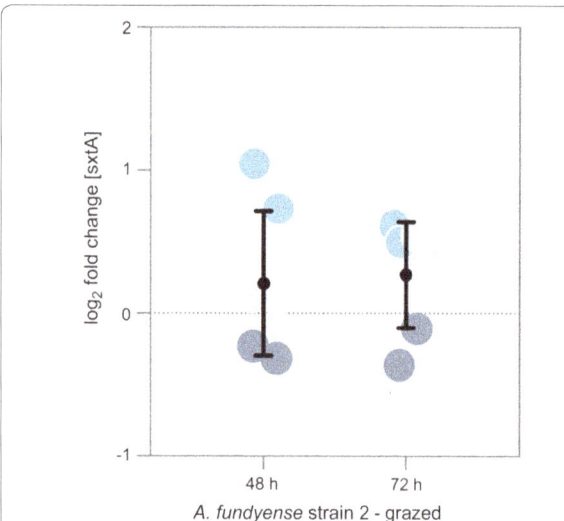

Fig. 4 Expression of the *sxt*A gene fragment. The log2 expression ratio for the putative *sxt*A gene fragment as determined for Alex2 from the direct grazing experiment after 48 and 72 h. Quantification of the relative expression compared to the control treatments was done using a reference spike-in gene to normalize expression level of controls and treatments. The different *blue colours* represent biological replicates, *dots* with *same colours* represent technical replicates

The increase in toxin biosynthesis differs in its physiological characteristics between the two strains: Alex5 increased the amount of PSTs without changing the relative composition of individual PSTs; in Alex2 the PST profile changed to a higher proportion of the saxitoxin derivate (Fig. 3). The higher amount of PST and in particular saxitoxin in Alex2 did, however, at the measured time points not correlate with a higher expression of the putative *sxt*A gene [35] that initiate the saxitoxin biosynthesis pathway as verified via RT-qPCR (Fig. 4). The *sxt*A gene product might not be the limiting step in the biosynthesis of saxitoxin or might have been upregulated already at the onset of the induced defense response prior to the observed morphological and physiological changes. Further, the biosynthesis of saxitoxin could be regulated by other means e.g. post-transcriptional, by the regulation of other enzymes participating in this pathway [36] or by the provision of precursors and reduction equivalents. The observed differences in the two strains may also point towards a convoluted biosynthesis pathway where several routes could lead to an increase in intracellular PST content. An increase in PSTs could therefore be as well associated with changes in other traits where beneficial trait associations led to a selective advantage. For example, we cannot rule out that the PST increase per biovolume is purely due to a decrease of cell size in our experiment. However, previous studies with the same genotype Alex5 showed increased PST contents after 48 h exposure to copepods on a per cell basis, irrespective of the biolvolume [17]. Selander et al. showed that

after removal of copepods, grazer-induced PST contents in *A. fundyense* returned to control levels after approximately 5 cell divisions, whereas the average biovolume returned to control levels after one to two cell divisions. Thus *A. fundyense* cells remained more toxic compared to controls with comparable biovolumes [16].

With the cage experiment design we demonstrated that the copepod cues alone are sufficient to induce the expression of defensive features. However, due to a potential signal dampening effect the signals alone lead to a weaker response compared to the direct grazing experiments (Figs. 1, 2) [17, 30, 37, 38]. The reduction of cell chains, however, remained at comparable levels to the direct grazing experiment in both strains (Figs. 1, 2). A meta-analysis on herbivore resistance in plants presumed that traits other than secondary metabolites (e.g. morphology, life history, phenology, primary chemistry and physiology) are more effective in defense [39]. As the effect of increased toxin content in *Alexandrium* on grazing rates strongly depends on the grazer abundance, species composition and local adaptations [37, 40–45], the significance of increased toxin content for defense is still unclear, particularly because intoxication of the grazer benefits the whole population and additionally potential competitors, unless the more toxic cells are rejected by grazers and directly benefit from their inducible toxicity [15, 37, 41]. The observed morphological changes are preventive by reducing encounter rates with grazers [13] and may therefore directly contribute increased fitness to responding cells.

Conclusions
In conclusion, we showed that the presence of copepods has the potential to induce the formation of a comparable phenotype from initially different genotypes which highlights the selective pressure that emanates from grazers in the pelagic realm. We could further show that the morphological changes might be stronger correlated to the induced defense response than changes in PST-content are. The induced defense response will reduce the trait variability for natural selection to operate on; the costs for its induction however can be different for each strain on several levels i.e. in terms of physiological and ecological costs. Selection on different genotypes can therefore occur in the presence of plasticity due to different costs and trade-offs that arise for the respective genotype. The occurrence of multiple environmental cues to respond to in nature however might equalize the effects of strain-specific trade-offs within populations.

Abbreviation
PST(s): paralytic shellfish toxin(s).

Authors' contributions
Conceived and designed the experiments: SW ES UJ. Performed the experiments: SW ES. Analyzed the data: SW ES. Wrote the paper: SW ES UJ. All authors read and approved the final manuscript.

Author details
[1] Department of Ecological Chemistry, Alfred Wegener Institute, Helmholtz Centre for Polar and Marine Research, 27570 Bremerhaven, Germany. [2] Department of Biological and Environmental Sciences, University of Gothenburg, Vasaparken, 40530 Gothenburg, Sweden. [3] Helmholtz Institute for Functional Marine Biodiversity (HIFMB), 26111 Oldenburg, Germany.

Acknowledgements
The authors thank Anke Stüken for her help with setting up the RT-qPCR analysis for the sxtA gene. Further, the authors thank the two anonymous reviewers for their constructive comments.

Competing interests
The authors declare that they have no competing interests.

Funding
Financial support was provided by the PACES research program of the Alfred-Wegener-Institute Helmholtz-Zentrum für Polar-und Meeresforschung, by the German Research Foundation (DFG) Priority Programme DynaTrait (1704; JO 702/7-1) and by FORMAS research Grant No 223-2012-693 to Erik Selander.

Role of funding: PACES: study design, data analysis and interpretation, manuscript writing; DFG: data analysis and interpretation, manuscript writing; FORMAS: study design, data analysis and interpretation, manuscript writing.

References
1. Bolnick DI, Amarasekare P, Araújo MS, Bürger R, Levine JM, Novak M, Rudolf VHW, Schreiber SJ, Urban MC, Vasseur DA. Why intraspecific trait variation matters in community ecology. Trends Ecol Evol. 2011;26(4):183–92.
2. Pfennig DW, Wund MA, Snell-Rood EC, Cruickshank T, Schlichting CD, Moczek AP. Phenotypic plasticity's impacts on diversification and speciation. Trends Ecol Evol. 2010;25(8):459–67.
3. Litchman E, Klausmeier CA. Trait-based community ecology of phytoplankton. Annu Rev Ecol Evol Syst. 2008;39(1):615–39.
4. Hansen B, Bjornsen PK, Hansen PJ. The size ratio between planktonic predators and their prey. Limnol Oceanogr. 1994;39(2):395–403.
5. Van Donk E, Ianora A, Vos M. Induced defences in marine and freshwater phytoplankton: a review. Hydrobiologia. 2011;668:3–19.
6. Bergkvist J, Thor P, Jakobsen HH, Wängberg S-Å, Selander E. Grazer-induced chain length plasticity reduces grazing risk in a marine diatom. Limnol Oceanogr. 2012;57(1):318–24.
7. Corno G, Jürgens K. Direct and indirect effects of protist predation on population size structure of a bacterial strain with high phenotypic plasticity. Appl Environ Microbiol. 2006;72(1):78–86.
8. Falkowski PG, Katz ME, Knoll AH, Quigg A, Raven JA, Schofield O, Taylor FJR. The evolution of modern eukaryotic phytoplankton. Science. 2004;305:354–60.
9. Calbet A. Mesozooplankton grazing effect on primary production: a global comparative analysis in marine ecosystems. Limnol Oceanogr. 2001;46:1824–30.
10. Calbet A, Landry MR. Phytoplankton growth, microzooplankton grazing, and carbon cycling in marine systems. Limnol Oceanogr. 2004;49:51–7.
11. Cyr H, Pace ML. Allometric theory-extrapolations from individuals to communities. Ecology. 1993;74:1234–45.
12. Verschoor AM, Vos M, van der Stap I. Inducible defences prevent strong population fluctuations in bi- and tritrophic food chains. Ecol Lett. 2004;7:1143–8.
13. Selander E, Jakobsen HH, Lombard F, Kiørboe T. Grazer cues induce stealth behavior in marine dinoflagellates. Proc Natl Acad Sci. 2011;108:4030–4.
14. Selander E, Kubanek J, Hamberg M, Andersson MX, Cervin G, Pavia H. Predator lipids induce paralytic shellfish toxins in bloom-forming algae. Proc Natl Acad Sci. 2015;112(20):6395–400.
15. Selander E, Thor P, Toth G, Pavia H. Copepods induce paralytic shellfish toxin production in marine dinoflagellates. Proc R Soc B-Biol Sci. 2006;273:1673–80.
16. Selander E, Fagerberg T, Wohlrab S, Pavia H. Fight and flight in Dinoflagellates? Kinetics of simultaneous grazer-induced responses in *Alexandrium tamarense*. Limnol Oceanogr. 2012;57:58–64.
17. Wohlrab S, Iversen MH, John U. A molecular and co-evolutionary context for grazer induced toxin production in *Alexandrium tamarense*. PLoS ONE. 2010;5:e15039.
18. Yang I, Selander E, Pavia H, John U. Grazer-induced toxin formation in dinoflagellates: a transcriptomic model study. Eur J Phycol. 2011;46:66–73.
19. Senft-Batoh CD, Dam HG, Shumway SE, Wikfors GH. A multi-phylum study of grazer-induced paralytic shellfish toxin production in the dinoflagellate *Alexandrium fundyense*: a new perspective on control of algal toxicity. Harmful Algae. 2015;44:20–31.
20. Tillmann U, Hansen P. Allelopathic effects of *Alexandrium tamarense* on other algae: evidence from mixed growth experiments. Aquat Microb Ecol. 2009;57:101–12.
21. Alpermann TJ, Tillmann U, Beszteri B, Cembella AD, John U. Phenotypic variation and genotypic diversity in a planktonic population of the toxigenic marine dinoflagellate *Alexandrium tamarense* (Dinophyceae). J Phycol. 2010;46:18–32.
22. Tillmann U, Alpermann TL, da Purificação RC, Krock B, Cembella A. Intrapopulation clonal variability in allelochemical potency of the toxigenic dinoflagellate *Alexandrium tamarense*. Harmful Algae. 2009;8:759–69.
23. Wohlrab S, Tillmann U, Cembella A, John U. Trait changes induced by species interactions in two phenotypically distinct strains of a marine dinoflagellate. ISME. 2016.
24. John U, Litaker RW, Montresor M, Murray S, Brosnahan ML, Anderson DM. Formal revision of the *Alexandrium tamarense* Species Complex (Dinophyceae) taxonomy: the introduction of five species with emphasis on molecular-based (rDNA) classification. Protist. 2014;165(6):779–804.
25. Keller MD, Selvin RC, Claus W, Guillard RRL. Media for the culture of oceanic ultraphytoplankton. J Phycol. 1987;23:633–8.
26. Frost BW. Effects of size and concentration of food particles on the feeding behavior of the marine planktonic copepod *Calanus pacificus*. Limnol Oceanogr. 1972;17:805–15.
27. Asp TN, Larsen S, Aune T. Analysis of PSP toxins in Norwegian mussels by a post-column derivatization HPLC method. Toxicon. 2004;43:319–27.
28. Freitag M, Beszteri S, Vogel H, John U. Effects of physiological shock treatments on toxicity and polyketide synthase gene expression in *Prymnesium parvum* (Prymnesiophyceae). Eur J Phycol. 2011;46:193–201.
29. Pfaffl MW. A new mathematical model for relative quantification in real-time RT-PCR. Nucleic Acids Res. 2001;29:e45.
30. Harvell DC, Tollrian R. Why inducible defenses? In: Tollrian R, Harvell DC, editors. The ecology and evolution of inducible defenses. Princeton: Princeton University Press; 1999. p. 3–9.
31. Jonsson PR, Pavia H, Toth G. Formation of harmful algal blooms cannot be explained by allelopathic interactions. Proc Natl Acad Sci USA. 2009;106:11177–82.
32. John U, Tillmann U, Hülskötter J, Alpermann TJ, Wohlrab S, Van de Waal DB. Intraspecific facilitation by allelochemical mediated grazing protection within a toxigenic dinoflagellate population. Proc R Soc Lond B: Biol Sci. 2015;282(1798):20141268.
33. Ellner SP. Rapid evolution: from genes to communities, and back again? Funct Ecol. 2013;27(5):1087–99.

34. Bjærke O, Jonsson PR, Alam A, Selander E. Is chain length in phytoplankton regulated to evade predation? J Plankton Res. 2015;37:1110–9.

35. Stüken A, Orr RJS, Kellmann R, Murray SA, Neilan BA, Jakobsen KS. Discovery of nuclear-encoded genes for the neurotoxin saxitoxin in dinoflagellates. PLoS ONE. 2011;6:e20096.

36. Perini F, Galluzzi L, Dell'Aversano C, Iacovo E, Tartaglione L, Ricci F, Forino M, Ciminiello P, Penna A. SxtA and sxtG gene expression and toxin production in the Mediterranean *Alexandrium minutum* (Dinophyceae). Marine Drugs. 2014;12(10):5258.

37. Senft-Batoh CD, Dam HG, Shumway SE, Wikfors GH, Schlichting CD. Influence of predator–prey evolutionary history, chemical alarm-cues, and feeding selection on induction of toxin production in a marine dinoflagellate. Limnol Oceanogr. 2015;60(1):318–28.

38. Bergkvist J, Selander E, Pavia H. Induction of toxin production in dinoflagellates: the grazer makes a difference. Oecologia. 2008;156:147–54.

39. Carmona D, Lajeunesse MJ, Johnson MTJ. Plant traits that predict resistance to herbivores. Funct Ecol. 2011;25:358–67.

40. Turner JT, Tester PA. Toxic marine phytoplankton, zooplankton grazers, and pelagic food webs. Limnol Oceanogr. 1997;42:1203–14.

41. Teegarden GJ. Copepod grazing selection and particle discrimination on the basis of PSP toxin content. Mar Ecol Prog Ser. 1999;181:163–76.

42. Turner TJ, Doucette JG, Powell LC, Kulis MD, Keafer AB, Anderson MD. Accumulation of red tide toxins in larger size fractions of zooplankton assemblages from Massachusetts Bay. USA. Mar Ecol Prog Ser. 2000;203:95–107.

43. Colin SP, Dam H. Testing for resistance of pelagic marine copepods to a toxic dinoflagellate. Evol Ecol. 2004;18:355–77.

44. Teegarden GJ, Campbell RG, Anson DT, Ouellett A, Westman BA, Durbin EG. Copepod feeding response to varying *Alexandrium* spp. cellular toxicity and cell concentration among natural plankton samples. Harmful Algae. 2008;7:33–44.

45. Dam HG, Haley ST. Comparative dynamics of paralytic shellfish toxins (PST) in a tolerant and susceptible population of the copepod *Acartia hudsonica*. Harmful Algae. 2011;10:245–53.

8

Insect herbivory in a mature *Eucalyptus* woodland canopy depends on leaf phenology but not CO$_2$ enrichment

Andrew N. Gherlenda[1*] [iD], Ben D. Moore[1], Anthony M. Haigh[2], Scott N. Johnson[1] and Markus Riegler[1*]

Abstract

Background: Climate change factors such as elevated atmospheric carbon dioxide concentrations (e[CO$_2$]) and altered rainfall patterns can alter leaf composition and phenology. This may subsequently impact insect herbivory. In sclerophyllous forests insects have developed strategies, such as preferentially feeding on new leaf growth, to overcome physical or foliar nitrogen constraints, and this may shift under climate change. Few studies of insect herbivory at elevated [CO$_2$] have occurred under field conditions and none on mature evergreen trees in a naturally established forest, yet estimates for leaf area loss due to herbivory are required in order to allow accurate predictions of plant productivity in future climates. Here, we assessed herbivory in the upper canopy of mature *Eucalyptus tereticornis* trees at the nutrient-limited *Eucalyptus* free-air CO$_2$ enrichment (EucFACE) experiment during the first 19 months of CO$_2$ enrichment. The assessment of herbivory extended over two consecutive spring—summer periods, with a first survey during four months of the [CO$_2$] ramp-up phase after which full [CO$_2$] operation was maintained, followed by a second survey period from months 13 to 19.

Results: Throughout the first 2 years of EucFACE, young, expanding leaves sustained significantly greater damage from insect herbivory (between 25 and 32 % leaf area loss) compared to old or fully expanded leaves (less than 2 % leaf area loss). This preference of insect herbivores for young expanding leaves combined with discontinuous production of new foliage, which occurred in response to rainfall, resulted in monthly variations in leaf herbivory. In contrast to the significant effects of rainfall-driven leaf phenology, elevated [CO$_2$] had no effect on leaf consumption or preference of insect herbivores for different leaf age classes.

Conclusions: In the studied nutrient-limited natural *Eucalyptus* woodland, herbivory contributes to a significant loss of young foliage. Leaf phenology is a significant factor that determines the level of herbivory experienced in this evergreen sclerophyllous woodland system, and may therefore also influence the population dynamics of insect herbivores. Furthermore, leaf phenology appears more strongly impacted by rainfall patterns than by e[CO$_2$]. e[CO$_2$] responses of herbivores on mature trees may only become apparent after extensive CO$_2$ fumigation periods.

Keywords: Arthropod, Climate change, Eucalypt, FACE, Plant–insect interaction

Background

Climate change and its drivers can have a significant impact on the physiology, abundance and distribution of insect herbivores [1–3]. Elevated CO$_2$ concentrations (e[CO$_2$]) often reduce the growth and survival of insect herbivores as a plant-mediated effect influenced by the decrease in leaf nitrogen concentrations [4–7] and an increase in secondary metabolites, such as phenolic compounds [8–10] generally observed at e[CO$_2$]. Furthermore, some studies suggest modulation of plant hormone signalling and induced plant defence at e[CO$_2$] [9, 10]. Climate change may also alter the timing and amount of precipitation, and this can potentially impact insect

*Correspondence: a.gherlenda@westernsydney.edu.au; m.riegler@westernsydney.edu.au
[1] Hawkesbury Institute for the Environment, Western Sydney University, Locked Bag 1797, Penrith, NSW 2751, Australia
Full list of author information is available at the end of the article

abundance and phenology both directly, and indirectly as a consequence of changes in plant phenology and productivity [11–13]. It has been demonstrated that e[CO_2] can increase plant net primary production (NPP) [14, 15] and this could potentially benefit insect herbivores as a result of greater resource availability. However, this increase in NPP may be constrained or even reduced in nutrient-limited [16, 17] or water-limited forests [18], or also due to changing herbivory patterns. Furthermore, the measurement of NPP in field experiments may be underestimated if the impacts of herbivory are not measured, in particular the failure of new leaves to expand, due to herbivory on meristems and very young expanding leaves.

Eucalyptus (Myrtaceae) is both an ecologically and economically important tree genus in many parts of the world [19, 20]. *Eucalyptus* species are often characterised by sclerophyllous leaves and are often associated with low-fertility soils common in Australia [21–23]. As for all plants, the chemical and physical properties of *Eucalyptus* leaves change with age; young leaves typically have higher nitrogen concentration and moisture content, and reduced toughness compared to older leaves [11, 24, 25]. These factors increase palatability of young foliage to many herbivorous insects, and this can result in enhanced insect performance when feeding on young compared to older leaves [26–28]. Changes in the amount and occurrence of rainfall events may alter the relationship of insect herbivores with leaf phenology, potentially affecting diversity and abundance of insects within these forests. Furthermore, many plants, including *Eucalyptus*, invest heavily in secondary defence compounds [29, 30], and the production of these secondary compounds may vary throughout leaf development [29, 31, 32]. Herbivore induced plant defence, however, does not appear to occur in *Eucalyptus* [33] but see [34].

Increased consumption of leaves, or compensatory feeding, is often observed in herbivorous insects as a response to plants grown under e[CO_2]—this is to compensate for the dilution of leaf nitrogen [5]. Leaf consumption by herbivorous insects at e[CO_2] may result in an additional 17–40 % of leaf damage compared to current levels [7, 35]. Despite the potential for compensatory feeding at e[CO_2], the survival of herbivorous insects may be reduced while developmental time is often increased [5, 36–38]. Therefore, e[CO_2] may increase leaf damage caused by individual insects to forest trees, however, the abundances of insects in these forests may be reduced due to the negative effects of e[CO_2] on insect survival and development. Overall this may result in no net change of leaf damage to trees at e[CO_2].

It has previously been demonstrated that the production of insect herbivore excrements (or frass), a crude proxy for insect abundance and herbivory, increased after large rainfall events during the spring and summer at the *Eucalyptus* free-air CO_2 enrichment (EucFACE) experiment [11]. This increase in insect activity coincided with an increase in leaf area index (LAI) at EucFACE [18], suggesting a direct link between the abundance of *Eucalyptus*-feeding insects and leaf phenology. However no e[CO_2] effects on frass deposition or LAI changes within the first 2 years of EucFACE were found, suggesting that insect herbivory and canopy processes may not be impacted by early stages of [CO_2] fumigation at EucFACE.

The measurement of frass deposition onto the woodland floor does not reveal which leaf age class experiences most damage from herbivory. Furthermore, e[CO_2] may change leaf phenology and thereby resource availability for herbivores. It may also alter the preference of insect herbivores for different leaf stages if the relative palatability of young expanding versus fully expanded (mature) or old leaves changes under e[CO_2]. Any change in the consumption of young expanding leaves may therefore affect the recruitment of new leaves in the forest canopy, and place stress on plants. For new leaves, LAI measurement methods may struggle to discriminate between insect removal of leaf area and reductions in NPP. This can result in an incorrect estimate and under-evaluation of NPP of forests, particularly if climate change factors alter the herbivory of new leaf production.

This study investigated the relationship between insect herbivory and leaf phenology of *Eucalyptus tereticornis* Sm., and the impacts of e[CO_2] and rainfall patterns on these processes in a mature, evergreen canopy of this tree species forming a naturally established woodland at the EucFACE experimental site. We hypothesised that rates of insect herbivory would respond to new leaf production which again would vary across time based on rainfall. It has previously been demonstrated that rainfall is the key driver of *Eucalyptus* leaf phenology, including at the study site [18, 39, 40]. The aims of this study were to: (1) compare the monthly levels of insect leaf herbivory under ambient and e[CO_2] conditions within a mature *Eucalyptus* canopy forming a woodland for two spring and summer periods at which herbivore activity was observed to be highest in the first and second year of EucFACE [11]; (2) provide estimates of leaf damage for different leaf age classes (young, mature, old) during the same two time periods which included the major new leaf production events of *E. tereticornis* [18], and (3) determine whether specific leaf age classes were preferred by insect herbivores and if this preference was altered under e[CO_2] during the first 2 years of EucFACE.

Methods
Study site
This study was conducted at the *Eucalyptus* free-air CO_2 enrichment (EucFACE) experiment located within a native Cumberland Plain woodland remnant [41, 42] in Richmond, NSW, Australia (33°37'S, 150°44'E). The vegetation at EucFACE has been undisturbed for at least 75 years and retains old-growth trees mixed with some re-growth. The vegetation community within the study site is characterised as Cumberland Shale Plains Woodland [41], with mature *E. tereticornis* as the only canopy forming tree species. The site has an open canopy, approximately 600 trees ha^{-1} [43], with a low density of forbs and occasional shrubs in the understorey, together with a diverse community of grasses. The site is on a loamy sand soil of the Richmond Formation [44], which is phosphorus-poor and limits tree growth at the site [45]. The average monthly temperature at the site during the time period of this study was 20 °C with an average monthly rainfall of 73 mm (Additional file 1: Fig. S1).

Six large 25 m diameter rings with a height of 28 m above ground, extending above the tree canopy, were constructed amongst the vegetation of the site. Adjacent to each ring stands a high canopy crane with a person basket that allows access to the canopy from above. Each ring also contains a central scaffold tower. Three rings were fumigated diurnally with CO_2 enriched air via a proportional-integral-derivative control algorithm [46], while the three remaining rings were control rings, fumigated with ambient air. Beginning in September 2012, the target [CO_2] in treatment rings was increased by 30 µmol mol^{-1} every month until February 2013; thereafter diurnal [CO_2] targets within the treatment rings were 150 µmol mol^{-1} above ambient levels of ~400 µmol mol^{-1}. Rainfall was recorded using automated tipping bucket gauges (Tipping Bucket Rain gauge TB4, Hydrological Services Pty Ltd, Liverpool, NSW, Australia) located 23.5 m above the ground on the central tower in three rings. Data from these sensors were logged every 15 min using CR3000 data loggers (Campbell Scientific, Townsville, Australia).

Leaf herbivory and leaf production measurement
From each of the six EucFACE rings three trees were randomly selected and marked in the first year, and a different set of three trees per ring was selected and marked in the second year. The upper canopy (approximately 17 m above-ground) was accessed using the canopy cranes to establish herbivory observation points on each selected tree. For each year, 14 branches per tree were tagged. We expected that the herbivory measured in the upper canopy was representative for the entire tree canopy because it had previously been demonstrated

that *Eucalyptus* trees display a homogeneous pattern of herbivory throughout the crown [47]. Leaves on each branch were numbered sequentially from the base to the proximal end. A black permanent marker was used to mark the abaxial leaf surface near the petiole, and this has previously been demonstrated not to alter leaf formation or herbivory [48]. New leaves were marked as they emerged behind the shoot tip. Branches and leaves were initially marked in October 2012 and branches were then monitored monthly until February 2013 during the CO_2 ramp-up phase (year 1). Three different trees per ring were selected, marked and observed monthly in the second monitoring period from August 2013 to March 2014 (year 2).

The surveyed periods coincided with the majority of chewing insect herbivore activity, as measured by frass deposition to the woodland floor [11] and the growth period of *E. tereticornis* during the austral spring and summer as measured by changes in LAI [18]. For each of the 2 years, the initial measurements of leaf area in October 2012 and August 2013 were used as a baseline to measure subsequent leaf consumption. For this purpose, approximately 100 leaves across the 14 branches per tree were marked for the monitoring throughout the consecutive months. Branches were selected for ease of access with crane and away from scaffolding to reduce risk of mechanical leaf damage or loss. Branches were then surveyed once each month for a period of four months (year 1) and seven months (year 2) and assessed for leaf damage that can be attributed to insect herbivores due to feeding marks and new leaf emergence. Leaf damage due to insect herbivores was recorded in two different ways: firstly leaves were classified into three age classes (see below) and monthly leaf damage was measured as damage within each of these age classes following the formula below; secondly cumulative leaf damage was calculated for individual leaves throughout their development during the two observation periods in year 1 and year 2 of EucFACE. Thus, cumulative leaf damage refers to the total amount of damage occurring within each monitoring period and not over the life of a particular leaf.

Based on size, colour, shape and texture, leaves were assigned to one of three age classes for each canopy survey point: young (new expanding leaves), mature (fully expanded) and old leaves [48]. Age class-specific herbivory was then calculated as average damage to leaves of each age class throughout the observation periods of each year. For each survey month, scaled digital photographs were taken of leaves that were still attached to branches. For this purpose, leaves were flattened between a scaled white board and a clear non-reflective plastic sheet [49]. Photographs from each month were then compared to photographs taken in the previous month.

For each month, existing leaf area (LA$_e$) was quantified using Adobe Photoshop CS5 (Adobe Systems Incorporated, California, USA) by manually tracing the leaf using scaled photographs. Potential leaf area (LA$_p$) i.e. the extent of the leaf area if herbivory had not occurred, was determined by manually drawing and digitally reconstructing the leaf [50]. Skeletonising, mining, and leaf rolling were rarely observed on marked leaves, and therefore disregarded in this study. The recorded damage was exclusively due to removal of leaf area by chewing insects. Monthly increments of leaf consumption were determined with the formula [48]:

$$Lc_{(n+1)} = \left(1 - \left(\left(\frac{LA_e}{LA_p}\right) - Lc_n\right)\right) * 100$$

where $Lc_{(n+1)}$ is the proportion (% missing) of leaf area consumed within the observation period of one month; LA_e is the actual leaf area recorded for that month; LA_p is the potential leaf area if herbivory had not occurred; and Lc_n is the proportion of herbivory that had occurred in the previous month.

Total leaf consumption at the end of each monitoring period in years 1 and 2 was determined by the sum of the monthly leaf consumption for each respective leaf age class per branch per tree for the monitored period. Loss of entire leaves due to herbivory was distinguished from leaf loss that may occur as a consequence of leaf senescence and abscission, or due to wind. A leaf that had completely disappeared was considered as lost due to herbivory if it had signs of herbivory in the previous month, or if an entire leaf that was undamaged in the previous month had disappeared except for its petiole. A leaf without signs of herbivory in the previous month was considered lost due to senescence and abscission, or due to wind, if it had disappeared together with its petiole. In this case no value of herbivory was assigned. This approach to assign complete loss of leaves due to herbivory is a conservative measure as it only covers known herbivory. Leaf production was determined by the average number of young new expanding leaves present per branch divided by the total number of leaves per branch and averaged per tree.

Statistical analysis

Linear mixed effects models were constructed using *nlme* [51] in R [v3.2.2, 52]. The fixed model contained [CO$_2$], month and their interaction. The random model included ring with tree as a nested factor to account for repeated measures. An autocorrelation function was used in order to test for temporal autocorrelation within years, and an autoregressive moving average (ARMA) correlation structure was employed to model dependence among observations of leaf consumption and leaf production across months using a first-order autoregressive structure (AR1) [53]. The number of expanding leaves present was log + 1 transformed to normalise the model-standardised residuals. The relationships of monthly leaf damage with both the average number of young leaves per branch and with rainfall, were modelled using linear mixed effects models and R^2 values were obtained using the *r.squaredGLMM* function in the *MuMIn* R package [54, 55].

Results

Approximately 3000 *E. tereticornis* leaves were measured for herbivory in each year across three different age classes. Leaf age classes had a significant effect on leaf consumption both in year 1 ($F_{2,30} = 245.654$, $P < 0.001$; Fig. 1a) and in year 2 ($F_{2,30} = 286.435$, $P < 0.001$; Fig. 1b). Young leaves incurred approximately ten times more leaf damage than either mature or old leaves (averages ranged between 25 and 32 % loss in leaf area for young leaves versus less than 2 % loss for mature or old leaves) in both years (Fig. 1). No significant CO$_2$ treatment effect on leaf consumption was observed across leaf age classes (year 1: $F_{1,4} = 0.399$, $P = 0.562$; year 2: $F_{1,4} = 0.042$, $P = 0.848$; Fig. 1).

Significant temporal variation in the amount of leaf consumption during the monitoring periods was observed in both years (year 1: $F_{3,48} = 10.108$, $P < 0.001$; year 2: $F_{6,92} = 30.998$, $P < 0.001$; Fig. 2). Monthly leaf consumption peaked in December in the first year and in January in the second year. No significant difference in monthly leaf consumption was observed between CO$_2$ treatments in either year (year 1: $F_{1,4} = 3.992$, $P = 0.116$; year 2: $F_{1,4} = 0.028$, $P = 0.876$; Fig. 2). Total cumulative leaf consumption observed during the monitoring periods did not differ between CO$_2$ treatments (year 1: $F_{1,4} = 6.341$, $P = 0.066$; year 2: $F_{1,4} = 1.681$, $P = 0.265$; Table 1). The loss of young leaf production between the two years was nearing significance, with less young leaf production being lost in the second year ($F_{1,4} = $, $P = 0.051$; Table 1).

The average number of young leaves present per branch in the first year peaked in November ($F_{3,48} = 21.999$, $P < 0.001$; Fig. 2c) while production of young leaves was highest in January and February of year 2 ($F_{6,92} = 68.570$, $P < 0.001$; Fig. 2d). No significant differences in young expanding leaf production were observed between CO$_2$ treatments in either year (year 1: $F_{1,4} = 7.075$, $P = 0.056$; year 2 $F_{1,4} = 0.114$, $P = 0.753$). Furthermore, e[CO$_2$] did not alter the timing of flush production between year 1 and 2 ($F_{1,167} = 3.004$, $P = 0.085$). Strong positive correlations were observed between the number of expanding leaves present and leaf consumption (R$^2 = 0.303$, d.f. $= 161$, $P < 0.001$; Fig. 3a). Rainfall that occurred two

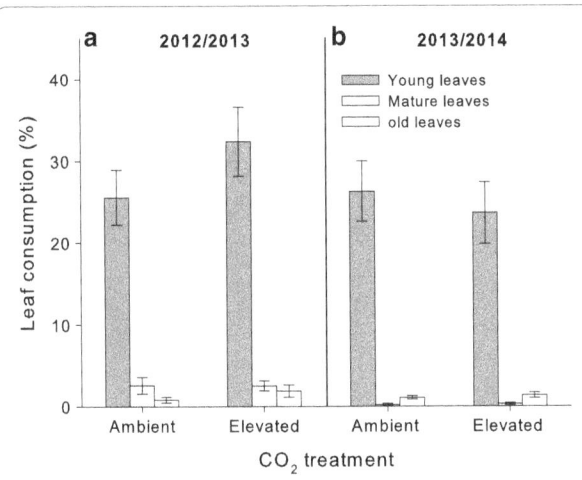

Fig. 1 Leaf consumption during three different leaf age classes; young (expanding leaves, *dark grey bars*), mature (fully expanded, *light grey bars*), and old leaves (*open bars*) of mature *Eucalyptus tereticornis* trees grown at ambient or elevated [CO$_2$] across two time periods, in year 1 (**a**) and year 2 (**b**). The *figure inset* indicates leaf age

months prior to the monitoring data points was positively correlated with leaf production (R^2 = 0.317, d.f. = 161, $P < 0.001$; Fig. 3b). Temperature was also positively correlated with leaf production (R^2 = 0.137, d.f. = 161, $P < 0.001$; Fig. 3c).

The number of new young leaves produced by trees and lost to herbivory did not differ as a result of CO$_2$ treatment in either year (year 1: $F_{1,4}$ = 0.385, P = 0.569; year 2: $F_{1,4}$ = 0.148, P = 0.720; Table 1). Overall, approximately 37 % of new young leaves produced by trees escaped any form of chewing damage (Table 1). CO$_2$ treatments did not affect this percentage of young leaves escaping any form of chewing damage in either years (year 1: $F_{1,4}$ = 0.043, P = 0.846; year 2: $F_{1,4}$ = 1.060, P = 0.361).

Discussion

Herbivory by chewing insects was measured over the first 2 years of the EucFACE experiment. In the first year, e[CO$_2$] was gradually increased from ambient conditions to +150 μmol mol^{-1}, while in the second year, e[CO$_2$] was maintained at 550 μmol mol^{-1}. The two monitoring periods included two major leaf production events, one in each year during spring and summer. Independent of [CO$_2$], consumption of young expanding leaves was very high and decreased to very low levels once leaves were fully expanded. The drastically higher levels of herbivory on young leaves drove monthly variations in overall leaf damage by insect herbivores due to the variation in leaf production. Furthermore, e[CO$_2$] did not affect damage due to herbivory and total leaf consumption. This was in line with our expectation that we would not detect

compensatory feeding because of a previous study in which we detected that concentrations of foliar nitrogen and total phenolics at the EucFACE site were not affected by CO$_2$ fumigation over the first 2 years of EucFACE [11]. Such a lack in plant responses may be due to the capacity of mature trees to compensate short term e[CO$_2$] exposure by retrieving nutrient reserves [11].

We found that herbivory by leaf chewing insects increased with the production of new young leaves suggesting that insect feeding is synchronised with the emergence of new young leaves. Insects generally prefer young expanding leaves over mature and old leaves as a result of higher nutrient content and lower physical defences in young leaves, in particular in sclerophyllous *Eucalyptus* leaves [27, 36, 48, 56]. We have previously demonstrated that foliar nitrogen concentration was higher in flush growth than mature leaves at EucFACE, independent of [CO$_2$] [11]. Herbivory decreased to very low levels in fully expanded leaves, suggesting that leaf flush chewers rather than senescent leaf chewers were the dominant feeding guild during the monitoring periods of this study. Frass production over the same time period [11] displayed a similar pattern of increased deposition around the periods of new leaf production [18]. This highlights the importance of new leaf production not only to insect herbivores but also for insect-mediated nutrient cycling within forests [11].

Herbivory on *E. tereticornis* in a native woodland observed in this study is within the range observed in other Australian systems. For example, in an Australian rainforest insect herbivore damage on young expanding leaves ranged between 10 and 30 % depending on tree species, and this fell to less than 5 % once leaves matured [48]. Similarly, Moles and Westoby [57] reported that leaf damage to expanding leaves from 51 woody dicotyledonous species in a coastal dry sclerophyll forest ranged between 0 and 51 %. Irrespective of leaf age, reported total leaf herbivore damage in Australian forests ranges between 5 and 44 % [23, 47, 58, 59]. Overall, at EucFACE we observed total herbivore consumption of between 9 and 17 % of leaf area, which is in the lower range of previous studies conducted in Australian forests. No leaf mining or leaf rolling was observed; therefore the damage measured in this study was exclusively due to leaf chewing insects. This is often the most common type of leaf damage observed in forests, when compared to other insect feeding types [59–61]. The feeding guild of leaf chewers is also the most likely impacted by e[CO$_2$] [7]. A previous study of herbivory on *E. tereticornis* at the study site also identified leaf chewers as the dominant feeding guild [62].

The responses to eCO$_2$ of two key leaf chewing insect herbivores found at the EucFACE site had previously

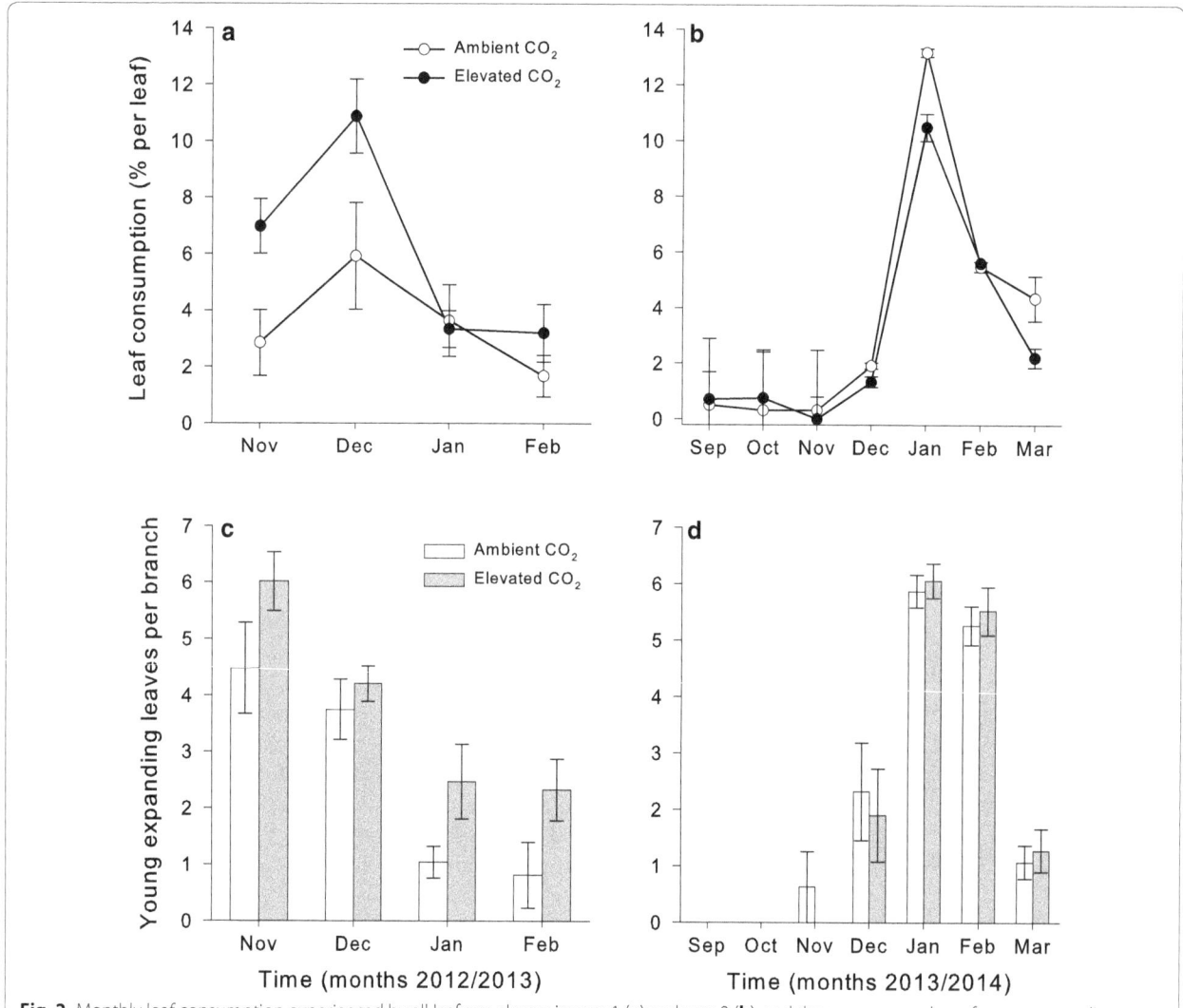

Fig. 2 Monthly leaf consumption experienced by all leaf age classes in year 1 (**a**) and year 2 (**b**), and the average number of young expanding leaves observed per branch in year 1 (**c**) and year 2 (**d**) on mature *Eucalyptus tereticornis* trees exposed to ambient (*open circles* or *bars*) or elevated (*closed circles* or *bars*) [CO_2] at the EucFACE site. The *figure insets* indicate CO_2 treatment

Table 1 Mean percentage (±SE) of total cumulative leaf damage, young leaf production completely lost to herbivory and young leaves which remained undamaged during the expansion stage on mature *E. tereticornis* under ambient or elevated [CO_2] at the EucFACE site over 2 years

CO_2 treatment	Year 1		Year 2	
	Ambient	Elevated	Ambient	Elevated
Total cumulative leaf consumption (%)	9.9 ± 1.7	16.6 ± 1.1	13.5 ± 2.5	9.0 ± 0.7
Young leaf production completely lost to herbivory (%)	21.0 ± 5.6	17.3 ± 3.5	9.7 ± 4.3	8.0 ± 3.7
Young leaf production remaining undamaged (%)	39.3 ± 5.9	27.5 ± 3.5	43.7 ± 3.7	38.0 ± 4.2
Young leaf production damaged (%)	39.7 ± 4.4	55.2 ± 3.6	48.7 ± 3.8	54.0 ± 4.4

The value of young leaf production that was damaged is complementary to the lost and undamaged new leaf production values

been tested on young *E. tereticornis* trees in greenhouse experiments. The cup moth *Doratifera quadriguttata* (Lepidoptera: Limacodidae) and the leaf beetle *Paropsis* *atomaria* (Coleoptera: Chrysomelidae), both experienced negative effects of e[CO_2] such as increased mortality, extended developmental times and reduced pupal

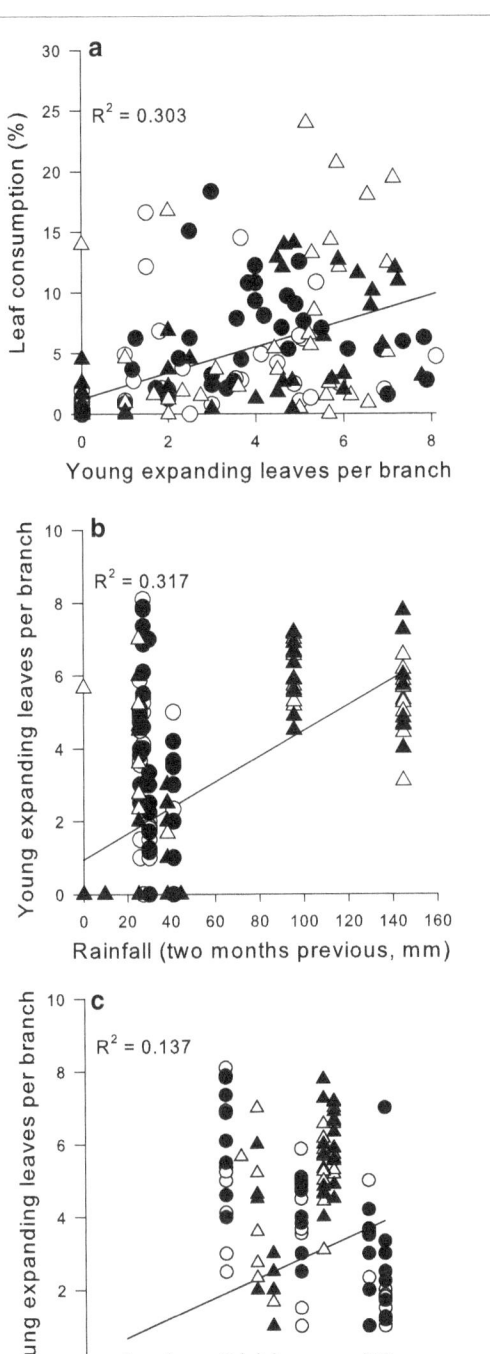

Fig. 3 Linear mixed effect model regression of young leaf production per branch and monthly leaf consumption (**a**), rainfall from two months prior and leaf production per branch (**b**), and temperature and leaf production per branch (**c**) observed on *Eucalyptus tereticornis* trees at the EucFACE site under ambient (*open symbols*) or elevated (*filled symbols*) [CO_2] in year 1 (*circles*) and year 2 (*triangles*)

weights, while both species also displayed signs of compensatory feeding [36–38]. However, we did not observe any differences in the amount of leaf damage occurring to mature *E. tereticornis* trees at EucFACE under CO_2 enrichment. In other forests undergoing CO_2 fumigation e[CO_2] has been observed to either decrease [63, 64], cause no change [65], or increase [66, 67] the amount of leaf damage. These differences in herbivory responses to e[CO_2] across a variety of forest types may be due to the complexity in interactions between biotic and abiotic factors which impact insect herbivores and therefore herbivory.

The different timing of the major flush production events between the 2 years at EucFACE may have affected the composition of herbivore communities within the woodland. Shifts in the timing of flush growth may have detrimental effects on herbivores that depend on flush with the potential outcome of an altered insect herbivore community structure. Although we did not directly assess insect populations our data indirectly suggests that such potential changes in chewing insect herbivore populations may have occurred. It appeared that less of the newly produced leaves were consumed in the second year, potentially due to the altered timing of flush growth. This may also indicate that there was a shift away from flush leaf chewers. In another study that assessed deposition of lerp (small covers produced by plant sap-feeding psyllids) to the woodland floor, a rapid outbreak succession by two psyllid species, *Glycaspis* sp. and *Cardiaspina fiscella* was detected on *E. tereticornis* after March 2014 [68]. This suggests that a rapid change in the overall canopy insect community composition must have occurred directly after the leaf phenology and herbivory surveys presented here. Our previous study on frass deposition by leaf chewing insects also demonstrated that less frass was produced in 2013/2014 (year 2) than in 2012/2013 (year 1) [11].

Our leaf herbivory study only focussed on leaf-chewing insects, and sap-feeding insects were not considered here. This focus on the leaf-chewing feeding guild may underestimate the impacts of herbivory occurring in the canopy and the true level of canopy biomass lost to insect herbivores in our study site. Significant numbers of plant sap-feeding psyllids were detected at EucFACE after March 2014 [68]. Some psyllid species can be significant leaf defoliators of *Eucalyptus*. For example, *Cardiaspina* sp. has caused area-wide defoliation on *Eucalyptus moluccana* but not any other *Eucalyptus* species in the Cumberland Plain Woodlands since 2009 [69]. Similarly, *C. fiscella* caused significant defoliation of the EucFACE site at the end of 2014 [68], after the end of the study presented here. This defoliation occurred because *Cardiaspina* can induce leaf senescence and defoliation [34]. However, throughout the surveying period of our herbivory study at EucFACE (from October 2012 until March 2014) we did not observe any high abundance of leaf defoliating psyllids [68]. After the ramp-up phase (from March 2013 to December 2014) the abundance of lerps deposited to the woodland floor was reduced for three psyllid species at e[CO_2], and signs for compensatory feeding were detected in one of these three species, *Glycaspis* sp. [68]. These differences were recorded despite the absence of any measureable differences in N concentration in leaves [11].

The loss of young expanding leaves, as found in our study, may be more detrimental to plants than the loss of mature leaves, because the energy and resources invested in the production of new leaves have yet to be recovered [48]. This may impact forest growth and the estimation of carbon storage potential within forests and highlights the need to consider the role of herbivorous insects in ecosystem functioning [66]. Removal of leaf material by insect herbivores is often not accounted for in models of plant productivity as common techniques used to measure productivity, such as LAI [70, 71], fail to account for the loss of newly produced plant material to insect herbivory. This can complicate estimates of NPP and may underestimate true productivity of forests and utilisation of forest resources. Furthermore, plant NPP often increases at e[CO_2] as a result of a carbon fertilisation effect [72, 73]. However, in the early stages of CO_2 fumigation at the EucFACE site we did not observe an increase in leaf production. This may indicate that mature *E. tereticornis* trees within the site are limited by water and nutrient availability [18, 45], and increased photosynthesis due to e[CO_2] may also result in increased respiration in the ecosystem [43] rather than the production of more leaves.

Conclusions

Predictions about damage inflicted by insect herbivores at e[CO_2] are difficult to make under field conditions owing to the complex interactions between plants, insect herbivores and their antagonists. This uncertainty in insect herbivore responses hinders the ability to accurately determine model-specifications of their response to climate change [74]. Contrary to our original expectations, CO_2 fumigation at the EucFACE site for the first 2 years did not affect total leaf consumption by herbivores or their leaf age preference, in part likely due to the lack of e[CO_2] responses in concentrations of foliar nitrogen and total phenolics in insect frass [11]. However, it is clear that new and expanding leaves were

heavily damaged while fully expanded leaves were not. Damage on young foliage is often not accounted for in estimates of forest productivity, yet this can amount to substantial underestimates of true forest productivity. Rainfall-mediated production of new leaves is an important regulator of insect herbivory in sclerophyllous forests due to the physical barriers to consumption present in mature leaves [75], and this will require further attention in climate change studies. Shifts in rainfall patterns, a potential outcome of climate change [76], can have significant effects on insect community composition, herbivory and insect frass deposition. This may have detrimental or positive outcomes for ecosystems and humans, for example by stimulating pest populations, or regulating ecosystem services that insects provide in forests and managed plantations. Understanding how rainfall may interact with e[CO_2] in altering insect herbivory and herbivore abundances is important in predicting the impacts of climate change variables on insect herbivore population dynamics in forests and plantations.

Abbreviations

e[CO_2]: elevated carbon dioxide concentrations; EucFACE: *Eucalyptus* free-air CO_2 enrichment; LA$_e$: existing leaf area; LAI: leaf area index; LA$_p$: potential leaf area; Lc_n: proportion of herbivory that had occurred in the previous month; NPP: net primary production.

Authors' contributions

ANG and MR conceived and designed the experiments with input by BDM, AMH and SNJ. ANG performed the experiments. ANG analysed the data. ANG and MR primarily wrote the manuscript; other authors helped with drafting the text and interpreting the data. All authors have read and approved the final manuscript.

Author details

[1] Hawkesbury Institute for the Environment, Western Sydney University, Locked Bag 1797, Penrith, NSW 2751, Australia. [2] School of Science and Health, Western Sydney University, Locked Bag 1797, Penrith, NSW 2751, Australia.

Acknowledgements

We thank Vinod Kumar and Steven Wohl for technical assistance at the EucFACE site.

Competing interests

The authors declare that they have no competing interests.

Funding

EucFACE is supported by the Australian Commonwealth Government in collaboration with Western Sydney University. EucFACE was built as an initiative of the Australian Government as part of the Nation-building Economic Stimulus Package. This research was also supported by an Australian Postgraduate Award to ANG and DP1095972 of the Australian Research Council. The funding body had no role in the design of the study, collection, analysis, interpretation of the data and in writing of the manuscript.

References

1. Cornelissen T. Climate change and its effects on terrestrial insects and herbivory patterns. Neotrop Entomol. 2011;40(2):155–63.
2. Tylianakis JM, Didham RK, Bascompte J, Wardle DA. Global change and species interactions in terrestrial ecosystems. Ecol Lett. 2008;11(12):1351–63.
3. Buckley LB, Kingsolver JG. The demographic impacts of shifts in climate means and extremes on alpine butterflies. Funct Ecol. 2012;26(4):969–77.
4. Ainsworth EA, Rogers A. The response of photosynthesis and stomatal conductance to rising [CO_2]: mechanisms and environmental interactions. Plant Cell Environ. 2007;30(3):258–70.
5. Robinson EA, Ryan GD, Newman JA. A meta-analytical review of the effects of elevated CO_2 on plant–arthropod interactions highlights the importance of interacting environmental and biological variables. New Phytol. 2012;194(2):321–36.
6. Hunter MD. Effects of elevated atmospheric carbon dioxide on insect–plant interactions. Agric For Entomol. 2001;3(3):153–9.
7. Stiling P, Cornelissen T. How does elevated carbon dioxide (CO_2) affect plant–herbivore interactions? A field experiment and meta-analysis of CO_2-mediated changes on plant chemistry and herbivore performance. Glob Chang Biol. 2007;13(9):1823–42.
8. Ryan GD, Rasmussen S, Newman JA. Global atmospheric change and trophic interactions: are there any general responses? In: Baluska F, Ninkovic V, editors. Plant communication from an ecological perspective. Berlin: Springer-Verlag; 2010. p. 179–214.
9. Zavala JA, Casteel CL, DeLucia EH, Berenbaum MR. Anthropogenic increase in carbon dioxide compromises plant defense against invasive insects. Proc Natl Acad Sci USA. 2008;105(13):5129–33.
10. DeLucia EH, Nabity PD, Zavala JA, Berenbaum MR. Climate change: resetting plant-insect interactions. Plant Physiol. 2012;160(4):1677–85.
11. Gherlenda AN, Crous KY, Moore BD, Haigh AM, Johnson SN, Riegler M. Precipitation, not CO_2 enrichment, drives insect herbivore frass deposition and subsequent nutrient dynamics in a mature Eucalyptus woodland. Plant Soil. 2016;369:29–39.
12. Whitfield TJS, Novotny V, Miller SE, Hrcek J, Klimes P, Weiblen GD. Predicting tropical insect herbivore abundance from host plant traits and phylogeny. Ecology. 2012;93(sp8):S211–22.
13. Lee MA, Manning P, Walker CS, Power SA. Plant and arthropod community sensitivity to rainfall manipulation but not nitrogen enrichment in a successional grassland ecosystem. Oecologia. 2014;176(4):1173–85.
14. Norby RJ, DeLucia EH, Gielen B, Calfapietra C, Giardina CP, King JS, Ledford J, McCarthy HR, Moore DJP, Ceulemans R. Forest response to elevated CO_2 is conserved across a broad range of productivity. Proc Natl Acad Sci USA. 2005;102(50):18052–6.
15. Zak DR, Pregitzer KS, Kubiske ME, Burton AJ. Forest productivity under elevated CO_2 and O_3: positive feedbacks to soil N cycling sustain decade-long net primary productivity enhancement by CO_2. Ecol Lett. 2011;14(12):1220–6.
16. Norby RJ, Warren JM, Iversen CM, Medlyn BE, McMurtrie RE. CO_2 enhancement of forest productivity constrained by limited nitrogen availability. Proc Natl Acad Sci USA. 2010;107(45):19368–73.
17. Johnson DW. Progressive N limitation in forests: review and implications for long-term responses to elevated CO_2. Ecology. 2006;87(1):64–75.
18. Duursma RA, Gimeno TE, Boer MM, Crous KY, Tjoelker MG, Ellsworth DS. Canopy leaf area of a mature evergreen Eucalyptus woodland does not respond to elevated atmospheric [CO_2] but tracks water availability. Glob Chang Biol. 2016;22:1666–76.
19. Paine TD, Steinbauer MJ, Lawson SA. Native and exotic pests of Eucalyptus: a worldwide perspective. Annu Rev Entomol. 2011;56:181–201.
20. FAO. Global forest resources assessment 2010. Rome: Food and Agriculture Organization of the United Nations; 2010.
21. Ohmart C, Edwards P. Insect herbivory on Eucalyptus. Annu Rev Entomol. 1991;36(1):637–57.
22. Keith DA. Ocean shores to desert dunes: the native vegetation of New South Wales and the ACT. Hurstville: New South Wales Department of Environment and Conservation; 2004.
23. Landsberg J, Ohmart C. Levels of insect defoliation in forests: patterns and concepts. Trends Ecol Evol. 1989;4(4):96–100.
24. Loney P, McArthur C, Sanson G, Davies N, Close D, Jordan G. How do soil nutrients affect within-plant patterns of herbivory in seedlings of Eucalyptus nitens? Oecologia. 2006;150(3):409–20.
25. Close DC, McArthur C, Hagerman AE, Fitzgerald H. Differential distribution of leaf chemistry in eucalypt seedlings due to variation in whole-plant nutrient availability. Phytochemistry. 2005;66(2):215–21.
26. Rapley LP, Allen GR, Potts BM. Genetic variation in Eucalyptus globulus in relation to susceptibility from attack by the southern eucalypt leaf beetle. Chrysophtharta agricola. Aust J Bot. 2004;52(6):747–56.
27. Nahrung HF, Duffy MP, Lawson SA, Clarke AR. Natural enemies of Paropsis atomaria Olivier (Coleoptera: Chrysomelidae) in south-eastern Queensland eucalypt plantations. Aust J Ecol. 2008;47(3):188–94.
28. Boege K, Marquis RJ. Facing herbivory as you grow up: the ontogeny of resistance in plants. Trends Ecol Evol. 2005;20(8):441–8.
29. McKiernan AB, Hovenden MJ, Brodribb TJ, Potts BM, Davies NW, O'Reilly-Wapstra JM. Effect of limited water availability on foliar plant secondary metabolites of two Eucalyptus species. Environ Exp Bot. 2014;105:55–64.
30. Moore BD, Wallis IR, Wood JT, Foley WJ. Foliar nutrition, site quality, and temperature influence foliar chemistry of tallowwood (Eucalyptus microcorys). Ecol Monogr. 2004;74(4):553–68.
31. Silvestre AJD, Cavaleiro JAS, Delmond B, Filliatre C, Bourgeois G. Analysis of the variation of the essential oil composition of Eucalyptus globulus Labill. from Portugal using multivariate statistical analysis. Ind Crops Prod. 1997;6(1):27–33.
32. Goodger JQD, Choo TYS, Woodrow IE. Ontogenetic and temporal trajectories of chemical defence in a cyanogenic eucalypt. Oecologia. 2007;153(4):799–808.
33. Henery ML, Wallis IR, Stone C, Foley WJ. Methyl jasmonate does not induce changes in Eucalyptus grandis leaves that alter the effect of constitutive defences on larvae of a specialist herbivore. Oecologia. 2008;156(4):847–59.
34. Steinbauer MJ, Burns AE, Hall A, Riegler M, Taylor GS. Nutritional enhancement of leaves by a psyllid through senescence-like processes: insect manipulation or plant defence? Oecologia. 2014;176(4):1061–74.
35. Coley PD. Possible effects of climate change on plant/herbivore interactions in moist tropical forests. Clim Change. 1998;39(2–3):455–72.
36. Gherlenda AN, Haigh AM, Moore BD, Johnson SN, Riegler M. Responses of leaf beetle larvae to elevated [CO_2] and temperature depend on Eucalyptus species. Oecologia. 2015;177(2):607–17.
37. Murray TJ, Ellsworth DS, Tissue DT, Riegler M. Interactive direct and plant-mediated effects of elevated atmospheric [CO_2] and temperature on a eucalypt-feeding insect herbivore. Glob Chang Biol. 2013;19:1407–16.
38. Murray TJ, Tissue DT, Ellsworth DS, Riegler M. Interactive effects of pre-industrial, current and future [CO_2] and temperature on an insect herbivore of Eucalyptus. Oecologia. 2013;171(4):1025–35.
39. Myers BA, Williams RJ, Fordyce I, Duff GA, Eamus D. Does irrigation affect leaf phenology in deciduous and evergreen trees of the savannas of northern Australia? Aust J Ecol. 1998;23(4):329–39.
40. Pook EW, Gill AM, Moore PHR. Long-term variation of litter fall, canopy leaf area and flowering in a Eucalyptus maculata forest on the south coast of New South Wales. Aust J Bot. 1997;45(5):737–55.
41. Tozer M. The native vegetation of the Cumberland Plain, western Sydney: systematic classification and field identification of communities. Cunninghamia. 2003;8(1):1–75.
42. Benson DH. The natural vegetation of the Penrith 1:100 000 map sheet. Cunninghamia. 1992;2(4):541–96.
43. Drake JE, Macdonald CA, Tjoelker MG, Crous KY, Gimeno TE, Singh BK, Reich PB, Anderson IC, Ellsworth DS. Short-term carbon cycling responses of a mature eucalypt woodland to gradual stepwise enrichment of atmospheric CO2 concentration. Glob Chang Biol. 2015;22:380–90.
44. Bannerman SM, Hazelton PA. Soil landscapes of the Penrith 1:100,000 sheet map and report. Sydney: Soil Conservation Service of NSW; 1990.
45. Crous KY, Ósvaldsson A, Ellsworth DS. Is phosphorus limiting in a mature Eucalyptus woodland? Phosphorus fertilisation stimulates stem growth. Plant Soil. 2015;391:293–305.

46. Lewin KF, Nagy J, Nettles WR, Cooley DM, Rogers A. Comparison of gas use efficiency and treatment uniformity in a forest ecosystem exposed to elevated [CO_2] using pure and prediluted free-air CO_2 enrichment technology. Glob Chang Biol. 2009;15(2):388–95.

47. Lowman MD. Herbivory in Australian forests-a comparison of dry sclerophyll and rain forest canopies. Proc Linn Soc NSW. 1995;115:77–87.

48. Lowman MD. Temporal and spatial variability in insect grazing of the canopies of five Australian rainforest tree species. Aust J Ecol. 1985;10(1):7–24.

49. Paul GS, Montagnini F, Berlyn GP, Craven DJ, Van Breugel M, Hall JS. Foliar herbivory and leaf traits of five native tree species in a young plantation of Central Panama. New For. 2012;43(1):69–87.

50. Unsicker SB, Mody K. Influence of tree species and compass bearing on insect folivory of nine common tree species in the West African savanna. J Trop Ecol. 2005;21(2):227–31.

51. Pinheiro J, Bates D, DebRoy S, Sarkar D. *nlme*: Linear and nonlinear mixed effects models. R Package Version 3.1-122. 2015. https://cran.r-project.org/package=nlme. Accessed 19 Aug 2015.

52. R Development Core Team. R. A language and environment for statistical computing. Vienna: R Foundation for Statistical Computing. 2015. https://www.R-project.org. Accessed 3 July 2015.

53. Bader MKF, Leuzinger S, Keel SG, Siegwolf RTW, Hagedorn F, Schleppi P, Körner C. Central European hardwood trees in a high-CO_2 future: synthesis of an 8 year forest canopy CO_2 enrichment project. J Ecol. 2013;101(6):1509–19.

54. Johnson PCD. Extension of Nakagawa and Schielzeth's R^2 GLMM to random slopes models. Methods Ecol Evol. 2014;5(9):944–6.

55. Barton K. MuMIn: Multi-Model Inference. R package version 1.15.1. 2015. https://cran.r-project/package=MuMIn. Accessed 3 July 2015.

56. Nahrung HF, Dunstan PK, Allen GR. Larval gregariousness and neonate establishment of the eucalypt-feeding beetle *Chrysophtharta agricola* (Coleoptera: Chrysomelidae: Paropsini). Oikos. 2001;94(2):358–64.

57. Moles AT, Westoby M. Do small leaves expand faster than large leaves, and do shorter expansion times reduce herbivore damage? Oikos. 2000;90(3):517–24.

58. Fox LR, Morrow PA. Estimates of damage by herbivorous insects on *Eucalyptus* trees. Aust J Ecol. 1983;8(2):139–47.

59. Nooten SS, Hughes L. Patterns of insect herbivory on four Australian understory plant species. Aust J Ecol. 2013;52(4):309–14.

60. Landsberg J, Gillieson DS. Regional and local variation in insect herbivory, vegetation and soils of eucalypt associations in contrasted landscape positions along a climatic gradient. Aust J Ecol. 1995;20(2):299–315.

61. Adams JM, Zhang Y. Is there more insect folivory in warmer temperate climates? A latitudinal comparison of insect folivory in eastern North America. J Ecol. 2009;97(5):933–40.

62. Lopaticki G. The response of eucalypt arthropod communities to varying resource availability and habitat complexity BSc Honours thesis. Penrith: University of Western Sydney; 2010.

63. Knepp RG, Hamilton JG, Mohan JE, Zangerl AR, Berenbaum MR, DeLucia EH. Elevated CO_2 reduces leaf damage by insect herbivores in a forest community. New Phytol. 2005;167(1):207–18.

64. Stiling P, Cattell M, Moon DC, Rossi A, Hungate BA, Hymus G, Drake B. Elevated atmospheric CO_2 lowers herbivore abundance, but increases leaf abscission rates. Glob Chang Biol. 2002;8(7):658–67.

65. Hamilton JG, Zangerl AR, Berenbaum MR, Pippen J, Aldea M, DeLucia EH. Insect herbivory in an intact forest understory under experimental CO_2 enrichment. Oecologia. 2004;138(4):566–73.

66. Couture JJ, Meehan TD, Kruger EL, Lindroth RL. Insect herbivory alters impact of atmospheric change on northern temperate forests. Nature Plants. 2015;1(3):150–60.

67. Meehan TD, Couture JJ, Bennett AE, Lindroth RL. Herbivore-mediated material fluxes in a northern deciduous forest under elevated carbon dioxide and ozone concentrations. New Phytol. 2014;204(2):397–407.

68. Gherlenda AN, Esveld JL, Hall AAG, Duursma RA, Riegler M. Boom and bust: rapid feedback responses between insect outbreak dynamics and canopy leaf area impacted by rainfall and CO_2. Glob Chang Biol. 2016. doi:10.1111/gcb.13334.

69. Hall AAG, Gherlenda AN, Hasegawa S, Johnson SN, Cook JM, Riegler M. Anatomy of an outbreak: the biology and population dynamics of a *Cardiaspina* psyllid species in an endangered woodland ecosystem. Agric For Entomol. 2015;17:292–301.

70. Reich PB. Key canopy traits drive forest productivity. Proc Biol Sci. 2012;279:2128–34.

71. Walker AP, Hanson PJ, De Kauwe MG, Medlyn BE, Zaehle S, Asao S, Dietze M, Hickler T, Huntingford C, Iversen CM. Comprehensive ecosystem model-data synthesis using multiple data sets at two temperate forest free-air CO_2 enrichment experiments: model performance at ambient CO_2 concentration. J Geophys Res: Biogeo. 2014;119(5):937–64.

72. Ainsworth EA, Long SP. What have we learned from 15 years of free-air CO_2 enrichment (FACE)? A meta-analytic review of the responses of photosynthesis, canopy properties and plant production to rising CO_2. New Phytol. 2005;165(2):351–72.

73. De Kauwe MG, Medlyn BE, Zaehle S, Walker AP, Dietze MC, Wang YP, Luo Y, Jain AK, El-Masri B, Hickler T. Where does the carbon go? A model-data intercomparison of vegetation carbon allocation and turnover processes at two temperate forest free-air CO_2 enrichment sites. New Phytol. 2014;203(3):883–99.

74. Boulanger Y, Gray DR, Cooke BJ, De Grandpré L. Model-specification uncertainty in future forest pest outbreak. Glob Chang Biol. 2016;22(4):1595–607.

75. Steinbauer MJ. Specific leaf weight as an indicator of juvenile leaf toughness in Tasmanian bluegum (*Eucalyptus globulus* ssp. *globulus*): implications for insect defoliation. Austral For. 2001;64(1):32–7.

76. IPCC (ed.). Climate change 2013: The physical science basis. Contribution of working group I to the fifth assessment report of the Intergovernmental Panel on Climate Change. Cambridge: Cambridge University Press; 2013.

77. Gherlenda AN, Moore BD, Haigh AM, Johnson SN, Riegler M. Data from: insect herbivory in a mature *Eucalyptus* woodland canopy depends on leaf phenology but not CO_2 enrichment. BMC Ecol. 2016. doi:10.4225/35/57e49ec6dd3eb.

9

Cold spell *en route* delays spring arrival and decreases apparent survival in a long-distance migratory songbird

Martins Briedis[1]* ⓘ, Steffen Hahn[2] and Peter Adamík[1,3]

Abstract

Background: Adjusting the timing of annual events to gradual changes in environmental conditions is necessary for population viability. However, adaptations to weather extremes are poorly documented in migratory species. Due to their vast seasonal movements, long-distance migrants face unique challenges in responding to changes as they rely on an endogenous circannual rhythm to cue the timing of their migration. Furthermore, the exact mechanisms that explain how environmental factors shape the migration schedules of long-distance migrants are often unknown.

Results: Here we show that long-distance migrating semi-collared flycatchers *Ficedula semitorquata* delayed the last phase of their spring migration and the population suffered low return rates to breeding sites while enduring a severe cold spell *en route*. We found that the onset of spring migration in Africa and the timing of Sahara crossing were consistent between early and late springs while the arrival at the breeding site depended on spring phenology at stopover areas in each particular year.

Conclusion: Understanding how environmental stimuli and endogenous circannual rhythms interact can improve predictions of the consequences of climate changes on migratory animals.

Keywords: Circannual rhythm, Climate change, Geolocator, Long-distance migrant, Phenology, Weather extremes

Background

Over the course of the 20th century, the Earth's near-surface temperature has increased, [1] and many species have advanced their phenology as a response to this climate warming [2]. Among those, various migratory birds have advanced their spring migration and breeding schedules [3], with stronger responses in short-distance compared to long-distance migrants [4].

Long-distance migrants spend the non-breeding period in the areas where they often have limited possibilities to assess the climatic conditions at their distant breeding grounds, thus limiting their ability to time the spring migration accordingly. Current theory suggests that long-distance migratory birds depend on endogenously controlled circannual rhythms to cue their spring migration

[5, 6]. Photoperiod and environmental factors may serve as *Zeitgeber* to fine-tune the timing of departure [7–9]. While the mechanisms regulating the onset of spring migration are not yet fully understood, even less is known about the processes modifying migration rates and decision making *en route* [10]. Thus, the specific factors that determine the observed advances in spring arrival of long-distance migrants remain unknown.

The understanding how animals respond to the changing environment is of special importance with respect to increasing frequency of extreme weather events [11]. Inability to respond to a rapidly changing environment can have severe consequences on population demography and viability. If long-distance migrants rely solely on endogenous signals to time the entire spring migration, this could result in suboptimal arrivals at the breeding sites, possibly leading to mismatches of food peak availability and food demand [12].

*Correspondence: martins.briedis@upol.cz
[1] Department of Zoology, Palacký University, tř. 17. listopadu 50, 77146 Olomouc, Czech Republic
Full list of author information is available at the end of the article

Here we examine how long-distance migrating semi-collared flycatchers *Ficedula semitorquata* respond to contrasting climatological conditions encountered in two consecutive spring migrations. Flycatchers' peak arrival period at their breeding range extends from the end of March to the beginning of April [13]. In Southeastern Europe in 2014, this period was the warmest on record since 2000, followed by an exceptional cold spring in 2015 with temperatures well below the long-term average (Fig. 1). Such extreme and opposing conditions present an ideal opportunity to study phenotypic plasticity in a natural setting. We were particularly interested to test whether this obligatory long-distance migrant is capable of adjusting its migration rate based on environmental cues *en route* to fine-tune arrival at the breeding site.

Methods

Study site and geolocators

Our study site is located in eastern Bulgaria (42°55′N, 27°48′E) approximately 8 km from the Black Sea coast at 120–150 masl. Habitat at the breeding site is oak woodland dominated by Hungarian oak *Quercus frainetto* with very little undergrowth. A population of approximately 100 pairs of semi-collared flycatchers breeds in nest boxes.

During the breeding season of 2013 and 2014 we equipped 40 (17 males, 23 females) and 49 (27 males, 22 females) adults with geolocators (GDL2.0, Swiss Ornithological Institute; weight including the harness: 0.6 g) which were fitted on birds' backs using elastic leg-loop silicone harnesses. The geolocators on average constituted 4.6 ± 0.3% (SD) of the bird's body mass. There was no difference in the average load of the geolocator

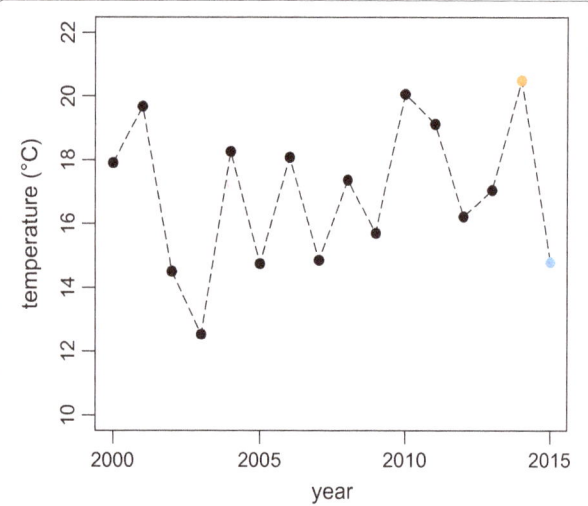

Fig. 1 Average land surface temperature in Bulgaria, Greece and Turkey from 22 March to 7 April from 2000 to 2015 [Data available from the U.S. Geological Survey (http://www.usgs.gov/)]

between the birds that returned and those that did not return (average ± SD; returned: 4.7 ± 0.3%, n = 18; not returned: 4.6 ± 0.3%, n = 71; t test: t_{87} = −1.19, p = 0.25).

We did an extensive recapturing of the tagged birds upon their arrival at the breeding site. Birds were captured using mist-nets and traps inside the nest boxes before the initiation of nest building. All adult breeders were captured later in the season when feeding nestlings, allowing for additional geolocator retrieval from the birds not captured earlier. In total we recovered 18 geolocators (2014: n = 11, 2015: n = 7); however, due to technical problems, we obtained spring migration data from only 5 [2 females, 3 males (1 incomplete)] and 6 (2 females, 4 males) devices in 2014 and 2015, respectively.

In addition, we acquired spring migration passage dates of flycatchers from the Antikythira Bird Observatory, Greece (35°51′N, 23°18′E, [14]) from 2007 to 2015.

Data analysis

We processed the light recording data using the R-package 'GeoLight' v2.0 [15], having determined sunrise and sunset times with 'Geolocator' software (Swiss Ornithological Institute) beforehand. We filtered the datasets for outlaying sun events using the 'loessFilter' function (k value = 2). We determined departure from the non-breeding site and arrival at the breeding site using the 'changeLight' function (probability of change q = 0.8). Minimum stationary period duration was set to 3 days. We determined Sahara crossing time according to the procedure described by Adamík et al. [16]. In short, during the Sahara crossing days geolocator's light sensor records uninterrupted maximal light intensities throughout the day, suggesting that birds cross the ecological barrier with a non-stop flight or at least prolonging the typical nocturnal flight for several hours into the following day. We adjusted the probability of change in the 'changeLight' function for each individual starting from q = 0.8, so that the function detects Sahara crossing time as a movement period. Annual timing of key migration phases are given as median date plus interquartile range (IQR) throughout.

To test for differences in apparent local survival rates between 2013–2014 and 2014–2015, we used a Chi squared goodness-of-fit test without Yates correction.

Weather data acquisition

We obtained land surface temperature data (data set: MOD11A2) and leaf area index (MOD15A2) data during the spring migration period (10 February–7 April) from MODIS terra and aqua satellites, accessed from the Land Processes Distributed Active Archive Center (LP DAAC) at the US Geological Survey (USGS) Earth

Resources Observation and Science (EROS) Center (https://lpdaac.usgs.gov/). We obtained wind data for the 850 mb pressure level (approximately 1500 masl) from the National Center for Environmental Prediction (NCEP)/National Center for Atmospheric Research (NCAR) Reanalysis dataset [17] using R-package 'RNCEP' [18]. Data were gathered across a 2.5° grid for every 6 h period in 2014, 2015 and annually averaged across the whole spring migration period (10 February–7 April). Winds at the 850 m bar pressure level are largely free of orographic distortion and, thus, are frequently used for describing wind patterns experienced by migratory birds [19].

Results

Weather patterns

The average land surface temperature during the spring migration period across Bulgaria, Greece and Turkey—countries on the species flyway—from 22 March–7 April was 20.5 °C in 2014, while in 2015 it was only 14.8 °C (Fig. 2a, b). This was the largest such difference in air temperature for over a decade (Fig. 1). Plant phenology, measured by leaf development, was delayed by approximately 29 days in 2015 compared to 2014 (Fig. 2c). Along other parts of the flycatchers' migratory flyway of the flycatchers, the prevailing winds and temperatures were similar between the two study years (Fig. 3).

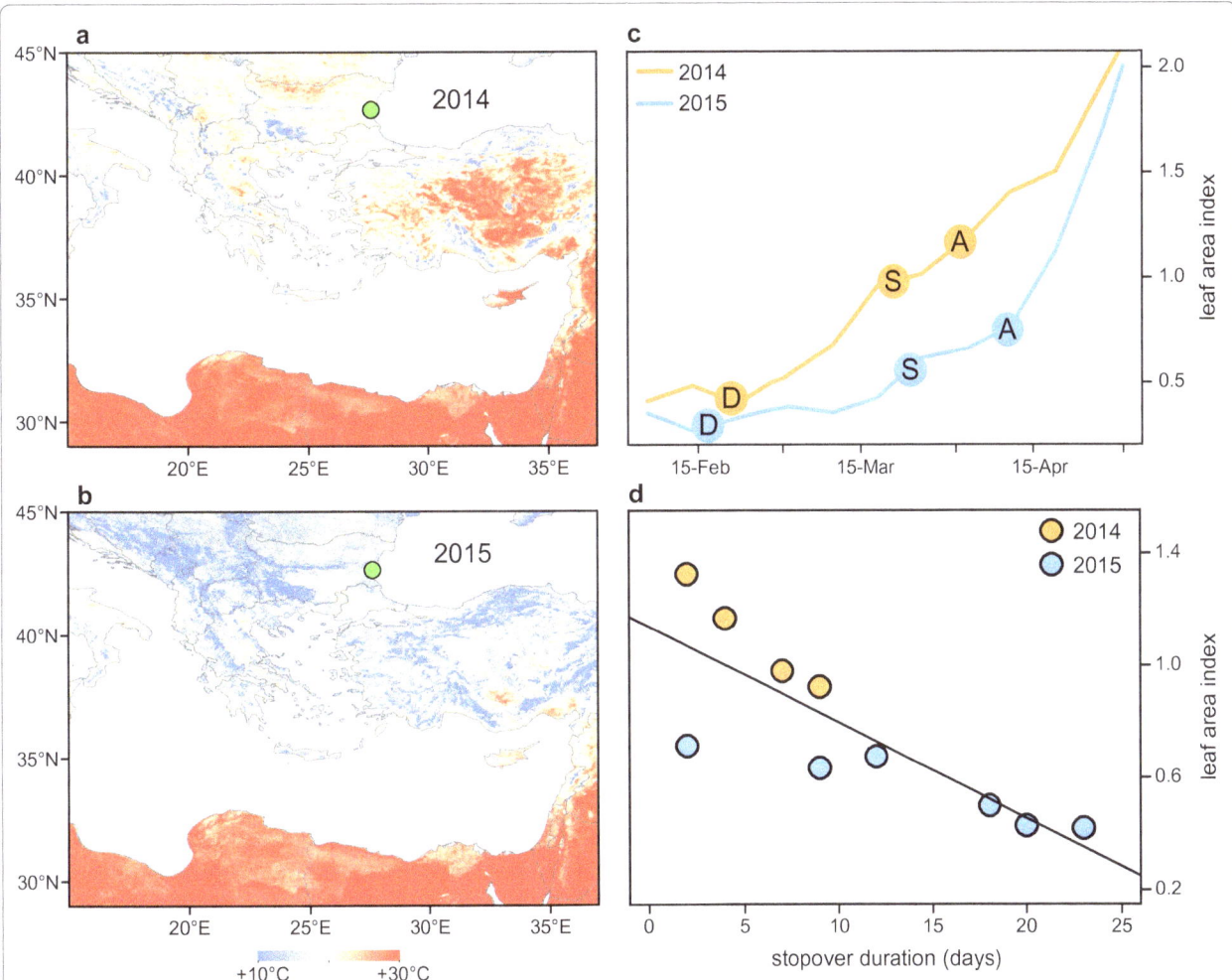

Fig. 2 Annual differences in weather conditions and the corresponding migration phenology of semi-collared flycatchers. Land surface temperatures (°C) from 22 March–7 April in **a** 2014 and **b** 2015. **c** Leaf area index (m² of leaf area per m² ground area) progression from 6 February–1 May at the flycatcher's breeding site in 2014 (*orange*) and 2015 (*blue*) and the related flycatcher migration phenology in each year, including (D) departure from the non-breeding site, (S) Sahara crossing, and (A) arrival at the breeding site. **d** Stopover duration north of the Sahara in relation to leaf area index at the breeding site at the time of Sahara crossing [The background maps in **a** and **b** made were made from data available from the U.S. Geological Survey (http://www.usgs.gov/); maps were created in ArcMap 10.1 (http://www.esri.com/)]

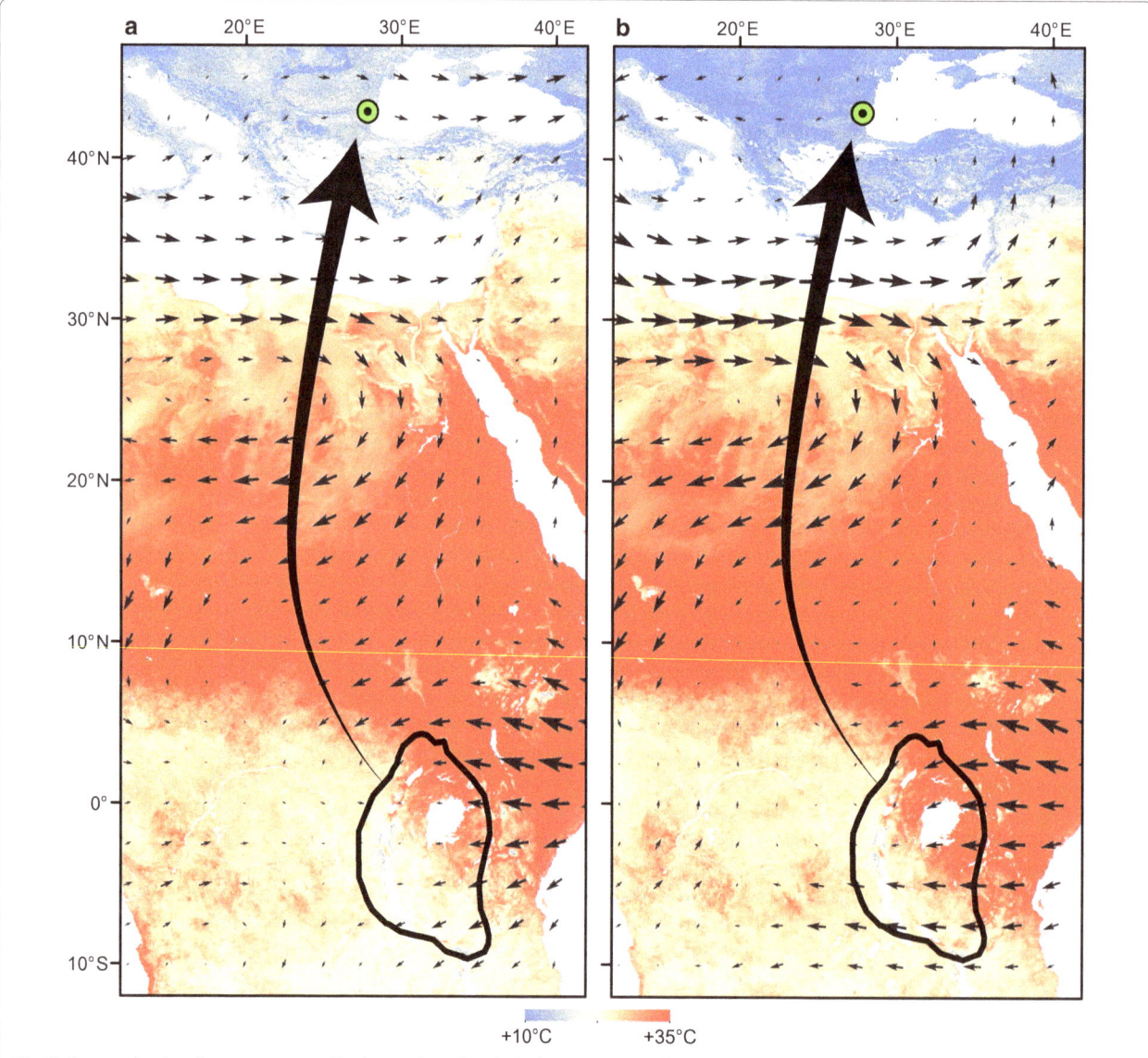

Fig. 3 Average land surface temperature (*background map*) and wind patterns at 850 kPa pressure level (*small arrows*) during semi-collared flycatcher spring migration from 10 February to 7 April in **a** 2014 and **b** 2015. *Black shape* outlines semi-collared flycatcher non-breeding range (BirdLife International and NatureServe 2011) and *large arrows* indicate spring migration routes [Temperature data available from the U.S. Geological Survey (http://www.usgs.gov/); wind data available from National Oceanic and Atmospheric Administration (http://www.noaa.gov/). Maps were created in ArcMap 10.1 (http://www.esri.com/)]

Responses of migrants

During both years, flycatchers departed from their non-breeding grounds in Eastern-Central Africa in the second half of February {median date 2014: 21 February [interquartile range (IQR) = 17–22 Feb], 2015: 16 February (11–19 Feb), Fig. 2c} and crossed the Sahara desert in late March [2014: 23 March (17–30 Mar), 2015: 27 March (21 Mar–5 Apr)]. After crossing the Sahara, the birds stayed in the Mediterranean Basin for 5 days (3.5–7.1) in 2014 before arriving at the breeding site on 2 April (29 Mar–7 Apr). In 2015 birds spent three times longer

(mean 15 days, IQR 9.8–19.5) in the Mediterranean Basin and arrived at the breeding site on 10 April (9–11 Apr, see Additional file 1). We found a negative relationship between the time spent in the Mediterranean Basin and leaf development at the breeding site (Pearson's one-tailed correlation: r = −0.82, n = 10, p = 0.002; Fig. 2d). The median spring migration passage times of flycatchers at Antikythira Bird Observatory in 2014 and 2015 were within the species' typical long-term passage period (2014: 17 Apr; 2015: 14 Apr; 2007–2015: 15 Apr, IQR = 12–18 Apr).

We also observed prominent sex differences in migration timing, with males crossing the Sahara and arriving at the breeding site earlier than females in both years. The distinct protandry resulted in stronger delays in males' migration schedule than in females'. In 2015 males

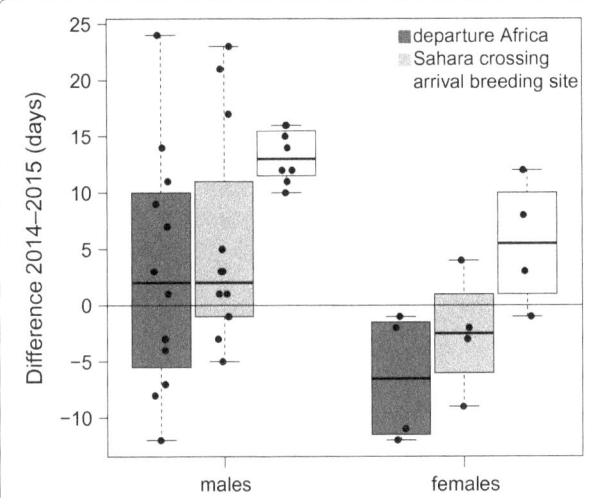

Fig. 4 Sex-specific difference in individual migration schedules between a cold (2015) and a warm spring (2014). The delay in arrival is larger for the earlier migrating males than for later migrating females. Individual data points represent difference between every possible pair of two individuals tracked in different years

arrived at the breeding site on average 13 days later than in 2014, while the difference for females was only 5.5 days (Fig. 4).

The cold spell of 2015 also had severe consequences on apparent local survival. Return rates of geolocator-tagged and ringed-only control birds were approximately two times lower in 2015 (Table 1), with males and older individuals (more than 3 years old) affected more severely.

Discussion

Our findings show that a cold spell encountered *en route* delayed spring arrival and decreased local apparent survival in a trans-Equatorial migrant. After reaching the temperate climatic zone where environmental cues of spring phenology become available, tracked semi-collared flycatchers flexibly adjusted their migration rate and advanced (in the warm spring of 2014) or delayed (in the cold spring of 2015) their arrival at the breeding site depending on local conditions (e.g. temperature and leaf development).

The typical passage times of semi-collared flycatchers at Antikythira Bird Observatory range from the end of March to the end of April [14], with most birds passing in the second decade of April. Median passage times in the second decade of April may imply that the flyway through Antikythira is used by different populations than ours, and those populations migrating through Antikythira

Table 1 Differences in return rates of semi-collared flycatchers between 2014 and 2015, and between control group and geolocator-tagged group

	2014	2015	χ^2	p value
Control				
Males	58.8% (30/51)	23.8% (19/80)	6.29	*0.01*
Females	38.1% (24/63)	21.7% (15/69)	1.78	0.18
2cy	64.3% (18/28)	51.4% (19/37)	0.11	0.74
2cy+	41.9% (36/86)	13.4% (15/112)	10.90	*<0.001*
Total	47.4% (54/114)	22.8% (34/149)	7.87	*0.005*
Tagged				
Males	47.1% (8/17)	18.5% (5/27)	1.31	0.25
Females	13.0% (3/23)	9.1% (2/22)	1.4e−30	1
2cy	37.5% (6/16)	21.4% (3/14)	0.11	0.75
2cy+	20.8% (5/24)	11.4% (4/35)	0.23	0.63
Total	27.5% (11/40)	14.3% (7/49)	0.99	0.32
	Control	**Tagged**	χ^2	**p value**
2014–2015				
Males	37.4% (49/131)	29.5% (13/44)	0.24	0.63
Females	29.5% (39/132)	11.1% (5/45)	3.22	0.07
2cy	56.9% (37/65)	30.0% (9/30)	1.68	0.20
2cy+	25.8% (51/198)	15.3% (9/59)	1.39	0.24
Total	33.5% (88/263)	20.2% (18/89)	2.69	0.10

Significant differences are given in italics

do not pass there until after the prolonged stopovers of our tracked birds in 2015. Indeed in 2014, birds from our study population arrival at the breeding site earlier than the median passage time at Antikythira, supporting this idea.

So far, contrasting results have been reported in long-distance migrants regarding their ability to use environmental signals to cue spring migration [5]. Nearctic-Neotropical long-distance migrants have been shown to use environmental cues to some extent to adjust their migration rate in spring [20, 21]. On the contrary, pied flycatchers *Ficedula hypoleuca* were not able to adjust the arrival time proportionally to the increasing spring temperatures suggesting a tight endogenous routine controlling phenology of spring migration in that population [22]. Recent tracking studies confirm these findings, showing that breeding site arrival date in pied flycatchers largely depends on the onset of spring migration, rather than birds making adjustments *en route* [23]. In the closely related collared flycatcher *F. albicollis*, spring arrival at different breeding sites is related to local phenology, and timing of the onset of spring migration seems to be less important [24]. This coincides with our findings in semi-collared flycatchers. The differences between these three *Ficedula* species may be related to the migratory flyway they use during the spring migration. Resource availability and ecological barriers encountered *en route* can influence on the rate and timing of bird migration [25, 26]. Species that encounter ecological barriers along the migratory flyway and have larger migratory distance show a greater degree of variation in their migratory behaviour and ability to adjust migration rate in response to the environment. In spring, pied flycatchers migrate along the western Afro-Palearctic flyway, while collared and semi-collared flycatchers migrate along the central Afro-Palearctic flyway. Migrants using the central Afro-Palearctic flyway encounter larger ecological barriers (e.g. the distance to cross the Sahara Desert is larger) and harsher conditions compared to the species using the western Afro-Palearctic flyway.

To date there seems to be no general consensus on where along a migration route the changing conditions should have the largest effect on the timing of bird arrival [10]. Tøttrup et al. [27] demonstrated that drought in the Horn of Africa delayed spring arrival of Afro-Palearctic migrants, as birds prolonged their stopovers in this area. This, when considered with our results suggests that prolonged stopovers due to adverse weather conditions could occur at any place along the migratory route (in the tropics and temperate regions alike), and can cause delayed arrival at the breeding sites.

As a consequence of adverse weather, increased mortality rates have previously been reported across different taxa [28]. Our finding of low apparent survival of flycatchers in a year with adverse weather conditions likely indicates increased mortality. Alternatively, birds may have acted opportunistically and settled for breeding elsewhere along the migratory route or exhibited a higher degree of breeding dispersal compared to the previous year. In our study, males showed lower return rates than females in the colder spring of 2015. By arriving earlier, males are exposed to a more hostile environment, including lower food availability, than later arriving females. Similarly, older flycatchers usually arrive at the breeding site earlier than younger ones and would therefore undergo similar consequences to those of males versus females. In cliff swallows *Petrochelidon pyrrhonota* higher mortality of older individuals was found as a result of a cold spell, coinciding with our findings of low return rates [29].

Geolocator attachment has been shown to negatively affect return rates of birds [30]. However the recent evidence is ambiguous, with a number of studies showing no apparent effect on return rates of the tagged birds [e.g. 31, 32], while some report negative influence [33] including delayed breeding site arrival time and decreased breeding success in the year following the geolocator deployment [34]. Furthermore, the differences in return rates between tagged and control birds seem to vary among sites within the same species [24, 35]. Therefore, having a control group of ringed only individuals within a study population is recommended in order to evaluate the impact of the attached devices on the animals. It may be that the limited sample size of tagged birds restricted our ability to detect a significant negative effect on individual apparent survival associated with carrying the geolocator, despite the fact that return rates of the geolocator-tagged individuals in our study were lower than for ringed only birds (see Table 1). However, we have no reason to believe that the extra weight of the geolocators influenced the migration speed and stopover behaviour of our study birds, as our field observations show simultaneous arrival of the tagged and ringed-only birds.

Conclusions

Our tracked flycatchers prolonged their stopovers in the Mediterranean region when confronting a cold spell, while the population as a whole suffered increased mortality. One must keep in mind that tracking by geolocator only provides data from recaptured, surviving individuals. Individuals differ in their response to abiotic stressors [36], and those not returning may have died due to an inappropriate response strategy. Because of spatial and temporal differences in climate change [1], long-distance migrants might be particularly challenged in their responses. For migratory birds the ability to combine external and internal stimuli appears to be essential for successful organization of the annual cycle. Understanding how species, populations, and even individuals respond to the changing climate and its associated weather extremes can help to predict the consequences for their population dynamics. Large phenotypic plasticity is likely to play a crucial role for population viability under the rapidly changing environment.

Additional files

Additional file 1. Video of semi-collared flycatcher spring migration progression tracked by light-level geolocators in relation to temperature anomalies in 2014 and 2015.

Additional file 2. Raw sunrise and sunset data recorded by the geolocators.

Authors' contributions
MB and PA carried out field work. MB analysed the data and was a major contributor in writing the manuscript. All authors actively commented during the writing of the manuscript. All authors read and approved the final manuscript.

Author details
[1] Department of Zoology, Palacký University, tř. 17. listopadu 50, 77146 Olomouc, Czech Republic. [2] Department of Bird Migration, Swiss Ornithological Institute, Seerose 1, 6204 Sempach, Switzerland. [3] Museum of Natural History, nám. Republiky 5, 77173 Olomouc, Czech Republic.

Acknowledgements
Thanks to Johan for providing access to the study area. We thank M. Král, T. Koutný, S. Peev and M. Ilieva for their help in the field and S. Bauer, N. Friedman, and two anonymous reviewers from Axios Reviews for helpful comments on an earlier version of the manuscript. This is contribution n. 22 from Antikythira Bird Observatory—Hellenic Ornithological Society/BirdLife Greece. Antikythira Bird Observatory is funded from the A.G. and A.P. Leventis Foundation.

Competing interests
The authors declare that they have no competing interests.

Funding
This study was funded by Czech Science Foundation (grant #13-06451S to PA, MB) and in part by Palacky University grant scheme (IGA_PRF to MB).

References
1. Easterling DR, Karl TR, Gallo KP, Robinson DA, Trenberth KE, Dai A. Observed climate variability and change of relevance to the biosphere. J Geophys Res. 2000;105:101–14.
2. Walther G, Post E, Convey P, Menzel A, Parmesan C, Beebee TJC, et al. Ecological responses to recent climate change. Nature. 2002;416:389–95.
3. Parmesan C. Ecological and evolutionary responses to recent climate change. Annu Rev Ecol Syst. 2006;37:637–69.
4. Pearce-Higgins JW, Green RE. Birds and climate change: impacts and conservation responses. Cambridge: Cambridge University Press; 2014.
5. Knudsen E, Lindén A, Both C, Jonzén N, Pulido F, Saino N, et al. Challenging claims in the study of migratory birds and climate change. Biol Rev. 2011;86:928–46.
6. Berthold P. Genetic control of migratory behaviour in birds. Trends Ecol Evol. 1991;6:254–7.
7. Berthold P. Control of bird migration. London: Chapman & Hall; 1996.
8. Kok OB, Van Ee CA, Nel DG. Daylength determines departure date of the spotted flycatcher (*Muscicapa striata*) from its winter quarters. Ardea. 1990;79:63–6.
9. Bauer S, Nolet BA, Giske J, Chapman JW, Åkesson S, Hedenström A, et al. Cues and decision rules in animal migration. In: Milner-Gulland EJ, Fryxell JM, Sinclair AR, editors. Animal migration—a synthesys. Oxford: Oxford University Press; 2011. p. 68–87.
10. Lindström Å, Chapman BB, Jonzén N, Klaassen M. Movement and migration in a changing world. In: Hansson L-A, Åkesson S, editors. Animal movment across scales. Oxford: Oxford University Press; 2014. p. 36–50.
11. Meehl GA, Karl T, Easterling DR, Changnon S, Pielke R, Changnon D, et al. An introduction to trends in extreme weather and climate events: observations, socioeconomic impacts, terrestrial ecological impacts, and model projections. Bull Am Meteorol Soc. 2000;81:413–6.
12. Thomas DW, Blondel J, Perret P, Lambrechts MM, Speakman JR. Energetic and fitness costs of mismatching resource supply and demand in seasonally breeding birds. Science. 2001;291:2598–600.
13. Briedis M, Träff J, Hahn S, Ilieva M, Král M, Peev S, et al. Year-round spatiotemporal distribution of the enigmatic semi-collared flycatcher *Ficedula semitorquata*. J Ornithol. 2016;157:895–900.
14. Barboutis C, Evangelidis A, Akriotis T, Fransson T. Spring migration phenology and arrival conditions of the Eastern Bonelli's Warbler and the Semi-collared Flycatcher at a small Greek island. Ringing Migr. 2013;28:39–42.
15. Lisovski S, Hewson CM, Klaassen RHGG, Korner-Nievergelt F, Kristensen MW, Hahn S, et al. Geolocation by light: accuracy and precision affected by environmental factors. Methods Ecol Evol. 2012;3:603–12.
16. Adamík P, Emmenegger T, Briedis M, Gustafsson L, Henshaw I, Krist M, et al. Barrier crossing in small avian migrants: individual tracking reveals prolonged nocturnal flights into the day as a common migratory strategy. Sci Rep. 2016;6:21560.
17. Kalnay E, Kanamitsu M, Kistler R, Collins W, Deaven D, Gandin L, et al. The NCEP/NCAR 40-year reanalysis project. Bull Am Meteorol Soc. 1996;77:437–71.
18. Kemp MU, Emiel van Loon E, Shamoun-Baranes J, Bouten W. RNCEP: global weather and climate data at your fingertips. Methods Ecol Evol. 2012;3:65–70.
19. Erni B, Liechti F, Bruderer B. The role of wind in passerine autumn migration between Europe and Africa. Behav Ecol. 2005;16:732–40.
20. Marra PP, Francis CM, Mulvihill RS, Moore FR. The influence of climate on the timing and rate of spring bird migration. Oecologia. 2005;142:307–15.
21. Kelly JF, Horton KG, Stepanian PM, de Beurs KM, Fagin T, Bridge ES, et al. Novel measures of continental-scale avian migration phenology related to proximate environmental cues. Ecosphere. 2016;7:e01434.
22. Both C, Visser ME. Adjustment to climate change is constrained by arrival date in a long-distance migrant bird. Nature. 2001;411:296–8.
23. Ouwehand J, Both C. African departure rather than migration speed determines variation in spring arrival in pied flycatchers. J Anim Ecol. 2017;86:88–97.
24. Briedis M, Hahn S, Gustafsson L, Henshaw I, Träff J, Král M, et al. Breeding latitude leads to different temporal but not spatial organization of the annual cycle in a long-distance migrant. J Avian Biol. 2016;47:743–8.
25. La Sorte FA, Fink D. Migration distance, ecological barriers and en-route variation in the migratory behaviour of terrestrial bird populations. Glob Ecol Biogeogr. 2017;26:216–27.
26. Thorup K, Tøttrup AP, Willemoes M, Klaassen RHG, Strandberg R, Vega ML, et al. Resource tracking within and across continents in long-distance bird migrants. Sci Adv. 2017;3:E1601360.

27. Tøttrup AP, Klaassen RHG, Kristensen MW, Strandberg R, Vardanis Y, Lindström Å, et al. Drought in Africa caused delayed arrival of European songbirds. Science. 2012;338:1307.

28. Bailey LD, van de Pol M. Tackling extremes: challenges for ecological and evolutionary research on extreme climatic events. J Anim Ecol. 2016;85:85–96.

29. Brown CR, Brown MB. Intense natural selection on body size and wing and tail asymmetry in cliff swallows during severe weather. Evolution. 1998;52:1461–75.

30. Costantini D, Møller AP. A meta-analysis of the effects of geolocator application on birds. Curr Zool. 2013;59:697–706.

31. Wijk RE, Souchay G, Jenni-Eiermann S, Bauer S, Schaub M. No detectable effects of lightweight geolocators on a Palaearctic-African long-distance migrant. J Ornithol. 2016;157:255–64.

32. Briedis M, Beran V, Hahn S, Adamík P. Annual cycle and migration strategies of a habitat specialist, the Tawny Pipit *Anthus campestris*, revealed by geolocators. J Ornithol. 2016;157:619–26.

33. Rodríguez-Ruiz J, Parejo D, de la Puente J, Valera F, Calero-Torralbo MA, Bermejo A, et al. Short- and long-term effects of tracking devices on the European Roller *Coracias garrulus*. Ibis. 2016;158:179–83.

34. Arlt D, Low M, Pärt T. Effect of geolocators on migration and subsequent breeding performance of a long-distance passerine migrant. PLoS ONE. 2013;8:e82316.

35. Weiser EL, Lanctot RB, Brown SC, Alves JA, Battley PF, Bentzen R, et al. Effects of geolocators on hatching success, return rates, breeding movements, and change in body mass in 16 species of Arctic-breeding shorebirds. Mov Ecol. 2016;4:12.

36. Breuner CW, Delehanty B, Boonstra R. Evaluating stress in natural populations of vertebrates: total CORT is not good enough. Funct Ecol. 2013;27:24–36.

10

Host plant range of a fruit fly community (Diptera: Tephritidae): does fruit composition influence larval performance?

Abir Hafsi[1,2], Benoit Facon[1,3], Virginie Ravigné[1], Frédéric Chiroleu[1], Serge Quilici[1], Brahim Chermiti[2] and Pierre-François Duyck[1,4*] iD

Abstract

Background: Phytophagous insects differ in their degree of specialisation on host plants, and range from strictly monophagous species that can develop on only one host plant to extremely polyphagous species that can develop on hundreds of plant species in many families. Nutritional compounds in host fruits affect several larval traits that may be related to adult fitness. In this study, we determined the relationship between fruit nutrient composition and the degree of host specialisation of seven of the eight tephritid species present in La Réunion; these species are known to have very different host ranges *in natura*. In the laboratory, larval survival, larval developmental time, and pupal weight were assessed on 22 fruit species occurring in La Réunion. In addition, data on fruit nutritional composition were obtained from existing databases.

Results: For each tephritid, the three larval traits were significantly affected by fruit species and the effects of fruits on larval traits differed among tephritids. As expected, the polyphagous species *Bactrocera zonata*, *Ceratitis catoirii*, *C. rosa*, and *C. capitata* were able to survive on a larger range of fruits than the oligophagous species *Zeugodacus cucurbitae*, *Dacus demmerezi*, and *Neoceratitis cyanescens*. Pupal weight was positively correlated with larval survival and was negatively correlated with developmental time for polyphagous species. Canonical correspondence analysis of the relationship between fruit nutrient composition and tephritid survival showed that polyphagous species survived better than oligophagous ones in fruits containing higher concentrations of carbohydrate, fibre, and lipid.

Conclusion: Nutrient composition of host fruit at least partly explains the suitability of host fruits for larvae. Completed with female preferences experiments these results will increase our understanding of factors affecting tephritid host range.

Keywords: Tephritidae, Nutriments composition, Carbohydrate, Food webs, Host range

Background

Arthropods constitute the most diverse group of animals on Earth, and a large fraction of Arthropod species are phytophagous [1]. Phytophagous insects range from those that are strictly monophagous, i.e., that are able to develop on only one host plant, to those that are extremely polyphagous, i.e., that are able to develop on

hundreds of plant species belonging to numerous families [2]. Most phytophagous insect species, however, are specialised for feeding on a small range of host plants, and this specialisation is thought to have contributed to the huge diversification of insects that consume plants [1, 3, 4].

The diet breadth of phytophagous insect is shaped by many evolutionary and ecological processes [3, 5, 6]. In most species, females locate, recognise and accept host plants for oviposition, eggs hatch on the host plants, and larvae develop to the adult stage by consuming various parts of the host plants. Diet breadth is therefore

*Correspondence: duyck@cirad.fr
[4] UMR « Peuplements Végétaux et Bio-agresseurs en Milieu Tropical », CIRAD Pôle de Protection des Plantes, 7 chemin de l'Irat, 97410 Saint Pierre, La Réunion, France
Full list of author information is available at the end of the article

the product of female choice and larval performance [1, 7]. Both traits have been shown to evolve rather rapidly [8]. As a result of their joint evolution, female preferences generally but not always correlate with larval performances [9]. The lack of correlation between female preference and larval performance has been previously reported in some studies [10–13]. In particular, the link between female preference and larval performance varies with the degree of diet specialization [6] and depends on the taxonomic diversity of studied host plants [6, 12]. Correlations between female preference and larval performance are more often observed for monophagous insects than for polygophagous insects and across and within plant families than among genotypes within a plant species [6, 12]. Balgawi et al. [10] suggested that host preferences may not be the result of optimization of the preference-performance relationship but linked with other behavioural, physio-chemical and/or physiological associations between insects and their larval hosts. In the field, availability and abundance of host plants and predation may also modify this relationship [13, 14]. While females very rarely prefer a plant that does not support good performance of their offspring [6], larvae may often develop on a larger set of host plants than that selected by females [11, 15, 16]. Hence, larval performance can provide insight into the potential diet breadth of a species [17].

Larval performance is influenced by a number of factors related to the ecological context, i.e., which plants are present in the environment and their availability for insects [18, 19], and to plant intrinsic value for larval development and survival [10, 20–22]. Plants furnish phytophagous insects with an array of vital resources. Moreover, plants in most families contain toxic compounds, including secondary metabolites [23, 24]. Secondary metabolites have been widely described in the vast literature on insect-plant coevolution [25–27]. Chemicals can alter the nutritional value of plant tissues by making them poisonous [28, 29] especially to polyphagous species that feed on vegetative organs or inflorescences. Some polyphagous insects, however, feed and develop in the pulp of fleshy fruits from a wide range of species and families. Frugivorous Tephritidae, for example, feed on many kinds of ripening and commercially grown fruits whose defensive compounds disappear when fruits are ripe [30].

Various nutritional compounds in host plants affect the life history traits of phytophagous larvae [31, 32]. Carbohydrates and lipids in particular greatly affect larval performance, while mineral nutrition may be more crucial for adult fecundity [25]. As a consequence, plants that differ in nutrient content often differ in their effects on insect fitness [21, 22, 26]. Understanding how larval

performance relates to plant nutrient composition should therefore shed light on the determinants of insect host range.

Plant availability for insects depends on plant abundance and phenology but also on biotic interactions. In most if not all natural or agronomic landscapes, the same plants can serve as hosts for several phytophagous insect species, whose host ranges may overlap at different scales [33, 34]. Understanding how phytophagous insect communities organize on a given plant community requires the study of the determinants of diet breadth for the different insect species that share a large part of their host range in a given environment.

The tephritids of La Réunion Island are ideal for studying the effects of fruit composition on a community of phytophagous insects. The tephritids, also called true fruit flies, are economically important because many species in this family attack important fruit and vegetable crops in tropical and subtropical regions worldwide. La Réunion hosts eight tephritid species (two indigenous and six exotic) that infest a large number of cultivated and wild host plants in the same, rather small area; the island occupies 2512 km^2. These eight species differ in diet breadth. Four species (*Ceratitis catoirii*, *C. capitata*, *C. rosa*, and *Bactrocera zonata*) are polyphagous [35]. *Ceratitis rosa*, for example, is found on 60 host species belonging to 20 families in La Réunion [36]. In contrast, the four other species are considered to be oligophagous, i.e., they have a limited host range; *Dacus demmerezi*, *D. ciliatus*, and *Zeugodacus cucurbitae* are found mostly on Cucurbitaceae, and *Neoceratitis cyanescens* is mostly associated with Solanaceae [36]. In spite of these important differences, the diet breadths of the eight species do overlap in that the polyphagous species are able to infest Cucurbitaceae or Solanaceae and the oligophagous species may infest some hosts belonging to families other than the Cucurbitaceae or Solanaceae.

Laboratory studies with *C. catoirii*, *C. capitata*, *C. rosa*, and *B. zonata* [37] and other studies with the tephritids *B. dorsalis* [38] and *C. fasciventris* [39] have indicated that larval performances can differ drastically on different host species. The nutrient content of a larval diet greatly affects larval growth, developmental time, and survival and also affects the number and the fitness of the adult fruit flies produced [38, 40–42]. For instance, the low concentrations of carbohydrate in the upper parts of papaya and orange fruits reduce the development of *C. capitata*, while the high concentrations in the lower parts of the fruits enhance larval development [43]. Similarly, Nash and Chapman [44] found that *C. capitata* developmental time and larval survival are affected by protein quantity and quality as well as by carbohydrate quantity and quality.

Table 1 List of host species studied

Family	Species	Common name
Anacardiaceae	*Mangifera indica*	Mango
Annonaceae	*Annona reticulata*	Custard apple
Caricaceae	*Carica papaya*	Papaya
Combretaceae	*Terminalia catappa*	Indian almond
Cucurbitaceae	*Citrullus lanatus*	Water melon
	Cucumis melo	Melon
	Cucumis sativus	Cucumber
	Cucurbita maxima	Pumpkin
	Cucurbita pepo	Zucchini
	Sechium edule	Chayote
Moraceae	*Ficus carica*	Fig
Myrtaceae	*Psidium cattleyanum*	Strawberry guava
	Psidium guajava	Guava
Oxalidaceae	*Averrhoa carambola*	Star fruit
Rosaceae	*Eriobotrya japonica*	Loquat
	Prunus domestica	Plum
	Prunus persica	Peach
Rutaceae	*Citrus reticulata Blanco*	Mandarin
Solanaceae	*Capsicum frutescens*	Chilli
	Solanum betacea	Tree tomato
	Solanum lycopersicum	Tomato
	Solanum melongena	Eggplant

The studies cited in the previous paragraph suggested that fruit nutrient composition may be an important factor influencing the organization of tephritid communities. Determining whether this is true requires basic information on the relationship between fruit nutrient composition and the performance of individual tephritid species. In this study, we documented the potential host range of seven of the eight tephritid species in La Réunion by measuring larval performance on 22 host plants belonging to 11 plant families (Table 1), representing the most common crops and wild plants affected by tephritids on the island. For these 22 host plants, we gathered information on fruit nutrient composition from the literature and examined the extent to which fruit biochemical composition was associated with the observed differences in tephritid larval performance.

Methods
Tephritids
This study of tephritids in La Réunion included four polyphagous species (*C. catoirii*, *C. capitata*, *C. rosa*, and *B. zonata*) and three oligophagous species (*D. demmerezi* and *Z. cucurbitae* are mostly associated with cucurbit hosts, and *N. cyanescens* is mostly associated with Solanaceae fruits). *Dacus ciliatus* could not be included because of difficulties in its rearing. The seven species

were reared from specimens initially obtained from rose-apple (*Syzygium jambos*) for *C. catoirii*, beach naupaka (*Scaevola taccada*) for *C. capitata*, rose-apple for *C. rosa*, Indian almond (*Terminalia catappa*) for *B. zonata*, cucumber (*Cucumis sativus*) for *D. demmerezi*, zucchini (*Cucurbita pepo*) for *Z. cucurbitae*, and bugweed (*Solanum mauritianum*) for *N. cyanescens*. No permissions were required to collect these samples from the field. Larvae of *C. catoirii*, *C. capitata*, *C. rosa*, and *B. zonata* were subsequently reared on an artificial diet [45, 46] for generations F147–F160, F4–F10, F1–F3, and F118–F131, respectively. Larvae of *D. demmerezi* and *Z. cucurbitae* were reared on zucchini for generations F10–F18 and F54–F69, respectively, and larvae of *N. cyanescens* were reared on potato (*Solanum tuberosum*) for generations F10–F18. The species were not reared on the same food substrate because each was placed on the adequate artificial diet or host plant that enabled optimal development under laboratory conditions. In addition, some of these species are difficult to find in the field or to maintain in the laboratory, which explains the variability in the numbers of generations among species. Populations were maintained at several thousands of individuals per generation. Laboratory rearing was conducted under constant environmental conditions (25 ± 1 °C; 70 % relative humidity; L:D 12:12 photoperiod supplemented with natural light to maintain twilight conditions). Eggs were collected using an inverted perforated plastic cup swabbed with the flesh of host fruits; citrus for *C. catoirii*, *C. capitata*, *C. rosa*, and *B. zonata*; pumpkin for *D. demmerezi* and *Z. cucurbitae*; and potato for *N. cyanescens*. Eggs were placed in a 2 % Nipagine/sodium benzoate solution and were incubated in environmental chambers at a constant temperature of 25 °C for 30 h for *Z. cucurbitae*, 55 h for *N. cyanescens*, and for 48 h for the other five species.

Experimental setup
An important technical challenge in measuring larval performance on intact fruits of many different plant species is the heterogeneity in fruit decay. We developed simplified diets that contained 250 g of fruit pulp without peel or seeds, 4 g of agar–agar (to provide a suitable texture), and 10 ml of a 4 % Nipagine/sodium benzoate solution (to prevent fungal and bacterial growth). While these diets differed from fresh fruits in term of physical texture, they allowed measuring individual fitness traits, following a high number of homogenous replicates, and obtaining comparable measurements of larval performance with a focus on fruit nutrient value.

Components of the diet were blended and placed in individual 5-ml plastic cups, each containing 5 g of mixture. Each combination of tephritid species and fruit species was represented by 50 replicate cups, giving a total of 7700 cups. One neonate larva (<2 h old) was placed in each cup. Each cup was set in a shallow pan with sand to allow pupation. Because temperature greatly affects tephritid larval development, all experiments were carried out in environmental chambers at a constant temperature of 25 °C, which is favourable for the seven species [45–48].

Larval performance

Three indicators of larval performance were assessed: survival rate, developmental time, and pupal weight. Pupal weight is a good indicator of female fecundity and therefore of individual female fitness in Tephritidae [49, 50]. Every 48 h during 60 days, all cups were examined and pupae were collected. Larval survival was recorded as the number of pupae recovered from each host. Developmental time was recorded as the time from placement in the cup to pupation. Each pupa was weighed (Sartorius® Germany, precision: 10^{-4} g).

Fruit nutrient composition

To study the relationship between tephritid larval performance and fruit nutrient composition, we gathered data on nutrient composition from the United States Department of Agriculture (USDA) National Nutrient Database for Standard Reference, the ANSES (French Agency for Food, Environmental and Occupational Health Safety) French Food Composition Table, and published studies (data used and references are presented in Additional file 1). For the 22 plant species, data on the content of water, proteins, total carbohydrates, fat, fibres, sodium, magnesium, potassium, phosphorus, calcium, iron, and vitamin C are presented in a data set available in Additional file 1. Nutrients that could not be documented for all 22 plant species were excluded from the analysis to avoid problems caused by missing data. While nutrient content obtained from the literature probably did not strongly differ from the diets used for trait measurements, water content could have been more affected during the processing of diets. In order to check if water content in database and diets were representative of fresh fruits, we measured water contents in diets and fresh fruits for six fruit species (*Annona reticulate, Carica papaya, Cucumis sativus, Cucurbita maxima, Psidium gujava,* and *Solanum lycopersicum*). A strong linear relationship was observed between the water contents of artificial diets and the water contents estimates from database ($R^2 = 0.87$; P < 0.001; Y = 0.90X + 12.13) and between water contents in fresh fruits and water contents in artificial diets (P < 0.0001; $R^2 = 0.97$, Y = 0.96X + 6.28) (Additional file 1). While artificial diets tend to contain more water than expected from literature data (intercept different from 0), differences between fruits were conserved (slope close to 1).

Statistical analyses

All analyses were performed with R-3.2.2-win [51]. Larval survival was analysed using a general linear model (GLM) with a binomial error as a function of plant species, tephritid species, and interaction. Developmental time and pupal weight were analysed by analysis of variance (ANOVA) as a function of plant species, tephritid species, and interaction. The relationship between the three indicators of larval performance was investigated using linear models.

Canonical correspondence analysis (CCA) [52] was used to study the relationship between nutrient composition of the 22 host fruits and larval survival of the seven tephritid species using the function *pcaiv* in the *ade4* package [53]. The CCA analysis consist of carrying out a factorial correspondence analysis (FCA) on the fitted variables (larval survival of seven tephritid species on 22 fruit species) after the regression on the instrumental variables (chemical composition of 22 fruit species). The significance of CCA was tested by a Monte-Carlo test [54] that evaluated the significance and the stability of decomposition of the total inertia of larval survival with only permutation of the rows of biochemical composition table.

Results of CCA analyses (presented in Additional file 2) showed that the first axis of the factorial space explained 64.5 % of the global variability. The variability in larval survival of the seven tephritid species was mostly explained (64.9 %) by the contents of 12 nutrients in the 22 fruit species. A Monte-Carlo test showed that the inertia projected by 1000 permutations was not significant (P = 0.141). Water, lipid, carbohydrate, and fibre were the nutrients most correlated with the first axis (correlations >0.5). A second CCA analysis restricted to these four nutrients was carried out to clarify their influence on larval survival and to increase the stability of the decomposition of inertia.

Results

Influence of host fruit on larval survival

Larval survival rates differed significantly among the fruit species (ΔDev$_{21, 1119}$ = 7948, P < 0.001) and among the tephritid species (ΔDev$_{6, 639}$ = 7942, P < 0.001); the interaction between tephritid and fruit species was also significant (ΔDev$_{125, 2120}$ = 7817, P < 0.001) (Fig. 1). Larvae of the polyphagous species *B. zonata, C. catoirii, C. capitata,* and *C. rosa* survived on a wide range of hosts (on 17–20 of the 22 fruit species tested). On these host fruits, the four polyphagous tephritid species had a larval survival rate ranging from 2 to 100 %, with the highest larval survival exceeding 60 % in tree tomato, guava, mango, and Indian almond. *Dacus demmerezi* and *Z. cucurbitae,* which are considered specialised on Cucurbitaceae fruits,

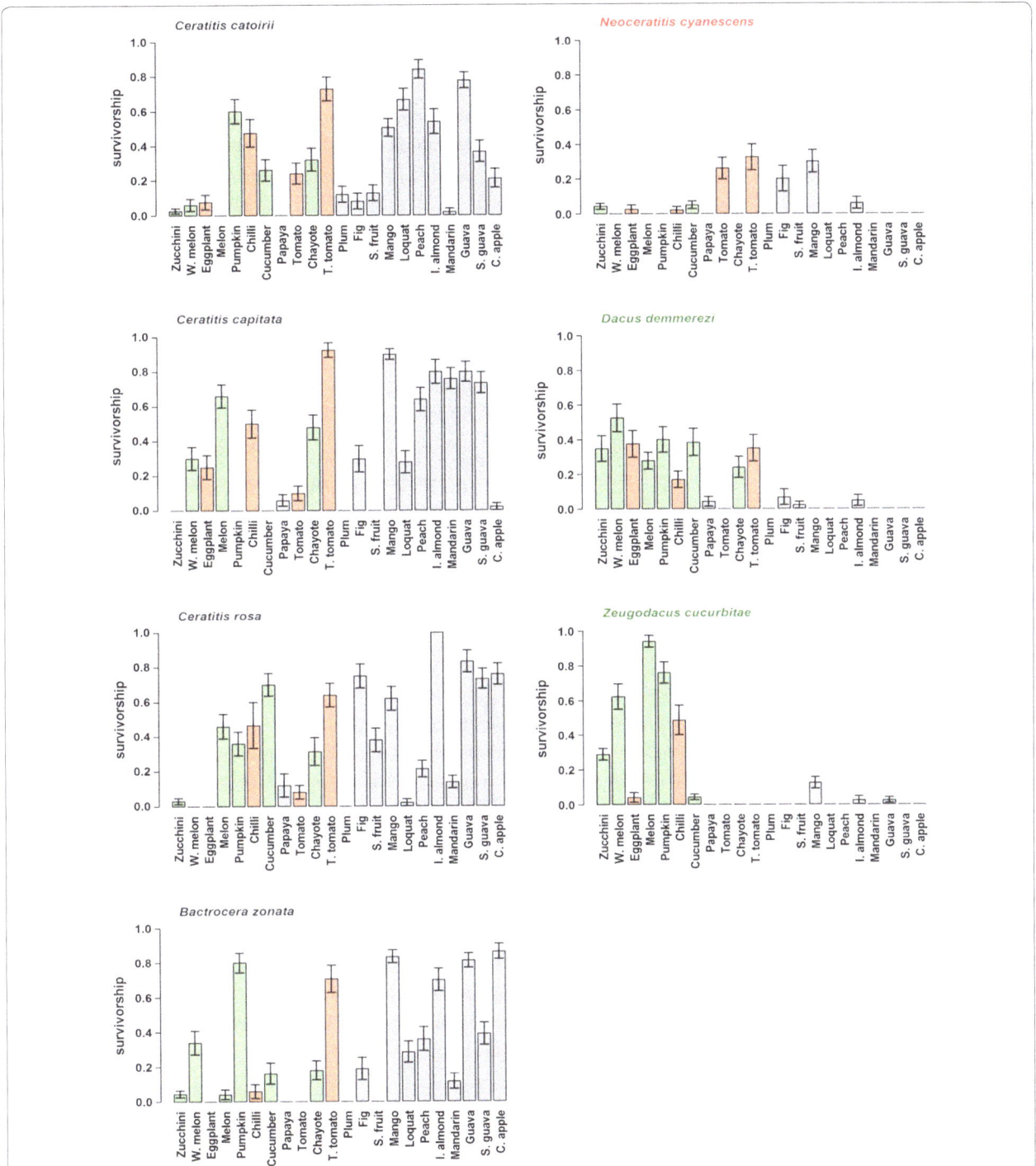

Fig. 1 Larval survival for seven tephritid species reared on 22 host fruit species occurring in La Reunion. Values are mean ± SE. Host fruits belonging to the Cucurbitaceae and Solanaceae are indicated by *green* and *red*, respectively. Host fruits are ordered by coordinate of the first axis of CCA analyses (see "Methods" section)

survived on 13 and 10 fruit species, respectively; survival of *D. demmerezi* was highest on six of the cucurbit species, and survival of *Z. cucurbitae* was highest on five of the cucurbit species. *Dacus demmerezi* and *Z. cucurbitae*

also survived on fruits from the Solanaceae (eggplant and chilli), the Anacardiaceae (mango), and the Combretaceae (Indian almond). *Neoceratitis cyanescens*, which is considered specialised on the Solanaceae family, had the highest survival rate on fruit species belonging to the Solanaceae but also on mango, and was able to survive, albeit with lower survival rates, on cucumber and zucchini fruits.

Influence of host fruit on pupal weight
Pupal weight differed significantly among fruit species ($F_{21, 24593} = 51.1$, $P < 0.001$) and among tephritid species ($F_{6, 198010} = 418.7$, $P < 0.001$); the interaction between tephritid and fruit species was also significant ($F_{76, 4457} = 9.4$, $P < 0.001$) (Fig. 2). Across fruit species, pupal weight was lowest for *C. capitata* and highest for *D. demmerezi*. For the four polyphagous tephritids, pupal weight was highest on peach and tree tomato for *C. catoirii*; on guava, mango, and tree tomato for *C. capitata*; on guava for *C. rosa*; and on custard apple for *B. zonata*. For the three oligophagous tephritids, pupal weight was highest on cucumber for *D. demmerezi*, on zucchini and cucumber for *Z. cucurbitae*, and on tree tomato for *N. cyanescens*.

Relationship between pupal weight, larval survival, and developmental time
Developmental time differed significantly among fruit species ($F_{21, 965} = 31.2$, $P < 0.001$) and among tephritid species ($F_{6, 6410} = 207.2$, $P < 0.001$); the interaction between tephritid and fruit species was also significant ($F_{76, 451} = 14.6$, $P < 0.001$) (Additional file 3). On all fruits that supported larval survival, pupal weight was positively correlated with larval survival for *C. catoirii* ($P < 0.001$), *C. capitata* ($P < 0.001$), *C. rosa* ($P = 0.0013$), and *B. zonata* ($P < 0.001$), and was negatively correlated with developmental time for *C. catoirii* ($P < 0.001$), *C. capitata* ($P < 0.001$), *B. zonata* ($P < 0.001$), and *Z. cucurbitae* ($P = 0.022$) (Table 2; Additional file 4). Similar but statistically insignificant relationships were also observed for the other species.

Relationship between fruit nutrient composition and larval survival
CCA analysis showed that the first eigenvalue, which explained 74.33 % of the total variation, was overwhelmingly larger than the second one, which explained only 14.26 % of the total variation. Most of the common structure of the two data matrices is therefore contained in the first axis. Water, lipid, carbohydrate, and fibre explained 30.05 % of the variability in survival of larvae of the seven tephritid species reared on 22 fruit species. A Monte-Carlo test showed significant inertia projected for 1000 permutations ($P_{value} = 0.028$).

CCA analysis (Fig. 3a) showed that all tested fruits fell into two distinct groups, with species in the Solanaceae (tomato, tree tomato, eggplant, and chili) and

Cucurbitaceae (pumpkin, zucchini, cucumber, and water melon) in the first group, and the other species in the second group. Shorter arrows (like those for zucchini, melon, guava, mango, Indian almond, and strawberry guava) indicate a better concordance between fruit nutrient composition and survival of tephritid species larvae.

The direction of the vector for water concentration was opposite to the direction for the vectors for carbohydrate, fibre, and lipid concentration (Fig. 3b), which indicated a negative correlation between these components. The length of these vectors indicates the importance of the components in explaining the structure of the two data matrices; importance decreases with vector length. Vectors of larval survival of *D. demmerezi* and *Z. cucurbitae* (Fig. 3c) were pointed in the same direction as the vector for water and in the opposite direction as the vectors for carbohydrate, fibre, and lipid (Fig. 3b), indicating that survival of *D. demmerezi* and *Z. cucurbitae* larvae was positively correlated with the concentration of water in fruits and was negatively correlated with the concentration of carbohydrate, fibre, and lipid in fruits. Larval survival of *B. zonata*, *C. catoirii*, *C. capitata*, and *C. rosa* was positively correlated with the concentration of carbohydrate, fibre, and lipid in fruits and negatively correlated with the concentration of water in fruits. Larval survival of *N. cyanescens* depended slightly on the components represented by the first axis of the CCA analyses, and was positively correlated with concentration of water and lipid in host fruits.

Discussion
The present study aimed to evaluate the contribution of available host plants to the performance of tephritid larvae occurring in sympatry. We established that host identity had a marked influence on larval survival, developmental time, and pupal weight. In general, these three fitness traits were positively correlated with each other, i.e., larvae reared on fruits supporting high survival had high pupal weights and short developmental times. More specifically, pupal weight was positively correlated with larval survival for *B. zonata*, *C. catoirii*, *C. capitata*, and *C. rosa* and was negatively correlated with developmental time for *B. zonata*, *C. catoirii*, *C. capitata*, and *Z. cucurbitae*. Similar relationships between the three traits were also observed for *D. demmerezi* and *N. cyanescens* but at a non-significant level, probably because the statistical power was limited by the low number of host fruits that supported survival of these oligophagous species. Pupal weight has previously been shown to be positively correlated with female fecundity in several tephritids [37, 40, 49]. This absence of trade-offs between fitness components suggests that hosts differ in nutritional value for tephritid development [55]. Some fruits have sufficient

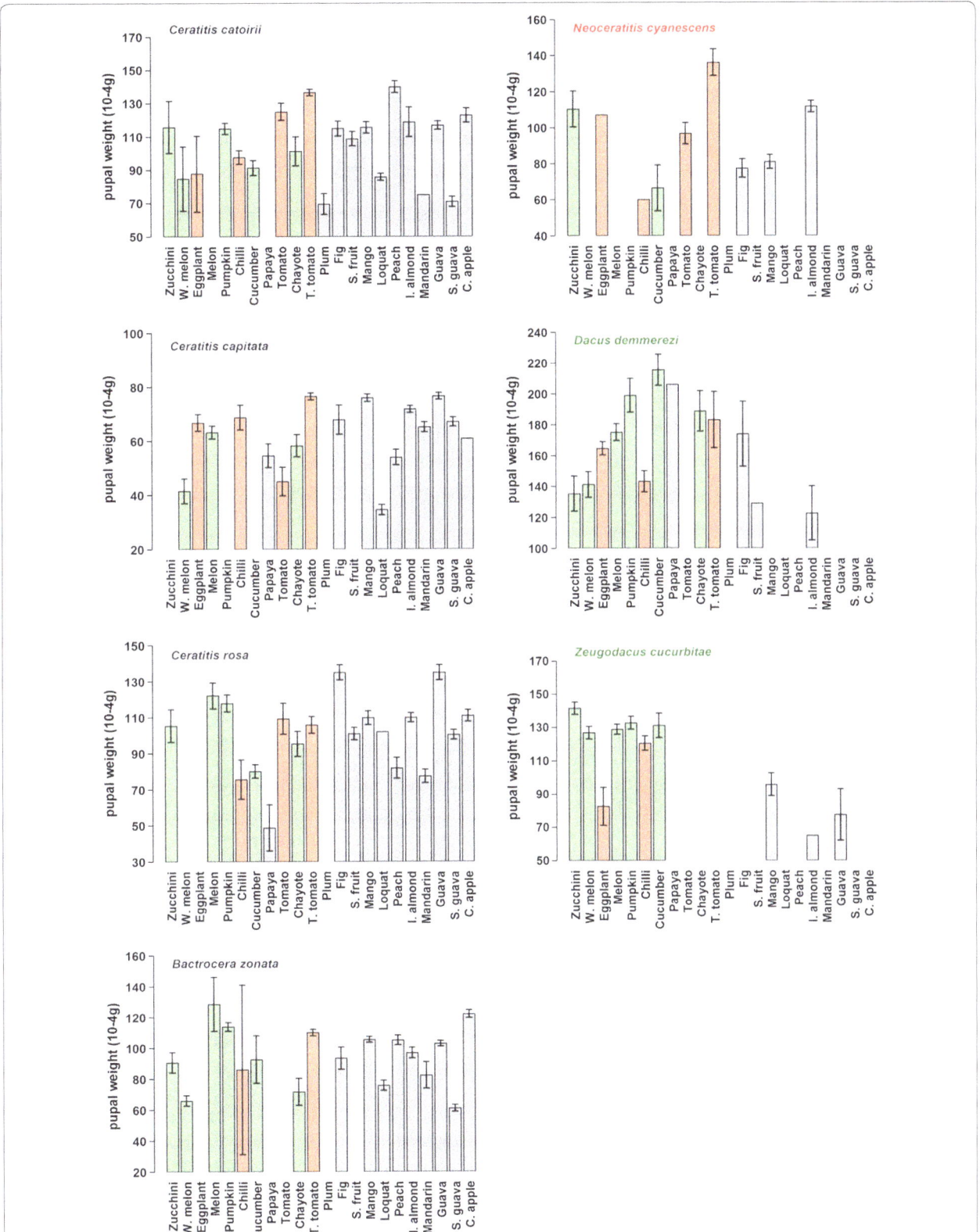

Fig. 2 Pupal weight (10^{-4} g) for the seven tephritid species reared on 22 host fruit species occurring in La Reunion. Values are mean ± SE. Host fruits belonging to the Cucurbitaceae and Solanaceae are indicated by *green* and *red*, respectively. Host fruits are ordered by coordinate of the first axis of CCA analyses (see "Methods" section). Pupal weight scales among species were changed for better representation

Table 2 Analyses by linear models of pupal weight as a function of larval survival and developmental time for seven Tephritidae species reared on 22 fruit species

Tephritidae species	Effect	Residuals; Df	Estimate	F value	P value
Ceratitis catoirii	Survival	1; 393	6.71	86.06	<0.001
	DT	1; 393	−2.49	290.38	<0.001
Ceratitis capitata	Survival	1; 413	30.40	161.40	<0.001
	DT	1; 413	−0.45	15.95	<0.001
Ceratitis rosa	Survival	1; 288	22.91	10.43	0.0013
	DT	1; 288	−0.09	0.29	0.5880
Bactrocera zonata	Survival	1; 406	24.40	188.58	<0.001
	DT	1; 406	−2.27	126.27	<0.001
Neoceratitis cyanescens	Survival	1; 590	68.42	3.37	0.0714
	DT	1; 590	−0.51	0.99	0.3245
Dacus demmerezi	Survival	1; 141	−9.04	0.05	0.8152
	DT	1; 141	−0.15	0.03	0.8568
Zeugodacus cucurbitae	Survival	1; 193	5.59	2.35	0.1265
	DT	1; 193	−0.66	5.32	0.0221

DT developmental time, *Survival* larval survival

Significant effects (P < 0.05) are indicated in italic

nutritional value to maximize all three of the measured performance traits. This is the case for guava, mango, and Indian almond, which supported relatively high survival (up to 60 %) and high pupal weights for the four polyphagous tephritids. In other geographical areas, these fruit species are the most utilized by the *Bactrocera dorsalis* complex of fruit flies [56]. Interestingly, the host range measured in the laboratory for the oligophagous tephritids like *Z. cucurbitae* included fruit species that are not hosts in the field in La Réunion but are hosts in other areas. *Zeugodacus cucurbitae* has been reported to cause damage on mango fruit in Africa and on papaya fruit in India [57, 58], confirming that *Z. cucurbitae* may use these plants in other ecological contexts.

Larval performance measured in tephritid colonies reared in laboratory for several generations may represent an important limitation of this study. Reports of differences between laboratory reared and wild insect strains and populations have gained much attention because of the relevance of such changes in behaviour and life history traits of phytophagous insects [59–62]. However, female oviposition preferences are generally found to change more rapidly than larval performance that is considered more conservative. For example in the bruchid beetle, after 11 generations of natural and artificial selection, genetic changes occurred in the behavioural response to host change, but not in the physiological performance of larvae [63]. Our study has other important limitations. First, we did not assess

some potentially important components of specialisation. Female choice, for example, was not measured but can greatly affect fruit fly abundance [13] and represent an important factor that facilitates host-plant adaptation [64]. Second, while symbiotic bacteria in the insect gut can impact larval development and adult fitness [65], these effects were not measured in our study. Symbiotic bacteria assemblages play a major role in detoxification processes thus making plant tissues edible for phytophagous insects and promote adaptation between phytophagous insects and host plants [66, 67]. Finally, we did not assess interactions among fruit fly species. Intra- and interspecific competition among larvae of different species within fruits and between adult females for egg laying sites are potentially important factors affecting host plant use by phytophagous insects in the field [68, 69]. So additional research is needed to determine how the results of current study relate to community dynamics.

Multivariate analyses (CCA) indicated that the water, lipid, carbohydrate, and fibre content of fruits may explain 30 % of the variability in the larval performance of the seven tephritid species. Larval survival of the polyphagous species *B. zonata*, *C. catoirii*, *C. capitata*, and *C. rosa* was positively correlated with carbohydrate, lipid, and fibre contents and negatively correlated with water content. In contrast, larval survival of the two species *D. demmerezi* and *Z. cucurbitae* associated with cucurbit hosts was positively correlated with water content and negatively correlated with carbohydrate and lipid content. Larval survival of *N. cyanescens*, which was the one species associated with Solanaceae fruits, was positively correlated with water and lipid contents.

Carbohydrates were previously identified as potentially important determinants of fruit nutritive value for tephritid larvae. Zucoloto [70], and Fernandes Da Silva and Zucoloto [43] showed that larvae of *C. capitata* moved to the part of the fruit containing the highest quantities of carbohydrate. These and current results are consistent with those reported for *C. capitata* [44], Acrididae [71], Drophila [72], and Noctuidae [73].

Behmer [74] suggested that the larvae of some phytophagous insect species prefer high-carbohydrate diets whereas others prefer high-protein diets. Our results suggest that polyphagous tephritids differ from oligophagous tephritids in terms of nutrient requirements. The performance of polyphagous species was strongly associated with carbohydrate, lipid, and fibre contents and was not associated with protein content. This contradicts previous experimental findings that decreases in protein quantity and quality reduced larval development and survival for *C. capitata* [44]. An explanation may be that the protein concentration of host fruit is generally low and invariant among fruits, whereas carbohydrate

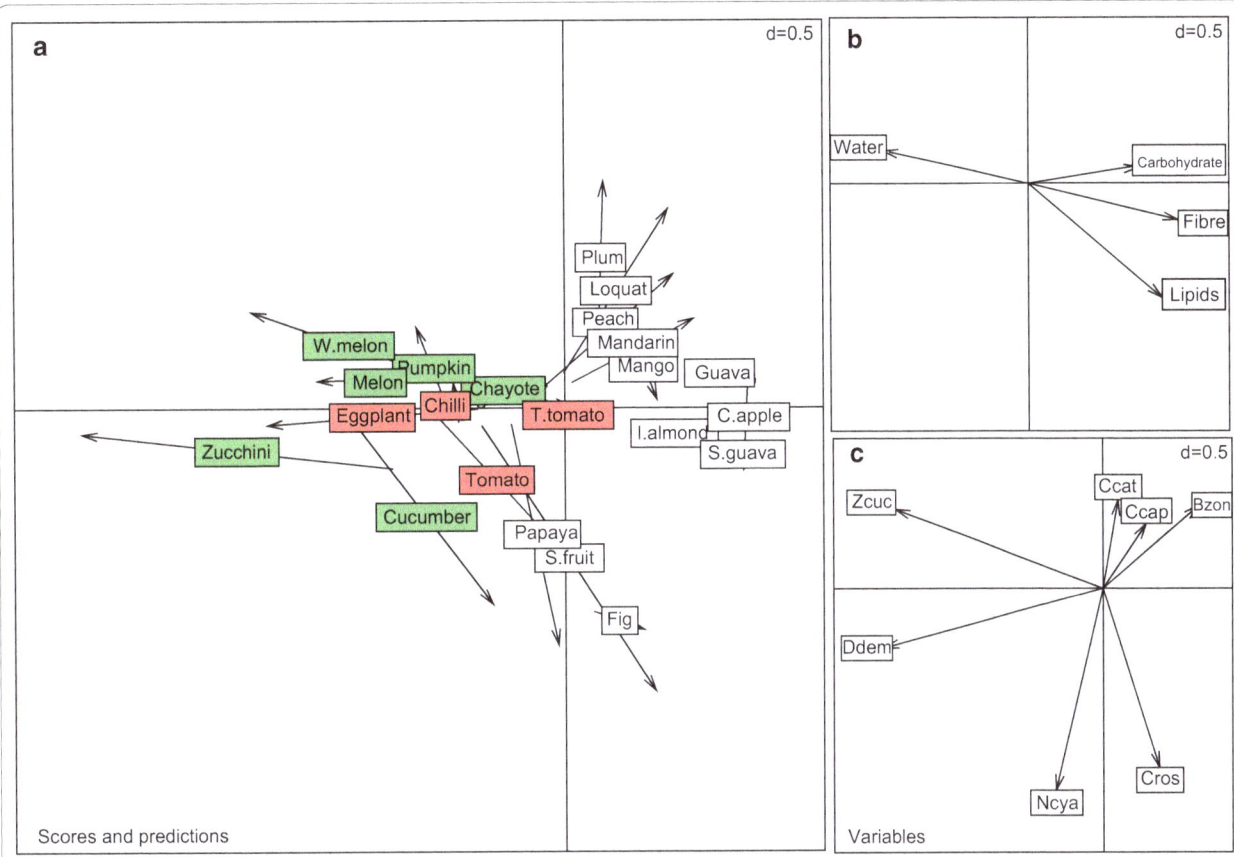

Fig. 3 CCA analysis of the relationship between fruit composition and survival of seven tephritid species. **a** Position of the different fruit species in the factorial space of the CCA analysis with contribution of nutrient composition and larval survival on the studied fruits. **b** Contribution of the nutrient composition to the factorial space. **c** Position of larval survival of the seven tephritid species in the factorial space. Host fruits belonging to Cucurbitaceae and Solanaceae families are indicated by *green* and *red*, respectively. Ccat, *C. catoirii*; Ccap, *C. capitata*; Cros, *C. rosa*; Bzon, *B. zonata*; Zcuc, *Z. cucurbitae*; Ddem, *D. demmerezi*; Ncya, *N. cyanescens*

concentration in fruits is higher and more variable [75]. Carbohydrate concentration but not protein content was retained in our canonical correspondence analysis. In contrast to polyphagous species, the oligophagous tephritids *D. demmerezi* and *Z. cucurbitae* performed better on fruits with high water content than on fruits with high carbohydrate, fibre, and lipid contents. The results of this study are consistent with the hypothesis that polyphagous and oligophagous insects differ in their nutritional requirements [76]. Because pulpy fruits have less structural diversity than stems, leaves, or inflorescences, fruit-feeding tephritids have better prospects to evolve wide host ranges than those tephritids that feed on other plant parts [77].

In the current study, we used those nutrients whose contents in all 22 of the studied fruits were available in published data bases. It follows that some chemical compounds were not included. While nutritional quality of fruit pulp of the plant greatly influences larval performance, host specialisation often cannot be predicted

solely by classical nutritional measures. It is also affected by defensive chemicals (secondary metabolites and volatile compounds) and plant physical characteristics (fruit and peel texture) [13, 78, 79]. Plant secondary metabolites are commonly thought to directly or indirectly deter the fecundity or the oviposition of phytophagous insects by being toxic or by reducing nutrient assimilation [25]. Erbout et al. [80] found that larvae of the polyphagous tephritid *C. fasciventris* did not survive in fruits containing high alkaloid concentrations. Among insect herbivores, oligophagous species are less affected than polyphagous species by defensive chemicals in the tissues that they typically consume [81]. This is well illustrated by *N. cyanescens*, which is the only tephritid in La Réunion Island able to survive on the fruits of most species belonging to the Solanaceae; such fruits often contain toxic compounds [49]. Volatile compounds of the fruit peel may determine the attractivity or non-attractivity of host plants to Tephritidae, and thus enable flies to discriminate between host and non-host plants [13, 82].

Hardness of the pericarp/peel determine the ability of the female to ovipositor [79, 83]. While physical and chemical properties of the peel may also affect larval survival, this was not taken into account in our study. For example early instar larvae may suffer from heavy mortality as a consequence of flavedo essential oils and gum secretion in *Citrus* [84–86], of the formation of hardened calluses around egg cavities and of peel mechanical resistance that prevent larvae to reach the fruit pulp [84].

Conclusions

Our results suggest that nutrient composition at least partly explains the suitability of host fruits for larvae of the seven tephritids in La Réunion Island. From an applied perspective, information on the performance of phytophagous larvae on potential hosts is essential for predicting future host range expansion, population size, and plant damage [32, 87]. Future studies should also investigate female preference to increase our understanding of the factors driving tephritid host range.

Additional files

Additional file 1. Nutrient contents obtained from the literature for 22 fruit species (g or mg per 100 g of pulp).

Additional file 2. CCA analysis of the relationship between fruit composition and survival of the larvae of seven tephritid species. (a) Position of the fruit species in the factorial space of CCA analyses with contribution of nutrient composition and larval survival on the studied fruits. (b) Contribution of the nutrient composition to the factorial space. (c) Position of larval survival of the seven tephritid species in the factorial space. Host fruits belonging to the Cucurbitaceae and Solanaceae families are in green and red, respectively. Ccat: *C. catoirii*, Ccap: *C. capitata*, Cros: *C. rosa*, Bzon: *B. zonata*, Zcuc : *Z. cucurbitae*, Ddem: *D. demmerezi*, Ncya: *N. cyanescens*.

Additional file 3. Duration (days) of the larval stage of seven tephritid species reared on 22 different host fruits occurring in La Reunion. Values are means ± SE. Host fruits belonging to the Cucurbitaceae and Solanaceae families are in green and red, respectively. Host fruits are ordered by coordinate of the first axis of the CCA analyses (see "Methods" section).

Additional file 4. Relationship between pupal weight (10^{-4} g) and larval duration (days) for seven tephritid species reared on 22 different host fruits. Determination coefficients (R^2) and regression lines are given when relationships are significant ($P < 0.05$).

Abbreviations

USDA: United States Department of Agriculture; ANSES: French Agency for Food, Environmental and Occupational Health Safety; CCA: canonical correspondence analysis; ANOVA: analysis of variance; GLM: general linear model.

Authors' contributions

PFD, SQ, BC and AH conceived and designed the experiments. AH performed the experiments. AH, FC, BF, VR and PFD analyzed the data. AH, BF, VR and PFD wrote the manuscript. All authors read and approved the final manuscript.

Author details
[1] CIRAD, UMR PVBMT, 97410 Saint Pierre, France. [2] Institut Supérieur Agronomique de Chott-Mariem, Laboratoire d'Entomologie et de Lutte Biologique, Université de Sousse, 4042 Sousse, Tunisia. [3] UMR « Centre de Biologie pour la Gestion des Populations », INRA-SPE, 755 avenue du Campus, Agropolis, CS 30016, 34988 Montferrier sur Lez, Cedex, France. [4] UMR « Peuplements Végétaux et Bio-agresseurs en Milieu Tropical », CIRAD Pôle de Protection des Plantes, 7 chemin de l'Irat, 97410 Saint Pierre, La Réunion, France.

Acknowledgements
We thank Jim Payet and Serge Glénac for their assistance in rearing larvae and in finding the different fruit species. We are grateful to Agathe Allibert for her help with statistical analyses, B. Jaffee for revising the English, and two anonymous reviewers for their comments on the manuscript. This work was funded by the European Union: European regional development fund (ERDF), by the Conseil Régional de la Réunion and by the Centre de coopération Internationale en Recherche Agronomique pour le Développement (CIRAD). Abir Hafsi PhD scholarship and travel funds were provided from the Ministère de l'Enseignement supérieur et de recherche scientifique de la Tunisie. The authors greatly acknowledge the Plant Protection Platform (3P, IBISA). Special thanks to Serge Quilici, who initiated this study and who died on 1 March 2015.

Competing interests
The authors declare that they have no competing interests.

References
1. Jaenike J. Host specialization in phytophagous insects. Annu Rev Ecol Syst. 1990;21:243–73.
2. Jermy T. Evolution of insect-host plant relationships. Am Nat. 1984;124(5):609–30.
3. Futuyma DJ, Moreno G. The evolution of ecological specialization. Annu Rev Ecol Syst. 1988;19:207–33.
4. Schoonhoven LM, Jermy T, van Loon JJA. Insect-plant biology: from physiology to evolution. London: Chapman & Hall; 1998.
5. Abrams PA. Adaptive change in the resource-exploitation traits of a generalist consumer: the evolution and coexistence of generalists and specialists. Evolution. 2006;60(3):427–39.
6. Gripenberg S, Mayhew PJ, Parnell M, Roslin T. A meta-analysis of preference-performance relationships in phytophagous insects. Ecol Lett. 2010;13(3):383–93.
7. Ravigné V, Dieckmann U, Olivieri I. Live where you thrive: joint evolution of habitat choice and local adaptation facilitates specialization and promotes diversity. Am Nat. 2009;174(4):141–69.
8. Craig TP, Itami JK. Evolution of preference and performance relationships. In: Tilmon KJ, editor. Specialization, speciation, and radiation The evolutionary biology of herbivorous insects. London: University of California Press; 2008. p. 20–8.
9. Keeler MS, Chew FS. Escaping an evolutionary trap: preference and performance of a native insect on an exotic invasive host. Oecologia. 2008;156(3):559–68.
10. Wiklund C. Host plant suitability and the mechanism of host selection in larvae of *Papilio machaon*. Entomol Exp Appl. 1973;16(2):232–42.
11. Thompson JN. Evolutionary ecology of the relationship between oviposition preference and performance of offspring in phytophagous insects. Entomol Exp Appl. 1988;47(1):3–14.
12. Balagawi S, Drew RA, Clarke AR. Simultaneous tests of the preference-performance and phylogenetic conservatism hypotheses: is either theory useful? Arth Plant Int. 2013;7(3):299–313.
13. Fitt GP. The roles of adult and larval specialisations in limiting the occurrence of five species of *Dacus* (Diptera: Tephritidae) in cultivated fruits. Oecologia. 1986;69(1):101–9.
14. Cunningham J. Can mechanism help explain insect host choice? J Evol Biol. 2012;25(2):244–51.
15. Mayhew PJ. Adaptive patterns of host-plant selection by phytophagous insects. Oikos. 1997;79(3):417–28.

16. Price PW. Patterns in the population dynamics of insect herbivores. In: Leather SR, Watt AD, Mills NJ, Walters KFA, editors. Individuals, populations and patterns in ecology. Andover: Intercept; 1994. p. 109–17.

17. Futuyma DJ, Keese MC. Evolution and coevolution of plants and phytophagous arthropods. In: Gerald AR, May RB, editors. Herbivores their interaction with secondary plant metabolites ecological and evolutionary processes. II ed. London: Academic Press Limited; 1992. p. 437–65.

18. Kuussaari M, Singer M, Hanski I. Local specialization and landscape-level influence on host use in an herbivorous insect. Ecology. 2000;81(8):2177–87.

19. Fahrig L, Paloheimo J. Effect of spatial arrangement of habitat patches on local population size. Ecology. 1988;69(2):468–75.

20. Friberg M, Wiklund C. Host plant preference and performance of the sibling species of butterflies *Leptidea sinapis* and *Leptidea reali*: a test of the trade-off hypothesis for food specialisation. Oecologia. 2009;159(1):127–37.

21. Murphy SM. The effect of host plant on larval survivorship of the Alaskan swallowtail butterfly (*Papilio machaon aliaska*). Entomol Exp Appl. 2007;122(2):109–15.

22. Scriber JM, Feeny P. Growth of herbivorous caterpillars in relation to feeding specialization and to the growth form of their food plants. Ecology. 1979;60(4):829–50.

23. Cornell HV, Hawkins BA. Herbivore responses to plant secondary compounds: a test of phytochemical coevolution theory. Am Nat. 2003;161(4):507–22.

24. Mithöfer A, Boland W. Plant defense against herbivores: chemical aspects. Annu Rev Plant Biol. 2012;63:431–50.

25. Awmack CS, Leather SR. Host plant quality and fecundity in herbivorous insects. Annu Rev Entomol. 2002;47:817–44.

26. Chapman R. Foraging and food choice in phytophagous insects. In: Hardege JD, editor. Chemical ecology. Oxford: Eolss Publishers; 2009. p. 99–141.

27. Chapman R. The chemical inhibition of feeding by phytophagous insects: a review. Bull Entomol Res. 1974;64(03):339–63.

28. Howe GA, Jander G. Plant immunity to insect herbivores. Annu Rev Plant Biol. 2008;59:41–66.

29. Schoonhoven L, Meerman J. Metabolic cost of changes in diet and neutralization of allelochemics. Entomol Exp Appl. 1978;24(3):689–93.

30. Aluja M, Mangan RL. Fruit fly (Diptera: Tephritidae) host status determination: Critical conceptual, methodological, and regulatory considerations. Annu Rev Entomol. 2008;53:473–502.

31. Bernays E, Chapman R. Host-plant selection by phytophagous insects. In: Bernays E, Chapman R, editors. Behavior: the process of host-plant selection, vol. 2. New York: Springer; 1994. p. 95–165.

32. Scriber J, Slansky JF. The nutritional ecology of immature insects. Annu Rev Entomol. 1981;26(1):183–211.

33. Koerner SE, Burkepile DE, Fynn RW, Burns CE, Eby S, Govender N, Hagenah N, Matchett KJ, Thompson DI, Wilcox KR. Plant community response to loss of large herbivores differs between North American and South African savanna grasslands. Ecology. 2014;95(4):808–16.

34. Ritchie M, Olff H. Herbivore diversity and plant dynamics: compensatory and additive effects. In: Olff H, Brown VK, Drent RH, editors. Herbivores: between plants and predators. Oxford: Blackwell science; 1999. p. 175–204.

35. White IM, Elson-Harris MM. Fruit flies of economic significance: their identification and bionomics. Wallingford: CAB International; 1992.

36. Quilici S, Jeuffrault E. Plantes-hôtes des mouches des fruits: Maurice, Réunion, Seychelles. La Réunion: Graphica Saint-André; 2001.

37. Duyck PF, David P, Pavoine S, Quilici S. Can host-range allow niche differentiation of invasive polyphagous fruit flies (Diptera: Tephritidae) in La Réunion? Ecol Entomol. 2008;33(4):439–52.

38. Ekesi S, Nderitu PW, Chang CL. Adaptation to and small-scale rearing of invasive fruit fly *Bactrocera invadens* (Diptera: Tephritidae) on artificial diet. Ann Entomol Soc Am. 2007;100(4):562–7.

39. Ekesi S, Mohamed SA, Chang CL. A liquid larval diet for rearing *Bactrocera invadens* and *Ceratitis fasciventris* (Diptera: Tephritidae). Int J Trop Insect Sci. 2014;34(S1):S90–8.

40. Krainacker D, Carey J, Vargas R. Effect of larval host on life history traits of the Mediterranean fruit fly, *Ceratitis capitata*. Oecologia. 1987;73(4):583–90.

41. Vargas R, Mitchell S, Hsu C-L, Walsh WA. Laboratory evaluation of diets of processed corncob, torula yeast, and wheat germ on four developmental stages of Mediterranean fruit fly (Diptera: Tephritidae). J Econ Entomol. 1994;87(1):91–5.

42. Kaspi R, Mossinson S, Drezner T, Kamensky B, Yuval B. Effects of larval diet on development rates and reproductive maturation of male and female Mediterranean fruit flies. Physiol Entomol. 2002;27(1):29–38.

43. Fernandes Da Silva PG, Zucoloto FS. The influence of host nutritive value on the performance and food selection in *Ceratitis capitata* (Diptera, Tephritidae). J Insect Physiol. 1993;39(10):883–7.

44. Nash WJ, Chapman T. Effect of dietary components on larval life history characteristics in the medfly (*Ceratitis capitata*: Diptera, Tephritidae). PLoS ONE. 2014;9(1):1–9.

45. Duyck PF, Quilici S. Survival and development of different life stages of three Ceratitis spp. (Diptera: Tephritidae) reared at five constant temperatures. Bull Entomol Res. 2002;92(06):461–9.

46. Duyck PF, David P, Quilici S. A review of relationships between interspecific competition and invasions in fruit flies (Diptera: Tephritidae). Ecol Entomol. 2004;29(5):511–20.

47. Brévault T, Quilici S. Relationships between temperature, development and survival of different life stages of the tomato fruit fly, *Neoceratitis cyanescens*. Entomol Exp Appl. 2000;94(1):25–30.

48. Vayssières JF, Carel Y, Coubes M, Duyck PF. Development of immature stages and comparative demography of two cucurbit-attacking fruit flies in Reunion Island: *Bactrocera cucurbitae* and *Dacus ciliatus* (Diptera Tephritidae). Environ Entomol. 2008;37(2):307–14.

49. Brévault T, Duyck PF, Quilici S. Life-history strategy in an oligophagous tephritid: the tomato fruit fly, *Neoceratitis cyanescens*. Ecol Entomol. 2008;33(4):529–36.

50. Krainacker D, Carey J, Vargas R. Size-specific survival and fecundity for laboratory strains of two tephritid (Diptera: Tephritidae) species: implications for mass rearing. J Econ Entomol. 1989;82(1):104–8.

51. R Development Core Team. R: a language and environment for statistical computing. Vienna: R foundation for statistical computing; 2008.

52. Ter Braak CJ. Canonical correspondence analysis: a new eigenvector technique for multivariate direct gradient analysis. Ecology. 1986;67(5):1167–79.

53. Dray S, Dufour A-B. The ade4 package: implementing the duality diagram for ecologists. J Stat Softw. 2007;22(4):1–20.

54. Rubinstein RY, Kroese DP. Simulation and the Monte Carlo method, vol. 707. New York: Wiley; 2011.

55. Stearns SC. The evolution of life histories, vol. 249. Oxford: Oxford University Press; 1992.

56. Clarke AR, Armstrong KF, Carmichael AE, Milne JR, Raghu S, Roderick GK, Yeates DK. Invasive phytophagous pests arising through a recent tropical evolutionary radiation: the *Bactrocera dorsalis* complex of fruit flies. Annu Rev Entomol. 2005;50:293–319.

57. Vayssières JF, Rey JY, Traoré L. Distribution and host plants of *Bactrocera cucurbitae* in West and Central Africa. Fruits. 2007;62(6):391–6.

58. Kapoor V, Agarwal M, Cavalloro R. Fruit flies and their increasing host plants in India: In: Proceedings of the CEC/IOBC international symposium on fruit flies of economic importance, Athens, Greece. Rotterdam: AA Balkema, Published; 1982. p. 252–7.

59. Vargas RI, Carey JR. Comparison of demographic parameters for wild and laboratory-adapted Mediterranean fruit fly (Diptera: Tephritidae). Ann Entomol Soc Am. 1989;82(1):55–9.

60. Genc H. Adaptation process of wild population of olive fruit fly (*Bactrocera Oleae* (Rossi) (Diptera: Tepritidae)) into the laboratory. In: International conference on biological, civil and environmental engineering (BCEE-2014). Dubai (UAE). Published; 2014.

61. Vaníčková L, do Nascimento RR, Hoskovec M, Ježková Z, Břízová R, Tomčala A, Kalinová B. Are the wild and laboratory insect populations different in semiochemical emission? The case of the medfly sex pheromone. J Agric Food Chem. 2012;60(29):7168–76.

62. Richerson JV, Cameron EA. Differences in pheromone release and sexual behavior between laboratory-reared and wild gypsy moth adults. Environ Entomol. 1974;3(3):475–81.

63. Wasserman SS, Futuyma DJ. Evolution of host plant utilization in laboratory populations of the southern cowpea weevil, *Callosobruchus maculatus* Fabricius (Coleoptera: Bruchidae). Evolution. 1981;35(4):605–17.

64. West-Eberhard MJ. Phenotypic plasticity and the origins of diversity. Annu Rev Ecol Syst. 1989;20:249–78.

65. Augustinos AA, Kyritsis GA, Papadopoulos NT, Abd-Alla AM, Cáceres C, Bourtzis K. Exploitation of the medfly gut microbiota for the enhancement of sterile insect technique: use of *Enterobacter* sp. in larval diet-based probiotic applications. PloS ONE. 2015;10(9):e0136459.

66. Tsuchida T, Koga R, Fukatsu T. Host plant specialization governed by facultative symbiont. Science. 2004;303(5666):1989.

67. Frago E, Dicke M, Godfray HCJ. Insect symbionts as hidden players in insect—plant interactions. Trends Ecol Evol. 2012;27(12):705–11.

68. Feder JL, Reynolds K, Go W, Wang EC. Intra-and interspecific competition and host race formation in the apple maggot fly, *Rhagoletis pomonella* (Diptera: Tephritidae). Oecologia. 1995;101(4):416–25.

69. Duyck PF, David P, Junod G, Brunel C, Dupont R, Quilici S. Importance of competition mechanisms in successive invasions by polyphagous tephritids in La Réunion. Ecology. 2006;87(7):1770–80.

70. Zucoloto FS. Feeding habits of *Ceratitis capitata* (Diptera: Tephritidae): can larvae recognize a nutritionally effective diet? J Insect Physiol. 1987;33(5):349–53.

71. Joern A, Behmer ST. Importance of dietary nitrogen and carbohydrates to survival, growth, and reproduction in adults of the grasshopper *Ageneotettix deorum* (Orthoptera: Acrididae). Oecologia. 1997;112(2):201–8.

72. Lee KP, Simpson SJ, Clissold FJ, Brooks R, Ballard JW, Taylor PW, Soran N, Raubenheimer D. Lifespan and reproduction in *Drosophila*: new insights from nutritional geometry. PNAS. 2008;105(7):2498–503.

73. Roeder KA, Behmer ST, Davidowitz G. Lifetime consequences of food protein-carbohydrate content for an insect herbivore. Funct Ecol. 2014;28(5):1135–43.

74. Behmer ST. Insect herbivore nutrient regulation. Annu Rev Entomol. 2009;54:165–87.

75. Finglas P, Roe M, Pinchen H, Berry R, Church S, Dodha S, Farron Wilson M, Swan G. McCance and Widdowson's the composition of foods. Cambridge: Royal Society of Chemistry; 2015.

76. Raubenheimer D, Simpson S. Nutrient balancing in grasshoppers: behavioural and physiological correlates of dietary breadth. J Exp Biol. 2003;206(10):1669–81.

77. Zwolfer H, Cavalloro R. Life systems and strategies of resource exploitation in tephritids. In: Proceedings of the CEC/IOBC international symposium on fruit flies of economic importance, Athens, Greece, 16–19 November 1982. Rotterdam: AA Balkema, Published; 1983. p. 16–30.

78. Renwick JAA. Variable diets and changing taste in plant–insect relationships. J Chem Ecol. 2001;27(6):1063–76.

79. Bateman M. The ecology of fruit flies. Annu Rev Entomol. 1972;17(1):493–518.

80. Erbout N, De Meyer M, Vangestel C, Lens L. Host plant toxicity affects developmental rates in a polyphagous fruit fly: experimental evidence. Biol J Linnean Soc. 2009;97(4):728–37.

81. Ali JG, Agrawal AA. Specialist versus generalist insect herbivores and plant defense. Trends Plant Sci. 2012;17(5):293–302.

82. Balagawi S, Vijaysegaran S, Drew RA, Raghu S. Influence of fruit traits on oviposition preference and offspring performance of *Bactrocera tryoni* (Froggatt) (Diptera: tephritidae) on three tomato (*Lycopersicon lycopersicum*) cultivars. Aust J Entomol. 2005;44(2):97–103.

83. Díaz-Fleischer F, Aluja M. Clutch size in frugivorous insects as a function of host firmness: the case of the tephritid fly *Anastrepha ludens*. Ecol Entomol. 2003;28(3):268–77.

84. Greany P, Styer S, Davis P, Shaw P, Chambers D. Biochemical resistance of citrus to fruit flies. Demonstration and elucidation of resistance to the Caribbean fruit fly *Anastrepha suspensa*. Entomol Exp Appl. 1983;34(1):40–50.

85. Papachristos DP, Papadopoulos NT, Nanos GD. Survival and development of immature stages of the Mediterranean fruit fly (Diptera: Tephritidae) in citrus fruit. J Econ Entomol. 2008;101(3):866–72.

86. Papachristos DP, Papadopoulos NT. Are citrus species favorable hosts for the Mediterranean fruit fly? A demographic perspective. Entomol Exp Appl. 2009;132(1):1–12.

87. Smirle M. Larval performance of two leafroller species on known and potential hosts. Entomol Exp Appl. 1993;67(3):223–31.

11

Effects of salinity on nest-building behaviour in a marine fish

Topi K. Lehtonen[1,2,3,4], Bob B. M. Wong[4] and Charlotta Kvarnemo[3*]

Abstract

Background: Parental allocation and reproductive success are often strongly influenced by environmental factors. In this respect, salinity is a key factor influencing species distributions and community structure in aquatic animals. Nevertheless, the effects of salinity on reproductive behaviours are not well known. Here, we used the sand goby (*Pomatoschistus minutus*), a small fish inhabiting a range of different salinities, to experimentally assess the effects of changes in salinity on nesting behaviour, a key component of reproduction in sand gobies and many other taxa.

Results: We found that salinity levels influenced some aspects of male nesting behaviour (i.e. nest entrance size) but not others (i.e. latency to build a nest, choice of nest site, sand on top of nest) and that small and large individuals were differently affected. In particular, the importance of body size in adjustment of nest entrance depended on the salinity level.

Conclusion: The results support the prediction that geographically widespread aquatic species, such as sand gobies, are able to perform well under a range of salinity levels. The phenotype by environment interaction found between male size and behavioural responses to salinity can, in turn, help to explain the notable variation observed in nest-building (and other) behaviours closely linked to reproduction.

Keywords: Body size, Environmental change, Nest-building, Parental care, Phenotypic plasticity, Salinity, Sand goby

Background

Parental allocation and reproductive success are often strongly influenced by environmental factors [1]. This is especially true in species that rear their eggs or young in purpose-built nests [2–5], with nest builders often adjusting their choice of nesting site or nest architecture according to local environmental conditions [6–8]. Such adjustments, in turn, can affect the costs of nest building and nest maintenance, as well as the suitability of the conditions in the nest for the developing offspring [2, 9–13] (but see also [14]). Furthermore, nest characteristics may also act as extended phenotypic signals that reveal important information about the quality of the builder [15], with the value of this information often influenced by environmental context [16, 17].

For aquatic animals, salinity is a key factor influencing species distributions and community structure [18–21]. For instance, salinity can affect metabolic costs and growth rates of both adults and juveniles—even in species capable of surviving under a range of salinity levels [22–25]. Aside from such metabolic and physiological effects, salinity can also affect the costs and benefits of parental behaviours. For example in the flagfish, *Jordanella floridae*, salinity influences the benefits of egg-care and nest-directed behaviours [26–28]. Indeed, the effects of salinity on reproductive behaviours may be particularly pertinent in environments where salinity levels vary both spatially and temporally, such as the Baltic Sea. For instance, egg development of many marine species in the Baltic Sea is affected by the lower salinity levels [29–32], with gametes of commercially important taxa—such as cod, *Gadus morhua*, and flatfishes—being close to the limit of their salinity tolerance, while also showing local adaptation to salinity [19, 30, 33, 34].

*Correspondence: lotta.kvarnemo@bioenv.gu.se
[3] Department of Biological and Environmental Science, University of Gothenburg, Box 463, 40530 Gothenburg, Sweden
Full list of author information is available at the end of the article

The sand goby, *Pomatoschistus minutus*, is a small marine fish with a widespread distribution across low and high salinity environments of coastal Europe, including those of the Baltic Sea [35], where salinities range from <3 ppt in the Northern Baltic to conditions close to fully marine/oceanic near the mouth of the Sea [36–39]. Male sand gobies typically build nests under empty mussel shells or rocks by excavating sand under the substrate and piling it on top of the shell or rock, leaving a single narrow opening. The size of the nesting site (also known as 'nesting resource' sensu [40]) and the characteristics of the nest itself can have a direct influence on male reproductive success and offspring survival. For example, nests with considerable amounts of sand on top (i.e. those that are well-covered) have been found to protect eggs and nest-tending males from predation [41]. Indeed, the amount of sand males use when building their nests can be substantial (Fig. 1), with the weight of sand piled on top of the nest sometimes exceeding 100 times the body mass of the nest-builder [40, 42]. Earlier results also suggest that food supplemented males invest more in nest building than less well-fed, control males [43]. The size of the entrance of a sand goby nest, in turn, is likely to be relevant in terms of both ventilation of the eggs and avoidance of egg predation, with ventilation being facilitated by a large nest entrance and predator defence aided by a small entrance [44, 45]. Interestingly, females prefer builders of elaborate nests in many [46–48] but not all [48–50] environmental settings. Regarding the size of the nesting resource, large nests accommodate more eggs [51, 52], while potentially also being more costly—not

only to build but also to defend against nest take-overs and egg predation attempts [53, 54].

Here, we expected males to adjust both nest-building behaviour and nest architecture to key environmental conditions (see [40, 42, 44, 55]), in this case salinity levels. In particular, not only can salinity impact metabolic costs to adults and developing young, but eggs at lower salinity levels are also more susceptible to microbial infections from pathogens, such as *Saprolegnia* water moulds [56, 57]. Yet, despite the potential importance of salinity on reproductive success, very little is known about the effects of salinity on nest choice, nest-building behaviours or nest architecture. This is an important knowledge gap because we expect such behavioural adjustments to also affect the capacity of sand gobies—and other nest-building fish—to colonize new habitats and to cope with changes in salinity levels, such as those predicted to take place in the Baltic Sea [29–31].

In the current study, we experimentally assessed the effects of salinity on nest building behaviour and nest architecture in the sand goby, by focussing on a population from the low salinity environment (salinity: 5.5 ppt) of the northern Baltic Sea. We considered four mutually exclusive hypotheses. First, sand gobies are adapted to local environmental conditions. If this is the case, we predict that the latency or ability to build a nest, the size of the chosen nest site, and/or the extent of nest elaboration should differ depending on salinity, with non-native salinities being linked to costs that have a negative impact on nest building behaviour. Second, if an isotonic environment (~ 9–12 ppt) results in energy

Fig. 1 A sand goby nest constructed by piling sand on *top* of, and excavating under, a halved flowerpot

savings, as shown for some aquatic organisms [22], the nesting behaviours could, instead, be positively influenced by an intermediate salinity level. Third, because the ancestral population of sand gobies that colonised the Baltic Sea several thousand years ago lived in high salinity conditions (see [58, 59]), as do most of the modern sand goby populations outside of the Baltic Sea [38], nest building may have evolved to peak (in terms of building motivation/latency and nest elaboration) in high salinity conditions. Finally, given their geographically widespread distribution [38], it is also feasible that sand gobies may be able to perform equally well under a range of different salinity levels. In that case, we may not expect to see any differences in nesting behaviour or nest architecture.

In addition to evaluating the effects of salinity, we also assessed the relationship between the nest-related behaviours and male body size. We considered that body size may be important because recent studies have suggested that individuals with different body sizes vary in their responses to environmental conditions [40, 42, 60]. In particular, these studies found the association between male size, the level of nest elaboration and motivation to spawn successfully to be positive only in the absence of environmental disturbance, such as predation risk or water turbidity. Moreover, smaller individuals may, due to their larger surface-to-volume ratio, be less tolerant of suboptimal salinities.

Methods

Fish collection and housing

The study was carried out at the Tvärminne Zoological Station (59°50.7′ N; 23°15.0′ E) of the University of Helsinki in 2014 during the sand goby breeding season, which, in this population, peaks from late May to early July. With permission from the field station, we collected sand goby males for the experiment in a nature reserve located near the station and owned by the University of Helsinki. Besides using a hand trawl for catching gobies [57], we also placed artificial nesting resources (10 × 10 cm ceramic tiles) in a nearby shallow bay and waited for males to start building their nests. The nesting males were then caught using dip nets and transported to the field station. Males were first kept for a short period (less than a week) in aquaria of ~100 l, fed *ad libitum* with live mysid shrimp (*Neomysis integer*) and supplied with a continuous through flow of natural brackish seawater, pumped straight from the Baltic Sea. All stocking, acclimatisation and experimental tanks (see below) were placed in a green house that was subject to natural day/night rhythm (with the length of the day being on average 18.5 h during the time of the study).

Acclimatisation

Before the experiment, focal males were acclimatised to the appropriate salinity treatments, i.e. 6, 12 and 24 ppt (see below). To achieve this, we haphazardly distributed male gobies into acclimatisation tanks (length × width × height of water level: 70 × 25 cm × approx. 25 cm) containing a 2 cm layer of sand as substrate. Concurrently, we had 1–2 acclimatisation tanks per salinity treatment, each housing initially 15–25 males, with new acclimatisation runs being initiated when needed. The tanks were continuously aerated and placed within larger tanks with continuously renewed seawater to ensure that the temperature of the acclimation tanks followed natural conditions and was identical to the temperature in the stock tanks.

All of the acclimatisation tanks initially contained water that was maintained at a salinity level of 6 ppt, which was achieved using a commercial sea salt mix ('Instant Ocean', Spectrum Brands Inc.) added to deionised water. This salinity level is slightly higher than the salinity level experienced by gobies in this part of the Baltic (~5.5 ppt). The initial salinity levels in the acclimation tanks were then adjusted over time depending on treatment. For the 'high salinity treatment' (24 ppt), we gradually increased the salinity level over a 7-day period until we reached a salinity level of 24 ppt. For our 'medium salinity treatment' (12 ppt), salinity was increased at a similar rate, this time over a 3–4 day period, until we reached the target level of 12 ppt. For the 'low salinity treatment' (6 ppt), no additional salt was added to the water. To help maintain high water quality in the acclimatisation tanks, we performed a 50 % water change on day 7 in all of the tanks, whereby water removed from the tanks was replaced with clean water of the appropriate salinity level. Apart from this, tanks were checked regularly to ensure that any water that had been lost to evaporation was replaced with deionised water to maintain the target salinity levels in the tanks. For all three treatments, males entered the experiment (see below) 7–21 days after the start of the acclimatisation period. Here, we wanted to minimise treatment differences in the time fish spent in the target salinity prior to experimentation, while also running a similar number of replicates of each treatment at any given time. During the stocking and acclimation periods, males were fed with live mysid shrimp and frozen chironomid larvae.

Experimental design

The aim of the experiment was to investigate whether changes in salinity levels affect investment of sand goby males into nest-building. Specifically we were interested in measuring (1) the time taken for males to begin nest building (as a measure of their motivation to nest), (2)

the size of the preferred nest site, and (3) the characteristics of the nest itself (i.e. nest architecture in terms
of nest elaboration and nest entrance size). Before the
start of each replicate, the male was weighed using an
electronic balance and its total length was measured
to the nearest 0.5 mm using a measuring board with a
grid scale. Each focal male was then placed into an individual experimental tank measuring 75 × 25 × 20 cm
(length × width × height of water level), the bottom of
which was covered with a 4 cm layer of fine sand. Each
tank also contained three halved clay flowerpots for the
focal male to potentially select to build his nest. These
flowerpots differed in size, representing small (diameter
of the mouth of the pot = 4 cm), medium (6.5 cm) and
large (9.5 cm) nest sites. The three nest sites were randomly assigned to the left, right and centre of each tank,
all with their entrances facing the front of the aquarium.

Water in these experimental tanks was prepared as
above. Each tank was aerated by a pump, with an airstone
being placed in the back of the tank behind the middle
nest site. As with the acclimatisation tanks, experimental tanks were placed within a larger aquarium that was
supplied with a through flow of fresh seawater to ensure
that temperature was the same as the stock tanks. Males
from acclimatisation tanks were only transferred to
experimental tanks of matching salinities. Six replicates
(n_{low} = 2; n_{medium} = 1; n_{high} = 3) were discontinued
because the male showed signs of distress (e.g. erratic
swimming behaviour). The experiment was successfully
replicated 36 times in each treatment, with male total
lengths [mean ± SE] being 52.0 ± 0.8 cm, 51.9 ± 0.9 cm,
and 52.1 ± 0.9 cm, and weights 1.03 ± 0.05 grams,
1.05 ± 0.06 g, and 1.04 ± 0.05 g in the low, medium and
high salinity treatments, respectively (with one missing
set of body size values in the high salinity treatment due
to human error).

For each replicate, males were given up to 60 h to start
building a nest. During this time, tanks were checked
~7 times daily between 07:30 and 22:30 to record male
behaviour, male location and any signs of nest building.
A male was considered to have initiated nest-building
when he had started to pile sand on top of, and excavate
under, the pot [40, 61, 62]. When a male was observed
to have started building a nest, the time it took to initiate nest building was recorded as the time point that
was midway between the check in which the onset of
nest building was observed and the previous check [63].
The male was left in the tank for another 11 h before we
recorded which one of the three pots it was occupying.
In many replicates, the male had, at least partially, built
a nest under more than one of the three nest sites (as
also observed by Japoshvili et al. [61]). In such cases, we
determined the male's principal choice being the nest site

inside which it had been most often observed after the
onset of nest-building.

For each constructed nest, we measured two ecologically relevant nest attributes (see [61]), namely the level
of nest elaboration (measured as the amount of sand the
male had displaced to cover the flower pot) and the width
of the entrance to the nest. The level of nest elaboration
(sensu Lehtonen and Wong [50]) was assessed by carefully lifting the flowerpot into a tray and collecting the
sand that had been piled on top of the pot. Due to the
shape of the flowerpots, only the sand placed directly on
the ridge of the pot was collected. Our visual assessment
suggests that this sand sample had a high correlation
with the total amount of sand the male had placed on the
nest and it was at least as reliable an estimate of the total
amount of sand as, for example, the height of the sand
layer on the ridge of the pot [61]. The collected sand was
then dried in an oven for 36 h at 60 °C, and weighed on
an electric balance [50]. The width of the nest entrance
was measured by taking a digital photograph of the front
of the nest using an Olympus XZ-1 digital camera, with
a ruler placed next to the nest entrance as a scale (Fig. 1).
This scale was later used for calibration in the image analysis software ImageJ (U.S. National Institutes of Health,
Bethesda, Maryland, USA) to enable us to measure the
nest entrance width.

After the experiments, most of the remaining experimental males were retained for other behavioural
research or, wherever possible, gradually acclimated to
local seawater conditions before being returned to the
sea. In cases where re-acclimatisation was not possible
due to logistical constraints, animals were euthanized
(four males from the 24 ppt treatment).

Statistical analyses

To test whether the time males spent in the experimental tank before starting to build a nest differed between
treatments, we applied a Cox proportional hazards survival analysis with salinity treatment and male body mass
as explanatory variables. Any males that did not commence nest-building within the allocated 60 h period
were 'right censored' for the purpose of the survival analysis [64]. To investigate the focal males choice between
small, medium and large nesting sites, we used an ordinal
logistic regression, again with treatment and body mass
as explanatory variables. Finally, the amount of sand on
the top of the nest site, as well as the nest entrance width
(both log-transformed for improved normality), were
analysed in two separate linear models, with salinity and
male body mass as fixed effects, and the size of the chosen nest site and the time until the onset of nest building
(in hours, log-transformed) as covariates. We then simplified the models by removing the salinity × body size

interaction, if it was found to be non-significant [65]. All analyses were run using R 3.1.0 software (R Development Core Team).

Results

Time to nest building
Of the 36 males in each treatment, 34, 33 and 33 built a nest within 60 h in the low, medium and high salinity treatments, respectively, indicating no difference between the salinity treatments (G test of independence with Williams' correction, $G = 0.0102$, $df = 2$, $p = 0.99$). Similarly, neither salinity (Cox proportional hazards model, salinity treatment effect, $\chi^2 = 3.111$, $df = 2$, $p = 0.21$), male weight ($\chi^2 = 1.403$, $df = 1$, $p = 0.24$), nor their interaction ($\chi^2 = 1.356$, $df = 2$, $p = 0.51$) had a significant effect on time from the beginning of the experiment until the onset of nest building. It is worth noting that the main effects were also non-significant in a model fitted without the interaction term.

Choice of the size of nest site
In the low salinity (6 ppt), none of the males chose the small pot, 23 chose the medium-sized pot, and 11 the large pot. In the medium salinity (12 ppt), the numbers of chosen pots were 2, 19 and 12 for small, medium and large pots, respectively. Finally, in high salinity (24 ppt), males chose 5, 19 and 9 small, medium, and large pots, respectively. Hence, the treatments did not significantly differ in the distribution of chosen nest sites (ordinal logistic regression, $z = 0.721$, $df = 2$, $p = 0.47$). However, male body mass had a significant effect (ordinal logistic regression, $z = 2.58$, $p < 0.01$), with larger males choosing larger nest sites (Fig. 2). Overall, males showed a strong preference towards medium-sized nest sites (61

medium nests chosen in 100 trials; binomial distribution with the H_0 being that a medium-sized nest would be chosen every third time, $p < 0.001$), indifference towards large nest sites (32 large nests chosen out of 100 trials; binomial distribution, $p = 0.87$) and avoidance of small nest sites (7 small nest chosen; binomial distribution, $p < 0.001$).

Nest characteristics
For the amount of sand piled on top of their preferred nest site, the interaction between salinity and body mass was found to be non-significant (general linear model, $F_{2,90} = 0.2947$, $p = 0.75$) and we refitted the model without the interaction term. The simplified model showed that salinity level did not have a significant effect ($F_{2,92} = 1.306$, $p = 0.28$), whereas male body mass did ($F_{1,92} = 21.64$, $p < 0.001$). Specifically, the amount of sand piled on top of the nest was positively associated with male size (Fig. 3). Similarly, both of the covariates were significant: nests with a later onset had less sand piled on top of them ($F_{1,92} = 16.41$, $p < 0.001$) and more sand was piled on larger nests ($F_{2,92} = 10.26$, $p < 0.001$). Nevertheless, there was notable overlap in the amount of sand on the ridge of different sized nests (small: mean $= 1.8$ g, range 0.2–12.1 g, $n = 7$; medium: mean $= 5.8$ g, range 0.7–33.3 g, $n = 61$; large: mean $= 13.7$ g, range 0.2–28.9 g, $n = 32$).

In terms of nest entrance width, we found a significant interaction between salinity and male body mass (general linear model, $F_{2,90} = 7.834$, $p < 0.001$). In particular, nest

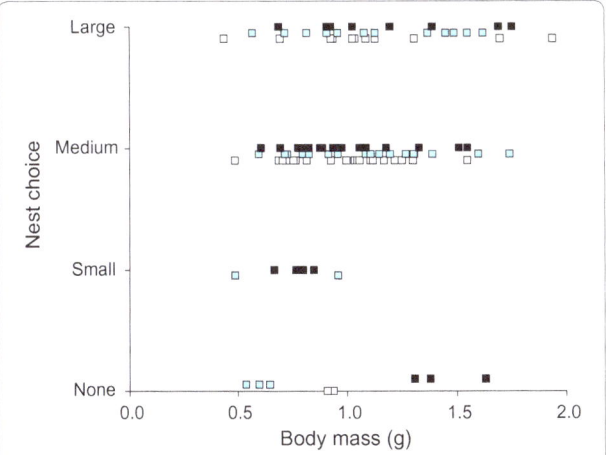

Fig. 2 Choice of nesting site size in relation to salinity treatment and male body mass. *White boxes* low salinity, *light blue boxes* medium salinity, *dark blue boxes* high salinity

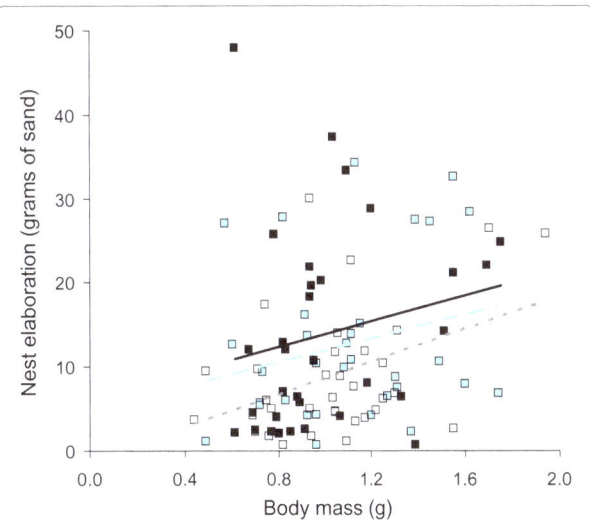

Fig. 3 Degree of nest elaboration, measured as grams of sand on ridge of nesting site, relative to salinity and male body mass. *White boxes + dotted trend line* = low salinity, *light blue boxes + dashed trend line* = medium salinity, *dark blue boxes + solid trend line* = high salinity

entrance width was positively associated with male body size in low and medium salinity but negatively associated in high salinity (Fig. 4). As with the sand on the nest, both the onset time ($F_{1,90} = 5.996$, p = 0.016) and the size of the chosen nest ($F_{2,90} = 12.02$, p < 0.001) had a significant effect, with larger entrances in nests of later onset and an extensive variation in nest entrance width for each of the three nest site sizes: (small: mean = 25 mm, range 15–33 mm, n = 7; medium: mean = 25 mm, range 9–41 mm, n = 61; large: mean = 36 mm, range 18–71 mm, n = 32).

Discussion

We found that male sand gobies from a low salinity population took comparable amounts of time to begin nest construction, irrespective of the salinities to which they were experimentally exposed. This suggests that the motivation for males to reproduce was unaffected by the increase in salinity.

An ability to perform reproductive behaviours under a range of different salinities is concordant with the expansive geographic distribution of sand gobies across coastal Europe, which encompasses both high and low salinity habitats. Indeed, the high salinity treatment in our study is the closest to the marine conditions experienced by the ancestral population of sand gobies that colonised the Baltic Sea several thousand years ago (see [58, 59]) and which most of the modern sand goby populations (outside the Baltic Sea) presently inhabit [38]. This suggests that even though the sand goby population used in our study presently inhabits a brackish water environment,

they have nevertheless retained their eagerness to reproduce under higher salinity levels.

Similarly to previous work [54, 61], we found that the choice of nest site was dependent on male size, that is, larger males preferred to nest in larger nest sites. However, male choice of nesting site was not affected by salinity. Previous studies on multiple fish species have shown that the size of the nest site can have a direct bearing on male reproductive success by, for example, acting as a physical limit to the number of eggs a male is able to receive [52, 66, 67]. Larger broods, however, are also more energetically demanding to look after e.g. because they require more fanning [68]. Because metabolic demands on both the adult and the offspring are expected to vary with salinity [22–25], we might have expected sand goby males to adjust their choice of nesting sites at different salinities in response to differences in the costs of care, e.g. by choosing smaller sites at higher salinities. However, this was not the case. One possible reason is that the costs of caring for broods is unaffected by differently sized nest sites, as found in a study where male size and brood size were kept constant [69], or that the costs of caring for differently sized broods is unaffected by different salinity levels. Alternatively, males may be able to adjust their behaviours in some other way to counter any associated changes in the cost of care (e.g. through subsequent changes to parental behaviours). If either hypothesis is true, males may not need to adjust their choice of nesting site per se. Another possibility is that choice of nest size in this population is adapted for a low salinity environment and that gobies, when exposed to higher salinities, are making suboptimal choices with respect to the size of their nesting site. In such a situation, inappropriate behavioural responses can cause animals to make poor nesting decisions, with potentially negative consequences to offspring fitness, as has been shown, for example, in birds [70, 71].

Increases in salinity had different effects on the two measures of nest architecture examined in our study, amount of sand piled on top of the nest and nest entrance width. The amount of piled sand was not affected by salinity level. This may not be surprising if the main reason for sand piling is to help conceal the nest against potential predators [41], since the value of having a well-constructed (i.e. concealed) nest should be important irrespective of salinity. Alternatively, the amount of sand piled on top of the nest may also act as an extension of the male's phenotype [42, 43] by revealing important information about the quality of the nest builder to choosy females (i.e. by serving as an extended phenotypic signal; sensu Schaedelin and Taborsky [15]). Hence, in the context of the current study, the degree of nest elaboration can also be important if the trait is condition

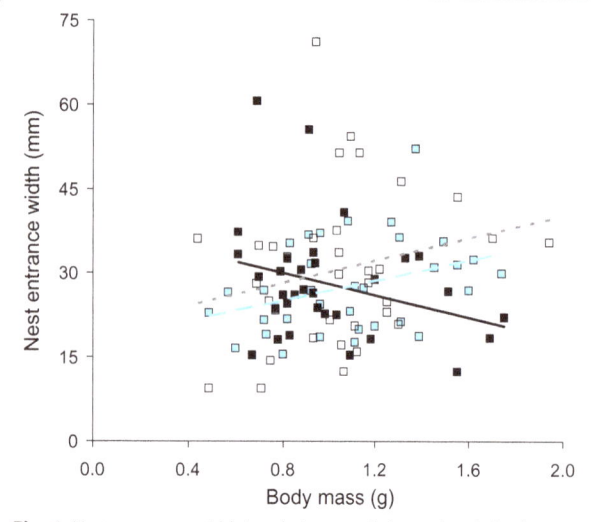

Fig. 4 Nest entrance width in relation to salinity and male body mass. *White boxes + dotted trend line* = low salinity, *light blue boxes + dashed trend line* = medium salinity, *dark blue boxes + solid trend line* = high salinity

dependent and if male condition is important to offspring survival. However, in a recent study, Lehtonen and Wong [50] found that male condition may be a poor predictor of hatching success in this particular sand goby population. Moreover, that same study revealed that the degree of nest elaboration and male condition was temporally unstable, making the amount of sand piled on top of the nest a potentially unreliable signal of male quality.

In contrast to the results we observed for the degree of nest elaboration, we found that the salinity treatment—in interaction with other factors—did affect nest entrance size. The size of the nest entrance is likely to be relevant to the nest holder's fitness in a wide range of taxa [72], as recently shown for instance in birds [73, 74]. In fish, nest entrance size has been linked to offspring survival, with studies showing that larger nest entrances improve the flow of oxygenated water into the nest, which is critical to embryo development [10, 11]. Nests with larger entrances, however, are also more difficult to defend against predators and nest challengers. Not surprisingly, male sand gobies are known to adjust the size of their nest entrance according to changes in environmental conditions, such as oxygen levels, water temperature, and presence of egg predators [43–45]. In the current study, we found that the salinity effects on nest entrance size were dependent on the size of the male. Specifically, male size was positively associated with nest entrance width in low and medium salinities, but not in the high salinity treatment. Why might this be the case?

Individuals often adjust their behaviours in different ways depending on environmental context, with evidence that variation in individual responses could be adaptive [75, 76]. Body size, in particular, can be an important source of individual behavioural variation in a wide range of taxa [77, 78]. In sand gobies, several studies have previously reported differences in male reproductive behaviour linked to body size, with adjustment of nesting behaviour to environmental factors (e.g. water turbidity, predation risk, intrasexual competition), contingent upon the size of the male [40, 42, 60]. It has been suggested that this may be due to differences in the costs and benefits incurred by small and large males in response to different environmental conditions [42]. In this respect, small males may be less able or willing to build tighter (i.e. smaller) nest entrances in higher salinity environments due to additional costs to small males under such conditions (for example as a consequence of sensitivity to osmotic stress as a result of their higher surface to volume ratio; see [79]). Furthermore, nest entrance size (and associated rate of water flow) could also be linked to salinity-dependent fitness costs by influencing the susceptibility of eggs to infection at lower salinities (e.g. by

Saprolegnia water moulds [56, 57]). Indeed, small and large males may differ in their ability to protect their developing eggs from infection (with large males having larger sperm duct glands [80] that produce an antimicrobial mucus, which parental males apply around the eggs [81]), to attract mates to fill their nests with eggs [17, 82], and to defend their nests against potential egg predators [40, 83]. Such factors could conceivably influence the way small and large males adjust their nest entrances in response to variation in salinity.

Conclusion

We found that some aspects of male nesting behaviour were influenced by salinity, while others were not, and that the impacts of salinity were different for small and large individuals. In particular, although the level of salinity change did not affect the eagerness of sand gobies to commence nest building, it did influence the importance of body size in adjustment of nest entrance size, which is an important characteristic of the nest. Hence, our results in sand gobies suggest that even populations that complete their whole life cycle in a brackish water environment can retain the ability to perform normal reproduction-related behaviours under a range of salinity levels. The phenotype by environment interaction between male size and behavioural response to salinity may, in turn, help to explain the notable variation observed in nest-building (and other) behaviours closely linked to reproduction.

Abbreviations
ppt: parts per thousand; cm: centimetres; g: grams; °C: degrees of celsius; df: degrees of freedom.

Author contributions
TKL conceived and designed the study, performed the experiment and led writing of the manuscript. BBMW provided advice for data collection and interpretation, and participated in writing the manuscript. CK provided advice for designing the study and analysing the data, performed the image measurements, and edited the manuscript. All authors read and approved the final manuscript.

Author details
[1] Department of Biosciences, Åbo Akademi University, Tykistökatu 6, 20520 Turku, Finland. [2] Section of Ecology, Department of Biology, University of Turku, 20014 Turku, Finland. [3] Department of Biological and Environmental Science, University of Gothenburg, Box 463, 40530 Gothenburg, Sweden. [4] School of Biological Sciences, Monash University, Melbourne, VIC 3800, Australia.

Acknowledgements
We thank the Tvärminne Zoological Station and Kai Lindström for logistic support, Andreas Lindén for statistical advice, and Santeri Lehtonen and Anniina Saarinen for assistance at the station. The study was also associated with the Linnaeus Centre for Marine Evolutionary Biology, CeMEB. The funding was provided by the Department of Biology, University of Turku (TKL), The Australian Research Council (BBMW) and The Swedish Research Council (CK).

Competing interests
The authors declare that they have no competing interests.

Animal ethics
Our work complies with the international animal care guidelines of the Association for the Study of Animal Behaviour [84] and the ARRIVE guidelines [85]. The experimental procedures were approved by ELLA - the National Animal Experiment Board of Finland (decision number: ESAVI/4144/04.10.07/2014).

References

1. Carlisle TR. Brood success in variable environments: implications for parental care allocation. Anim Behav. 1982;30:824–36. doi:10.1016/S0003-3472(82)80156-5.
2. Canali E, Ferrante V, Todeschini R, Verga M, Carenzi C. Rabbit nest construction and its relationship with litter development. Appl Anim Behav Sci. 1991;31:259 66. doi:10.1016/0168 1591(91)90010 U.
3. Lukas JA, Orth DJ. Factors affecting nesting success of smallmouth bass in a regulated Virginia stream. T Am Fish Soc. 1995;124:726–35. doi:10.1577/1548-8659.
4. Suski CD, Ridgway MS. Climate and body size influence nest survival in a fish with parental care. J Anim Ecol. 2007;76:730–9. doi:10.1111/j.1365-2656.2007.01242.x.
5. Byrne PG, Keogh JS. Extreme sequential polyandry insures against nest failure in a frog. Proc R Soc Lond B. 2009;276:115–20. doi:10.1098/rspb.2008.0794.
6. Burton NHK. Nest orientation and hatching success in the tree pipit Anthus trivialis. J Avian Biol. 2006;37:312–7. doi:10.1111/j.2006.0908-8857.03822.x.
7. Eggers S, Griesser M, Nystrand M, Ekman J. Predation risk induces changes in nest-site selection and clutch size in the Siberian jay. Proc R Soc Lond B. 2006;273:701–6. doi:10.1098/rspb.2005.3373.
8. Rushbrook BJ, Head ML, Katsiadaki I, Barber I. Flow regime affects building behaviour and nest structure in sticklebacks. Behav Ecol Sociobiol. 2010;64:1927–35. doi:10.1007/s00265-010-1003-3.
9. Bult A. Lynch CB Nesting and fitness: lifetime reproductive success in house mice bidirectionally selected for thermoregulatory nest-building behavior. Behav Genet. 1997;2:231–40. doi:10.1023/A:1025610130282.
10. Jones JC, Reynolds JD. Oxygen and the trade-off between egg ventilation and brood protection in the common goby. Behaviour. 1999;136:819–32. doi:10.1163/156853999501586.
11. Jones JC, Reynolds JD. The influence of oxygen stress on female choice for male nest structure in the common goby. Animal Behav. 1999;57:189–96. doi:10.1006/anbe.1998.0940.
12. Takegaki T, Nakazono A. The role of mounds in promoting water-exchange in the egg-tending burrows of monogamous goby, Valenciennea longipinnis (Lay et Bennett). J Exp Mar Biol Ecol. 2000;253:149–63. doi:10.1016/S0022-0981(00)00251-3.
13. Petit C, Hossaert-McKey M, Perret P, Blondel J, Lambrechts MM. Blue tits use selected plants and olfaction to maintain an aromatic environment for nestlings. Ecol Lett. 2002;5:585–9. doi:10.1046/j.1461-0248.2002.00361.x.
14. Burhans DE, Thompson FR III. Effects of time and nest-site characteristics on concealment of songbird nests. Condor. 1998;100:663–72.
15. Schaedelin FC, Taborsky M. Extended phenotypes as signals. Biol Rev. 2009;84:293–313. doi:10.1111/j.1469-185X.2008.00075.x.
16. Jennions MD, Petrie M. Variation in mate choice and mating preferences: a review of causes and consequences. Biol Rev. 1997;72:283–327. doi:10.1111/j.1469-185X.1997.tb00015.x.
17. Lehtonen TK, Lindström K. Females decide whether size matters: plastic mate preferences tuned to the intensity of male–male competition. Behav Ecol. 2009;20:195–9. doi:10.1093/beheco/arn134.
18. Peterson MS, Ross ST. Dynamics of littoral fishes and decapods along a coastal river-estuarine gradient. Estuar Coast Shelf Sci. 1991;33:467–83. doi:10.1016/0272-7714(91)90085-P.
19. Westin L, Nissling A. Effects of salinity on spermatozoa motility, percentage of fertilized eggs and egg development of Baltic cod (Gadus morhua), and implications for cod stock fluctuations in the Baltic. Mar Biol. 1991;108:5–9. doi:10.1007/BF01313465.
20. Jassby AD, Kimmerer WJ, Monismith SG, Armor C, Cloern JE, Powell TM, Schubel JR, Vendlinski TJ. Isohaline position as a habitat indicator for estuarine populations. Ecol Appl. 1995;5:272–89. doi:10.2307/1942069.
21. Kimmerer WJ. Effects of freshwater flow on abundance of estuarine organisms: physical effects or trophic linkages? Mar Ecol Prog Ser. 2002;243:39–55. doi:10.3354/meps243039.
22. Bœuf G, Payan P. How should salinity influence fish growth? Comp Biochem Phys C. 2001;130:411–23. doi:10.1016/S1532-0456(01)00268-X.
23. Imsland AK, Foss A, Gunnarsson S, Berntssen MHG, FitzGerald R, Bonga SW, Ham EV, Nævdal G, Stefansson SO. The interaction of temperature and salinity on growth and food conversion in juvenile turbot (Scophthalmus maximus). Aquaculture. 2001;198:353–67. doi:10.1016/S0044-8486(01)00507-5.
24. Imsland AK, Gústavsson A, Gunnarsson S, Foss A, Árnason J, Arnarson I, Jónsson AF, Smáradóttir H, Thorarensen H. Effects of reduced salinities on growth, feed conversion efficiency and blood physiology of juvenile Atlantic halibut (Hippoglossus hippoglossus L.). Aquaculture. 2008;274:254–9. doi:10.1016/j.aquaculture.2007.11.021.
25. Glover DC, DeVries DR, Wright RA. Effects of temperature, salinity and body size on routine metabolism of coastal largemouth bass Micropterus salmoides. J Fish Biol. 2012;81:1463–78. doi:10.1111/j.1095-8649.2012.03385.x.
26. St Mary CM, Noureddine CG, Lindström K. Effects of the environment on male reproductive success and parental care in the Florida flagfish, Jordanella floridae. Ethology. 2001;107:1035–52. doi:10.1046/j.1439-0310.2001.00747.x.
27. St Mary CM, Gordon E, Hale RE. Environmental effects on egg development and hatching success in Jordanella floridae, a species with parental care. J Fish Biol. 2004;65:760–8. doi:10.1111/j.1095-8649.2004.00481.x.
28. Hale RE. Evidence that context-dependent mate choice for parental care mirrors benefits to offspring. Anim Behav. 2008;75:1283–90. doi:10.1016/j.anbehav.2007.09.034.
29. Nissling A, Westin L. Egg mortality and hatching rate of Baltic cod (Gadus morhua) in different salinities. Mar Biol. 1991;111:29–32. doi:10.1007/BF01986341.
30. Thorsen A, Kjesbu OS, Fyhn HJ, Solemdal P. Physiological mechanisms of buoyancy in eggs from brackish water cod. J Fish Biol. 1996;48:457–77. doi:10.1111/j.1095-8649.1996.tb01440.x.
31. Karås P, Klingsheim V. Effects of temperature and salinity on embryonic development of turbot (Scophthalmus maximus L.) from the North Sea, and comparisons with Baltic populations. Helgoländer Meeresun. 1997;51:241–7. doi:10.1007/BF02908710.
32. Nissling A, Dahlman G. Fecundity of flounder, Pleuronectes flesus, in the Baltic Sea—Reproductive strategies in two sympatric populations. J Sea Res. 2010;64:190–8. doi:10.1016/j.seares.2010.02.001.
33. Nissling A, Westin L, Hjerne O. Reproductive success in relation to salinity for three flatfish species, dab (Limanda limanda), plaice (Pleuronectes platessa), and flounder (Pleuronectes flesus), in the brackish water Baltic Sea. ICES J Mar Sci. 2002;59:93–108. doi:10.1006/jmsc.2001.1134.
34. Berg PR, Jentoft S, Star B, Ring KH, Knutsen H, Lien S, Jakobsen KS, André C. Adaptation to low salinity promotes genomic divergence in Atlantic cod (Gadus morhua L.). Genome Biol Evol. 2015;7:1644–63. doi:10.1093/gbe/evv093.
35. Miller PJ. Gobiidae. In: Whitehead PJP, editor. Fishes of the north-eastern Atlantic and the Mediterranean, vol. III. Paris: UNESCO; 1986. p. 1019–85.
36. Wiederholm AM. Distribution of Pomatoschistus minutus and Pomatoschistus microps (Gobiidae, Pisces) in the Bothnian Sea: importance of salinity and temperature. Memoranda Soc Fauna Flora Fennica. 1987;63:56–62.
37. Granéli E, Wallström K, Larsson U, Granéli W, Elmgren R. Nutrient limitation of primary production in the Baltic Sea area. Ambio. 1990;19:142–51.

38. Bonsdorff E. Zoobenthic diversity-gradients in the Baltic Sea: continuous post-glacial succession in a stressed ecosystem. J Exp Mar Biol Ecol. 2006;330:383–91. doi:10.1016/j.jembe.2005.12.041.

39. Zettler ML, Schiedek D, Bobertz B. Benthic biodiversity indices versus salinity gradient in the southern Baltic Sea. Mar Pollut Bull. 2007;55:258–70. doi:10.1016/j.marpolbul.2006.08.024.

40. Lehtonen TK, Lindström K, Wong BBM. Effect of egg predator on nest choice and nest construction in sand gobies. Anim Behav. 2013;86:867–71. doi:10.1016/j.anbehav.2013.08.005.

41. Lindström K, Ranta E. Predation by birds affects population structure in breeding sand goby, Pomatoschistus minutus, males. Oikos. 1992;64:527–32.

42. Lehtonen TK, Lindström K, Wong BBM. Body size mediates social and environmental effects on nest building behaviour in a fish with paternal care. Oecologia. 2015;178:673–83. doi:10.1007/s00442-015-3264-z.

43. Olsson KH, Kvarnemo C, Svensson O. Relative costs of courtship behaviours in nest-building sand gobies. Anim Behav. 2009;77:541–6. doi:10.1016/j.anbehav.2008.10.021.

44. Lissåker M, Kvarnemo C, Svensson O. Effects of a low oxygen environment on parental effort and filial cannibalism in the male sand goby, Pomatoschistus minutus. Behav Ecol. 2003;14:374–81. doi:10.1093/beheco/14.3.374.

45. Lissåker M, Kvarnemo C. Ventilation or nest defense: parental care trade- offs in a fish with male care. Behav Ecol Sociobiol. 2006;60:864–73. doi:10.1007/s00265-006-0230-0.

46. Svensson O, Kvarnemo C. The importance of sperm competition risk and nest appearance for male behavior and female choice in the sand goby, Pomatoschistus minutus. Behav Ecol. 2005;16:1042–8. doi:10.1093/beheco/ari085.

47. Lehtonen TK, Lindström K. Repeatability of mating preferences in the sand goby. Anim Behav. 2008;75:55–61. doi:10.1016/j.anbehav.2007.04.011.

48. Lehtonen TK, Wong BBM, Lindström K. Fluctuating mate preferences in a marine fish. Biol Lett. 2010;6:21–3. doi:10.1098/rsbl.2009.0558.

49. Lehtonen TK, Lindström K. Females decide whether size matters: plastic mate preferences tuned to the intensity of male–male competition. Behav Ecol. 2009;20:195–9. doi:10.1093/beheco/arn134.

50. Lehtonen TK, Wong BBM. Should females prefer males with elaborate nests? Behav Ecol. 2009;20:1015–9. doi:10.1093/beheco/arp091.

51. Lindström K. Male-male competition for nest sites in the sand goby, Pomatoschistus minutus. Oikos. 1988;53:67–73.

52. Lindström K. Female spawning patterns and male mating success in the sand goby Pomatoschistus minutus. Mar Biol. 1992;113:475–80. doi:10.1007/BF00349174.

53. Lindström K. The effect of resource holding potential, nest size and information about resource quality on the outcome of intruder-owner conflicts in the sand goby. Behav Ecol Sociobiol. 1992;30:53–8. doi:10.1007/BF00168594.

54. Kvarnemo C. Size-assortative nest choice in the absence of competition in males of the sand goby, Pomatoschistus minutus. Env Biol Fish. 1995;43:233–9. doi:10.1007/BF00005855.

55. Svensson O, Kvarnemo C. Sexually selected nest-building—Pomatoschistus minutus males build smaller nest-openings in the presence of sneaker males. J Evol Biol. 2003;16:896–902. doi:10.1046/j.1420-9101.2003.00591.x.

56. Lehtonen TK, Kvarnemo C. Infections may select for filial cannibalism by impacting egg survival in interactions with water salinity and egg density. Oecologia. 2015;178:699–706. doi:10.1007/s00442-015-3246-1.

57. Lehtonen TK, Kvarnemo C. Density effects on fish egg survival and infections depend on salinity. Mar Ecol Prog Ser. 2015;540:183–90. doi:10.3354/meps11517.

58. Björck S. A review of the history of the Baltic Sea, 13.0–8.0 ka BP. Quatern Int. 1995;27:19–40. doi:10.1016/1040-6182(94)00057-C.

59. Westman P, Sohlenius G. Diatom stratigraphy in five offshore sediment cores from the northwestern Baltic proper implying large scale circulation changes during the last 8500 years. J Paleolimnol. 1999;22:53–69. doi:10.1023/A:1008011511101.

60. Wong BBM, Järvenpää M, Lindström K. Risk-sensitive mating decisions in a visually compromised environment. Biol Lett. 2009;5:600–2. doi:10.1098/rsbl.2009.0350.

61. Japoshvili B, Lehtonen TK, Wong BBM, Lindström K. Repeatability of nest size choice and nest building in sand gobies. Anim Behav. 2012;84:913–7. doi:10.1016/j.anbehav.2012.07.015.

62. Wong BBM, Lehtonen TK, Lindström K. Male nest choice in sand gobies, Pomatoschistus minutus. Ethology. 2008;114:575–81. doi:10.1111/j.1439-0310.2008.01500.x.

63. Lindström K, Lehtonen TK. Mate sampling and choosiness in the sand goby. Proc R Soc B. 2013;280:20130983. doi:10.1098/rspb.2013.0983.

64. Lagakos SW. General right censoring and its impact on the analysis of survival data. Biometrics. 1979;35:139–56.

65. Crawley MJ. The R book. Chichester: John Wiley and Sons Ltd; 2007.

66. Hastings PA. Female choice and male reproductive success in the angel blenny, Coralliozetus angelica (Teleostei: Chaenopsidae). Anim Behav. 1988;36:115–24. doi:10.1016/S0003-3472(88)80254-9.

67. Marconato A, Bisazza A, Marin G. Correlates of male reproductive success in Padogobius martensi (Gobiidae). J Fish Biol. 1989;34:889–99. doi:10.1111/j.1095-8649.1989.tb03372.x.

68. Coleman RM, Fischer RU. Brood size, male fanning effort and the energetics of a nonshareable parental investment in bluegill sunfish, Lepomis macrochirus (Teleostei: Centrarchidae). Ethology. 1991;87:177–88. doi:10.1111/j.1439-0310.1991.tb00245.x.

69. Björk JR, Kvarnemo C. Mechanisms behind size-assortative nest choice by sand goby males in absence of intrasexual competition. Anim Behav. 2012;83:55–62. doi:10.1016/j.anbehav.2011.09.033.

70. Schlaepfer MA, Runge MC, Sherman PW. Ecological and evolutionary traps. Trends Ecol Evol. 2002;17:474–80. doi:10.1016/S0169-5347(02)02580-6.

71. Rodewald AD, Shustack DP, Jones TM. Dynamic selective environments and evolutionary traps in human-dominated landscapes. Ecology. 2011;92:1781–8. doi:10.1890/11-0022.1.

72. Clutton-Brock TH. The evolution of parental care. Princeton: Princeton University Press; 1991.

73. Wesolowski T, Rowinski P. The breeding performance of blue tits Cyanistes caeruleus in relation to the attributes of natural holes in a primeval forest. Bird Study. 2012;59:437–48. doi:10.1080/00063657.2012.722189.

74. Collias NE, Collias EC. Nest building and bird behaviour. Princeton: Princeton University Press; 2014.

75. Dingemanse NJ, Kazem AJN, Denis Réale D, Wright J. Behavioural reaction norms: animal personality meets individual plasticity. Trends Ecol Evol. 2010;25:81–9. doi:10.1016/j.tree.2009.07.013.

76. Dingemanse NJ, Wolf M. Between-individual differences in behavioural plasticity within populations: causes and consequences. Anim Behav. 2013;85:1031–9. doi:10.1016/j.anbehav.2012.12.032.

77. Blanckenhorn WU. The evolution of body size: what keeps organisms small? Q Rev Biol. 2000;75:385–407.

78. Hunt J, Breuker CJ, Sadowski JA, Moore AJ. Male–male competition, female mate choice and their interaction: determining total sexual selection. J Evol Biol. 2009;22:13–26. doi:10.1111/j.1420-9101.2008.01633.x.

79. Hildrew AG, Raffaelli DG, Edmonds-Brown R. Body size: the structure and function of aquatic ecosystems. Cambridge: Cambridge University Press; 2007.

80. Kvarnemo C, Manson W, Svensson O. Investment in testes, sperm-duct glands and lipid reserves differs between male morphs but not between early and late breeding season in Pomatoschistus minutus. J Fish Biol. 2010;76:1609–25. doi:10.1111/j.1095-8649.2010.02587.x.

81. Giacomello E, Marri L, Marchini D, Mazzoldi C, Rasotto MB. Sperm-duct gland secretion of the grass goby Zosterisessor ophiocephalus exhibits antimicrobial activity. J Fish Biol. 2008;73:1823–8. doi:10.1111/j.1095-8649.2008.02069.x.

82. Kvarnemo C, Forsgren E. The influence of potential reproductive rate and variation in mate quality on male and female choosiness in the sand goby. Pomatoschistus minutus. Behav Ecol Sociobiol. 2000;48:378–84. doi:10.1007/s002650000246.

83. Lindström K, Pampoulie C. Effects of resource holding potential and resource value on tenure at nest sites in sand gobies. Behav Ecol. 2005;16:70–4. doi:10.1093/beheco/arh132.

84. Guidelines for the treatment of animals in behavioural research and teaching. Anim Behav. 2012; 83:301–309. doi:10.1016/j.anbehav.2011.10.031.

85. Kilkenny C, Browne WJ, Cuthill IC, Emerson M, Altman DG. Improving bioscience research reporting: the ARRIVE guidelines for reporting animal research. PLoS Biol. 2010;8:e1000412. doi:10.1371/journal.pbio.1000412.

12

Reciprocal transplants support a plasticity-first scenario during colonisation of a large hyposaline basin by a marine macro alga

Daniel Johansson[1,2], Ricardo T. Pereyra[1,2], Marina Rafajlović[2,3] and Kerstin Johannesson[1,2*] ⓘ

Abstract

Background: Establishing populations in ecologically marginal habitats may require substantial phenotypic changes that come about through phenotypic plasticity, local adaptation, or both. West-Eberhard's "plasticity-first" model suggests that plasticity allows for rapid colonisation of a new environment, followed by directional selection that develops local adaptation. Two predictions from this model are that (i) individuals of the original population have high enough plasticity to survive and reproduce in the marginal environment, and (ii) individuals of the marginal population show evidence of local adaptation. Individuals of the macroalga *Fucus vesiculosus* from the North Sea colonised the hyposaline (≥2–3‰) Baltic Sea less than 8000 years ago. The colonisation involved a switch from fully sexual to facultative asexual recruitment with release of adventitious branches that grow rhizoids and attach to the substratum. To test the predictions from the plasticity-first model we reciprocally transplanted *F. vesiculosus* from the original population (ambient salinity 24‰) and from the marginal population inside the Baltic Sea (ambient salinity 4‰). We also transplanted individuals of the Baltic endemic sister species *F. radicans* from 4 to 24‰. We assessed the degree of plasticity and local adaptation in growth and reproductive traits after 6 months by comparing the performance of individuals in 4 and 24‰.

Results: Branches of all individuals survived the 6 months period in both salinities, but grew better in their native salinity. Baltic Sea individuals more frequently developed asexual traits while North Sea individuals initiated formation of receptacles for sexual reproduction.

Conclusions: Marine individuals of *F. vesiculosus* are highly plastic with respect to salinity and North Sea populations can survive the extreme hyposaline conditions of the Baltic Sea without selective mortality. Plasticity alone would thus allow for an initial establishment of this species inside the postglacial Baltic Sea at salinities where reproduction remains functional. Since establishment, the Baltic Sea populations have evolved adaptations to extreme hyposaline waters and have in addition evolved asexual recruitment that, however, tends to impede local adaptation. Overall, our results support the "plasticity-first" model for the initial colonisation of the Baltic Sea by *Fucus vesiculosus*.

Keywords: Common garden, Reciprocal transplant, Salinity, Asexual reproduction, *Fucus vecisulosus*, *Fucus radicans*, Baltic Sea

Background

Some species are able to establish populations in ecologically marginal habitats where the physical environment is radically different from the species' native environment.

If a marginal habitat is relatively local, a population may be established and sustained by continuous recruitment of individuals from a nearby population forming a "source-sink" relationship [1]. However, some marginal environments are larger than the typical dispersal distance of a species, and populations established in these areas must be self-sustained. If this requires new phenotypic traits or phenotypic buffering [2], plasticity and/or

*Correspondence: Kerstin.Johannesson@marine.gu.se
[1] Department of Marine Sciences, University of Gothenburg, Tjärnö, Strömstad, Sweden
Full list of author information is available at the end of the article

directional selection need to be involved in the tuning of traits so that the individuals survive the new environment. Plasticity may seem to be the ideal mechanism, as colonisation in the presence of plasticity can take place without significant losses of genetic variation [3]. However, various constraints including increased costs of plasticity may put a halt to colonising new environments [4]. Moreover, plasticity may not always be adaptive and establishing a population in an environment outside the range of the native environmental variation may increase the risk of plasticity being harmful. If genetic variation is already present in a population as standing genetic variation, local adaptation by means of directional selection may be an efficient and rapid alternative to plasticity [5]. In comparison to plasticity, however, selection will, most likely, be accompanied by much larger loss of genetic variation. If new mutations are required, adaptation will be limited by the waiting time for these, which is usually very long [6]. Furthermore, local adaptation may be counteracted by gene flow [1], and restricted by demographic characteristics of species, such as small population sizes and long generation times [7]. On the other hand, it has been shown that hybridisation and introgression may contribute new genetic variation that may support local adaptation [8].

The Baltic Sea is one of the world's largest brackish-water environments, and, as such, a truly marginal marine habitat. This postglacial semi-enclosed brackish water basin formed from a freshwater lake that opened to the sea about 8500 years ago [9]. Today the Baltic Sea has a surface salinity ranging from 2 to 3‰ in its innermost parts, 6–8‰ in the central parts, and towards the opening to the North Sea, through the Danish Straits, salinity changes rapidly from ~8 to ~20‰. Outside these straits, surface salinity successively increases to full marine salinity (>30‰) in the western part of the North Sea.

Since the opening, the Baltic Sea has been colonised by marine species, some of which have been introduced from other parts of the world [10], but the bulk of marine species have invaded the Baltic Sea from nearby areas of the North Sea. Following the North Sea-Baltic Sea transect, a majority of the marine species shows genetic clines that are steepest in the Danish Straits where the salinity gradient is strongest [11]. The steep clines are caused by local adaptation (e.g. [12, 13]), and isolation effects related to this.

Fucoid macroalgae are foundation species of rocky-shore ecosystems in temperate and subarctic waters. Of a handful of Atlantic species, one species, *Fucus vesiculosus*, has been able to establish populations deep inside the Baltic Sea including areas of strong hyposaline (2–4‰) waters, such as the inner parts of the Gulf of Riga, the Gulf of Finland and the Bothnian Sea [14,

15]. A most intriguing finding is that the colonisation of the Baltic Sea is paralleled by a switch from what seems to be obligate sexual recruitment of new attached thalli outside the basin, to facultative asexual recruitment of new attached and fully sexually reproduced thalli inside the basin [16]. It has been suggested that this switch is due to an increased risk of polyspermy and failure of sexual reproduction in hyposaline waters [17]. However, predominance of recruitment by cloning is not everywhere correlated to salinity [18]. During establishment of *F. vesiculosus* in the Baltic Sea, a separate species (*F. radicans*) diverged from the Baltic lineage of *F. vesiculosus* [19, 20]. *Fucus radicans* is endemic in the Baltic Sea. It is sympatric with *F. vesiculosus* over large parts of the Bothnian Sea and in Estonia [15, 21]. Both species are dioecious with both males and females being capable of asexual reproduction.

Asexual reproduction in Baltic Sea populations of both species is accomplished by the production of adventitious branches that come loose and reattach to the bottom by formation of rhizoids [16]. Adventitious branches are present also in thalli of *Fucus* outside the Baltic Sea, where they have most likely been formed after physical damage from grazers [22], but asexual formation of new thalli has never been reported outside the Baltic Sea.

We hypothesised that the establishment in the Baltic Sea by *F. vesiculosus* may have followed the plasticity-first model [3]. We investigated this by testing two predictions from the model: (i) individuals of the original population have high enough plasticity to tolerate the marginal environment, and (ii) individuals of the marginal population show evidence of local adaptation. The Baltic Sea populations of *F. vesiculosus* and *F. radicans* both descend from a common *F. vesiculosus* lineage originating in the eastern part of the North Sea, close to the entrance of the Baltic Sea [21]. Thus we used individuals from a population in this area to represent the ancestors from which the current Baltic Sea individuals of both species have derived. We used a reciprocal transplant experiment [23] to compare the development of key traits between North Sea and Baltic Sea populations in both native and non-native salinity. This allowed us to separate between locally adapted and plastic traits. To avoid confounding effects from the native environment we used small adventitious branches that were detached from the mother thalli and cultured during 6 months in the laboratory. We assessed survival and measured growth rate as proxies for physiological tolerance and general fitness. We also recorded formation of a second generation of small adventitious branches from the first generation branches and the formation of rhizoids from the primary branches. Both formation of adventitious branches and formation of rhizoids are necessary to accomplish asexual reproduction. Finally, we recorded formation of receptacles as a proxy for the allocation of resources into sexual reproduction.

Methods
Sampling and characterisation
We sampled *F. radicans* and *F. vesiculosus* (73 individuals in total) from a sympatric site on the Swedish coast of the northern Baltic Sea (Skagsudde, N 63°11′21″, E 19°0′13″; Fig. 1). In addition, we sampled 20 individuals of *F. vesiculosus* from a North Sea population on the Swedish west coast (Saltö, N 58°52′16″, E 11°7′11″; Fig. 1). The Baltic samples were from a depth of 3–6 m and an average salinity of 4‰, while the North Sea samples were from the intertidal, with fluctuating salinity around an average of 24‰ (range 15–30‰). Sampling was performed in July 2011 and fresh thalli were brought to the laboratory and stored in tanks with water of ambient salinity (4 and 24‰, respectively). We hereafter refer to each of these three samples (*F. radicans* from the Baltic Sea, *F. vesiculosus* from the Baltic Sea, and *F. vesiculosus* from the North Sea) as our three "populations". Formal identification of the individuals within each sample was done by D.J. on basis of morphological criteria and microsatellite genotype.

For the two Baltic Sea species we performed a first separation based on morphology and confirmed (or in a few cases, corrected) the identification using genotypes in 9

microsatellite markers. Extraction of DNA, PCR reactions and microsatellite analyses were done following the description in [24]. The software STRUCTURE [25] was used to assign individuals into genetically coherent groups, and GENCLONE [26] was used to identify clones among the Baltic Sea individuals. All individuals from the North Sea, on the other hand, were a priori assumed to be unique multi-locus genotypes as asexual reproduction has never been reported from outside the Baltic Sea [16, 27].

We assessed background phenotypic differentiation between Atlantic and Baltic Sea populations by measuring variation in morphological traits in all thalli sampled in the field using the traits earlier described to discriminate between the two species [19]. The measurements included distance between the two most distant dichotomies on a branch, frond width measured midway between the same two dichotomies (both these measurements were repeated in five branches per individual), stipe length measured as the distance between the holdfast and the first branching point, and total length measured as the distance between the holdfast and the most distal tip. We used principal components analysis to assess overall differences among the three populations in size and shape.

In a long-term (6 months) reciprocal transplant experiment North Sea and Baltic Sea adventitious branches detached from large thalli grew new vegetative tissue in both 4 and 24‰ salinity. In this experiment each individual was represented by six ≈1 cm long adventitious branches that we randomly chose from a large number of adventitious branches grown on each individual. We used 48 individuals of Baltic *F. radicans*, 25 of Baltic *F. vesiculosus* and 20 of North Sea *F. vesiculosus*. Three branches from each individual were acclimatised to the low salinity and three to the high, by adjusting the salinity gradually over 4 weeks for those transplanted to a non-native salinity. The adventitious branches were incubated upright, attached with elastic threads to holders in tanks with 40 L water and flow-pumps to circulate the water. We used two tanks per salinity and individuals were randomly distributed between these while keeping numbers of individuals from each population constant in each tank. The water was prepared by mixing tap water with "Instant Ocean" salt and nutrient medium. Water was changed once every 2nd month. The experiment was run in a thermo constant room at 13 °C and a 16:8 h light: dark cycle for 6 months. The length increment of each adventitious branch was measured at the end of the experiment. In addition, we counted the number of new adventitious branches ("secondary branches") formed from the starting branches. Rhizoids (used to attach the vegetative part to the substratum) were formed at the

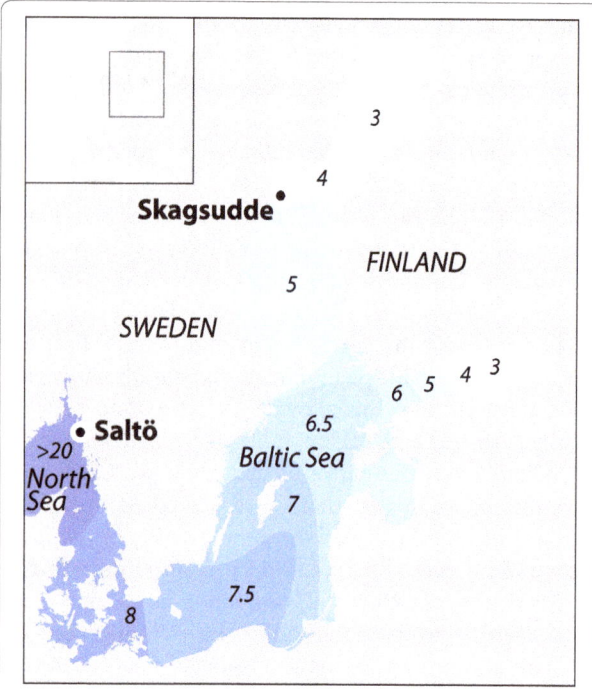

Fig. 1 Map showing sampling sites and the Baltic Sea salinity gradient. Source: Online Map Creation-Martin Weinelt, http://aquarius.ifm-geomar.de, visited 2009.05.01

basal end of some of the primary adventitious branches, and we noted presence or absence of rhizoids in any of the three replicate branches of the same individual after inspection using a stereomicroscope. Some of the primary adventitious branches also formed early stages of receptacles and in a similar way, we recorded presence or absence of these in each individual.

Statistical analysis

We conducted two separate analyses for each response parameter. In one analysis we compared the two sympatric populations of *F. radicans* and *F. vesiculosus* from the Baltic Sea. In a separate analysis we compared the Baltic and North Sea populations of *F. vesiculosus*. For all tests including *F. radicans* we performed analyses with both the full dataset, and datasets that were made balanced ($N = 20$) by random removal of individuals, using type-III sums of squares in the unbalanced cases. As results obtained in balanced and full data sets remained qualitatively the same, we only present the result of the balanced data set in figures and tables.

As adventitious branches of each individual were grown in both salinities (see above), we used a split-plot design for the analysis of variance (ANOVA). We assessed assumptions of normality and homogeneity of variances with box-plots and residual plots. Log transformation improved the growth data, which was analysed using the *aov* function in R (v. 2.15.2; [28]). Data on formation of secondary adventitious branches were analysed as a split-plot design with linear mixed-effects models using the *lme* function. We analysed occurrence (presence/absence) of rhizoids using Fisher's exact test, comparing species and origin separately. As rhizoids only formed in the high salinity in all three populations, the low salinity data was not included in the analysis. Receptacles were only formed by one population and in one salinity treatment, and these results were left without statistical evaluation. Confidence intervals for all means were calculated by multiplying standard error with the critical t-value at a confidence level of 95%.

Results

In the Baltic Sea site where both *F. radicans* and *F. vesiculosus* were sampled, the preliminary assignment of individuals to species, based on overall morphology, was mostly consistent with the result of the genetic analysis. Among the individuals sampled, however, four individuals were not correctly assigned from morphology and were moved to the correct taxa for all down-stream analyses after genotyping (Additional file 1). The genotyping further showed that the Baltic population of *F. vesiculosus* consisted of 10 unique multi-locus genotypes and 7 clones with 2–3 individuals per clone (in total 17

genotypes), while the population of *F. radicans* included 2 unique individuals and three clones with 9, 17 and 20 individuals each (in total 5 genotypes). Despite the replication of genotypes, in particular in *F. radicans*, we decided to include all individuals sampled in the experiment, in order to form a representative sample of a wild population from the Baltic Sea and to detect (if present) contributions from somatic mutations. Separate analyses were in addition done comparing traits among the three major clones of the *F. radicans* samples (see below).

The morphology of the adult thalli sampled in the wild revealed phenotypic variation among the three populations along the PC1 axis, reflecting population differences in overall size. Individuals of *Fucus vesiculosus* from the North Sea were, on average, larger than individuals of Baltic Sea *F. vesiculosus*, and individuals of *F. radicans* were the smallest (Fig. 2). There was no separation of the populations along the PC2 axis and this suggests that there were no major differences in shape among the three populations. The within population variation along the PC2 axis tended to be highest in the North Sea *F. vesiculosus* and lowest in the *F. radicans* population.

Among the adventitious branches transferred to a non-native salinity (from 4 to 24‰, or the reverse), 1–3 (and usually 3) of the replicate branches of each individual survived and formed new tissue during the 6 months experiment, indicating high plasticity in the physiological response to environmental salinity. However, comparing growth of Baltic and North Sea *F. vesiculosus*, we found a strong interaction between salinity and origin

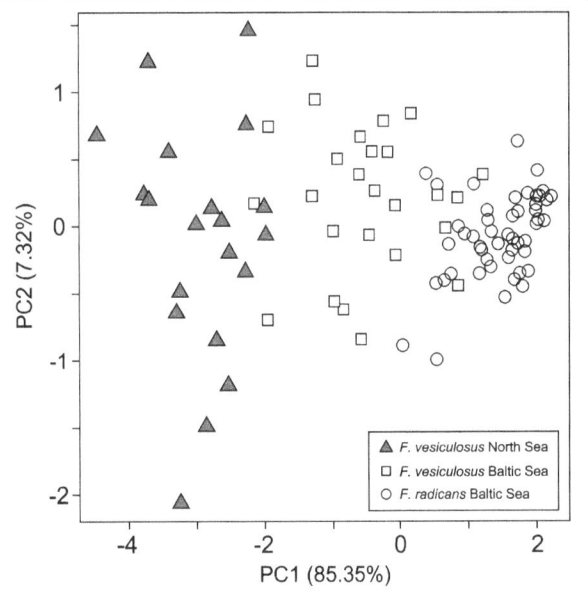

Fig. 2 Principal components analysis of four morphological characters (see text) in adult populations of *F. vesiculosus* from Saltö (North Sea), *F. vesiculosus* from Skagsudde (Baltic Sea) and *F. radicans* from Skagsudde (Baltic Sea)

($P < 0.001$; Table 1), indicating that both populations of *F. vesiculosus* grew better in their native salinity than in the other salinity. This supports that Baltic Sea populations have evolved local adaptation in traits that contribute to growth (Fig. 3a). Also the Baltic Sea population of *F. radicans* grew better in the hyposaline conditions of its native environment than in fully marine waters (Fig. 3a). Notably, *F. radicans* was less negatively impacted by high

salinity than was Baltic *F. vesiculosus*, resulting in a significant interaction be-tween salinity and species also in this comparison ($P = 0.033$; Table 1; Fig. 3a). We also compared growth rates among the three numerically dominant clones of *F. radicans* from the Baltic Sea and found a statistically significant effect of clone on growth with a similar trend of higher growth in the low salinity in all clones (Table 2; Fig. 4a), supporting the presence of genetic variation in this trait.

Secondary adventitious branches were formed from the basal part of the original adventitious branches in all three populations during the experimental period. When we compared the two populations of *F. vesiculosus* from the North Sea and the Baltic Sea, respectively, we found no significant effect of salinity and no interaction between salinity and origin, but a trend towards higher numbers of secondary adventitious branches in the Baltic population than in the North Sea population ($P = 0.073$; Table 3; Fig. 3b). Comparing the two sympatric populations from the Baltic Sea, *F. radicans* and Baltic *F. vesiculosus*, we found no significant interaction of species and salinity, and no significant effect of salinity alone. However, there was a strong difference between the two species ($P < 0.001$, Table 3), that is, *Fucus radicans* formed significantly more secondary adventitious branches than did sympatric individuals of *F. vesiculosus* in both salinities (Fig. 3b). Also in this trait we found differences among clones with one clone growing more adventitious branches than the other two clones, showing genetic variation being present in this trait (Table 4; Fig. 4b).

Table 1 ANOVA (split-plot design) analysing growth of adventitious branches during 6 months in a reciprocal transplant experiments as an effect of salinity (4 and 24‰), species and origin

Source	df	SS	MS	F	P
A. Baltic Sea *F. radicans* vs. *F. vesiculosus*					
Species	1	0.688	0.688	2.191	0.148
Residuals	36	11.312	0.314		
Salinity	1	4.782	4.782	35.67	<0.001
Species*salinity	1	0.662	0.662	4.939	0.033
Residuals	36	4.826	0.134		
B. Baltic Sea *F. vesiculosus* vs. North Sea *F. vesiculosus*					
Origin	1	<0.001	<0.001	<0.001	0.985
Residuals	36	10.62	0.295		
Salinity	1	0.052	0.052	0.375	0.544
Origin*salinity	1	7.689	7.689	55.757	<0.001
Residuals	36	4.964	0.138		

A. Comparison of growth rates between *F. radicans* and *F. vesiculosus* from a sympatric site in the Baltic Sea. B. Comparison of growth rates between *F. vesiculosus* from North Sea and Baltic Sea. See also Fig. 3a

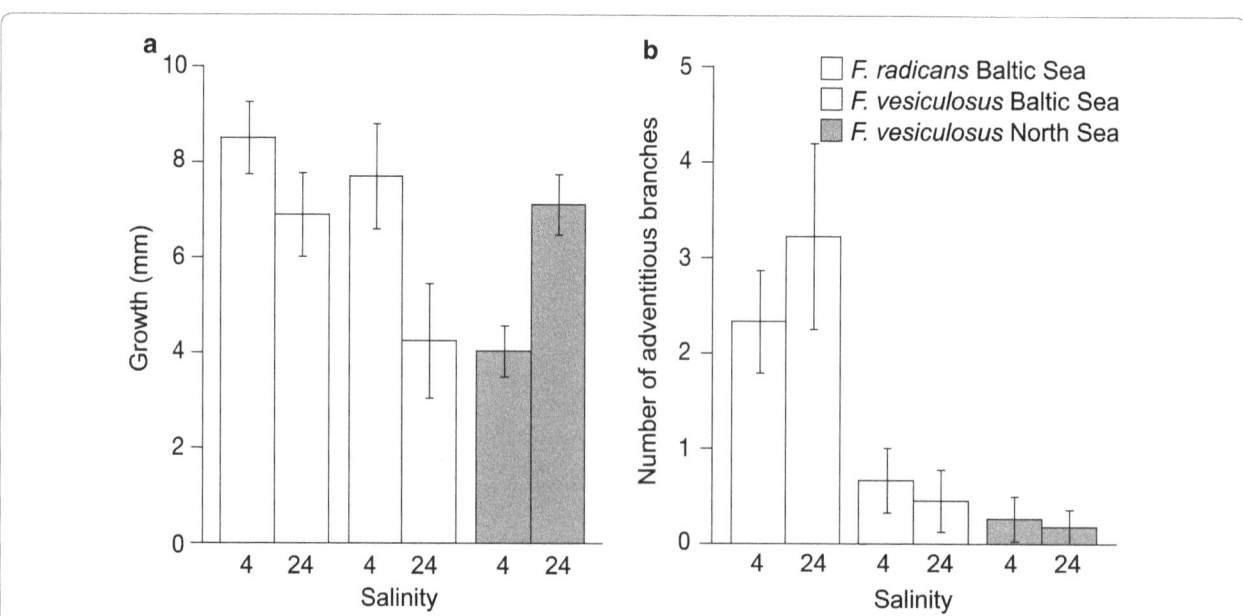

Fig. 3 Result of reciprocal transplants of North Sea and Baltic Sea *Fucus* during 6 months in 4 and 24‰ salinity. **a** Growth of 10 mm large adventitious branches. **b** Number of secondary branches formed per primary branch. Error bars show 95% CI

Table 2 ANOVA (split-plot design) analysing growth of adventitious branches during 6 months in a reciprocal transplant experiments as an effect of salinity (4 and 24‰), among three clones of *Fucus radicans* from Skagsudde (Baltic Sea), see also Fig. 4a

Source	df	SS	MS	F	P
Clone	2	167.9	83.97	7.57	0.002
Residuals	43	476.9	11.09		
Salinity	1	54.7	54.72	16.45	<0.001
Clone*Salinity	2	5.22	2.61	0.784	0.46
Residuals	43	143.1	3.33		

Some of the primary adventitious branches developed rhizoids and attached to the substratum, but unexpectedly rhizoids were only formed in the higher salinity treatment (24‰). The proportion of individuals that developed rhizoids was different among the three populations with 61% of *F. radicans* individuals and 20% of Baltic Sea *F. vesiculosus* forming rhizoids, but none of the North Sea *F. vesiculosus* did form rhizoids. In the sympatric Baltic Sea site, the difference between the two species was significant, and the difference between *F. vesiculosus* of Baltic Sea and North Sea origin was in addition marginally significant (Table 5). There was, however, no significant difference in how many individuals formed rhizoids among the three clones of *F. radicans* (results not shown). Although not quantified, we observed that most of the adventitious branches that formed rhizoids also attached to the substratum.

In 35% of the North Sea individuals the adventitious branches formed early stages of receptacles in the high salinity treatment. No receptacles were developed in the adventitious branches from the two Baltic populations in high or low salinity, or in adventitious branches of the North Sea population kept in low salinity.

Discussion

West-Eberhard [3] suggests that developmental plasticity is likely to precede the accumulation of genetic divergence during colonisation of a new environment. The mechanism is simply that a population in which all individuals survive the new environment by being plastic will avoid demographic bottlenecks during establishment caused by selective mortality. Following successful colonisation, assimilation of genetic differences that improves phenotypes and result in local adaptation of traits may follow [3]. Here we raise the question if the colonisation of the Baltic Sea by *F. vesiculosus* was initially made possible by a high degree of plasticity, but later followed by assimilation of genetic differences caused by directional selection that improved local adaptation and allowed for further colonisation up to the current range margin in extreme hyposaline waters. The plasticity-first model is supported if colonising individuals without prior adaptation will survive and reproduce in the new environment due to their plasticity. In contrast to a model where colonisation would only be possible under selective mortality, establishment aided by plasticity would give less serious demographic effects and probably less loss of genetic

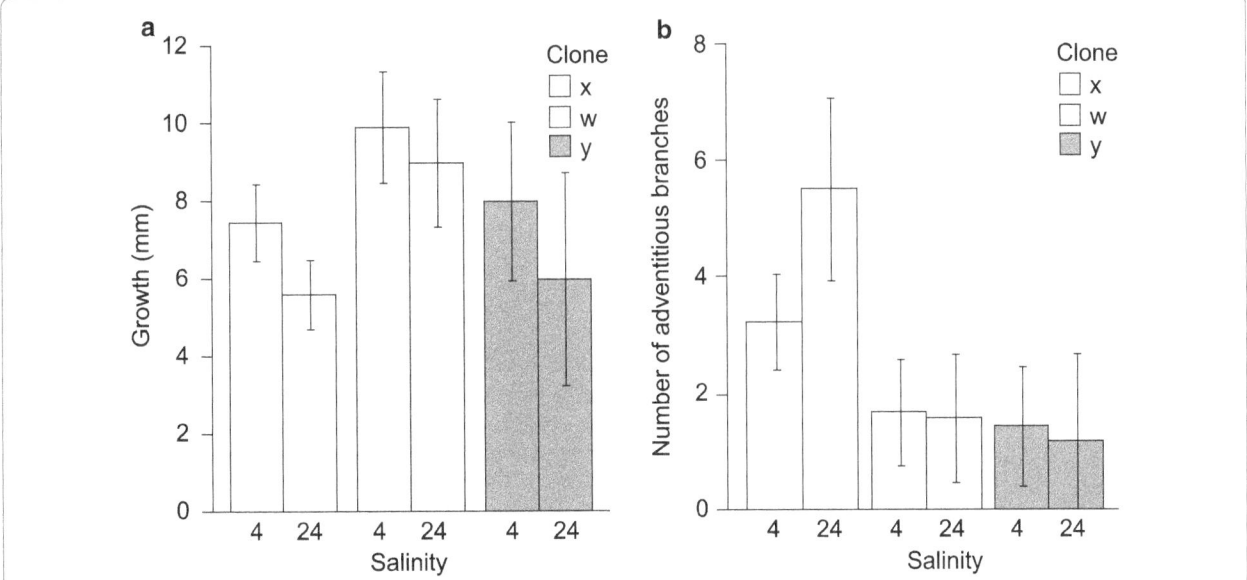

Fig. 4 The same experiment as in Fig. 3, comparing the results for the three large clones of *Fucus radicans*. N = 20 for clone X, 17 for W and 9 for Y. **a** Growth of 10 mm large adventitious branches. **b** Number of secondary branches formed per primary branch. Error bars show 95% CI. (For statistic evaluation see Tables 2 and 4.)

Table 3 ANOVA (linear mixed effects model, df = 1.34) analysing the number of secondary formed adventitious branches per primary branch, during 6 months in a reciprocal transplant experiments, as an effect of salinity (4 and 24‰), species and origin

Source	F	P
A. Baltic Sea *F. radicans* vs. *F. vesiculosus*		
Intercept	36.45	<0.001
Salinity	0.876	0.36
Species	15.65	<0.001
Salinity*species	2.648	0.11
B. Baltic Sea *F. vesiculosus* vs. North Sea *F. vesiculosus*		
Intercept	24.74	<0.001
Salinity	1.989	0.17
Origin	3.42	0.07
Salinity*origin	0.261	0.61

A. Analysing the effect of salinity and species (*F. radicans* and *F. vesiculosus*) from the same origin in the Baltic Sea. B. Analysing the effect of salinity and origin (North Sea and Baltic Sea) in populations of *F. vesiculosus*. See also Fig. 3b

Table 4 Linear mixed-effects model of number of secondary adventitious branches formed by the three clones of *Fucus radicans* from Skagsudde during growth in a common garden at two different salinities (see also Fig. 4b)

	df	F	P
Intercept	1.43	136.8	<0.001
Salinity	1.43	0.049	0.83
Clone	2.43	15.63	<0.001
Clone*salinity	2.43	3.24	0.049

Table 5 Proportion of individuals forming rhizoids in the three study populations of *Fucus*

Baltic Sea		North Sea
Fucus radicans	*F. vesiculosus*	*F. vesiculosus*
61%	20%	0%
P = 0.003		
	P = 0.053	

P values are derived from Fisher's exact test

variation by selection and drift, although plasticity would mask the genetic variation present in traits and may prevent improvements in fitness by local adaptation.

We here show that contemporary individuals of the North Sea population of *F. vesiculosus* have a high tolerance to low salinity and survive and grow in strongly hyposaline (4‰) waters. Assuming a similar capacity in the North Sea population during the time of the Baltic Sea invasion of *F. vesiculosus*, the individuals that colonised the Baltic Sea were capable of surviving at salinities currently found at the range margin of the Baltic *Fucus* distribution. However, for successful establishment in the Baltic Sea without prior local adaptation, also traits central to reproduction must remain functional in the low salinity. Indeed, an earlier study of sexual reproduction shows that gametes of *F. vesiculosus* from the North Sea remain functional down to salinities around 7–8‰ [17]. Thus high levels of plasticity in survival, growth and sexual function suggest that populations from the North Sea were able to rapidly colonise the southern part of the hyposaline Baltic Sea after its postglacial formation, or even more central parts as salinity during a period around 6000 years ago was higher than today [9]. Due to the lack of selective mortality this phase of the colonisation was presumably relatively rapid, and without major losses of genetic variation. Following the establishment in the southern and central parts of the Baltic Sea, the population seems to have improved its fitness by directional selection. The selection led to local adaptation and higher growth rate in 4‰ compared to 24‰ salinity, as shown in this study (and see [29]). Also local adaptation in sexual reproduction has taken place as current Baltic Sea populations maintain high sperm quality and rate of fertilisation below 8‰ [17]. There is also support for local adaptation in other traits of Baltic Sea *F. vesiculosus* and *F. radicans*. For example, tolerance to emersion stress is lost in Baltic Sea populations following the switch to a more subtidal distribution [30], and a majority of genotypes in *F. radicans* have reduced tolerance to stress from desiccation and freezing [31]. Thus following the initial phase of establishment in the Baltic Sea supported by plasticity, the second phase involving selection towards increased fitness in hyposaline waters seems also to have extended the distribution of the *F. vesiculosus* lineage (including the *F. radicans* branch) to its current range margins in extremely low salinities (3–4‰).

As part of the second phase of local adaptation, the new role of the adventitious branches is intriguing. The original (North Sea) role of these vegetative parts was probably to repair tissue damaged by grazers [22]. In the northern Baltic Sea, by contrast, adventitious branches form, fall off and develop rhizoids that reattach them to the substratum, and from there they grow clonal copies of the mother thallus. Formation of adventitious branches was most frequent in *F. radicans*, less frequent in the Baltic Sea *F. vesiculosus*, and least common or absent in North Sea *F. vesiculosus* in our experiments, which correlates to the prevalence of clones in these populations (this study and [32]). Asexual reproduction by means of re-attaching adventitious branches seems to be a unique trait to the Baltic Sea that in itself has promoted the spread and establishment in a new area. The reason for this may be that after its initial occurrence, asexual reproduction may have spread in the northern Baltic Sea by directional selection favouring individuals that invested more in asexual than in sexual reproduction. The observation of polyspermy in salinities below 5–6‰ [33] supports a selection-driven switch to asexual reproduction in hyposaline waters. Polyspermy may severely constrain sexual reproduction, at least in populations where gamete densities are high. Under risk of polyspermy, individuals that allocate more energy to formation of adventitious branches and less to building receptacles and forming gametes would be favoured by selection.

However, in the field also thalli recruited asexually form receptacles and produce gametes that are functional [16, 34]. This suggests that selection for asexual reproduction in hyposaline areas is not particularly strong. In addition, the observation that the prevalence of asexual recruitment is very variable and not correlated to salinity [18] suggests directional selection imposed by salinity being a less likely explanation for the spread of asexual recruitment inside the Baltic Sea. Alternatively, the distribution of clones may be a consequence of neutral processes during colonisation of the Baltic Sea (that is, random dispersal and demographic stochasticity). Indeed, according to recent modelling results, it seems highly likely that if asexual reproduction is an option, then fully neutral processes are likely to drive structuring of the population and determine the distribution of asexual populations during a colonisation event without involvement of selection on individual reproductive fitness [35].

Conclusions

Fucus vesiculosus is the only Atlantic lineage of fucoid algae that has been able to colonise deep into the Baltic Sea including environments with strong hyposaline conditions. As shown in this study, North Sea individuals of this species are tolerant enough to survive at least several months in salinities corresponding to the hyposaline conditions at the range margin of Baltic Sea populations. Moreover, as shown in an earlier study [17], sexual reproduction in North Sea *F. vesiculosus* is sustained down to salinities typical of the central Baltic Sea. Thus, plasticity in growth and sexual reproduction seemed strong enough to allow for an initial colonisation phase into the central Baltic Sea without selective mortality, corroborating the primary prediction of the plasticity-first model of West-Eberhard [3]. Furthermore, growth of North Sea individuals is already plastic enough to sustain survival down to strongly hyposaline waters at the current Baltic Sea range margins, and this trait was thus "preadapted" to colonise deeper into the Baltic Sea upon arrival of the

first populations. Sexual reproduction, on the other hand, was only enough plastic to sustain recruitment into the central Baltic Sea, but a dramatic switch to asexual reproduction most likely boosted further colonisation of more marginal areas. Both strong plasticity and dominance of asexual recruitment somewhat impede the action of selection. But, nevertheless, the Baltic Sea *Fucus* shows evidence of local adaption in several key traits supporting the second prediction of the plasticity-first hypothesis that the initial phase of colonisation is followed by improved fitness and local adaptation through directional selection [3].

Authors' contributions
DJ, RTP and KJ planned the study and sampled the material. DJ genotyped the material with the help of RTP, run the experiments and performed the statistical analyses. KJ and DJ wrote the manuscript with the help of MR. All authors read and approved the final manuscript.

Author details
[1] Department of Marine Sciences, University of Gothenburg, Tjärnö, Strömstad, Sweden. [2] Centre for Marine Evolutionary Biology, University of Gothenburg, Tjärnö, Strömstad, Sweden. [3] Department of Physics, University of Gothenburg, Gothenburg, Sweden.

Acknowledgements
We are grateful to Gerry Quinn and Jon Havenhand for advice on statistical analyses, and to Angelica Ardehed, Ellen Schagerström and Lena Kautsky for advices and practical support during field campaigns.

Competing interests
The authors declare that they have no competing interests.

Funding
Funding was provided by the BONUS project (Baltic Organisations Network for funding Science EEIG) to the BaltGene and Bambi projects, and by the Swedish Research Council VR, through a Linnaeus grant to the Centre for Marine Evolutionary Biology (CeMEB). The funding bodies had no role in the design of the study and collection, analysis, and interpretation of data, or in writing the manuscript.

References
1. Kawecki TJ. Adaptation to marginal habitats. Ann Rev Ecol Evol Syst. 2008;39:321–42.
2. Reusch TBH. Climate change in the oceans: evolutionary versus phenotypically plastic responses of marine animals and plants. Evol Appl. 2014;7:104–22.
3. West-Eberhard MA. Developmental plasticity and the origin of species differences. Proc Natl Acad Sci USA. 2005;102:6543–9.
4. Pigliucci M. Phenotypic plasticity: beyond nature and nurture. Baltimore: Johns Hopkins Univ. Press; 2001.
5. Lescak EA, Bassham SL, Catchen J, Gelmond O, Sherbick ML, von Hippel FA, Cresko WA. Evolution of stickleback in 50 years on earthquake-uplifted islands. Proc Natl Acad Sci USA. 2015;112:E7204–12.
6. Hoffmann AA, Sgrò CM. Climate change and evolutionary adaptation. Nature. 2011;470:479–85.
7. Chevin LM, Lande R, Mace GM. Adaptation, plasticity, and extinction in a changing environment: towards a predictive theory. PLoS Biol. 2010;8:e1000357.
8. Keller SR, Taylor DR. Genomic admixture increases fitness during a biological invasion. J Evol Biol. 2010;23:1720–31.

9. Zillén L, Conley DJ, Andren T, Andren E, Björck S. Past occurrences of hypoxia in the Baltic Sea and the role of climate variability, environmental change and human impact. Earth Sci Rev. 2008;91:77–92.

10. Ojaveer H, Jaanus A, MacKenzie BR, Martin G, Olenin S, Radziejewska T, Telesh I, Zettler ML, Zaiko A. Status of biodiversity in the Baltic Sea. PLoS ONE. 2010;5:e12467.

11. Johannesson K, André C. Life on the margin: genetic isolation and diversity loss in a peripheral marine ecosystem, the Baltic Sea. Mol Ecol. 2006;15:2013–29.

12. Luttikhuizen PC, Drent J, Peijnenburg KTCA, Van der Veer HW, Johannesson K. Genetic architecture in a marine hybrid zone: comparing outlier detection and genomic clines analysis in the bivalve *Macoma balthica*. Mol Ecol. 2012;21:3048–61.

13. Lamichhaney S, Barrio AM, Rafati N, Sundström G, Rubin C-J, Gilbert ER, Berglund J, Wetterbom A, Laikre L, Webster MT, Grabherr M, Ryman N, Andersson L. Population-scale sequencing reveals genetic differentiation due to local adaptation in Atlantic herring. Proc Natl Acad Sci USA. 2012;109:19345–50.

14. Bäck S, Ruuskanen A. Distribution and maximum growth depth of *Fucus vesiculosus* along the Gulf of Finland. Mar Biol. 2000;136:303–7.

15. Forslund H, Eriksson O, Kautsky L. Grazing and geographic range of the Baltic seaweed *Fucus radicans* (Phaeophyceae). Mar Biol Res. 2012;8:322–30.

16. Tatarenkov A, Bergström L, Jonsson RB, Serrão EA, Kautsky L, Johannesson K. Intriguing asexual life in marginal populations of the brown seaweed *Fucus vesiculosus*. Mol Ecol. 2005;14:647–51.

17. Serrão EA, Kautsky L, Brawley SH. Distributional success of the marine seaweed *Fucus vesiculosus* L. in the brackish Baltic Sea correlates with osmotic capabilities of Baltic gametes. Oecologia. 1996;107:1–12.

18. Ardehed A, Johansson D, Sundqvist L, Schagerström E, Zagrodzka Z, Kovaltchouk NA, Bergström L, Kautsky L, Rafajlovic M, Pereyra RT, Johannesson K. Divergence within and among seaweed siblings (*Fucus vesiculosus* and *F. radicans*) in the Baltic Sea. PLoS ONE. 2016;11:e0161266.

19. Bergström L, Tatarenkov A, Johannesson K, Jonsson RB, Kautsky L. Genetic and morphological identification of *Fucus radicans* sp Nov (Fucales, Phaeophyceae) in the brackish Baltic Sea. J Phycol. 2005;41:1025–38.

20. Pereyra RT, Bergström L, Kautsky L, Johannesson K. Rapid speciation in a newly opened postglacial marine environment, the Baltic Sea. BMC Evol Biol. 2009;9:70.

21. Pereyra RT, Huenchuñir C, Johansson D, Forslund H, Kautsky L, Jonsson PR, Johannesson K. Parallel speciation or long-distance dispersal? Lessons from seaweeds (*Fucus*) in the Baltic Sea. J Evol Biol. 2013;26:1727–37.

22. Van Alstyne KL. Adventitious branching as a herbivore-induced defense in the intertidal brown alga *Fucus distichus*. Mar Ecol Prog Ser. 1989;56:169–76.

23. Kawecki TJ, Ebert D. Conceptual issues in local adaptation. Ecol Lett. 2004;7:1225–41.

24. Johannesson K, Johansson D, Larsson KH, Huenchuñir CJ, Perus J, Forslund H, Kautsky L, Pereyra RT. Frequent clonality in fucoids (*Fucus radicans* and *Fucus vesiculosus*; Fucales, Phaeophyceae) in the Baltic Sea. J Phycol. 2011;47:990–8.

25. Pritchard JK, Stephens M, Donnelly P. Inference of population structure using multilocus genotype data. Genetics. 2000;155:945–59.

26. Arnaud-Haond S, Belkhir K. GENCLONE: a computer program to analyse genotypic data, test for clonality and describe spatial clonal organization. Mol Ecol Notes. 2007;7:15–7.

27. Tatarenkov A, Jonsson RB, Kautsky L, Johannesson K. Genetic structure in populations of *Fucus vesiculosus* (Phaeophyceae) over spatial scales from 10 m to 800 km. J Phycol. 2007;43:675–85.

28. R Core Team. R: A language and environment for statistical computing. R Foundation for Statistical Computing, Vienna, Austria. 2012. ISBN 3-900051-07-0, URL http://www.R-project.org/.

29. Bäck S, Collins JC, Russell G. Effects of salinity on growth of Baltic and Atlantic *Fucus vesiculosus*. Br Phycol J. 1992;27:39–47.

30. Pearson G, Kautsky L, Serrão E. Recent evolution in Baltic *Fucus vesiculosus*: reduced tolerance to emersion stresses compared to intertidal (North Sea) populations. Mar Ecol Prog Ser. 2000;202:67–79.

31. Johannesson K, Forslund H, Capetillo NÅ, Kautsky L, Johansson D, Pereyra RT, Råberg S. Phenotypic variation in sexually and asexually recruited individuals of the Baltic Sea endemic macroalga *Fucus radicans*: in the field and after growth in a common-garden. BMC Ecol. 2012;12:2.

32. Ardehed A, Johansson D, Schagerström E, Kautsky L, Johannesson K, Pereyra RT. Complex spatial clonal structure in the macroalgae *Fucus radicans* with both sexual and asexual recruitment. Ecol Evol. 2015;2015(5):4233–45.

33. Serrão EA, Brawley SH, Hedman J, Kautsky L, Samuelsson G. Reproductive success of *Fucus vesiculosus* (Phaeophyceae) in the Baltic Sea. J Phycol. 1999;35:254–69.

34. Forslund H, Kautsky L. Reproduction and reproductive isolation in *Fucus radicans* (Phaeophyceae). Mar Biol Res. 2013;9:262–7.

35. Rafajlović M, Kleinhans D, Gulliksson C, Fries J, Johansson D, Ardehed A, Sundqvist L, Pereyra RT, Mehlig B, Jonsson PR, Johannesson K. Neutral mechanisms forming large clones during colonisation of new areas. J Evol Biol. 2017. **(awaiting final decision)**.

13

Seasonal rainfall at long-term migratory staging sites is associated with altered carry-over effects in a Palearctic-African migratory bird

Marjorie C. Sorensen[1]*[iD], Graham D. Fairhurst[2], Susanne Jenni-Eiermann[3], Jason Newton[4], Elizabeth Yohannes[5] and Claire N. Spottiswoode[1,6]

Abstract

Background: An understanding of year-round habitat use is essential for determining how carry-over effects shape population dynamics in long-distance migratory songbirds. The recent discovery of long-term migratory staging sites in many species, prior to arrival at final wintering sites, adds complexity to efforts to decipher non-breeding habitat use and connections between sites. We investigated whether habitat conditions during migratory staging carry over to influence great reed warbler (*Acrocephalus arundinaceus*) body condition at final wintering sites in Zambia. We asked whether the presence/absence and strength of such carry-over effects were modified by contrasting rainfall conditions during 2 years.

Results: First, we found that individuals staging in a dry year had higher corticosterone ($CORT_f$) and stable nitrogen isotope values (suggesting higher aridity) than birds staging in a wet year, indicating that regional weather affected staging conditions. Second, we found that carry-over effects from staging habitat conditions (measured via carbon and nitrogen isotopes) to final winter site body condition (measured via scaled mass index and β-hydroxybutyrate) were only present in a dry year, suggesting that environmental factors have consequences for the strength of carry-over effects. Our results also suggest that wet conditions at final winter sites may buffer the effects of poor staging conditions, at least in the short term, since individuals that staged in a dry year had higher scaled mass indices in Zambia than individuals that staged in a wet year.

Conclusions: This study provides a first insight into the connections between long-term migratory staging sites and final wintering sites, and suggests that local environmental factors can modify the strength of carry-over effects for long-distance migratory birds.

Keywords: Great reed warblers, Long-term staging, Plasma metabolites, Migration, Stable isotopes

Background

The population dynamics of migratory animals may be affected by factors operating at multiple discrete sites throughout the annual cycle. Long-distance migrants must contend with extreme variation in weather conditions, habitat types and habitat quality, imposed by sites thousands of kilometers apart. This variation can be exacerbated by anthropogenic impacts that disrupt links between specific habitats and crucial periods of the annual cycle [1]. Breeding conditions are well known to have important consequences for population dynamics, but evidence is growing that conditions experienced during non-breeding seasons may be at least as important [2], such that conditions during migratory staging may carry over to affect future breeding success [3]. However,

*Correspondence: marjoriesorensen@gmail.com
[1] Department of Zoology, University of Cambridge, Cambridge CB2 3EJ, UK
Full list of author information is available at the end of the article

identifying such carry-over effects (defined as processes in one season influencing the success of an individual in the following season) between migratory staging sites has been challenging owing to difficulties in tracking small-bodied migrants across multiple sites throughout the annual cycle [3].

There is strong evidence that winter weather conditions, specifically rainfall effects on food availability, play an important role in phenology and reproductive success of migratory birds, such that low precipitation during the winter has long-term effects on population dynamics [4–7]. However, understanding individual carry-over effects from wintering ground conditions requires knowledge of movements within the non-breeding season and the specific habitats thus encountered. Recently, new light has been shed on the largely unknown winter migratory patterns of long-distance migratory songbirds, using miniature light-level geolocators [8]. Surprisingly, geolocator data show that rather than single wintering sites or "itinerant" winter movements [9], the use of long-term staging sites en route to core wintering grounds may be widespread among migratory songbirds [10–14]. Long-term staging sites complicate our understanding of year-round habitat use for migratory birds because variation in habitat quality among staging sites may have important consequences for population dynamics via carry-over effects on individual survival and condition.

Studying how conditions at multiple sites cumulatively affect year-round individual success is particularly important for Palearctic-African species conservation, management, and effective reserve design, as sub-Saharan wintering migrants are in a state of severe population decline [15, 16]. To begin addressing these shortcomings we studied the great reed warbler (*Acrocephalus arundinaceus*), a 30 g long-distance Palearctic-African migratory songbird that winters widely across sub-Saharan Africa, in southern Zambia. During autumn migration, great reed warblers pause for 2–3 months to moult their flight feathers before continuing on to their final winter sites further south where moult is completed [17]. We focused on rainfall as a potential driver of carry-over effects because its effect on food availability makes it a critical environmental factor for insectivorous birds [18]. We studied great reed warblers during 2 years with contrasting rainfall conditions at staging areas (estimated from isotopic data; see "Results"). Carry-over effects are unlikely to be uniform across years and our understanding of the processes that are capable of modifying their strength remains limited. Natural variation between our study years allowed us to investigate whether the presence and strength of carry-over effects vary depending on environmental conditions.

We first established that differences in rainfall between years affected the conditions birds experienced while staging. To do so, we utilised the fact that great reed warblers replace their flight feathers during staging [19], such that individuals arriving in Zambia carry an isotopic and hormonal signature of the conditions they experienced previously during staging. Specifically, we analysed stable isotope ratios of nitrogen (δ^{15}N) and carbon (δ^{13}C), and corticosterone ($CORT_f$) levels in flight feathers grown on staging grounds (Fig. 1). Both δ^{15}N and δ^{13}C isotope ratios have been used previously to assess habitat conditions experienced by migratory birds [20–22], trophic position in food webs, or marine versus freshwater ecosystems [23]. δ^{15}N increases with greater aridity, especially in Africa [24–26], and δ^{13}C decreases as C_3-dominated plant communities, which are associated with moister and cooler environments, become less common [27]. Corticosterone is a hormone involved in energy management and high levels can indicate compromised individual condition, reflecting a response to stressors such as reduced food availability and poor habitat quality [28–30]. $CORT_f$ is an integrated measure of the hormone and reflects conditions experienced over the feather growth period ([31, 32]; for a review, see [33]). We expected that if birds indeed experienced drier conditions during staging in 2011, this would be recorded in their flight feathers as higher δ^{15}N values (reflecting higher aridity) and higher $CORT_f$ (reflecting a physiological response to poor conditions); however, we expected no difference in δ^{13}C values since carbon isotopes are relatively stable between years [34].

Second, we asked whether variation in environmental conditions experienced during staging affected the presence and strength of carry-over effects. We used several physiological metrics to estimate condition on wintering grounds. Plasma metabolites are indicators of energetic state: high circulating levels of triglycerides in plasma (hereafter 'TRIG') indicate fat deposition and foraging rates over short time-scales [35, 36], whereas high circulating levels of β-hydroxybutyrate (hereafter 'BUTY') in plasma indicate dietary fasting, as BUTY is synthesized from fatty acids and largely replaces glucose in fueling metabolism during fasting [37, 38]. We also used scaled mass index [39] to estimate the amount of stored fat. Finally, to gain insight into stress levels on the final wintering site, we measured $CORT_f$ levels in body feathers which, unlike flight feathers, are replaced in Zambia after staging (Fig. 1). Since poorer staging conditions should exacerbate differences in quality between individuals, we predicted that in 2011, a year with below-average rainfall, carry-over effects would be stronger than in 2012, a year with above-average rainfall (Fig. 2).

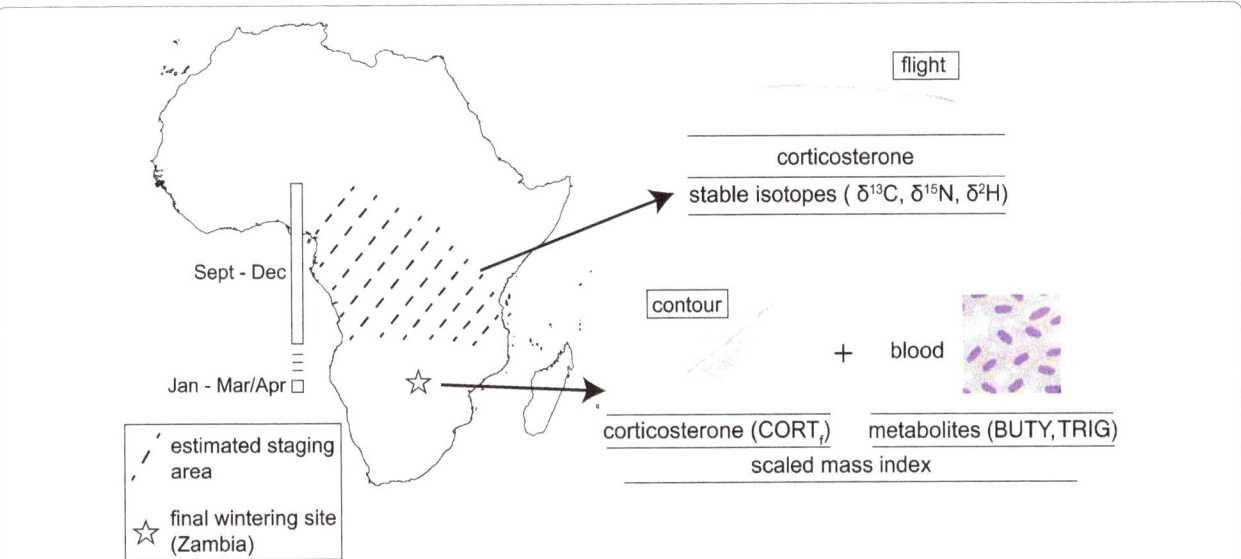

Fig. 1 Feather collection, blood sampling, and associated measures of condition for great reed warblers captured in Zambia. We tested whether the presence and strength of carry-over effects varied depending on habitat conditions. $\delta^{15}N$, $\delta^{13}C$, and $CORT_f$ were analysed in flight feathers moulted during staging to assess staging habitat quality, and δ^2H was added for use in a triple-isotope cluster analysis to estimate staging locations. Body condition at the Zambian winter site was assessed using plasma metabolites in blood, scaled mass index, and $CORT_f$ from contour feathers grown at this site

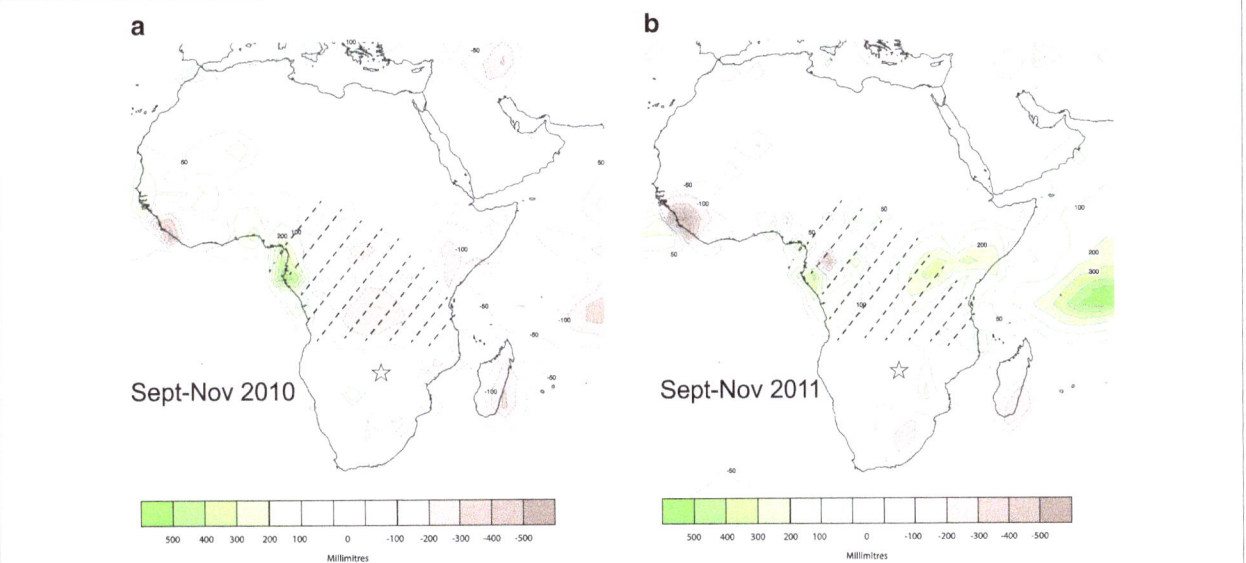

Fig. 2 Rainfall conditions experienced by great reed warblers during staging in **a** 2010 (2011 wintering season in Zambia) and **b** 2011 (2012 wintering season in Zambia). *Hashed lines* indicate estimated staging areas from triple-isotope cluster analysis (see "Results" section). Colour range (*brown to green*) indicates the range of rainfall anomalies (below average = *brown*, above average = *green*; NOAA data [42]. The Zambian study site is indicated by a *star*

Results

Estimated location of great reed warbler staging sites

We captured 26 great reed warblers in 2011 and 38 in 2012. The triple isotope cluster analysis assigned the majority of great reed warbler staging locations to cluster 2 (79 %), corresponding to Angola, Zambia, Zimbabwe, and southern portions of the Democratic Republic of the Congo. Some birds were assigned to cluster 1 (8 %) and cluster 3 (13 %), but none were assigned to cluster 4 which is associated with the Horn of Africa and the north-eastern Sahel region (Table 1; see [40] for a cluster map of Africa). A recent geolocator study of Swedish

Table 1 Estimated staging areas for great reed warblers caught in Zambia in 2011 and 2012

Cluster	2011	2012	Years combined
1	3 (12 %)	2 (5 %)	5 (8 %)
2	18 (72 %)	32 (84 %)	50 (79 %)
3	4 (16 %)	4 (11 %)	8 (13 %)
4	0 (0 %)	0 (0 %)	0 (0 %)

Cluster 1: Congo Basin, north Africa, and pockets of west Africa, southern Africa, and Madagascar; cluster 2: Angola, Zambia, Zimbabwe, and southern portions of the Democratic Republic of the Congo but also to extreme west Africa; cluster 3: a broad arc spanning the Atlantic coast of Senegal, south and east toward the Congo Basin, along Africa's eastern shoreline and westward into Botswana and Namibia; cluster 4: horn of Africa and the north-eastern Sahel region; see Fig. 3 in [40] for details. Staging clusters were estimated via a triple isotope cluster analysis, using isotope values (δ^{15}N, δ^{13}C, δ^{2}H) from tail feathers moulted during staging. The number of birds assigned to each cluster is included along with the corresponding proportion

breeding great reed warblers staging in West Africa found no further migration to southern Africa after staging and moult [17], which concurs with ringing data [41] and suggests that birds wintering in southern Africa do not stage in West Africa prior to reaching their final wintering grounds, and originate from different European breeding grounds. Lemke et al. [17] also found that the mean distance moved after staging was 678 km. Taken together, these results suggest that the most common staging sites for our study population are within 678–2000 km north of the Zambian study site (extent of most of cluster 2 excluding West African component of the cluster), and do not incorporate cluster 4 (East Africa) or West Africa.

Staging site rainfall in 2011 and 2012
Rainfall within estimated staging sites during autumn migration differed between the 2 years of this study. Birds wintering in Zambia in 2011 experienced lower than average rainfall during autumn migration staging, whereas birds wintering in 2012 experienced higher than average rainfall (Fig. 2).

Does staging rainfall affect conditions experienced by great reed warblers?
In support of the NOAA rainfall data (reported in Fig. 2; [42]), isotopic results also suggest that birds indeed experienced drier conditions in 2011 than in 2012: δ^{15}N values during staging were higher in 2011, suggesting that more arid conditions were experienced by birds in 2011 than in 2012 (t test: $t'_{54.4} = 2.9$, p = 0.006; 2011: mean = 9.9 ‰, SE ± 0.26; 2012: mean = 8.9 ‰, SE ± 0.22). In addition, as predicted, staging $CORT_f$ (feather corticosterone) was higher for 2011 than for 2012 (t test on ranked data: $t'_{52.9} = 6.9$, p < 0.0001; 2011: mean = 83.1 pg/mm, SE ± 2.3; 2012: mean = 65.0 pg/mm, SE ± 3.3; Fig. 3)

while $CORT_f$ from feathers grown on the final wintering grounds in Zambia was similar between years (t test on ranked data: $t'_{48.5} = -0.9$, p = 0.35; 2011: mean = 21.7 pg/mm, SE ± 1.86; 2012: mean = 21.7, SE ± 0.68; Fig. 3). As expected there was also no difference between years in δ^{13}C values from feathers grown during staging (δ^{13}C = t test: $t'_{47.5} = -0.3$, p = 0.75; 2011: mean = −15.7 ‰, SE ± 0.76; 2012: mean = −15.4 ‰, SE ± 0.53).

Does yearly variation in rainfall affect the strength of carry-over effects?
With scaled mass index (on the Zambian final winter site) as the response variable, interaction terms included in the top model set were δ^{13}C * year, and, δ^{15}N *year. Other models within 2 AIC_c scores included staging $CORT_f$, with combinations of year or δ^{15}N (Additional file 1: Table S1). The interaction term between δ^{13}C and year was significant in the multivariate model selected using AIC_c, whereas the interaction between δ^{15}N and year was not (Table 2). To better understand the relationship between δ^{13}C and year, we regressed scaled mass index against δ^{13}C separately for 2011 and 2012. The relationship between scaled mass index and δ^{13}C was negative in 2011 but not in 2012 (2011: $r^2 = 0.17$, p = 0.03; 2012: $r^2 = 0.03$, p = 0.37; Fig. 4a), indicating that in 2011 birds staging in areas with fewer C_3 plants (associated with moister, cooler habitats) had lower scaled mass indices

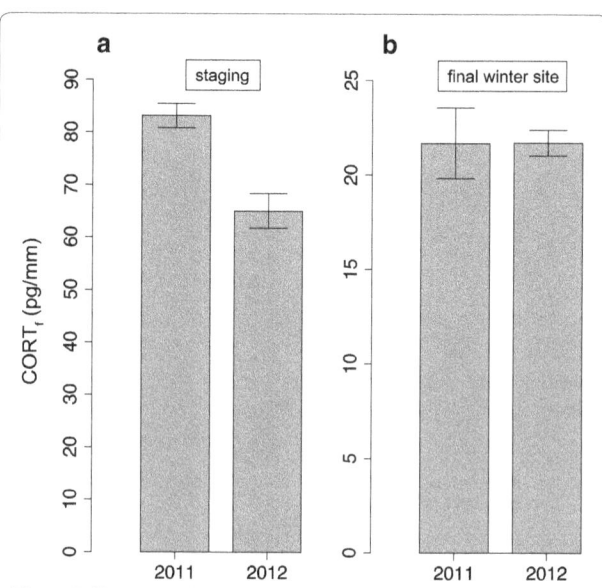

Fig. 3 Difference in the hormonal signatures of feathers ($CORT_f$) grown on **a** staging (t test on ranked data: $t'_{52.9} = 6.9$, p < 0.0001; 2011: n = 21, 2012: n = 34) and **b** final wintering grounds (t test on ranked data: $t'_{48.5} = -0.9$, p = 0.35; 2011: n = 24, 2012: n = 36) of great reed warblers in a year with below-average rainfall (2011), and a year with above-average rainfall (2012). *Whiskers* indicate ranges

Table 2 Summary of models (in the top 2 AIC$_c$ scores) with an interaction term included (see "Methods" section)

Model parameters	Model statistics			Interaction term statistics	
	F	r^2	p	t	p
Scaled mass index					
Year, carbon, staging CORT$_f$, carbon*year	2.75	0.19	0.04	2.19	0.03
Year, nitrogen, staging CORT$_f$, nitrogen*year	2.29	0.17	0.07	1.62	0.11
BUTY					
Year, nitrogen, scaled mass index, time at capture, nitrogen*year	2.83	0.22	0.03	2.89	0.005
Year, nitrogen, carbon, scaled mass index, time at capture, nitrogen*year	2.35	0.23	0.05	2.83	0.007
TRIG					
Year, carbon, scaled mass index, time at capture, carbon*year	2.98	0.24	0.02	1.48	0.15

Parameters were chosen using AIC$_c$ for each response variable separately (scaled mass index, B-hydroxybutyrate BUTY, triglyceride TRIG, final winter site feather corticosterone CORT$_f$. Scaled mass index and time at capture were added as additional covariates for BUTY and TRIG models (see "Methods" section). All covariates were standardized prior to analyses such that effect sizes were comparable

in Zambia. δ^{15}N was not related to scaled mass index in either 2011 (r^2 = 0.04, p = 0.32) or 2012 (r^2 = 0.009, p = 0.60; Fig. 4b).

With BUTY as the response variable, every model in the top model set included the interaction term δ^{15}N * year (Additional file 1: Table S1). The interaction between δ^{15}N and year was significant in both of the selected multivariate models (Table 2). When years were analysed separately and scaled mass index and time at capture were included as covariates (see "Methods" section), the relationship between δ^{15}N and BUTY was weakly positive in 2011 but not in 2012 (2011: t = 1.83, p = 0.08; 2012: t = −1.34, p = 0.2; Fig. 4c). This may indicate that in 2011, individuals staging in habitats with higher δ^{15}N values (corresponding to higher aridity) had higher BUTY levels (corresponding to increased fat catabolism and fasting) in Zambia.

With TRIG as the response variable, the interaction between δ^{13}C * year was included in the top model set. Other models within 2 AIC$_c$ scores included year, with combinations of δ^{13}C, δ^{15}N or staging CORT$_f$ (Additional file 1: Table S1). The interaction between δ^{13}C and year was not significant in the multivariate general linear model included in the top model set (Table 2). When years were analysed separately with scaled mass index and time at capture included as covariates (see "Methods" section), δ^{13}C was negatively related to TRIG in 2011 but not in 2012; however, these relationships were not significant (2011: t = −1.7, p = 0.1; 2012: t = 0.34, p = 0.74; Fig. 4d).

With winter CORT$_f$ as the response variable, no interaction terms were included in the top model set (Table 2). Models within 2 AIC$_c$ scores included combinations of carbon, nitrogen, and staging CORT$_f$ (Additional file 1: Table S1).

The relationships between staging conditions and final winter site conditions in Zambia did not differ between males and females in either 2011 or 2012 (Additional file 1: Table S3).

Zambian final winter site rainfall and body condition

We tested whether body condition at the final wintering site was better predicted by environmental conditions experienced during staging, or experienced at the final wintering site. Local rainfall conditions at our study site in Zambia were higher in 2011 (December–February total = 736 mm) than in 2012 (595 mm), which was the reverse of rainfall conditions on the staging grounds (Fig. 2). We found that winter CORT$_f$, BUTY, and TRIG did not differ between years (winter CORT$_f$: t test on ranked data t$'_{48.5}$ = −0.92, p = 0.36; BUTY: ANCOVA F$_{1,51}$ = 0.08, p = 0.79; TRIG: ANCOVA F$_{1,48}$ = 1.06, p = 0.31); however, the great reed warbler population in Zambia in 2012 had a lower scaled mass index (an estimate of stored fat) than the population in 2011 (t test t$'_{56.3}$ = 2.4, p = 0.02). This may suggest that in these 2 years body condition was more strongly related to the most recently experienced rainfall conditions (i.e. at final wintering grounds) rather than those experienced during staging. See Additional file 2 for all supporting data.

Discussion

Recent research has found that migratory songbirds wintering in sub-Saharan Africa frequently use long-term staging sites before reaching their final destination winter sites. Our results support the hypothesis that local conditions during staging carry over to influence individual condition on the final wintering grounds, but only in years when rainfall conditions are poor. We found that in a drier-than-average year on staging sites (2011),

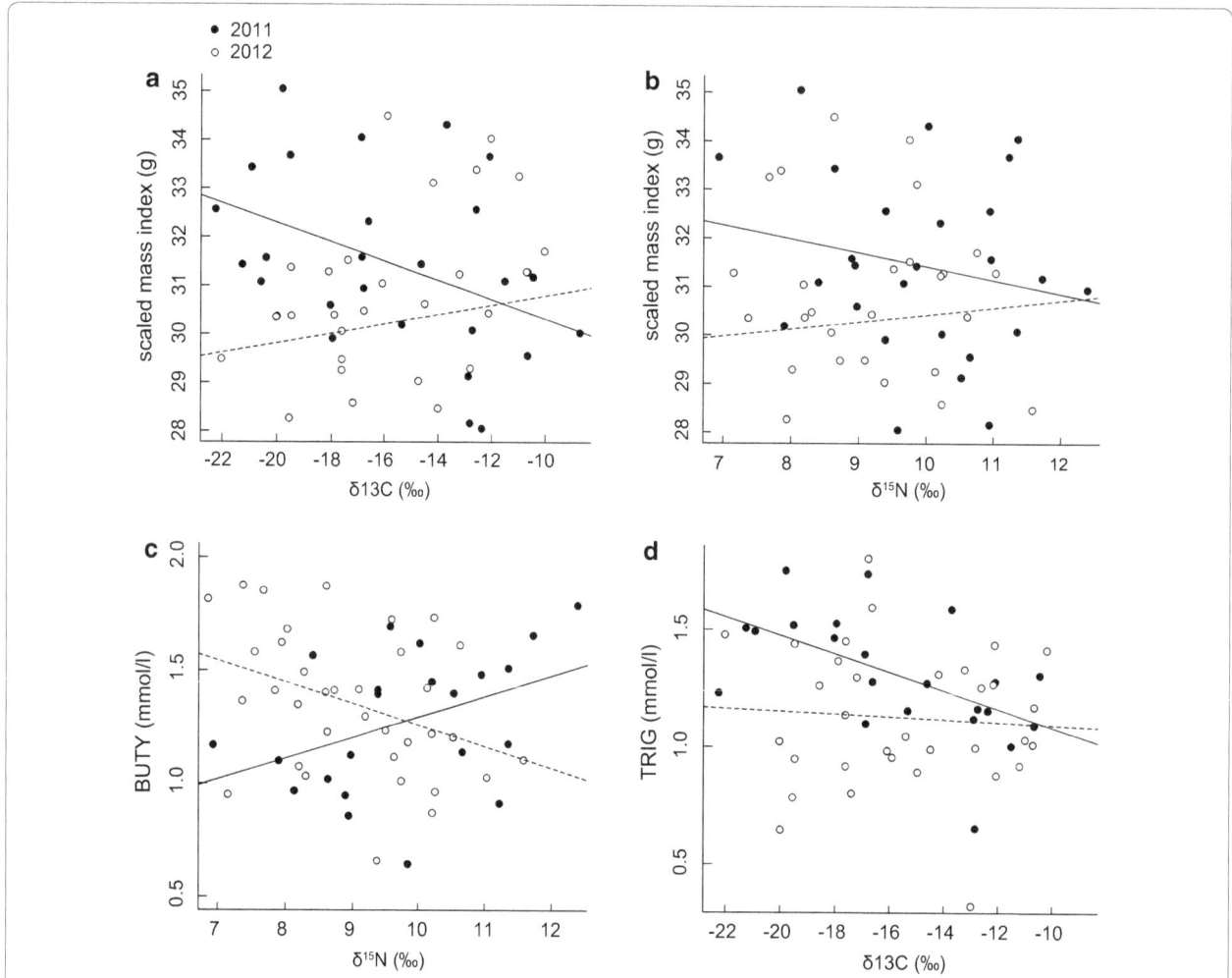

Fig. 4 Exploratory analysis for interaction terms that were included in the top model set (top 2 AIC_c scores; Table 2); the response variable and parameter of interest were regressed separately for each year. **a** scaled mass index: $\delta^{13}C$ * year (2011: $r^2 = 0.17$, p = 0.03; 2012: $r^2 = 0.03$, p = 0.37), **b** scaled mass index: $\delta^{15}N$ * year (2011: $r^2 = 0.04$, p = 0.32; 2012: $r^2 = 0.009$, p = 0.6); **b** BUTY: $\delta^{15}N$ * year (2011: t = 1.83, p = 0.08; 2012: t = −1.34, p = 0.19); **c** TRIG: $\delta^{13}C$ * year (2011: t = −1.7, p = 0.1; 2012: t = 0.34, p = 0.74). Time at capture and scaled mass index were included as covariates in **c** BUTY and **d** TRIG models (see "Methods" section), *plots* show the raw data

birds had higher $CORT_f$ levels and $\delta^{15}N$ values, respectively reflecting exposure to stronger (or more pervasive) stressors and higher aridity, than birds staging in a year with above-average rainfall (2012; Fig. 3). This indicated that rainfall during staging did indeed affect the local conditions experienced by the population. We then found that individual variation in staging conditions experienced by birds (estimated by $\delta^{13}C$ and $\delta^{15}N$) in 2011 carried over to affect their winter condition in Zambia, as estimated by scaled mass index and BUTY, both of which are likely to influence their overwinter survival and subsequent success on northwards migration. However, in a wetter-than-average year (2012), such individual variation in staging conditions was unrelated to over-winter condition in Zambia, suggesting that when conditions

were good, all individuals had access to high quality habitats. Our results demonstrate that differences in local long-term staging conditions may be important for species demography [1]. Moreover, our findings emphasise the challenges in conserving multi-site migration routes and understanding how annual variation in local conditions determine the strength of carry-over effects.

The data support the predictions of our hypothesis; however, they are based on a comparison between 2 years and additional years of study are needed to confirm the generality of the pattern they suggest. It is possible that other differences between years, in addition to rainfall, may have contributed to our observed results. However, rainfall is the most plausible driver given that its connection to food supply is known to be an important environmental driver for the success of wintering migratory birds [18, 43].

Can good conditions on final wintering sites buffer the effects of poor staging conditions?

Despite having experienced poorer habitat conditions during staging, birds wintering in Zambia in 2011 had higher scaled mass indices than in 2012. One explanation may be that local rainfall conditions at the winter study site were wetter (+141 mm) in 2011 than 2012, and that this difference was enough to counteract drier-than-average staging conditions. Alternatively, poor staging conditions in 2011 might have resulted in differential survival [43]. If only the highest quality birds were able to complete the second stage of migration to final winter sites, this would cause scaled mass index to be higher across the winter population despite the presence of carry-over effects from poor staging conditions. Either way, if rainfall underlies these patterns then good rainfall conditions in Zambia appear to be sufficient to counteract drier-than-average staging conditions, at least in the short term [6]. Whether and how carry-over effects from staging during dry years might affect long-term indicators of success (such as spring migration departure dates and success, survival, and fitness) is an important question for determining the consequences of staging site conditions, and should be a focus of future research exploiting continual improvements in tracking technology.

Migratory staging and connectivity

Determining the location of long-term staging sites is essential for understanding how staging habitat conditions may contribute to population dynamics. In this study, we used a triple isotope (δ^{15}N, δ^{13}C, δ^2H) cluster analysis to assign birds to staging locations. While this method can assign migrants to large clusters in Africa, it lacks the specificity that can be obtained from geolocator studies. Future work that combines geolocator and isotope data will help to improve the applicability of isotopic assignment in Africa. For species with strong migratory connectivity, determining specific staging sites may be essential for determining population responses to conditions during staging; for example, extreme local staging conditions may have specific and dramatic consequences later in the annual cycle. By contrast, for species with weak migratory connectivity, those effects will be diffuse across the species' range [44]. Great reed warblers seem likely to experience diffuse effects, since breeding and wintering populations appear to be loosely connected [17]. The discovery of multiple long-term staging sites in many species makes studies of migratory connectivity more difficult, since the degree of connectivity with breeding areas could change over the non-breeding season as birds shift locations. Determining how staging locations are connected to both breeding and wintering grounds is important for predicting the consequences of poor rainfall or drought conditions.

Conclusions

This study supports the hypothesis that habitat conditions on distant long-term staging sites can influence individual condition at final wintering sites, and suggests that these effects are likely to be strongest when conditions are relatively poor. Longitudinal studies are needed to confirm whether the latter pattern is general. Our results suggest that when conditions on final wintering sites are good, individuals may be able to buffer the effects of poor staging conditions in the short term; however, differential survival due to poor staging conditions may also account for this pattern. Research determining variation in the strength of carry-over effects between years has been limited (but see [18, 45, 46]), and to our knowledge this study is the first to assess carry-over effects between recently-discovered long-term staging sites and final wintering sites. Long-term staging sites add complexity to our understanding of population dynamics for declining migratory species especially in the context of global change, in which habitat loss (e.g. owing to agriculture) and unpredictable events outside the reproductive period such as food shortages (e.g. owing to drought) are likely to increase. Our results provide further incentive for logistically challenging endeavours that aim to quantify year-round habitat use and disentangle interactions between multiple environmental factors and sites.

Methods
Study site and field data

Fieldwork was conducted in southern Zambia on and around Muckleneuk Farm, near Choma, in ca. 280 ha of mesic grass and reed-bed habitat around each of two watercourses and associated dams (site 1: 16°39′S, 27°00′E; site 2: 16°37′S, 26°59′E), from January to April 2011 and 2012. The surrounding habitat is thornbush savannah and broad-leaved woodland with areas of tobacco, maize, and wheat cultivation. Great reed warblers are widely distributed across their European breeding grounds and sub-Saharan wintering grounds, and absent only from the southwestern half of South Africa and parts of Botswana and Namibia [47].

Great reed warblers were captured using 8–10 mist nets erected daily from 0530 to 1100 h. Birds were captured without the use of playbacks in order to avoid biases towards the most aggressive individuals. At the time of capture, all individuals were weighed with an electronic scale (± 0.01 g), and tarsus length, bill length, and wing length were measured with digital calipers (± 0.1 mm).

To estimate body condition, we calculated the scaled mass index from our morphometric data. This approach scales the mass of all individuals to that expected if they were all of identical body size, and is preferable to the more traditionally used residuals of mass on length [39]. With this approach, a single measure of size is recommended rather than a principal component analysis (PCA) of multiple measures [39], so we used wing length as our measure of structural size since it was correlated most strongly with weight (wing: r = 0.46, p < 0.001; bill: r = 0.35, p = 0.004; tarsus: r = −0.03, p = 0.85, n = 61). Within 10 min of capture, a 100 µl blood sample was taken for molecular sexing and metabolite analyses (see "Plasma metabolites" section below); thereafter, 5–8 contour feathers (collected from the upper breast) and the 5th right rectrix feather were sampled for isotope and CORT analyses (see below). Because great reed warblers are sexually monomorphic, sex was determined using a molecular technique. Primers for molecular sexing were modified from the primers 2550F and 2718R first described by [48]; see [49] for details. Juveniles (first-winter birds) may initiate staging moult on average 15 days later than adults (51 day total moult duration; [19]), which could cause differences in isotope signatures of feathers; however, this effect is likely to be minimal given the total length of moult. Juveniles were separated from adults by using tongue spot distinctiveness, eye, and tarsus colouration (for details see [50]). Indeed, we found no difference in isotope signatures between adults and juveniles in our data (δ^{15}N: t test $t'_{24.1} = 0.69$, p = 0.5; δ^{13}C: t test $t'_{23.8} = 0.13$, p = 0.9). Therefore, all individuals (2011 = 7 juveniles, 19 adults; 2012 = 9 juveniles, 28 adults) were pooled for analysis.

Rainfall measures

Figure 2 shows precipitation during migratory staging (September–November averages) in the northern-hemisphere autumn of 2010 (hereafter '2011' as this is when sampling took place in Zambia) and 2011 (hereafter '2012'). To quantify differences between years, we used 3 month precipitation anomalies from the NOAA Climate Prediction Center's CAMS_OPI dataset (http://www.ncdc.noaa.gov/cag/mapping/global; Accessed May 15 2014; [42]), which merges observations from rain gauges with precipitation estimates from a satellite algorithm. Anomalies are expressed as the deviation from the 1979–2000 mean value. September–November averages are presented in order to include rainfall just prior to autumn arrival on staging sites (end of September to early October; [17], since rainfall during this time is likely to be important given the time lag observed between precipitation and primary productivity [51]. We used farm records (Muckleneuk Farm, 16°38′S, 27°00′E) to assess

local rainfall at the final winter site in Zambia; rainfall was recorded daily in mm using a rain gauge (from December to February 2011 and 2012).

Isotope analysis
Carbon and nitrogen

Prior to all isotope analyses, feathers were washed in 2:1 chloroform:methanol solution for 24 h then rinsed with distilled water and left to air dry for 24 h. Sub-samples of 0.8–0.9 mg were weighed into tin capsules. Samples were combusted in a Pyrocube (Elementar, Hanau) elemental analyser. The resulting CO_2 and N_2 were separated by gas chromatography and admitted into the inlet of a Micro-mass (Manchester, UK) Isoprime isotope ratio mass spectrometer (IRMS) for determination of ^{13}C/^{12}C and ^{15}N/^{14}N ratios. Measurements are reported in δ-notation (δ^{13}C and δ^{15}N, respectively) relative to the Pee Dee Belemnite (PDB) for carbon and atmospheric N_2 for nitrogen in parts per thousand deviations (‰) using the formula δ (‰) = 1000 × [R_{sample}/$R_{standard}$−1]. Two sulfanilamides (Isoprime internal standards) and two Casein standards (in house standard) were used for every seven unknowns in sequence. Internal laboratory standards indicated measurement errors (SD) of ±0.05 ‰ for δ^{13}C, 0.12 ‰ for δ^{15}N.

Hydrogen

Cleaned samples of approximately 0.2 mg were weighed into silver capsules and pyrolysed in an Elementar (Hanau, Germany) Pyrocube elemental analyser over glassy carbon. The resulting H_2 was admitted into the source inlet of a Thermo (Bremen, Germany) XP Plus mass spectrometer. Measurements are reported in δ-notation relative to the international standard SMOW (Standard Mean Ocean Water). Organic materials involving H not bonded to carbon will readily exchange H with water vapour [52]. Subtraction of this exchangeable hydrogen is attained using standards with known non-exchangeable (i.e. indigenous) hydrogen isotope compositions. These are CFS (chicken feathers), BWB-II (bowhead whale baleen), and ISB (black kittiwake feathers); for CFS and BWB-II see [53]; for ISB see [54]. Replicate measurements of these standards imply measurement errors (SD) of around 2 ‰ for δ^2H.

Triple-isotope cluster analysis

We used the isotope signature of great reed warbler tail feathers, grown during staging, to determine the most likely staging locations in Africa. Specifically, we used a three-isotope (δ^2H, δ^{13}C, δ^{15}N) cluster analysis developed by [40] to assign birds to one of four clusters in Africa. This method uses isoscapes (i.e. maps of isotopic variation produced by iteratively applying models across regions of space) to estimate the origin of moulted feathers [55]. The clusters were developed by examining previously published plant $\delta^{13}C$ and $\delta^{15}N$ [56, 57] isoscapes and the long-term hydrologic δ^2H [55] isoscapes of Africa, to obtain information on geographic structure in order to find "natural" groupings in multivariate space (Fig. 2; see [40] for details). Linear discriminant function analysis was then used to derive algorithms that predict the posterior probability that a sample with a given multi-isotope composition could have originated from any given cluster within Africa, given the predicted ranges for feather δ^2H, $\delta^{13}C$, and $\delta^{15}N$ (see [40] for details). Error estimates were derived from [58]. All analyses were conducted within R version 2.15.0 (R Development Core Team 2012) applying algorithms derived from [40] by means of script provided by Steven Van Wilgenburg.

Plasma metabolites

Blood samples were collected into heparinized capillary tubes from the brachial vein and were centrifuged for 10 min (maximum speed of 8000 rpm) within 4 h of collection. The separated plasma was frozen at -20 °C until analysis. Samples were transported on ice to the Schweizerische Vogelwarte, Sempach, Switzerland, for analyses in May 2012. All metabolites were determined in the plasma using standard test-combinations, the Wako Auto-Kit for 3-hydroxybutyrate (cyclic enzymatic method; BUTY), and the enzymatic colorimetric test for triglycerides (TRIG) including free glycerol (Invicon HIT 917, PAP-method). Dilution curves of the great reed warbler plasma were run for both metabolites to test for linearity. The triglyceride assay was adapted to small volumes (5–10 µl). All samples were measured in duplicates. TRIG and BUTY values were linearly negatively related ($r^2 = 0.06$, $p = 0.05$); this suggests that the inclusion of free glycerol in our test of triglycerides did not obscure true triglyceride values [36].

Feather corticosterone

Corticosterone (CORT) was extracted from feathers using a methanol-based procedure [59] that has been replicated successfully in passerines [60, 61]. A single tail feather per bird was collected and CORT thus extracted. Because multiple contour feathers were collected from all individuals, two contour feathers per bird were extracted to ensure that $CORT_f$ measurements were well within limits of detectablility [62]. Following removal of the calamus, the length of each feather sample (proximal cut to distal tip of vane) was measured flat against the edge of a ruler. Feather samples were then placed in glass vials and cut into tiny pieces (c. 5 mm^2). Ten mL of methanol (HPLC grade; Fisher Scientific, Fair Lawn, NJ, USA) was added to each vial and samples were sonicated at room temperature for 30 min, followed by overnight incubation at 50 °C. Using a glass funnel fitted with polyester fibre, vacuum filtration was employed to separate the methanol extract from the feather pieces. Collected methanol extracts were evaporated and residues were reconstituted in 600 µL of phosphate-buffered saline (0.05 M, pH 7.6) and frozen at -20 °C until analysed by radioimmunoassay (RIA). Samples were extracted in two batches and the recovery efficiency of each extraction was assessed by spiking three sample extracts with approximately 5000 c.p.m. of ^3H-labelled CORT. The average recoveries were 98.6 and 93.2 %, and final values were adjusted for recoveries. Serial dilutions of feather extracts were parallel to the CORT standard curve. Samples were analysed in duplicate in two RIAs using a commercial antiserum (Sigma-Aldrich, St. Louis, MO, USA; product #C8784). The variability of each assay was assessed using six replicates of the same internal standard; average intra-assay coefficient of variation (CV) was 7.4 (SD = 1.1) % and inter-assay CV was 5.5 %. Average detection limit (%B/B_0) was 11.0 (SD = 1.5) pg CORT/100 µl of extract and all samples were above detection limits. $CORT_f$ measurements were corrected for total feather sample length, reflecting the time-dependent deposition of CORT into feathers, and were thus expressed as pg CORT/mm [31, 59, 63]. $CORT_f$ values are likely determined by the interaction between plasma CORT values and the dynamics of feather growth [33]. Although we cannot be sure that the higher $CORT_f$ values during staging in 2011 vs. 2012 were due to decreased feather growth rates, higher plasma CORT levels, or both, responses to poor environmental conditions would likely result in higher $CORT_f$ values in all of these scenarios [33], though future work is needed to fully understand the mechanism in our study species.

Statistical analysis

We analysed each response variable separately (scaled mass index, TRIG, BUTY, final winter site $CORT_f$). We used Akaike's Information Criterion corrected for small sample sizes (AIC_c; [64]) to select from the following predictors $\delta^{13}C$, $\delta^{15}N$, staging $CORT_f$, and interaction terms between each parameter and year. Because a difference between years was the a priori prediction for our hypothesis, if an interaction term was included in the top model set (within 2 AIC_c scores) we constructed multivariate general linear models of all top models including interaction terms (Table 2), and performed exploratory analysis by regressing the response variable and parameter of interest separately for each year (see "Results" section; Fig. 4). For TRIG and BUTY models, time at capture was included as a covariate since this has been shown to be related to plasma metabolite levels [65, 66]; in addition, scaled mass index was included as a covariate in order to assess plasma metabolites as a dynamic measure of change in body condition rather than a static measure of current state, such as scaled mass index [36]. All covariates were standardized prior to analyses such that effect sizes were comparable. All models residuals conformed to normality assumptions for parametric statistics; therefore, response variables were not transformed prior to analysis. Model averaging was not conducted, as sample sizes were not large enough for robust analysis [67].

Three TRIG data points had Cook's distances substantially higher than the mean (mean = 0.02 ± 0.008 SE; sample 1 = 0.21, sample 2 = 0.18, sample 3 = 0.34; [68]). All three outlying TRIG concentrations were measured within 2 weeks of the start of spring migration, such that these high TRIG concentrations were likely from individuals that had begun pre-migratory fattening and do not represent baseline winter feeding conditions. These three values were therefore removed from the analyses, but conclusions were unchanged if they were left in the dataset.

Sex was not included in the main models due to low residual degrees of freedom. Therefore, to determine whether carry-over effects differed between females and males, separate general linear models were instead run for each year, with an interaction term between sex and all three predictor variables ($\delta^{15}N$, $\delta^{13}C$, and staging $CORT_f$) included.

To test for differences between groups we used unequal variances (Welch's) t tests; when data did not meet assumptions of normality we ranked the data, following [69]. To test for differences in plasma metabolite values between years we used ANCOVA and included scaled mass index and time at capture as covariates.

Additional files

Additional file 1: Table S1. This file contains three tables: the full AIC_c ranked candidate model set (2AIC_c scores) of the relative importance of staging parameters on final winter site body condition ($\delta^{15}N$, $\delta^{13}C$, feather corticosterone, and interactions with year); **Table S2:** full results of the top multivariate general linear model with interaction terms included for each response variable (scaled mass index, BUTY, TRIG, and staging $CORT_f$); and **Table S3:** full results of models to test for sex differences in carry-over effects between years.

Additional file 2. The raw data collected from great reed warblers at the Zambian study site. Information includes year, sex, age (1 = juvenile, 2 = adult) time at capture (time.count = minutes since 0600 h), winter $CORT_f$ (feathercort.C), staging $CORT_f$ (feathercort.F), BUTY, TRIG, scaled mass index, $\delta^{13}C$, $\delta^{15}N$, δ^2H, and the estimated isotope cluster.

Abbreviations
CORT: corticosterone; $CORT_f$: feather corticosterone; $\delta^{15}N$: nitrogen isotope; $\delta^{13}C$: carbon isotope; TRIG: triglyceride; BUTY: β-hydroxybutyrate.

Authors' contributions
MCS and CNS designed the study, MCS conducted fieldwork, MCS, GDF, SJ-E, JN, and EY conducted lab assays, and MCS wrote the manuscript with contributions from CNS, GDF, SJ-E, JN and EY. All authors read and approved the final manuscript.

Author details
[1] Department of Zoology, University of Cambridge, Cambridge CB2 3EJ, UK. [2] Department of Biology, University of Saskatchewan, Saskatoon S7N 5E2, Canada. [3] Swiss Ornithological Institute, Sempach, Switzerland. [4] NERC Life Sciences Mass Spectrometry Facility, Scottish Universities Environmental Research Centre, Rankine Avenue, East Kilbride G75 0QF, UK. [5] Limnological Institute, University of Konstanz, Mainaustrasse 252, 78464 Constance, Germany. [6] DST-NRF Centre of Excellence at the FitzPatrick Institute, University of Cape Town, Cape Town, South Africa.

Acknowledgements
We thank Kenty Mudenda, Collins Moya, Jason Boyce, and Jane Jönsson for assistance in the field and lab, Steve Van Wilgenburg and Keith Hobson for providing isotopic cluster analysis code and guidance during analysis, Ian and Emma Bruce-Miller for their generous hospitality and for supplying rainfall data in Zambia, Hamish Ross for access to part of the study area, Staffan Bensch for generous access to lab space, and the Zambia Wildlife Authority for permits. We gratefully acknowledge Tracy Marchant for allowing us to use her endocrine lab at the University of Saskatchewan.

Competing interests
The authors declare that they have no competing interests.

Funding
MCS was funded by the Gates Cambridge Trust and the Natural Sciences and Engineering Research Council of Canada; CNS was supported by a Royal Society Dorothy Hodgkin Fellowship, a BBSRC David Phillips Fellowship (BB/J014109/1), and the DST-NRF Centre of Excellence at the FitzPatrick Institute. Isotope analyses at the LSMSF were funded by a NERC LSMSF Grant (EK206-16/12).

References
1. Newton I. Can conditions experienced during migration limit the population levels of birds? J Ornithol. 2006;147:146–66.
2. Newton I. Population limitation in migrants. Ibis. 2004;146:197–226.
3. Harrison XA, Blount JD, Inger R, Norris DR, Bearhop S. Carry-over effects as drivers of fitness differences in animals. J Anim Ecol. 2010;80:4–18.
4. Gordo O, Brotons L, Ferrer X, Comas P. Do changes in climate patterns in wintering areas affect the timing of the spring arrival of trans-Saharan migrant birds? Glob Change Biol. 2005;11:12–21.
5. Saino N, Szep T, Romano M, Rubolini D, Spina F, Moller AP. Ecological conditions during winter predict arrival date at the breeding quarters in a trans- Saharan migratory bird. Ecol Lett. 2004;7:21–5.
6. Angelier F, Tonra CM, Holberton RL, Marra PP. Short-term changes in body condition in relation to habitat and rainfall abundance in American redstarts Setophaga ruticilla during the non-breeding season. J Avian Biol. 2011;42:335–41.

7. Studds CE, Marra PP. Rainfall-induced changes in food availability modify the spring departure programme of a migratory bird. Proc R Soc B. 2011;278:3437–43.

8. McKinnon EA, Fraser KC, Stutchbury BJM. New discoveries in landbird migration using geolocators, and a flight plan for the future. Auk. 2013;130:211–22.

9. Newton I. The migration ecology of birds. London: Elsevier; 2010.

10. Stach R, Jakobsson S, Kullberg C, Fransson T. Geolocators reveal three consecutive wintering areas in the thrush nightingale. Anim Migr. 2012;1:1–7.

11. Delmore KE, Fox JW, Irwin DE. Dramatic intraspecific differences in migratory routes, stopover sites and wintering areas, revealed using light-level geolocators. Proc R Soc B. 2012;279:4582–9.

12. Jahn AE, Cueto VR, Fox JW, Husak MS, Kim DH, Landoll DV, et al. Migration timing and wintering areas of three species of flycatchers (*Tyrannus*) breeding in the Great Plains of North America. Auk. 2013;130:247–57.

13. Tøttrup AP, Klaassen RHG, Strandberg R, Thorup K, Kristensen MW, Jørgensen PS, et al. The annual cycle of a trans-equatorial Eurasian-African passerine migrant: different spatio-temporal strategies for autumn and spring migration. Proc R Soc B. 2011;279:1008–16.

14. Stutchbury BJM, Siddiqui R, Applegate K, Hvenegaard GT, Mammenga P, Mickle N, et al. Ecological causes and consequences of intratropical migration in temperate-breeding migratory birds. Am Nat. 2016;188:S28–40.

15. Vickery JA, Ewing SR, Smith KW, Pain DJ, Bairlein F, Škorpilová J, et al. The decline of Afro-Palaearctic migrants and an assessment of potential causes. Ibis. 2013;156:1–22.

16. Sanderson FJ, Donald PF, Pain DJ, Burfield IJ, van Bommel FPJ. Long-term population declines in Afro-Palearctic migrant birds. Biol Cons. 2006;131:93–105.

17. Lemke HW, Tarka M, Klaassen R, Akesson M. Annual cycle and migration strategies of a trans-Saharan migratory songbird: a geolocator study in the great reed warbler. PLoS One. 2013;8:e79209.

18. Studds CE, Marra PP. Linking fluctuations in rainfall to nonbreeding season performance in a long-distance migratory bird, *Setophaga ruticilla*. Clim Res. 2007;35:115–22.

19. Hedenström A, Bensch S, Hasselquist D, Lockwood M, Ottosson U. Migration, stopover and moult of the Great Reed Warbler *Acrocephalus arundinaceus* in Ghana, West Africa. Ibis. 1993;135:177–80.

20. Yohannes E, Hansson B, Lee RW, Waldenstrom J, Westerdahl H, Akesson M, et al. Isotope signatures in winter moulted feathers predict malaria prevalence in a breeding avian host. Oecologia. 2008;158:299–306.

21. Norris DR, Marra PP, Kyser TK, Sherry TW, Ratcliffe LM. Tropical winter habitat limits reproductive success on the temperate breeding grounds in a migratory bird. Proc R Soc B. 2004;271:59–64.

22. Evans KL, Newton J, Mallord JW, Markman S. Stable isotope analysis provides new information on winter habitat use of declining avian migrants that is relevant to their conservation. PLoS One. 2012;7:e34542.

23. Minagawa M, Wada E. Stepwise enrichment of ^{15}N along food chains: further evidence and the relation between δ^{15}N and animal age. Geochim Cosmochim Acta. 1984;48:1135–40.

24. Pardo LH, Nadelhoffer KJ. Using nitrogen isotope ratios to assess terrestrial ecosystems at regional and global scales. In: West JB, Bowen GJ, Dawson TE, Tu KP, editors. Isoscapes: understanding movements, pattern and process on Earth through isotope mapping. New York: Springer; 2009. p. 221–49.

25. Yohannes E, Hobson KA, Pearson DJ. Feather stable-isotope profiles reveal stopover habitat selection and site fidelity in nine migratory species moving through sub-Saharan Africa. J Avian Biol. 2007;38:347–55.

26. Heaton THE, Vogel JC, von La Chevallerie G, Collett G. Climatic influence on the isotopic composition of bone nitrogen. Nature. 1986;322:822–3.

27. Farquhar GD, Ehleringer JR, Hubick KT. Carbon isotope discrimination and photosynthesis. Annu Rev Plant Physiol Plant Mol Biol. 1989;40:503–37.

28. Wingfield JC, Ramenofsky M. Corticosterone and facultative dispersal in response to unpredictable events. Ardea. 1997;85:155–66.

29. Marra PP, Holberton RL. Corticosterone levels as indicators of habitat quality: effects of habitat segregation in a migratory bird during the non-breeding season. Oecologia. 1998;116:284–92.

30. Wingfield JC, Ramenofsky M. Hormones and the behavioral ecology of stress. In: Balm PHM, editor. Stress physiology in animals. Sheffield: Sheffield Academic Press; 1999. p. 1–51.

31. Jenni-Eiermann S, Helfenstein F, Vallat A, Glauser G, Jenni L. Corticosterone: effects on feather quality and deposition into feathers. Methods Ecol Evol. 2015;6:237–46.

32. Fairhurst GD, Marchant TA, Soos C, Machin KL, Clark RG. Experimental relationships between levels of corticosterone in plasma and feathers in a free- living bird. J Exp Biol. 2013;216:4071–81.

33. Romero LM, Fairhurst GD. Measuring corticosterone in feathers: strengths, limitations, and suggestions for the future. Comp Biochem Phys A. 2016. doi:10.1016/j.cbpa.2016.05.002.

34. Still CJ, Berry JA, Collatz GJ, DeFries RS. Global distribution of C$_3$ and C4 vegetation: carbon cycle implications. Global Biogeochem Cycles. 2003;17:1–14.

35. Guglielmo CG, Cerasale DJ, Eldermire C. A field validation of plasma metabolite profiling to assess refueling performance of migratory birds. Physiol Biochem Zool. 2005;78:116–25.

36. Jenni-Eiermann S, Jenni L. Plasma metabolite levels predict individual body mass- changes in a small long-distance migrant, the garden warbler. Auk. 1994;111:888–99.

37. Féry F, Plat L, Melot C, Balasse EO. Role of fat-derived substrates in the regulation of gluconeogenesis during fasting. Am J Physiol. 1996;270:E822–30.

38. Jenni-Eiermann S, Jenni L. Metabolic responses to flight and fasting in night- migrating passerines. J Comp Physiol B. 1991;161:465–74.

39. Peig J, Green AJ. New perspectives for estimating body condition from mass/length data: the scaled mass index as an alternative method. Oikos. 2009;118:1883–91.

40. Hobson KA, Van Wilgenburg SL, Wassenaar LI, Powell RL, Still CJ, Craine JM. A multi-isotope (δ^{13}C, δ^{15}N, δ^2H) feather isoscape to assign Afrotropical migrant birds to origins. Ecosphere. 2012;3:art44.

41. Yohannes E, Lee RW, Jochimsen MC. Stable isotope ratios in winter-grown feathers of great reed warblers *Acrocephalus arundinaceus*, clamorous reed warblers *A. stentoreus* and their hybrids in a sympatric breeding population in Kazakhstan. Ibis. 2011;153:502–8.

42. Janowiak JE, Xie P. CAMS-OPI: a global satellite–rain gauge merged product for real-time precipitation monitoring applications. J Climate. 1999;12:3335–42.

43. Szép T. Relationship between west African rainfall and the survival of central European sand martins *Riparia riparia*. Ibis. 2008;137:162–8.

44. Taylor CM, Norris DR. Population dynamics in migratory networks. Theor Ecol. 2009;3:65–73.

45. Legagneux P, Fast PLF, Gauthier G, Bety J. Manipulating individual state during migration provides evidence for carry-over effects modulated by environmental conditions. Proc R Soc B. 2012;279:876–83.

46. Harrison XA, Hodgson D, Inger R, Colhoun K, Gudmundsson G, McElwaine G, et al. Environmental conditions during breeding modify the strength of mass- dependent carry-over effects in a migratory bird. PLoS One. 2013;8:e77783.

47. Cramp S, Perrins CM. The birds of the Western Palearctic. Oxford: Oxford University Press; 1994.

48. Fridolfsson A, Ellegren H. A simple and universal method for molecular sexing of non-ratite birds. J Avian Biol. 1999;30:116.

49. Round PD, Hansson B, Pearson DJ, Kennerley PR, Bensch S. Lost and found: the enigmatic large-billed reed warbler *Acrocephalus orinus* rediscovered after 139 years. J Avian Biol. 2007;38:133–8.

50. Bensch S, Hasselquist D, Nielsen B, Hansson B. Higher fitness for philopatric than for immigrant males in a semi-isolated population of great reed warblers. Evolution. 1998.

51. Sinclair ARE. Factors affecting the food supply and breeding season of resident birds and movements of palaearctic migrants in a tropical african savannah. Ibis. 2008;120:480–97.

52. Schimmelmann A, Miller RF, Leavitt SW. Hydrogen isotopic exchange and stable isotope ratios in cellulose, wood, chitin, and amino compounds. In: Swart PK, Lohmann KC, McKenzie J, Savin S, editors. Climate change in continental isotopic records. Washington: Geophysical Monograph 68, American Geophysical Union; 1993. p. 367–74.

53. Hobson KA, Wassenaar LI. Tracking animal migration with stable isotopes. Amsterdam: Elsevier; 2008.

54. Fox TAD, Christensen TK, Bearhop S, Newton J. Using stable isotope analysis of multiple feather tracts to identify moulting provenance of vagrant birds: a case study of baikal teal *Anas formosa* in Denmark. Ibis. 2007;142:622–5.

55. Bowen GJ, Wassenaar LI, Hobson KA. Global application of stable hydrogen and oxygen isotopes to wildlife forensics. Oecologia. 2005;143:337–48.

56. Still CJ, Powell RL. Continental-scale distributions of vegetation stable carbon isotope ratios. Isoscapes: understanding movements, pattern and process on Earth through isotope mapping. New York: Springer; 2009. p. 179–93.

57. Craine JM, Elmore AJ, Aidar MPM, Bustamante M, Dawson TE, Hobbie EA, et al. Global patterns of foliar nitrogen isotopes and their relationships with climate, mycorrhizal fungi, foliar nutrient concentrations, and nitrogen availability. New Phytol. 2009;183:980–92.

58. Hobson KA, Bairlein F. Isotopic fractionation and turnover in captive garden warblers Sylvia borin: implications for delineating dietary and migratory associations in wild passerines. Can J Zool. 2003;81:1630–5.

59. Bortolotti GR, Marchant TA, Blas J. Corticosterone in feathers is a long-term, integrated measure of avian stress physiology. Funct Ecol. 2008;22:494–500.

60. Fairhurst GD, Vögeli M, Serrano D, Delgado A, Tella JL, Bortolotti GR. Can synchronizing feather-based measures of corticosterone and stable isotopes help us better understand habitat–physiology relationships? Oecologia. 2013;173:731–43.

61. Fairhurst GD, Dawson RD, van Oort H, Bortolotti GR. Synchronizing feather- based measures of corticosterone and carotenoid-dependent signals: what relationships do we expect? Oecologia. 2014;174:689–98.

62. Fairhurst GD, Bond AL, Hobson KA, Ronconi RA. Feather-based measures of stable isotopes and corticosterone reveal a relationship between trophic position and physiology in a pelagic seabird over a 153-year period. Ibis. 2015;157:273–83.

63. Bortolotti GR. Flaws and pitfalls in the chemical analysis of feathers: bad news- good news for avian chemoecology and toxicology. Ecol Appl. 2010;20:1766–74.

64. Burnham KP, Anderson DR. Model selection and multimodel inference: a practical information-theoretic approach. New York: Springer; 2002.

65. Guglielmo CG, O'Hara PD, Williams TD. Extrinsic and intrinsic sources of variation in plasma lipid metabolites of free-living western sandpipers Calidris mauri. Auk. 2002;119:437–45.

66. Jenni L, Jenni-Eiermann S. Metabolic responses to diurnal feeding patterns during the postbreeding, moulting and migratory periods in passerine birds. Funct Ecol. 1996;10:73–80.

67. Grueber CE, Nakagawa S, Laws RJ, Jamieson IG. Multimodel inference in ecology and evolution: challenges and solutions. J Evol Biol. 2011;24:699–711.

68. Fox J. Regression diagnostics: an introduction. Newbury Park: Sage; 1991.

69. Ruxton GD. The unequal variance t test is an underused alternative to student's t test and the Mann–Whitney U test. Behav Ecol. 2006;17:688–90.

70. Rollin BE, Kessel ML. Guidelines for the treatment of animals in behavioural research and teaching. Anim Behav. 2012;83:301–9.

14

Rodent-avoidance, topography and forest structure shape territory selection of a forest bird

Gilberto Pasinelli[1,2]*, Alex Grendelmeier[1,3], Michael Gerber[2,4] and Raphaël Arlettaz[3,5]

Abstract

Background: Understanding the factors underlying habitat selection is important in ecological and evolutionary contexts, and crucial for developing targeted conservation action in threatened species. However, the key factors associated to habitat selection often remain poorly known. We evaluated hypotheses related to abiotic and biotic factors thought to affect territory selection of the wood warbler *Phylloscopus sibilatrix*, a passerine living in an unpredictable environment owing to irregular rodent outbreaks and showing long-term declines particularly in Western Europe.

Results: Comparing breeding territories to unoccupied areas located close-by revealed that territory occupancy in north-western Switzerland was positively related to slope steepness (topographic hypothesis supported) as well as to numbers of tussocks and trees, respectively, while it showed a unimodal relationship to cover of herb layer (forest structure hypothesis supported). Furthermore, a strong negative correlation between breeding territory occupancy and rodent numbers was found, suggesting that wood warblers avoid areas with high rodent densities (rodent-avoidance hypothesis supported). Comparing breeding territories to abandoned territories showed that breeding territories were located on steeper slopes (topography hypothesis supported), at larger distance from the forest edge (anthropogenic disturbance hypothesis supported) and harboured more trees (forest structure hypothesis supported) than abandoned territories.

Conclusions: Aside from structural and topographic features of the habitat, wood warblers are affected by rodent numbers when settling, making habitat selection unpredictable from year to year. Forestry practices promoting relatively high tree densities, few bushes and an intermediate low-growing ground vegetation cover would enhance habitat quality for this declining passerine. In contrast, forestry practices aiming at increasing light in forests (selective thinning, group-felling) or keeping forest stands permanently covered with shrubs, bushes and trees of various sizes (continuous cover forestry) do not benefit the wood warbler.

Keywords: Habitat, Ecological niche, Forestry, AICc model selection, Aves, Passeriformes

Background

Understanding the mechanisms underlying habitat selection is important in ecological and evolutionary contexts as well as for the application of conservation measures in threatened species. For animals reproducing in seasonal environments, selecting a habitat to breed is a recurring annual task. Resident species can base breeding habitat selection on year-round interactions with their abiotic and biotic environments. In contrast, long-distance migratory species such as many songbirds, often spending most of the year outside the breeding grounds, have to select habitats shortly after returning to the breeding grounds. Here, we address patterns of habitat selection of the wood warbler *Phylloscopus sibilatrix*, a specialised songbird inhabiting the interior of European woodlands. These woodlands are subject to irregular rodent outbreaks arising from irregularly occurring seed mast of forest trees [1]. Wood warblers have been shown to respond to rodent numbers when settling in spring [2,

*Correspondence: gilberto.pasinelli@vogelwarte.ch
[1] Swiss Ornithological Institute, Sempach, Switzerland
Full list of author information is available at the end of the article

3], but how abiotic factors, structural habitat features and rodent numbers interact during habitat selection of this species has not yet been assessed.

We examined hypotheses proposed to be relevant both in the general context of habitat selection and in explaining population declines of the wood warbler in Western Europe. We compared current breeding territories to both unoccupied areas located nearby and to abandoned territories, that is, to previously but no longer occupied territories (see "Choice of abandoned territories"). According to the forest structure hypothesis, structural changes such as decreases in canopy cover or increases in understory vegetation over the last decades [e.g., 4] may have resulted in degradation and loss of many previously suitable forest habitats [5]. In contrast, in ecosystems like Białowieża National Park in Poland, a pristine environment mostly unaffected by humans, behaviour, ecology, breeding success and population trends of wood warblers did not significantly change over the past 25 years [2, 6]. While macro-habitat selection of the wood warbler has been subject to some studies [7–9], factors affecting habitat selection at the territory scale have received comparatively little attention [5, 10–12]. Based on habitat preferences established in previous studies, we expected breeding territories of wood warblers to be located in forest stands of medium age and to be characterized by relatively closed canopy, sparse undergrowth and sparse ground vegetation cover compared to unoccupied control areas and abandoned territories, respectively.

Predation risk has been shown to affect patterns of habitat selection in a variety of species [13–15]. In birds, nest predation is often the main reason for nest failure [16, 17] and can profoundly affect avian population dynamics. Predation has been shown to cause up to 95 % of all nest losses in the wood warbler [e.g., 6]. As suggested for other bird species [18, 19], avoiding areas of increased predation risk should thus be of central importance in breeding habitat selection of this species. Many potential wood warbler nest predators such as red fox (*Vulpes vulpes*) and marten (*Martes* spp.) feed on rodent species such as voles (*Microtus* spp.) and mice (*Apodemus* spp.). Increased density and activity of these predators in areas and years with high rodent populations might increase predation risk for wood warbler nests, in addition to the possible risk of direct predation by rodents. Numbers of wood warblers are lower in rodent outbreak years than in other years [2, 3]. High rodent density might indicate increased predation risk to wood warblers, which may thus avoid settling in such areas. We examined the hypothesis that wood warbler habitat selection at the territory scale was related to rodent density. According to this rodent-avoidance hypothesis, we expected that actual territories of wood warblers would have lower rodent densities than nearby unoccupied areas and that abandoned territories (see "Choice of abandoned territories" below) would show higher rodent densities than current breeding territories.

Disturbances due to increasing human activities can negatively affect breeding bird communities and population dynamics [20–23] and have been proposed to be a reason for wood warbler population declines in Switzerland [24]. Therefore, we evaluated the influence of disturbance-related variables on breeding and abandoned territories (referred to in the following as the "anthropogenic disturbance hypothesis"), expecting that abandoned territories would be located closer to areas exposed to human disturbance than breeding territories.

Finally, abiotic factors are part of a species' niche and are thus expected to affect habitat selection directly or indirectly [e.g. 25]. Accordingly, wood warblers have been found to prefer settling in inclined areas [5, 7, 26]. We thus expected breeding territories to be located on steeper slopes than control areas. Additionally, a preference for slopes with eastern to southerly aspects has been reported, while slopes with western and northern aspects appear to be avoided, this preference remaining unexplained [7, 26]. Breeding territories were therefore expected to exhibit more eastern to southern aspects than control areas. Because the wood warbler has disappeared from many parts of the Swiss lowlands, we furthermore expected breeding territories, compared to abandoned territories, to be located at higher elevation, on steeper and more east- to south-exposed slopes. The latter relationship was expected as wood warblers might today be restricted to the best available sites, i.e. to the most suitable aspects. Slope steepness, aspect and elevation (m above sea level, a.s.l.) are referred to as the topography hypothesis.

Throughout Western Europe, populations of this species have declined in the last three decades, while populations in Eastern Europe have remained relatively stable [6, 7, 27]. The causes of these differential population trajectories are unknown. The wintering grounds in tropical Africa do not appear to differ for birds from western and eastern populations [28], suggesting that changes in the breeding areas could underlie the diverging population trends. In Switzerland, the wood warbler is classified as vulnerable (VU) on the red list of breeding birds [29] and considered a priority species for the Swiss species recovery programme for birds [30]. Aside from evaluating habitat selection under unpredictable environmental conditions arising from irregular rodent outbreaks, an additional aim was thus to increase our understanding of the habitat requirements of this species to provide conservationists and foresters with management recommendations to ameliorate the species' habitat.

Methods

Study species

The wood warbler is a Eurasian, ground-nesting passerine wintering in tropical Africa and returning to the breeding grounds in mid-April. Unpaired males use a characteristic singing style and show a high singing activity from early morning throughout the day. After pairing, the singing style changes, and overall singing activity sharply drops to relatively low levels for the rest of the breeding season [7]. The changes in singing style use and singing activity allows distinguishing paired males from unpaired males.

Study areas

We searched for wood warbler territories in 15 study areas (Fig. 1; Additional file 1), which were chosen based on (a) the common breeding bird survey provided by the Swiss Ornithological Institute (the standardized Swiss national bird monitoring program, http://www.vogel-warte.ch/monitoring-common-breeding-birds.html), (b) the breeding bird atlas of the canton Zurich (www.birdlifezuerich.ch) and (c) http://www.ornitho.ch/ (the official birding exchange platform in Switzerland). In all study areas, we used the coordinates of previous sightings as rough starting points, from where we extensively searched for the species. The boundaries of the study areas were determined by natural circumstances like forest edge, strong changes in forest structure (e.g. young re-growths or coniferous plantations) or geographical features (e.g. deep valleys). Study areas were mostly located on steep, south-facing slopes within large deciduous forests dominated by beech *Fagus silvatica* and occasionally oaks *Quercus* spp. pine *Pinus silvestris* and fir *Picea abies* were intermixed to various degrees. No permissions were required to enter the forests. Size of the study areas is given in Additional file 1.

Territory mapping and nest searching

We started mapping territories in mid-April 2010–2012 by listening for the distinct wood warbler song. If no wood warbles were heard or observed, we played wood warbler songs for 10 s every 300 m to avoid overlooking territories. As soon as birds responded, we stopped the playback and noted the observations in a map. We checked each study area for the presence of wood warblers at least once a week until early July. A territory was

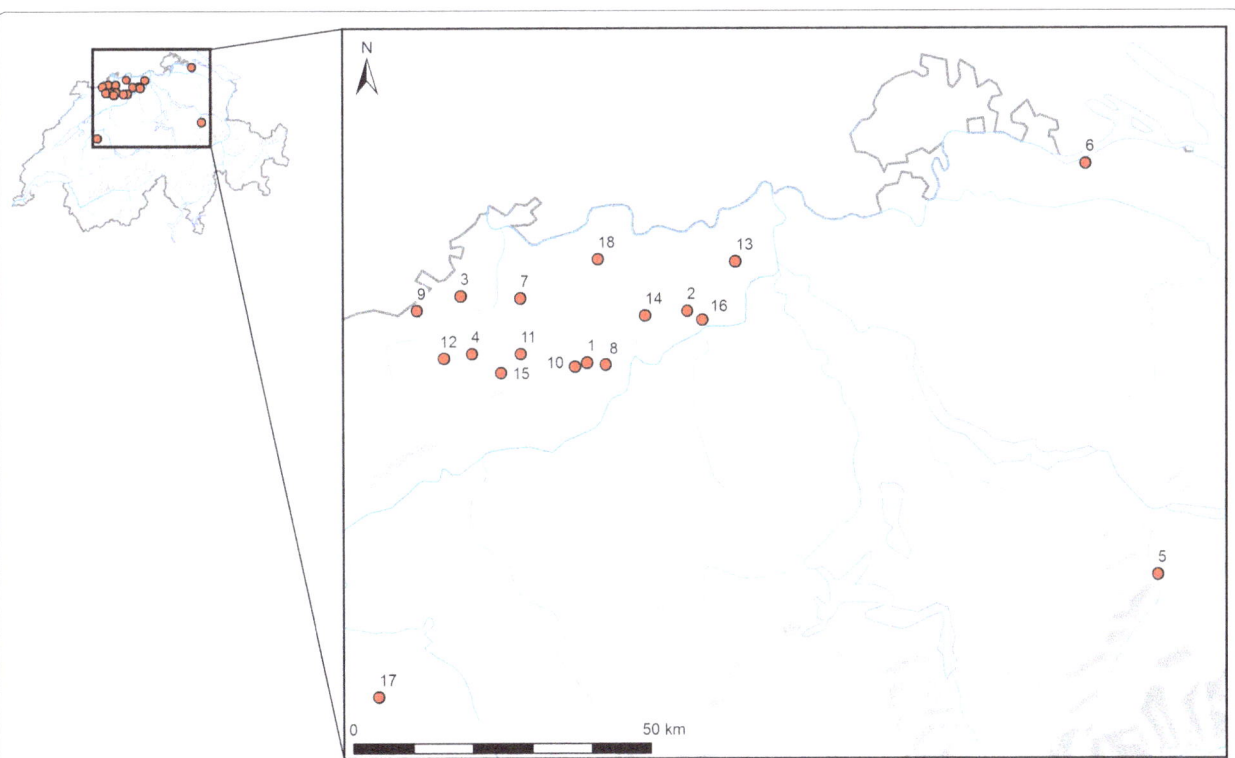

Fig. 1 Location of study areas in Switzerland. 1 = Belchen, 2 = Bänkerjoch, 3 = Blauen, 4 = Erschwil, 5 = Ennenda, 6 = Gündelhardt, 7 = Hochwald, 8 = Homberg, 9 = Kleinlützel, 10 = Langenbruck, 11 = Lauwil, 12 = Montsevelier, 13 = Mönthal, 14 = Oltingen, 15 = Scheltenpass, 16 = Staffelegg, 17 = Ueberstorf, 18 = Wintersingen. Study sites 13, 17 and 18 only used in the comparison between breeding territories vs. abandoned territories. See Additional file 1 for details on the study areas. Basemap © Institute of Cartography and Geoinformation, ETH Zurich, reproduced with permission from 11 April 2016

classified as occupied when (a) we observed a singing male twice in the same location with at least 7 days in between, (b) we observed a pair (two birds in the same territory showing no agonistic behaviour and observed at least twice on subsequent visits) or (c) we found the nest. Once territories had been established, they were regularly checked for the presence of females and nests. Due to the regular observer presence, a narrow search grid and the use of playback, it is highly probable that all territories (with and without nests) within a study area were found. Once the nest site was established, singing and general activity of both males and females concentrated to a small radius (mostly <25 m) around the nest, which corresponds to a circular area of about 1900 m^2, representing the upper end of the average breeding territory size after nest establishment [7]. For logistical reasons, it was impossible to record habitat variables and rodent density for all breeding territories. Therefore, territories were selected to get a representative number of successful and unsuccessful nests and a balanced sample from the different study areas. As we were interested in the patterns of habitat selection in breeding territories, territories without nests were not considered for habitat mapping.

Choice of control areas

To each breeding territory chosen for analysis we assigned a control area without wood warblers. We first defined eight possible control areas (i.e. X–Y-coordinates) 200–300 m from the nest of the respective territory in the four cardinal (N, E, S, W) and four inter cardinal (NE, SE, SW and NW) directions. To avoid trivial results, we ruled out control areas with habitats known to be unsuitable for wood warblers (non-forest areas, large clearings, purely coniferous forest patches, young re-growths and plantations). Also, control areas closer than 50 m to other breeding territories were excluded. This distance was based on the average breeding territory size (1200–1900 m^2) according to [7]. Of the non-excluded potential control areas, one was randomly selected. Absence of wood warblers in retained control areas was confirmed with playback.

Choice of abandoned territories

Abandoned territories were selected based on patterns of wood warbler occupancy over the past 10 years. Abandoned territories were defined as areas that had been deserted for the last 3 years, but that had been occupied at least three times in the 7 years before. This ensured that now-abandoned territories had not simply been in marginal habitats when they were occupied earlier and also accounted for the known nomadic behaviour of the species [2]. Based on data from the common breeding

bird survey of the Swiss Ornithological Institute, we selected and analysed 20 abandoned territories in 6 study areas (Additional file 1). This relatively low number of abandoned territories was a consequence of the need to know the exact location of the territory for recording habitat variables (see "Habitat variables" below). The centre of an abandoned territory was defined as the X–Y-coordinate averaged over the mapped observations (accuracy ~50 m) made during the three annual surveys of the Swiss common breeding birds monitoring scheme.

Habitat variables

Habitat variables were recorded after a nest was lost (predated or abandoned) or the nestlings had fledged to minimize disturbance at active nests. In the control areas, habitat variables were recorded at the same time as in their associated breeding territories. In abandoned territories, habitat variables were sampled towards the end of the breeding season.

For each breeding territory, control area and abandoned territory, respectively, we defined five quadratic sample areas of 50 m^2 each, as shown in Fig. 2. One square was centred on the territory centre (nest position, X/Y-coordinates in control areas and abandoned territories, see above), the centres of the other four squares were located 17 m from the territory centre on axes corresponding to

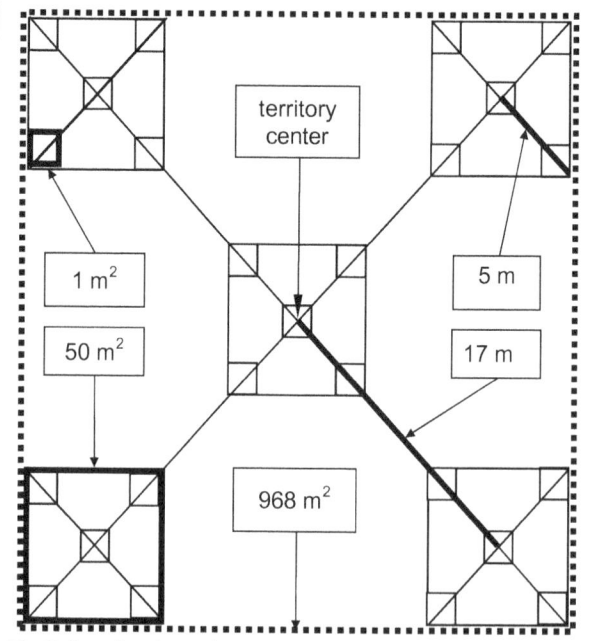

Fig. 2 Sampling design for assessing habitat structure variables and rodent numbers. Territory center is the nest position in occupied territories or X/Y-coordinates in control areas and abandoned territories (see text for details). *Bold lines* indicate distances, *bold squares* exemplify 1 m^2 and 50 m^2 squares, respectively. Adapted from [42]

the diagonals of the first square. Connecting the outermost points of these four squares results in an area covering roughly 1000 m² (Fig. 2), which corresponds to the lower end of the average breeding territory size of wood warblers [7]. Furthermore, we defined five 1-m²-squares located at the corners and the centre, respectively, in each of the five 50 m²-squares (Fig. 2).

Habitat characteristics were described with variables referring to forest structure, rodent abundance, anthropogenic disturbance and topography. Names, descriptions, calculations and values of variables are listed in Table 1. *Aspect*, a circular variable, was converted to the

variables *southness* [$-\cos(aspect \times \pi/180)$, where $1 = S$ and $-1 = N$] and *eastness* [$\sin(aspect \times \pi/180)$, where $1 = E$ and $-1 = W$].

Estimation of sky visibility

The percentage of sky visible from 1.5 m above ground level at each of the five 50 m²-squares was recorded from pictures of the crown canopy in order to estimate foliage density. We adopted the method described in [31], with the following adaptions and additions. We used a DSLR camera (Nikon D2Xs) with a standard zoom lens (18–70 mm f3.5-4.5G ED-IF AF-S DX Zoom Nikkor) at a

Table 1 Variable names and descriptions and associated hypotheses

Hypothesis	Variable	Description	Breeding	Control	Abandoned
Forest structure					
Ground variables	Cover of herb layer [a]	Percentage of ground covered by vegetation < 0.5 m, visually estimated	24.2, 10.9–42.6	14.8, 6.4–32.0	25.5, 12.2–29.2
	Number of tussocks [b]	Number of grass and sedge tussocks	325.5, 122.8–653.5	40.0, 3.0–216.0	28.5, 0–310.8
	Number of bushes [c]	Number of bushes > 0.5 m height and number of young trees with dbh < 8 cm	38.5, 17.8–70.8	69.0, 12.0–246.0	34.0, 6.8–137.8
Tree variables	Number of trees [c]	Number of trees with dbh > 8 cm	16.5, 13.0–22.0	13.0, 10.0–17.0	11.0, 7.8–14.5
	Number of trees branched < 4 m [c]	Number of trees with branches below 4 m	10.0, 6.0–14.0	7.0, 4.0–11.0	5.5, 3.8–10.3
	Number of trees branched < 10 m [c]	Number of trees with branches below 10 m	13.0, 9.0–18.0	9.0, 6.0–13.0	7.0, 6.0–12.0
	Tree dbh [d]	Average dbh of all trees with dbh > 8 cm	26.0, 22.3–30.0	31.0, 26.0–36.0	27.5, 24–37.5
	Tree species diversity [e]	Shannon's index of diversity based on tree species and dbh data	1.2, 0.7–1.5	0.9, 0.7–1.3	0.9, 0.6–1.2
	Sky visibility [d]	Percentage of sky visible (see "Estimation of sky visibility" section)	13.0, 9.0–19.0	14.0, 10.0–19.0	10.5, 9.0–21.8
	Proportion beech [e]	Number of beech trees divided by total number of trees	43.2, 20.3–59.3	50.0, 29.2–69.6	52.3, 33.3–58.5
	Proportion other deciduous trees [e]	Number of deciduous trees except beech divided by total number of trees	31.7, 18.6–47.1	25.0, 14.3–50	41.4, 18.6–53.1
	Proportion conifers [e]	Number of coniferous trees divided by total number of trees	13.1, 0–31.7	0, 0–29.4	0, 0–15.4
Rodent-avoidance	Rodent numbers	Number of rodents captured in the 25 traps per territory or control area	8.0, 0–15.8	13.0, 4.0–22.0	7.0, 1.0–13.3
Anthropogenic disturbance	Distance to paths [f]	Distance to paths, i.e. trails regularly used by humans	48.0, 15.0–75.8	–	37.5, 25.8–45.5
	Distance to forest edge [f]	Distance to edge of forest	148.5, 102.8–237.8	153.0, 72.0–224.0	98.5, 60.8–148.5
Topography	Elevation [f]	Elevation in m above sea level	698, 656–931	699, 610–920	575, 548–740
	Aspect [d]	Measured in degrees (°) with a compass in the centre of each 50-m²-square	174, 144–204	171, 127–227	162, 109–307
	Slope steepness [d]	Measured in degrees (°) with a compass in the centre of each 50-m²-square	31.5, 27.0–37.0	26.0, 21.0–33.0	21, 16.8–23.5

Shown are medians and interquartile (25–75 %) ranges

Dbh = diameter at breast height. N = 73 for breeding territories and control areas, respectively, and n = 20 for abandoned territories

[a] Averaged over the 25 1-m²-squares per breeding territory, per control area and per abandoned territory, respectively

[b] Summed over the 25 1-m²-squares per breeding territory, per control area and per abandoned territory, respectively

[c] Summed over the five 50-m²-squares per breeding territory, per control area and per abandoned territory, respectively

[d] Averaged over the five 50-m²-squares per breeding territory, per control area and per abandoned territory, respectively

[e] Calculated over the five 50-m²-squares per breeding territory, per control area and per abandoned territory, respectively

[f] Recorded for the centres of each breeding and abandoned territory and extracted from ecoGIS (www.ecogis.admin.ch)

focal length of 35 mm. To take a picture, the camera was held 1.5 m above ground in the centre of the respective 50 m²-square, lens pointing vertically up. Pictures were taken in camera RAW format and imported into Adobe Photoshop CS5 for editing. Import was performed with standard camera RAW settings. Brightness of green colours was reduced to the minimum and brightness of blue colours increased to the maximum in order to improve the contrast between sunlit green leaves and the blue sky. The pictures were edited as described in [31], downscaled to 1500 × 1000 pixels and transformed to black/white bitmaps before being processed by a self-written php-script to calculate the black/white pixel ratio.

Live-trapping of rodents

Rodents were captured with live-trapping in breeding territories, control areas and abandoned territories with permission nr. 410 issued by the Veterinary Office of the Canton Basel-Landschaft. For details about laws on animal experimentation in Switzerland see http://www.blv.admin.ch/themen/tierschutz/00777/index.html?lang=en. To avoid disturbance and for logistic reasons, we waited until at least three nests per study area had failed and/or had fledged. We used the same sampling layout for rodent trapping as for recording habitat variables (Fig. 2). In each of the five 50 m²-squares, five traps were placed near structures or, if found, near rodent trails, and covered with foliage. Thus, 25 traps were set up in each breeding territory, control area and abandoned territory, respectively. We used Longworth traps (Penlon Ltd., Abingdon, UK) made of steel or aluminium and Field Trip Trap Live Catch Trap made of plastic (Alana Ecology, Bishops Castle, UK). We provided commercial pet hay as bedding and apple pieces, oatmeal, peanut butter and hazel nuts as bait.

Traps were put out on the same day in a breeding territory and in its associated control area. Traps were active during 48 h. In 2010, traps were checked every 8 h, resulting in five capture occasions. Based on experiences from 2010, traps were checked every 12 h in 2011 and 2012, resulting in three capture occasions. Caught animals were put into a bag, classified to species or genus level (cryptic sibling species), marked by hair clipping and released immediately. We marked the animals using a nose hair trimmer, with the markings reflecting each capture occasion. From these markings, capture histories were later constructed to allow calculation of capture probabilities and rodent density.

Statistical analyses
Estimation of rodent density and rodent numbers

We analysed capture-recapture data using Program CAPTURE run through Program MARK v6.0 [32] and assumed a demographically closed population, since our trapping time frame only lasted 48 h. Even though we caught and identified several species, we pooled all captures to obtain a single estimate of rodent density per breeding territory, control area and abandoned territory, respectively. With Program CAPTURE, we computed estimates of capture probability and population density for the following models. (1) A null model of no time, behavior or heterogeneity effect (Mo), assuming all individuals of a population are equally at risk of capture on every trapping occasion. (2) A model of heterogeneity effects (Mh), assuming capture probabilities vary by individual animal. (3) A model with time effects (Mt), assuming capture probabilities vary with time. (4) A model of behavior effects (Mb), assuming capture probabilities vary by behavioral response to capture. (5) A model of behavior and heterogeneity effects (Mbh), assuming capture probabilities vary by individual animal and by behavioral response to capture. (6) A model of time and heterogeneity effects (Mth), assuming capture probabilities vary with time and by individual. (7) A model of time and behavior effects (Mtb), assuming capture probabilities vary with time and with behavioral effects (trap happiness, trap shyness). (8) A model of time, heterogeneity and behavior effects (Mtbh), for which however, there is currently no estimator. The first 7 models were then ranked by a model selection criterion between 0 and 1, where the most appropriate model scores a 1. We then used the rodent density estimate from the most appropriate model, calculated for each breeding territory, control area or abandoned territory for further analyses.

In addition to rodent density, we calculated the total number of trapped rodents by summing all captures over the 25 traps per territory, control area and abandoned territory, respectively ("rodent numbers"). Rodent density and rodent numbers turned out to be highly correlated (Spearman rank correlation, $r_s = 0.97$, n = 131). In the following, we only used rodent numbers, because estimation of rodent density was not possible for all territories in 2011 owing to very few rodent captures (rodent crash year).

Correlations among habitat variables

Strong (Spearman rank correlation coefficient $|r_s| > 0.7$) and positive correlations were only detected between the variables *number of trees* and *number of trees branched <4 m*, between *number of trees* and *number of trees branched <10 m*, and between *number of trees branched <4 m* and *number of trees branched <10 m* (see Additional file 2). In the data set used for comparing breeding territories and abandoned territories, two additional strong (and negative) correlations were found between the variables *tree dbh* (diameter at breast height) and *number of trees branched <4 m*, and between *proportion beech* and *proportion other deciduous trees*, respectively. In all subsequent analyses, we therefore never entered both variables of a strongly correlated variable pair into the same generalized linear mixed-effects model.

Model structure

Generalized linear mixed-effects models (GLMMs) with logit link and binomial error structure were used to assess the relationships between breeding territory selection and the habitat variables. The binary dependent variable always was the occupancy state of a site (0 for control areas and abandoned territories, respectively; 1 for breeding territories). The habitat variables potentially influencing breeding territory selection were defined as fixed effects. For the analysis of breeding territories vs. control areas, we used random effects with a hierarchical structure. The breeding territories were distributed over different study areas. Within the study areas, breeding territories and their control areas were always paired. Thus, we included two random effects: (1) *study area* to account for the dependency of breeding territories within the same study area and (2) *breeding territory-control area pairs nested within study area* to account for the paired structure of breeding territories and control areas. Given the nomadic behaviour and low philopatry of the wood warbler [2; own unpublished ringing data], the chance of observing the same individuals over multiple years was negligible, making the inclusion of a random effect to account for individual dependencies unnecessary. Likewise, locations of nest sites changed across years, so breeding territories were not repeatedly sampled over time. For the analysis of abandoned vs. breeding territories, we used no random effects for two reasons. (1) Abandoned territories were available from only 3 of the 15 study areas used to compare breeding territories vs. control areas and from 3 additional study areas (Additional file 1). (2) Abandoned territories were often located quite far away from breeding territories of the same study area, sometimes even in other forest stands; therefore, using a paired structure in the statistical analysis was not expedient.

Prior to the analyses, variables (all continuous) were standardized (mean = 0, standard deviation = 1). Analyses were performed in R (https://www.r-project.org/) using the packages lme4 [33] for model selection and AICcmodavg [34] for model-averaging. Note that model-averaged coefficients were very similar to those of the respective best across-hypothesis models [cf. 35]. Model fit was visually assessed with residual plots. Further, we evaluated the presence of spatial autocorrelation with

semi-variograms of the residuals [36]. Evidence for spatial autocorrelation was found in the analysis of breeding territories vs. abandoned territories, but not in the analysis of breeding territories vs. control areas. We thus included x- and y-coordinates of the territories and the interaction between x- and y-coordinates in all analyses of breeding territories vs. abandoned territories to account for spatial structure. Inspection of semi-variograms of the residuals following these analyses no longer indicated presence of spatial autocorrelation.

Modeling approach and model selection

The forest structure hypothesis included 12 habitat variables. To avoid over-parameterizing and convergence problems of models, variables of the forest structure hypothesis were assigned to three subgroups termed (1) "ground variables", which included *cover of herb layer, number of tussocks, number of bushes*; (2) "tree variables", which included *number of trees, number of trees branched <4 m, number of trees branched <10 m, tree dbh, tree species diversity* and *sky visibility*; and (3) "tree species composition", which included *proportion beech, proportion other deciduous trees* and *proportion conifers*. Models consisting of variables from each of these three subgroups were then separately evaluated (i.e. variables from the different subgroups were not jointly modelled in the first two steps, Fig. 3). The rodent-avoidance hypothesis only included the variable *rodent numbers*, the anthropogenic disturbance hypothesis (only evaluated for breeding vs. abandoned territories), *distance to paths* and *distance to forest edge*. The topography hypothesis included the variables *slope steepness, southness* and *eastness* (for *aspect*, see above) and, in the comparison of breeding vs. abandoned territories, *altitude* (Table 1). *Southness* and *eastness* were always jointly entered to or removed from models.

Candidate models (see Additional file 3) were compared with Aikake's Information Criterion corrected for small sample size AICc [37]. Models were ranked based on their AICc values, with the model having the lowest AICc being considered the best, given the data. Candidate models were evaluated as follows: models with ΔAICc < 2 compared to the best model were judged to have considerable support by the data. Competing models with ΔAICc < 2 compared to the best model, but differing by one parameter only, were evaluated with regard to their log-likelihood value. If the log-likelihood of a model containing the habitat variable A was almost equal to a model including A and habitat variable B, then B did not explain much additional variation in the data [37], and the model with both A and B was discarded in favour of the model with A only.

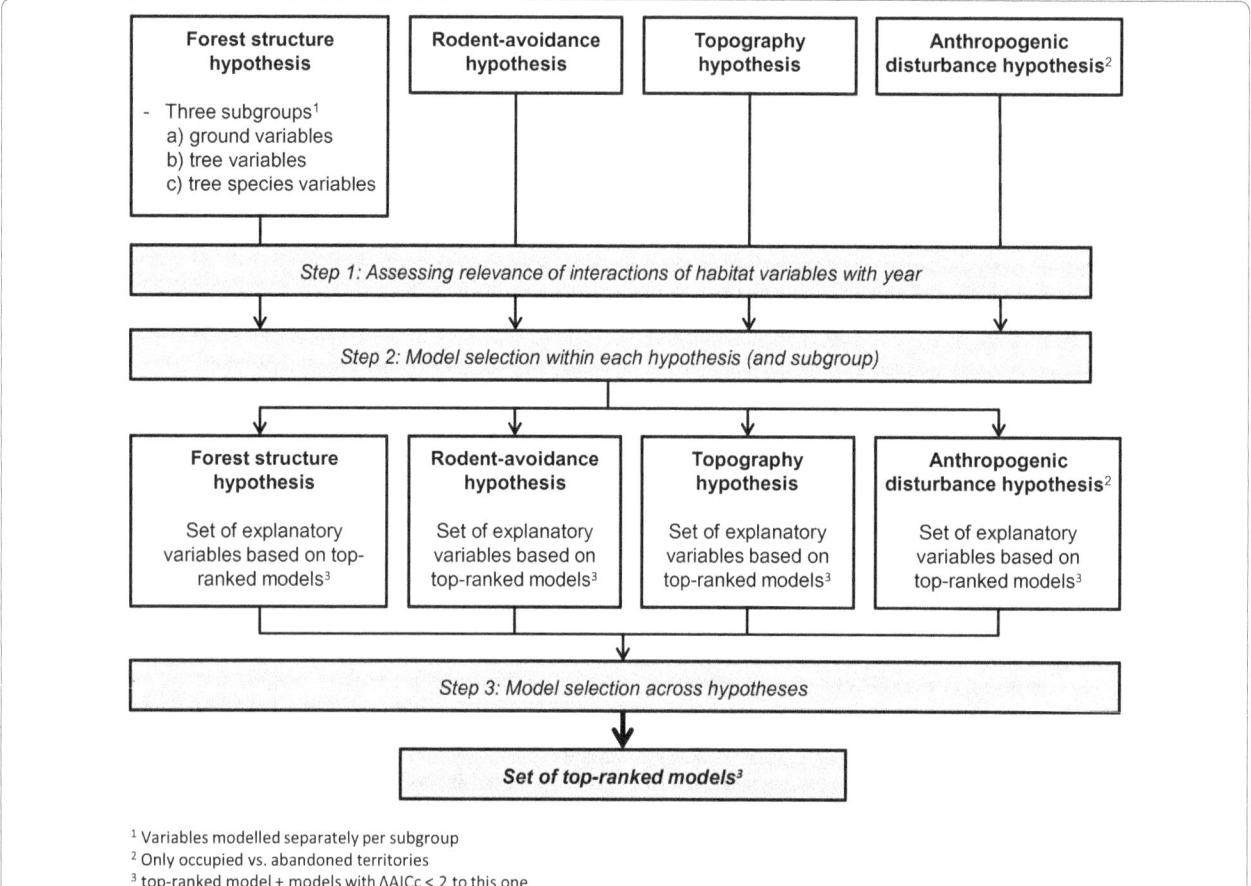

Fig. 3 Overview on the model selection design applied. Variables of the forest structure hypothesis were placed in three thematic subgroups to avoid over-parameterizing and convergence problems of models. For further details, see "Modeling approach and model selection" section and Additional file 3

Results

We only report results for analyses with *number of trees*. Analyses including *number of trees branched <4 m* or *number of trees branched <10 m*, both highly correlated to *number of trees* (see Additional file 2), gave almost identical results to analyses with *number of trees* and are presented in Additional file 4.

Breeding territories vs. control areas
Interactions with year (analysis step 1)

In the forest structure hypothesis, subgroup ground variables, two models including either the interaction of *cover of herb layer* and *year* or *number of bushes* and *year* were ranked highest and received very similar support (ΔAICc between models = 0.147). ΔAICc of the null model to the best model was 36.777. In the rodent-avoidance hypothesis, the model including the interaction between *rodent numbers* and *year* was ranked highest, with the null model having a ΔAICc of 5.057 to the highest ranking model. In the other hypotheses (and subgroups),

models with interactions with *year* generally received low support (ΔAICc always >4.3 to best models). We thus retained the interactions of *cover of herb layer* and *year*, *number of bushes* and *year* and *rodent numbers* and *year* for the next step.

Within-hypothesis analysis (analysis step 2)

Forest structure hypothesis Subgroup ground variables—Evaluation of models with all possible combinations of habitat variables showed that the model including *number of bushes*, *number of tussocks* and the quadratic effect of *cover of herb layer* was ranked highest, with ΔAICc of this model to the null model being >39 (Table 2). Three other models were within ΔAICc of 2 to the best model (details in Table 2). However, the fourth-ranked model including *number of bushes* and the quadratic effects of *number of tussocks* and *cover of herb layer* had almost the same log-likelihood value as the top model (Table 2). The quadratic effect of *number of tussocks* did thus not explain more variation in the data than the linear effect of *number*

Table 2 Model selection results of the analysis of breeding territories vs. control areas (n = 73 pairs)

Hypothesis	Variables in model	LL	K	AICc	ΔAICc	Weight
Forest structure						
(a) Ground variables	Number of bushes, number of tussocks, cover of herb layer^2	−77.238	7	169.288	0	0.232
	Number of bushes, year, number of bushes x year, number of tussocks	−75.155	9	169.633	0.345	0.195
	Number of bushes, number of tussocks	−80.208	5	170.845	1.557	0.106
	Number of bushes, number of tussocks^2, cover of herb layer^2	−77.089	8	171.230	1.942	0.088
	…					
	Null	−101.199	3	208.568	39.280	0.000
(b) Tree variables	Number of trees, tree dbh	−87.397	5	185.224	0	0.158
	Number of trees, tree dbh, tree species diversity^2	−85.854	7	186.520	1.297	0.083
	Number of trees, tree dbh, tree species diversity	−87.160	6	186.924	1.701	0.068
	…					
	Null	−101.199	3	208.568	23.345	0.000
(c) Tree species composition	Null	−101.199	3	208.568	0	0.114
	Proportion beech, propoprtion other deciduous trees, proportion conifers^2	−97.099	7	209.01	0.442	0.091
	Proportion beech, propoprtion other deciduous trees, proportion conifers	−98.227	6	209.059	0.491	0.089
	Proportion beech	−100.459	4	209.202	0.634	0.083
	Proportion beech^2, propoprtion other deciduous trees, proportion conifers^2	−96.542	8	210.134	1.566	0.052
Rodent-avoidance	Rodent numbers, year, rodent numbers x year	−93.230	8	203.511	0	0.498
	Rodent numbers	−98.100	4	204.483	0.972	0.306
	Null	−101.199	3	208.568	5.057	0.040
Topography	Slope steepness	−91.564	4	191.412	0	0.558
	…					
	Null	−101.199	3	208.568	17.156	0
Across hypotheses	Slope steepness, rodent numbers, number of tussocks, cover of herb layer^2, number of trees, number of bushes, tree dbh	−62.749	11	149.469	0	0.107
	Slope steepness, rodent numbers, number of tussocks, cover of herb layer^2, number of trees, number of bushes	−63.958	10	149.545	0.076	0.103
	Slope steepness, rodent numbers, number of tussocks, cover of herb layer^2, number of trees, tree dbh	−64.066	10	149.762	0.293	0.092
	Slope steepness, rodent numbers, number of tussocks, cover of herb layer^2, number of trees	−65.448	9	150.220	0.751	0.073
	Slope steepness, number of tussocks, cover of herb layer^2, number of trees, number of bushes	−65.976	9	151.275	1.806	0.043
	Slope steepness, rodent numbers, number of tussocks, cover of herb layer^2, number of trees, tree dbh, tree species diversity^2	−62.470	12	151.285	1.816	0.043
	Slope steepness, rodent numbers, number of tussocks, cover of herb layer^2, number of trees, tree species diversity^2	−63.658	11	151.287	1.817	0.043
	…					
	Null	−101.199	3	208.568	59.099	0.000

For each hypothesis, the top-ranked model (ΔAICc = 0), the models with ΔAICc < 2 to the top-ranked model and the null model (referred to as "null") are shown. "…" refers to additional models examined, but not listed in detail to avoid overlong table, as they were little informative

The quadratic effect of a variable x, composed of a linear and a quadratic component (x ± x^2), is denoted as x^2

LL log-likelihood, K number of parameters in the model (including random effects and intercept), *weight* Akaike weight (chance of the model to be the best one, given the candidate models)

of tussocks. In summary, we retained *number of bushes*, *number of tussocks*, the quadratic effect of *cover of herb layer* and the interaction of *number of bushes* and *year* for the subsequent across-hypotheses analysis.

Subgroup tree variables—The highest-ranked model included *number of trees* and *tree dbh* (ΔAICc of 23.34

to the null model, Table 2). Second-ranked was a model additionally including the quadratic effect of *tree species diversity*. The third-ranked model included the linear effect of *tree species diversity*, in addition to *number of trees* and *tree dbh*. However, log-likelihood of the third-ranked model was almost the same as for the highest-ranked model, suggesting that the inclusion of the linear effect of *tree species diversity* did not really contribute to explaining variation in the data. We thus retained *number of trees*, *tree dbh* and the quadratic effect of *tree species diversity* for the across-hypotheses analysis.

Subgroup tree species composition variables—The null model was ranked highest (Table 2). We thus did not retain any variable relating to tree species composition for the across-hypotheses analysis.

Rodent-avoidance hypothesis The model including the interaction between *rodent numbers* and *year* was ranked highest, followed by the model with *rodent numbers* only (Table 2). We thus retained both *rodent numbers* and the interaction between *rodent numbers* and *year* for the across-hypothesis analysis.

Topography hypothesis The model including *slope steepness* was ranked highest and the only one with support (ΔAIC to the next best model = 2.1, Table 2). We thus retained *slope steepness* for the across-hypotheses analysis.

Across-hypotheses analysis (analysis step 3)

Models including habitat variables retained from step 2 and their interactions with rodent numbers had ΔAICc > 2.2 to the highest-ranking model, which included the main effects *number of tussocks* and *rodent numbers* only (see Additional file 3 for explanation of modelling steps). Interactions with *rodent numbers* were thus not further considered.

The highest-ranked model (ΔAICc of 59.1 to the null model, Table 2) included the following variables: *slope steepness* (topography); *rodent numbers* (rodent avoidance); and, within forest structure: *number of tussocks*, *number of bushes* and the quadratic effect of *cover of herb layer* (subgroup ground variables); *number of trees* and *tree dbh* (subgroup tree variables). Six other models had ΔAICc values <2 to the highest-ranked model; they all included *slope steepness*, *number of tussocks*, *number of trees* and the quadratic effect of *cover of herb layer*. Five of the six high-ranking models further included *rodent numbers* (Table 2).

According to model-averaging (Table 3a), territory occupancy of the wood warbler was positively related to *number of tussocks*, *number of trees* and *slope steepness* (Fig. 4). On the other hand, territory occupancy showed a quadratic relationship with *cover of herb layer* and was negatively related to *rodent numbers*. *Number of bushes*, *tree dbh* and the quadratic effect of *tree species diversity*, the other three variables included in some of the five top models, appeared to be less important in explaining

territory occupancy, as the 95 % confidence intervals of their estimates included 0 (Table 3a).

Breeding vs. abandoned territories
Interactions with year (analysis step 1)

In the forest structure hypothesis, subgroup ground variables, one model including the interaction of *number of tussocks* and *year* ranked highest (ΔAICc to second-best model = 6.35 and to null model = 13.62), and this interaction was retained for the next step. For all other hypotheses (and subgroups), models including interactions with *year* received no support, and interactions with *year* were not further considered.

Within-hypothesis analysis (analysis step 2)

Forest structure hypothesis Subgroup ground variables—The highest-ranking model included the interaction between *number of tussocks* and *year* along with the respective main effects (ΔAICc to null model: > 12, Table 4). Four other models were within ΔAICc < 2 to the best model. All included *number of tussocks* and, in various combinations, the interaction between *number of tussocks* and *year*, *number of bushes* (linear and quadratic effects) and *cover of herb layer*. We thus retained *number of tussocks*, its interaction with *year*, *number of bushes* (linear and quadratic effects) and *cover of herb layer* (linear effect) for the across-hypotheses analysis.

Subgroup tree variables—Best-supported models consistently included *number of trees*, either as linear or quadratic effect (Table 4). Likewise, the quadratic effects of *tree species diversity* and *tree dbh*, respectively, were included in the best-supported model and in most models with ΔAICc < 2 to the best one. Finally, the linear or the quadratic effect of *sky visibility* was included in a few models with ΔAICc < 2 to the best model. We thus retained *number of trees* and *tree dbh* (for both variables as linear and quadratic effects) as well as *tree species diversity* (quadratic effect) and *sky visibility* (linear and quadratic effects) for the across-hypotheses analysis.

Subgroup tree species composition variables—*Proportion conifers* was included in the best-supported model. Log-likelihood values of two other supported models (ΔAICc < 2 to the best one) and the best-supported model were very similar, suggesting that the inclusion of *proportion beech* or *proportion other deciduous trees* did not contribute to explaining variation in the data (Table 4). We thus only retained the variable *proportion conifers* for the across-hypotheses analysis.

Rodent-avoidance hypothesis The rodent-avoidance hypothesis received no support, because the top-ranked model was the null model (Table 4). We thus did not

Table 3 Estimates, standard errors (SE), and 2.5–97.5 % confidence limits based on model-averaging from the across-hypotheses model selection of (A) breeding territories vs. control areas (n = 73 pairs) and (B) breeding territories (n = 56) vs. abandoned territories (n = 20)

Hypothesis	Variable	Estimate	SE	2.5 %	97.5 %
(a) Forest structure					
Ground variables	Cover of herb layer (lin. term)	0.98	0.45	0.09	1.87
	Cover of herb layer (quad. term)	−0.71	0.28	−1.25	−0.16
	Number of bushes	−0.58	0.37	−1.31	0.15
	Number of tussocks	1.78	0.81	0.18	3.37
Tree variables	Number of trees	0.94	0.31	0.32	1.55
	Tree dbh	−0.48	0.31	−1.09	0.13
	Tree species diversity (lin. term)	−0.22	0.27	−0.75	0.31
	Tree species diversity (quad. term)	0.27	0.21	−0.13	0.67
Topography	Slope steepness	0.91	0.28	0.35	1.46
Rodent-avoidance	Rodent numbers	−0.60	0.30	−1.19	−0.01
(b) Forest structure					
Tree variables	Number of trees	1.72	0.70	0.36	3.09
Disturbance	Distance to forest edge	3.58	1.52	0.60	6.57
Topography	Slope steepness	2.20	0.76	0.71	3.68

Shown are variables included in the highest ranking models and in models with ΔAICc < 2 to the highest ranking ones

Lin linear; *quad* quadratic

retain the variable *rodent numbers* for the across-hypothesis analysis.

Anthropogenic disturbance hypothesis The best-supported model (ΔAICc to null model: 22.96, Table 4) included the linear effect of *distance to forest edge* and the quadratic effect of *distance to path*. One other model had a ΔAICc < 2 to this model and included the quadratic effects of both *distance to forest edge* and *distance to path*. The linear and the quadratic effects of *distance to forest edge* and the quadratic effect of *distance to path* were kept for the across-hypotheses analysis.

Topographic hypothesis The highest-ranked model included the quadratic effects of *slope steepness* and *elevation* and the linear effects of *southness* and *eastness*, respectively (ΔAICc to null model: >21, Table 4). All other models had ΔAICc > 2.5 to the highest-ranking one. We retained the quadratic effects of both *slope steepness* and *elevation* and the linear effects of *southness* and *eastness* for the across-hypotheses analysis.

Across-hypotheses analysis (analysis step 3)

Because *rodent numbers* was not identified as relevant in step 2, interactions between *rodent numbers* and habitat variables were not analysed. The combined analysis of the variables, which were retained from the high-ranking models of the hypothesis-specific analyses above, was problematic because many models of the

across-hypotheses analyses had convergence problems. The problematic models always included the interaction between *number of tussocks* and *year*. We thus simplified the analysis by dropping this interaction and including instead *number of tussocks* as linear effect, and by including variables from only the top-ranked model per hypothesis. Because this simplification did not alleviate the convergence problems, we continued by only using linear effects instead of quadratic effects and by jointly including at most three habitat variables (plus the x- and y-coordinates and their interaction, see "Methods" section) in the different models. In the 176 candidate models based on the ten variables *distance to forest edge, number of tussocks, number of trees, slope steepness, elevation, aspect* (via *southness* and *eastness* jointly), *distance to path, tree species diversity, tree dbh* and *proportion conifers*, convergence problems no longer occurred.

ΔAICc of the highest-ranked model to the null model was 26.8. No other model was within ΔAICc 2 of the highest-ranked one (Table 4), which included *slope steepness, number of trees* and *distance to forest edge*.

Model-averaging across all models revealed that *slope steepness, number of trees* and *distance to forest edge* were the variables for which 95 % confidence intervals of their estimates did not include 0 (Table 3b). Thus, the topography hypothesis (via the variable *slope steepness*), the forest structure hypothesis (via *number of trees*) and the anthropogenic disturbance hypothesis (via *distance to forest edge*) were supported (Fig. 5).

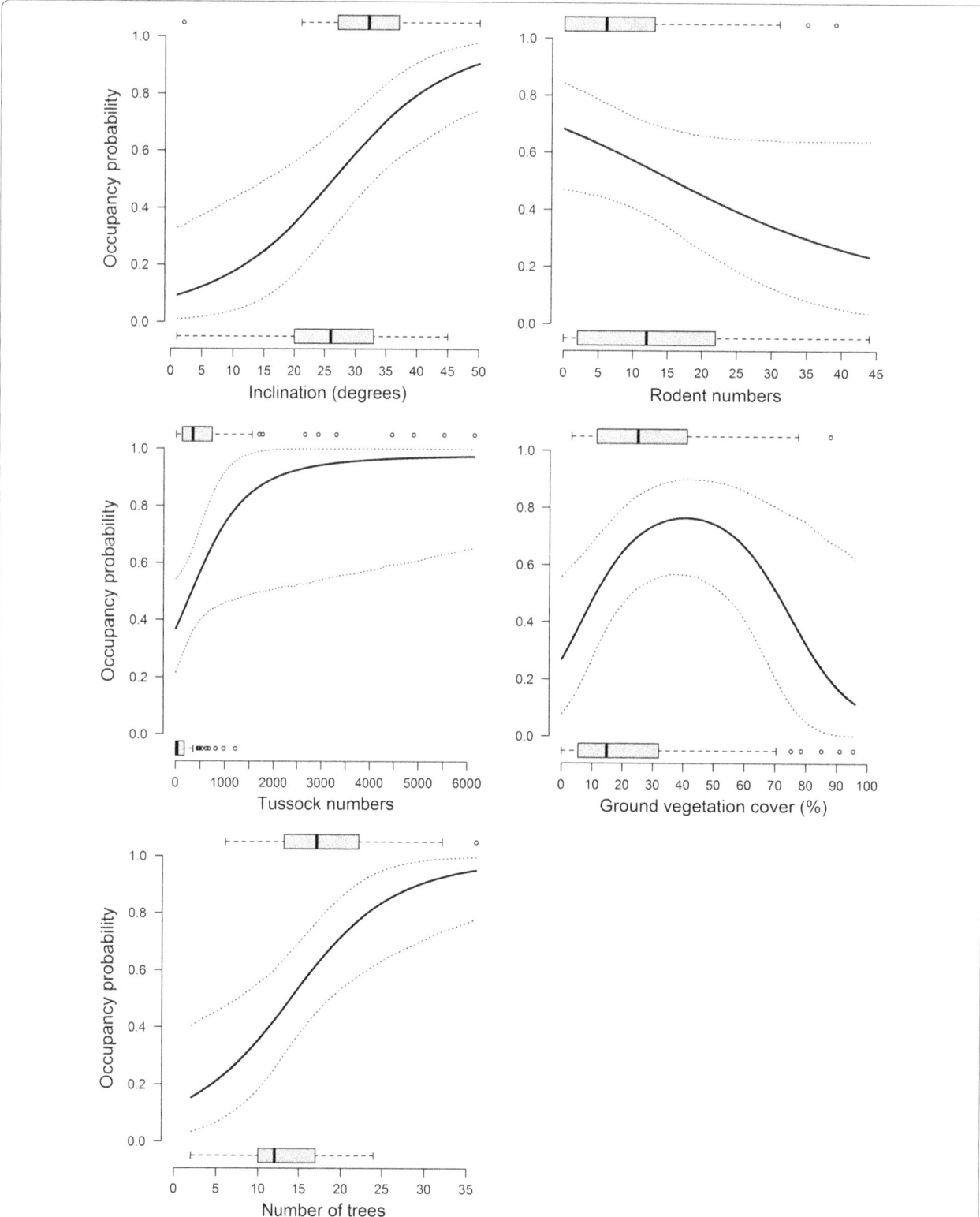

Fig. 4 Habitat variables discriminating breeding territories and control areas. Shown are plots of the five variables whose model-averaged coefficients did not include 0 (cf. Table 3). The *solid lines* are fitted values based on model-averaged coefficients of the seven top-ranked GLMMs of the across-hypotheses analysis (Table 2), the *dotted lines* show 95 % confidence limits. To calculate the fitted values, the variable of interest (x-axis) was varied within the observed range while the others were fixed on their average values. For each variable, *inset box plots* show median (*bold line*), 25–75 % range (*grey box*), range of data within 1.5 times the interquartile range from the corresponding quartile (*whiskers*) and observations beyond this range (*dots*) for occupancy probability equaling 0 (control areas) or 1 (breeding territories). $N_{territories} = 73$, $n_{control areas} = 73$

Table 4 Model selection results of the analysis of breeding territories (n = 56) vs. abandoned territories (n = 20)

Hypothesis	Variables in model	LL	K	AICc	ΔAICc	Weight
Forest structure						
(a) Ground variables	Number of tussocks, year, number of tussocks x year	−22.315	7	60.277	0	0.202
	Number of tussocks, number of bushes^2	−22.774	7	61.194	0.917	0.128
	Number of tussocks, number of bushes^2, cover of herb layer	−21.562	8	61.273	0.996	0.123
	Number of tussocks, year, number of tussocks x year, number of bushes	−21.798	8	61.745	1.468	0.097
	Number of tussocks, year, number of tussocks x year, cover of herb layer	−21.986	8	62.122	1.845	0.080
	...					
	Null	−31.868	4	72.299	12.022	0.000
(b) Tree variables	Number of trees^2, tree dbh^2, tree species diversity^2	−17.092	10	57.569	0	0.091
	Number of trees, tree species diversity^2	−21.187	7	58.020	0.452	0.072
	Number of trees, tree dbh^2, tree species diversity^2	−18.652	9	58.031	0.462	0.072
	Number of trees^2	−22.514	6	58.246	0.677	0.065
	Number of trees, tree species diversity^2, tree dbh	−20.060	8	58.269	0.701	0.064
	Number of trees, tree dbh^2, tree species diversity^2, sky visibility	−17.695	10	58.776	1.207	0.050
	Number of trees^2, tree dbh^2, tree species diversity^2, sky visibility	−16.336	11	58.797	1.229	0.049
	Number of trees^2, tree species diversity^2	−20.364	8	58.877	1.308	0.047
	Number of trees^2, tree dbh^2, tree species diversity^2, sky visibility^2	−15.234	12	59.420	1.852	0.036
	...					
	Null	−31.868	4	72.299	14.730	0.000
(c) Tree species composition	Proportion conifers	−29.723	5	70.303	0	0.265
	Proportion conifers, proportion beech	−29.533	6	72.283	1.981	0.098
	Proportion conifers, proportion other deciduous trees	−29.539	6	72.295	1.992	0.098
	...					
	Null	−31.868	4	72.299	1.996	0.098
Rodent-avoidance	Null	−31.868	4	72.299	0	0.575
	Rodent numbers	−31.023	5	72.903	0.605	0.425
Disturbance	Distance to forest edge, distance to path^2	−20.719	7	57.084	0	0.694
	Distance to forest edge^2, distance to path^2	−20.453	8	59.055	1.97	0.259
	...					
	Null	−31.868	4	72.299	15.214	0.000
Topography	Slope steepness^2, elevation^2, southness, eastness	−13.541	10	50.467	0	0.692
	...					
	Null	−31.868	4	72.299	21.832	0.000
Across hypotheses[a]	Slope steepness, distance to forest edge, number of trees	−14.985	7	45.457	0	0.734
	...					
	Null	−31.868	4	72.299	26.842	0.000

For each hypothesis, the top-ranked model (ΔAICc = 0), the models with ΔAICc < 2 to the top-ranked model and the null model (referred to as "null") are shown. "..." refers to additional models examined, but not listed in detail to avoid overlong table

LL log-likelihood; *K* number of parameters in the model (including intercept), *weight* Akaike weight (chance of the model to be the best one, given the candidate models)

The quadratic effect of a variable x, composed of a linear and a quadratic component (x $\pm$ x^2), is denoted as x^2

Each model included x- and y-coordinates (and their interaction) of territories to account for spatial autocorrelation

[a] Only linear terms of variables from best models per hypothesis and at most three habitat variables jointly used due to convergence problems with quadratic terms and more than three habitat variable per model

Discussion

Forest structure hypothesis

Forest structure in terms of both the ground vegetation (*number of tussocks, cover of ground vegetation, number of bushes*) and the tree layer (*number of trees* and *tree dbh*) was important for the selection of breeding territories in the wood warbler. Collectively, our findings suggest that wood warblers preferred to set up territories in forest stands of medium age (25–75 % quartiles of *number of trees* and *tree dbh*, respectively, in breeding territories:

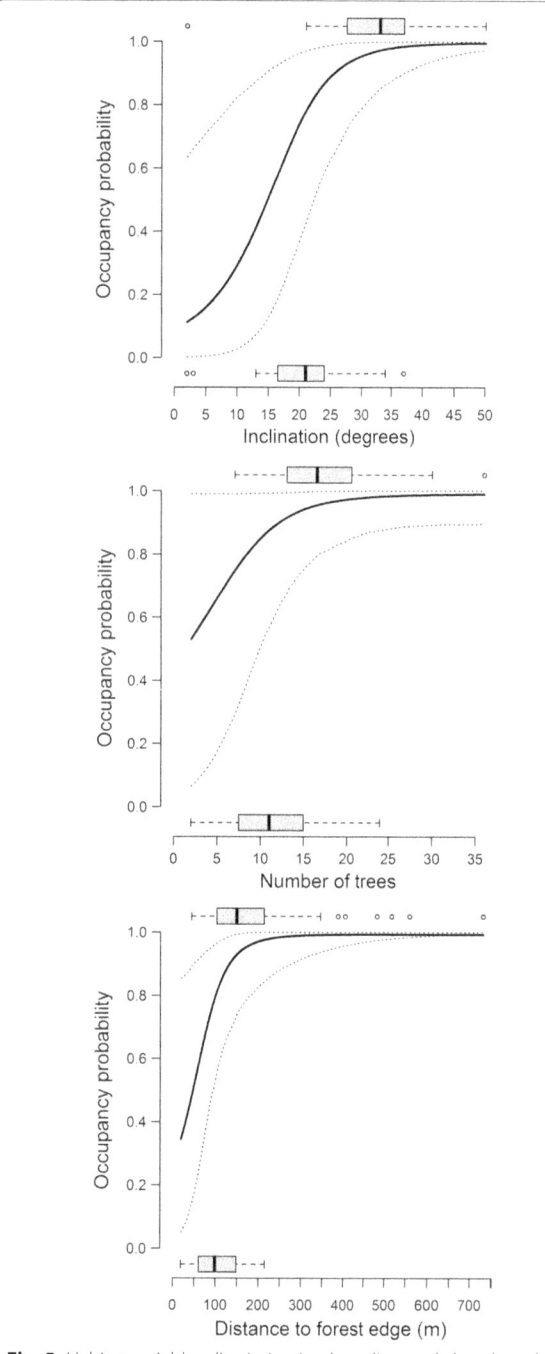

Fig. 5 Habitat variables discriminating breeding and abandoned territories. Shown are *plots* of the three variables whose model-averaged coefficients did not include 0 (Table 3). The *solid lines* are fitted values based on model-averaged coefficients of the three best-supported GLMs of the across-hypotheses analysis (Table 4), the *dotted lines* show 95 % confidence limits. To calculate the fitted values, the variable of interest (x-axis) was varied within the observed range while the others were fixed on their average values. For each variable, inset box plots show median (*bold line*), 25–75 % range (*grey box*), range of data within 1.5 times the interquartile range from the corresponding quartile (*whiskers*) and observations beyond this range (*dots*) for occupancy probability equaling 0 (abandoned territories) or 1 (breeding territories). $N_{occupied} = 56$, $n_{abandoned} = 20$

520–880 trees/ha and 22–30 cm, Table 1), corresponding to late pole wood as described in [38]. In the same study area as ours, but using remote sensing data, wood warbler territories were shown to be located in stands with relatively even-aged trees of medium height and low vertical diversity of canopy height [39]. Such stands are characterized by a relatively closed canopy and an open stem space with little branching below the canopy, resulting in relatively sparse ground (grass) vegetation cover and few bushes, i.e. little forest regeneration.

Our findings corroborate previous reports about associations of wood warblers with forest structure [5, 7–9, 11, 12, 40, 41]. Even though a preference for forests with a fairly closed canopy was not found in our study, median values for canopy closure were 87, 86 and 89.5 % in breeding territories, control areas and abandoned territories, respectively (estimated from sky visibility, Table 1) and thus within the range of values found in previous studies [60–90 %; e.g., 5, 9–11]. That canopy closure (i.e. sky visibility) did not differ between breeding territories, control areas and abandoned territories is in contrast to [39] who found canopy cover between 10 and 20 m above ground to be larger in breeding territories than in control areas in the same study area. These divergent results might be explained by the different methods used for assessing canopy closure. We measured canopy closure via an index of sky visibility (photographic camera pointed upwards) that included the foliage of all trees below the canopy while [39] assessed canopy closure between 10 m and 20 m height based on lidar signals.

Notwithstanding this methodological issue, wood warblers appear to favour forests with relatively little structural vertical diversity below a fairly closed canopy, as the number of bushes and young trees (dbh < 8 cm) was substantially lower in the breeding territories than in the control areas (Table 1). Aside from offering suitable conditions for nesting (see below), a relatively open under- and mid-storey may be particularly conducive to the wood warbler's courtship behaviour, which includes song-flights from low branches between tree trunks [7]. Note that openness in the under- and mid-storey does not simply arise from reduced tree density, as the number of trees (dbh > 8 cm) was lower in both control areas and abandoned territories than in breeding territories (Table 1). Fewer trees in a forest stand allow light to better penetrate the forest which in turn promotes growth of bushes and young trees, thereby reducing openness in the under- and mid-storey.

Breeding territory occupancy showed a quadratic relationship to ground vegetation cover (Fig. 4), corroborating previous findings about a preference for sparse ground vegetation typically around 20–30 % cover [5, 7, 9, 11]. Breeding territories also harboured markedly more grass and sedge tussocks than control areas and abandoned territories (Table 1). The occurrence of patches of ground vegetation cover appears to play a crucial role in the nesting ecology of the wood warbler. In Białowieża, 88.5 % of 156 nests were concealed among low (5–20 cm high) vegetation and under branches or spruce trees lying on the ground [49]. In our study, 87.7 % of 220 nests found between 2010 and 2014 were located in or very close to tussocks (own unpublished data). Not surprisingly, daily survival of wood warbler nests increased with nest concealment and with number of tussocks in our study areas [42].

An emergent property might arise from the combination of forest age structure and topographic conditions. On the one hand, relatively dense middle-aged forest stands situated on steep slopes might allow more direct sun radiation to reach the ground, favouring grasses and sedges, compared to similar forest stands in flat terrain. In steep forests, there is an increased probability of perpendicular incidence of sun rays onto the ground due to the spatial arrangement of trees. Canopy structure is in general measured vertically, either from above (lidar) or from the ground (photographic camera, this study), although most sun radiation is not vertical but diagonal. Interestingly, [39] found a positive relationship between wood warbler occurrence and potential direct solar radiation in March in the same study area. On the other hand, leaf litter might have a lower probability to accumulate on steep forest floors, inducing shallower soils and creating advantageous growth conditions for grasses and sedges compared to flat floors. This peculiar combination of features might explain the positive relationship of wood warbler breeding territories and inclination found in comparisons of breeding territories to both control areas and abandoned territories.

Rodent-avoidance hypothesis
Breeding territory occupancy was strongly and inversely related to rodent numbers. This agrees with findings from [2, 3], both showing that across years local wood warbler numbers were significantly negatively correlated to rodent numbers. Our study furthermore suggests that rodents might also influence wood warbler habitat selection at a much smaller, within-forest-stand scale. An explanation for the avoidance of areas with many rodents could be that wood warblers aim at reducing the probability of nest predation. Nest success in wood warblers has been found to range from 34 % [6] to 59 % [43], and nest predation typically accounts for the majority of nest losses (80–95 % in Białowieża, [6]; 37 % in Wales, [44]; 79 % in our study population, [42]). However, direct predation by rodents appears to be rare (own unpublished data based on nest cameras) or not existent [44]. In

Białowieża, nest loss rate was not related to rodent numbers, while the probability of nest failure was increased in rodent outbreak years [2, 6]. In our study area, nest survival was not related to rodent numbers either [42]. It thus seems that the avoidance of areas with many rodents is not due to direct predation of nests by rodents. Direct predation by rodents has been confirmed and implicated in territory selection in some other ground-breeding passerines (e.g., veerie *Catharus fuscescens* [19]; dusky warblers *Phylloscopus fuscatus* [18]). A more likely explanation for the avoidance of areas with many rodents could thus be that high prey densities attract rodent predators, thereby increasing the likelihood of (accidental) predation on wood warbler nests.

Breeding territories did not differ from abandoned ones in terms of rodent numbers, which rules out the possibility that territories were no longer occupied because of high rodent numbers. Thus, rodent-avoidance does not seem to be the reason for territory abandonment in our study area.

Topography hypothesis
According to the comparison of breeding territories and control areas, wood warblers prefer to settle in relatively steep terrain, such as forested slopes, as already evidenced [e.g., 5, 7, 26]. Steep forested slopes primarily occur along valley sides in our study area. Wood warblers also settle at little inclined slopes in otherwise largely flat wooded terrain elsewhere [26]. Reasons for the preference for steep slopes can be manifold. First, suitable habitat structure, particularly in terms of ground vegetation cover, could be more likely to occur at inclined than flat areas (see the emergent property mentioned above). Second, reduced or absent forest management in steep terrain due to unfavourable conditions for economic exploitation of timber would result in more extensive forest stands, i.e. more suitable wood warbler habitat than in the heavily managed lowland forests. Third, disturbance from recreational activities is likely to be reduced on less accessible steep slopes. In fact, all our study areas with wood warbler occurrence were relatively remote, which not only reduces human recreational disturbance but also represents an obstacle to timber exploitation, with the last intensive harvesting carried out 20–50 years ago [38] depending on study site. Finally, as nest entrances in wood warblers are oriented horizontally, nests on slopes, with entrances facing away from the slope, could allow wood warblers to easily escape from nests without getting entangled in the vegetation [45].

Breeding territories were also located on steeper slopes than abandoned ones. Even though we do not have data on forest structure prior to abandonment, abandoned territories were not located in marginal habitats as indicated by the patterns of wood warbler occurrence before abandonment. More likely, territories in flatter terrain have become abandoned due to structural changes owing to forest management, resulting in the lower tree density compared to still occupied territories along slopes. Another reason for the abandonment of territories located in flatter areas might be the absence of the emergent property described earlier on: steep slopes provide more opportunities for sun radiation to reach the forest floor, while there is less leaf litter accumulation on steep forest floor, which both may promote ground vegetation cover. In the end, a conjunction of different factors may play a role in breeding habitat selection by wood warblers: more appropriate habitat structure, less disturbance by humans and possibly predators, and less detrimental timber exploitation.

Anthropogenic disturbance hypothesis
Many studies have addressed the impact of disturbance by human recreational activities on wildlife, but general conclusions are hard to draw [46]. For instance, in grassland and forest habitats some species avoid the proximity of trails, while others, especially generalists, are attracted [20]. The influence of trails on breeding bird communities can be either due to increased edge effects, direct human disturbance, higher penetration of habitat by domestic and wild predators, or a combination thereof. In forests, it is especially ground-nesting species that seem to be negatively affected by disturbing visitors, increased predation and habitat change [22]. Under these premises, the wood warbler as a non-generalist, ground-nesting species is likely to be susceptible to disturbance [24].

The comparison of breeding territories vs. abandoned territories might indeed provide support for the disturbance hypothesis, but the difficulty remains to disentangle anthropogenic from predatory and/or habitat structure effects. Abandoned territories were located closer to the forest edge than current breeding territories (Fig. 5), while the two territory types did not differ in terms of distance to path or trails. Avoidance of edge-habitat could serve to reduce potential disturbance from increased human activities. However, avoidance of edge-habitat might equally likely result from increased presence of nest predators such as domestic cats or foxes, creating a "landscape of fear" [47], which appears to be widespread in wildlife [14, 15]. Finally, avoidance of edge-habitat could also be related to changes in habitat structure close to forest edges [e.g., 48], due to, for instance, altered light or microclimatic conditions promoting growth of under- and mid-storey vegetation, making the habitat unsuitable for wood warblers. Whether human

recreational activities, predation pressure and/or habitat conditions close to forest edges have recently changed in our study area cannot be answered yet.

Conclusions
Identifying and ranking cues associated to habitat selection, and linking these cues to fitness and population dynamics is critical for conserving and promoting high-quality habitats for threatened species. This study and a previous one [42] suggest, first, that grass and sedge tussocks are key habitat features for the wood warbler, affecting both territory selection and reproductive performance. Any forestry intervention promoting this type of ground vegetation is thus likely to enhance habitat quality for this declining passerine. Second, wood warblers prefer habitats with relatively high tree numbers, few bushes and an intermediate ground vegetation cover, for example in the form of tussocks. As these conditions are typically encountered in forest stands of middle age (i.e. pole wood), the wood warbler might be described as a coloniser of the middle stages of forest succession. However, the species also occurs in more mature woodland, such as old-growth forests [49], if they provide the structural features required (sparse and low-growing ground vegetation cover, relatively open stem space with few shrubs and bushes below a fairly closed canopy). Tree species do not appear to be decisive, as long as the required habitat structures are present [7; this study; T. Wesołowski pers. comm.].

Exploited forests might occasionally offer suitable conditions for the wood warbler, provided that management favours a high number of middle-aged trees that lead to stands with relatively closed canopy [39] and sufficient open space between tree trunks. In this sense, the current widespread small-scale thinning practice, which consists of removing few trees [i.e. single-tree selection and group-felling, 50] and favours light-demanding species, is clearly detrimental to the wood warbler as it creates too many gaps in the canopy. This practice boosts under-storey growth, particularly on fertile soils, and, due to competition, suppresses ground cover vegetation, especially grass and sedge tussocks.

In the mid-term, clear-felling of larger forest patches with subsequent re-growth leading to even-aged high forest stands would likely be more beneficial for the wood warbler than the current forestry practices that aim either at bringing more light into woodland by selective (and at times excessive) thinning, generating forest stands permanently covered with shrubs, bushes and trees of various sizes (continuous cover forestry), or at letting trees grow to climax in forest reserves. In the short-term, removal of shrubs and bushes in forest stands otherwise deemed suitable in structure for the wood warbler might perhaps provide temporary habitat for this endangered species. Future studies should more closely address (1) how forest structures suitable to the wood warbler can be achieved through targeted forest management, accounting for prevailing site conditions; (2) what are the mechanisms underlying avoidance of edge habitat (disturbance, predation, vegetation structure); and (3) why forest stands with low rodent densities are preferred by this European species.

Additional files

Additional file 1. Description of study areas. Study areas with coordinates, elevation, size, density and sample sizes for the comparisons of breeding territories vs. control areas (data from 2010 to 2012) and breeding territories vs. abandoned territories (data from 2010 and 2011).

Additional file 2. Spearman rank correlations among habitat variables. Coefficients of Spearman rank correlations among habitat variables.

Additional file 3. Modelling steps. Description of the three modelling steps (also see Fig. 3).

Additional file 4. Model selection results. Model selection results of the analyses of breeding territories vs. control areas and breeding territories vs. abandoned territories when using the variables number of trees <4 m and number of trees <10 m, respectively, instead of number of trees. All three variables are highly correlated (see Additional file 2).

Abbreviations

GLMM: generalized linear mixed-effects model; AICc: Aikake's information criterion corrected for small sample size.

Authors' contributions

GP conceived the study, participated in its design, performed statistical analyses and drafted the manuscript. AG participated in the design of the study, conducted fieldwork and helped drafting the manuscript. MG participated in the design of the study, conducted fieldwork, performed statistical analyses and drafted the manuscript. RA participated in the design of the study, provided the rodent traps and helped drafting the manuscript. All authors read and approved the final manuscript.

Author details

[1] Swiss Ornithological Institute, Sempach, Switzerland. [2] Department of Evolutionary Biology and Environmental Studies, University of Zurich, Zurich, Switzerland. [3] Division of Conservation Biology, Institute of Ecology and Evolution, University of Bern, Bern, Switzerland. [4] Schweizer Vogelschutz SVS/BirdLife Schweiz, Zurich, Switzerland. [5] Swiss Ornithological Institute, Valais Field Station, Sion, Switzerland.

Acknowledgements

We thank Christoph Bonetti, Roman Furrer and Thomas Vogler for their help in the field, Fränzi Korner-Nievergelt for statistical advice, Ueli Bühler, Pierre Mollet, Reto Spaar, Lechosław Kuczyński and three anonymous reviewers for comments to drafts of this manuscript.

Competing interests

The authors declare that they have no competing interests.

Funding

The study was supported by grants of the Hilfsfonds für die Schweizerische Vogelwarte Sempach, Lotteriefonds des Kantons Solothurn, PD-Stiftung der Universität Zürich, Stotzer-Kästli-Stiftung, Styner-Stiftung (all grants to GP) and the Swiss National Science Foundation (Grant number 31003A_143879/1 to GP and RA). The funding bodies had no role in the design of the study and collection, analysis, and interpretation of data and in writing the manuscript.

References

1. Pucek Z, Jedrzejewski W, Jedrzejewska B, Pucek M. Rodent population dynamics in a primeval deciduous forest (Bialowieza National Park) in relation to weather, seed crop, and predation. Acta Ther. 1993;38:199–232.
2. Wesołowski T, Rowinski P, Maziarz M. Wood warbler *Phylloscopus sibilatrix*: a nomadic insectivore in search of safe breeding grounds? Bird Study. 2009;56:26–33.
3. Szymkowiak J, Kuczynski L. Avoiding predators in a fluctuating environment: responses of the wood warbler to pulsed resources. Behav Ecol. 2015;26:601–8.
4. Amar A, Smith KW, Butler S, Lindsell JA, Hewson CM, Fuller RJ, Charman EC. Recent patterns of change in vegetation structure and tree composition of British broadleaved woodland: evidence from large-scale surveys. Forestry. 2010;83:345–56.
5. Mallord JW, Charman EC, Cristinacce A, Orsman CJ. Habitat associations of wood warblers *Phylloscopus sibilatrix* breeding in welsh oakwoods. Bird Study. 2012;59:403–15.
6. Wesołowski T, Maziarz M. Changes in breeding phenology and performance of Wood Warblers *Phylloscopus sibilatrix* in a primeval forest: a thirty-year perspective. Acta Ornithol. 2009;44:69–80.
7. Glutz von Blotzheim UN, Bauer KM. Handbuch der Vögel Mitteleuropas. Wiesbaden: Aula; 1991.
8. Mulhauser B. Ségrégation spatiale du Pouillot de Bonelli *Phylloscopus bonelli*, du Pouillot siffleur *Ph. sibilatrix* et du Pouillot véloce *Ph. collybita* dans un massif forestier du Val-de-Travers (canton de Neuchâtel, Suisse). Nos Oiseaux. 2000;47:221–8.
9. Marti J. Zur Habitatwahl des Waldlaubsängers *Phylloscopus sibilatrix* im Kanton Glarus. Ornithol Beob. 2007;104:45–52.
10. Quelle M, Lemke W. Strukturanalyse von Waldlaubsängerrevieren. Charadrius. 1988;24:196–213.
11. Hillig F. Verursachen Veränderungen im Brutgebiet den Bestandsrückgang des Waldlaubsängers (*Phylloscopus sibilatrix*)? Eine Untersuchung im Schwalm-Eder Kreis (Hessen) unter Berücksichtigung von Bruterfolg und Habitatveränderung. Fachhochschule Osnabrück: Diplomarbeit; 2009.
12. Reinhardt A, Bauer HG. Analyse des starken Bestandsrückgangs beim Waldlaubsänger *Phylloscopus sibilatrix* im Bodenseegebiet. Vogelwarte. 2009;47:23–40.
13. Creel S, Christianson D. Relationships between direct predation and risk effects. Trends Ecol Evol. 2008;23:194–201.
14. Cresswell W. Non-lethal effects of predation in birds. Ibis. 2008;150:3–17.
15. Lima SL. Predators and the breeding bird: behavioral and reproductive flexibility under the risk of predation. Biol Rev. 2009;84:485–513.
16. Ricklefs RE. An analysis of nesting mortality in birds. Smithsonian Contr Zool. 1969;9:1–48.
17. Thompson FR. Factors affecting nest predation on forest songbirds in North America. Ibis. 2007;149:98–109.
18. Forstmeier W, Weiss I. Adaptive plasticity in nest-site selection in response to changing predation risk. Oikos. 2004;104:487–99.
19. Schmidt KA, Ostfeld RS, Smyth KN. Spatial heterogeneity in predator activity, nest survivorship, and nest-site selection in two forest thrushes. Oecologia. 2006;148:22–9.
20. Miller SG, Knight RL, Miller CK. Influence of recreational trails on breeding bird communities. Ecol Appl. 1998;8:162–9.
21. Arlettaz R, Patthey P, Baltic M, Leu T, Schaub M, Palme R, Jenni-Eiermann S. Spreading free-riding snow sports represent a novel serious threat for wildlife. Proc R Soc Lond B Biol Sci. 2007;274:1219–24.
22. Kangas K, Luoto M, Ihantola A, Tomppo E, Siikamäki P. Recreation-induced changes in boreal bird communities in protected areas. Ecol Appl. 2010;20:1775–86.

23. Patthey P, Wirthner S, Signorell N, Arlettaz R. Impact of outdoor winter sports on the abundance of a key indicator species of alpine ecosystems. J Appl Ecol. 2008;45:1704–11.

24. Spaar R, Ayé R, Zbinden N, Rehsteiner U. Elemente für Artenförderungsprogramme Vögel Schweiz - Update 2011. Sempach: Schweizer Vogelschutz SVS/BirdLife Schweiz und Schweizerische Vogelwarte; 2012.

25. Begon M, Harper JL, Townsend CR. Ecology: individuals, populations, and communities. Oxford: Blackwell Science; 2005.

26. Bauer H-G, Hölzinger J. *Phylloscopus sibilatrix* (Bechstein 1793) Waldlaubsänger. In: Hölzinger J (editor) Die Vögel Baden-Württembergs. Band 3.1: Singvögel 1. Ulmer, Stuttgart (Hohenheim), p. 716–726. 1999.

27. Burfield IJ, van Bommel FPJ. Birds in Europe: population estimates, trends and conservation status. Cambridge: BirdLife International; 2004.

28. Hobson KA, Van Wilgenburg SL, Wesołowski T, Maziarz M, Bijlsma RG, Grendelmeier A, Mallord JW. A multi-isotope (δ^2H, and δ^{13}C, and δ^{15}N) approach to establishing migratory connectivity in palearctic-afrotropical migrants: an example using wood warblers phylloscopus SIBILATRIX. Acta Ornithol. 2014;49:57–69.

29. Keller V, Gerber A, Schmid H, Volet B, Zbinden N. Rote Liste Brutvögel. Gefährdete Arten der Schweiz, Stand 2010 Bundesamt für Umwelt, Bern, und Schweizerische Vogelwarte. Schweiz: Sempach; 2010.

30. Keller V, Ayé R, Müller W, Spaar R, Zbinden N. Die prioritären Vogelarten der Schweiz: Revision 2010. Ornithol Beob. 2010;107:265–85.

31. Coch T, Bertiller R, Trachsler B. Effiziente Erfassung der Kronendichte mit Hilfe fotografischer Senkrechtaufnahmen. Schweiz Z Forstwes. 2005;156:59–64.

32. White GC, Burnham KP. Program MARK: survival estimation from populations of marked animals. Bird Study. 1999;46:S120–39.

33. Bates D, Maechler M, Bolker B, Walker S. lme4: Linear mixed-effects models using Eigen and S4. R package version 1.0-5. 2013. http://CRAN.R-project.org/package=lme4.

34. Mazerolle MJ. AICcmodavg: Model selection and multimodel inference based on (Q)AIC(c). R package version 1.35. 2013. http://CRAN.R-project.org/package=AICcmodavg.

35. Cade BS. Model averaging and muddled multimodel inferences. Ecology. 2015;96:2370–82.

36. Korner-Nievergelt F, von Felten S, Roth T, Almasi B, Guélat J, Korner-Nievergelt P. Bayesian data analysis in ecology using linear models with R, BUGS, and stan. London: Academic Press; 2015.

37. Burnham KP, Anderson DR. Model selection and multi-model inference: a practical information-theoretic approach. New York: Springer; 2002.

38. Brändli U-B. Schweizerisches Landesforstinventar: Ergebnisse der dritten Erhebung 2004–2006. Eidgenössische Forschungsanstalt für Wald, Schnee und Landschaft WSL, Birmensdorf and Bundesamt für Umwelt, BAFU, Bern; 2010.

39. Huber N, Kienast F, Ginzler C, Pasinelli G. Using remote-sensing data to assess habitat selection of a declining passerine at two spatial scales. Landsc Ecol. 2016. doi:10.1007/s10980-016-0370-1.

40. Delahaye L, Vandevyvre X. Le Pouillot siffleur (*Phylloscopus sibiliatrix*) est-il une espèce indicatrice de la qualité des forêts feuillues ardennaises? Aves. 2008;45:3–14.

41. Wesołowski T. Polygyny in three temperate forest *Passerines* (with a critical reevaluation of hypotheses for the evolution of polygyny). Acta Ornithol. 1987;23:273–302.

42. Grendelmeier A, Arlettaz R, Gerber M, Pasinelli G. Reproductive performance of a declining forest passerine in relation to environmental and social factors: implications for species conservation. PLoS One. 2015;10:e0130954.

43. Moreau G. Etude d'une population de pouillot siffleur *Phylloscopus sibilatrix* dans une forêt du perche (Normandie). Alauda. 2001;69:103–10.

44. Mallord JW, Orsman CJ, Cristinacce A, Butcher N, Stowe TJ, Charman EC. Mortality of Wood Warbler *Phylloscopus sibilatrix* nests in Welsh Oakwoods: predation rates and the identification of nest predators using miniature nest cameras. Bird Study. 2012;59:286–95.

45. Piotrowska M, Wesołowski T. The breeding ecology and behaviour of the chiffchaff *Phylloscopus collybita* in primaeval and managed stands of Bialowieza Forest (Poland). Acta Ornithol. 1989;25:25–76.

46. Tablado Z, Jenni L. Determinants of uncertainty in wildlife responses to human disturbance. Biol Rev. 2015. doi:10.1111/brv.12224.

47. Laundré JW, Hernández L, Ripple WJ. The landscape of fear: ecological implications of being afraid. Open Ecol J. 2010;3:1–7.

48. Fuller RJ. Avian responses to transitional habitats in temperate cultural landscapes: woodland edges and young-growth. In: Fuller RJ, editor. Birds and habitats: relationships in changing landscapes. New York: Cambridge University Press; 2012. p. 125–49.

49. Wesołowski T. The breeding ecology of the wood warbler *Phylloscopus sibilatrix* in primaeval forest. Ornis Scand. 1985;16:49–60.

50. Fuller RJ. Bird life of woodland and forest. Cambridge: Cambridge University Press; 1995.

51. Pasinelli G, Grendelmeier A, Gerber M, Arlettaz R. Rodent-avoidance, topography and forest structure shape territory selection of a forest bird. Dryad. doi:10.5061/dryad.k20ng.

15

Collapse of an iconic conifer: long-term changes in the demography of *Widdringtonia cedarbergensis* using repeat photography

J. D. M. White[1,2], S. L. Jack[1,2], M. T. Hoffman[1,2*], J. Puttick[1,2], D. Bonora[1,2], V. Visser[3,4] and E. C. February[2]

Abstract

Background: Conifer populations appear disproportionately threatened by global change. Most examples are, however, drawn from the northern hemisphere and long-term rates of population decline are not well documented as historical data are often lacking. We use a large and long-term (1931–2013) repeat photography dataset together with environmental data and fire records to account for the decline of the critically endangered *Widdringtonia cedarbergensis*. Eighty-seven historical and repeat photo-pairs were analysed to establish 20th century changes in *W. cedarbergensis* demography. A generalized linear mixed-effects model was fitted to determine the relative importance of environmental factors and fire-return interval on mortality for the species.

Results: From an initial total of 1313 live trees in historical photographs, 74% had died and only 44 (3.4%) had recruited in the repeat photographs, leaving 387 live individuals. Juveniles (mature adults) had decreased (increased) from 27% (73%) to 8% (92%) over the intervening period. Our model demonstrates that mortality is related to greater fire frequency, higher temperatures, lower elevations, less rocky habitats and aspect (i.e. east-facing slopes had the least mortality).

Conclusions: Our results show that *W. cedarbergensis* populations have declined significantly over the recorded period, with a pronounced decline in the last 30 years. Individuals that established in open habitats at lower, hotter elevations and experienced a greater fire frequency appear to be more vulnerable to mortality than individuals growing within protected, rocky environments at higher, cooler locations with less frequent fires. Climate models predict increasing temperatures for our study area (and likely increases in wildfires). If these predictions are realised, further declines in the species can be expected. Urgent management interventions, including seedling out-planting in fire-protected high elevation sites, reducing fire frequency in higher elevation populations, and assisted migration, should be considered.

Keywords: Temperature, Conifer, Repeat photography, Population change, Climate change, Fire, Cederberg

Background

In a recent global review of forest tree species, climate-induced physiological stress driven by drought and high temperatures was found to be a major cause of mortality in the last 40 years [1]. Affected trees were also more vulnerable to climate-mediated processes such as insect or pathogen attack and wildfires. Conifers represented more than 40% of the 88 reviewed cases of mortality [1] and have recently been found to be particularly vulnerable to a warming climate [2].

The particularly rapid decline of the charismatic southern African endemic, *Widdringtonia cedarbergensis* Marsh [3] over the last century is representative of this global decline in conifers and has led the species to be listed as critically endangered under the current IUCN Red List of Plants [4]. While the diminishing number of *W. cedarbergensis* individuals have been anecdotally noted since the early 1800 s [5] and the focus of more

*Correspondence: timm.hoffman@uct.ac.za
[2] Department of Biological Sciences, University of Cape Town, Cape Town, South Africa
Full list of author information is available at the end of the article

systematic observation over the last 50 years (see [3]), there is as yet no quantitative evidence documenting the rate of decline and no plausible hypothesis for a cause.

The prevailing causal theories for the decline in *W. cedarbergensis* are (a) 18th and 19th century over-exploitation of the tree as a timber source which reduced and fragmented populations, thereby increasing 'edge-effects' and the likelihood of succumbing to fire [6–8], (b) late Quaternary climate change with warmer temperatures and less winter precipitation leading to a shift in the composition of co-occurring species with consequent changes in the fire regime [9, 10], (c) recent anthropogenic climate change, leading to temperature increases and aridity in the north-western areas of the fynbos biome, thereby increasing *W. cedarbergensis* mortality [11], and (d) the negative influence of insects and pathogens on reproductive output and survival [12].

The vegetation for the Cederberg Wilderness Area (CWA) is termed fynbos, a fire-prone vegetation type that includes many fire-adapted species with a fire return interval of approximately 11–15 years [13]. Fire is an important disturbance mechanism in fynbos and is thought to have had a strong influence in shaping the ecology and evolution of fynbos species [14]. *W. cedarbergensis*, however, is conspicuous in this environment in that it shows little or no adaptation to cope with fire. Mature trees are often killed by fire and display no re-sprouting ability, there is no canopy-stored seedbank, sapling growth rates are relatively slow, and the trees take much longer to reach reproductive maturity (ca. 40 years) than any co-occurring shrubs or trees [15]. Given such sensitivity to fire, an unfavourable fire regime can be very detrimental to this species. More recently there has been an increase in fire frequency with as many as six fires occurring in some parts of the Cederberg between 1977 and 2003 [16]. This suggests that the current fire return interval could be too short to allow for adequate establishment and reproductive output to maintain viable populations of the species [9, 13].

Widdringtonia cedarbergensis individuals are usually associated with rocky cliffs, outcrops and east facing slopes at high elevations and are rarely found at lower elevations, on flatter ground and deeper soils [3, 17]. This association with higher-lying rocky sites has not only been attributed to temperature amelioration [7] and protection from fire [3], but has also been suggested to give trees reliable access to available water trapped between bedding planes [11]. This water supply is reliant on regular rainfall, and therefore any change in the amount or seasonality of rainfall that may lead to a decline in plant available water will adversely affect the trees. Given that the CWA is projected to be one of the first areas in the Western Cape Province in South Africa to be affected by warming and drying due to anthropogenic climate change [18], the likely sensitivity of *W. cedarbergensis* to elevated temperatures and reduced moisture levels is concerning.

This study draws on a large collection of historical photographs of *W. cedarbergensis* (ca. 1931–1951 and 1960–1987) which were relocated and retaken (2007–2013), and represent the longest visual record of change in *W. cedarbergensis* populations to date. Here we (a) describe the pattern of recruitment and mortality of the species for the last 80 years, (b) evaluate the relationship between the degree of observed mortality and several likely environmental contributors to this mortality, and (c) speculate on the extent to which short-versus long-term changes in rainfall, temperature and fire regimes have affected the current distribution and demography. We use our findings to make recommendations regarding the future conservation of the species.

Methods

Study area and photograph sites

Widdringtonia cedarbergensis, commonly known as the Clanwilliam cedar, is endemic to the Cederberg Mountains of the Western Cape Province in South Africa. It has a patchy distribution between 900 and 1500 metres above sea level over approximately 250 km^2, and typically grows in quartz-derived sandstone soils [19]. The Cederberg experiences a Mediterranean-type climate with the majority of frontally-derived rainfall received during the cooler winter months (April–September), while summers are hot and dry. Cederberg vegetation is comprised mainly of Cederberg Sandstone Fynbos, which is a sclerophyllous and fire-prone vegetation type [20, 21].

At the elevations at which it grows, *W. cedarbergensis* is often the only tree species in the landscape. It is also uniquely different in growth form from the only other tree, *Protea nitida* and therefore easily distinguished. A historical collection of 87 photographs unambiguously showing *W. cedarbergensis* trees, taken between 1931 and 1987 by several photographers, were used in the analysis. Historical photographs were located in the Skerpioenspoort, Middelberg, Welbedacht and Heuningvlei regions and were repeated during three separate visits to the Cederberg in 2007, 2012 and 2013 (Fig. 1; see Additional file 1: Table S1 for further information on repeat photographs). In order to retake the historical photograph, sites were first located using a combination of topographical maps, notes accompanying photographs and landscape features. The exact position of the original camera was then determined in the field and, with the aid of a tripod for stability, the repeat photograph was taken using a high specification digital camera (Canon 5D MkII, Canon Inc., USA). Detailed site information, such as GPS co-ordinates, elevation, aspect, degree of

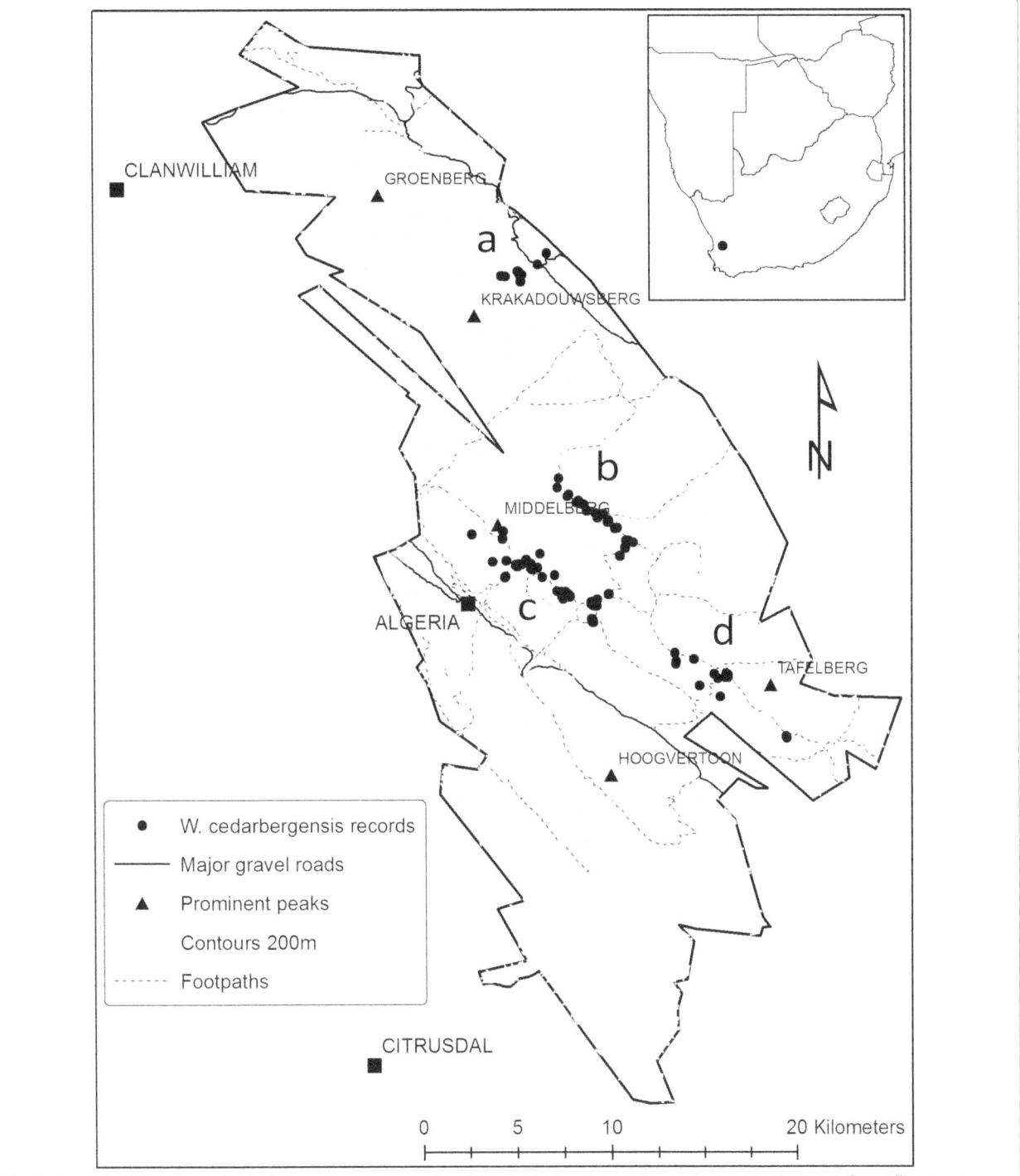

Fig. 1 Map of localities of sampled *Widdringtonia cedarbergensis* individuals. The locations of *W. cedarbergensis* individuals in four different regions in the Cederberg Wilderness Area, Western Cape, South Africa: *a* Heuningvlei, *b* Skerpioenspoort, *c* Middelberg and *d* Welbedacht. Features of interest include the weather stations at Algeria Forest Station and Clanwilliam

rockiness, general ecological description and description of major changes in *W. cedarbergensis* and other vegetation, were recorded for each site.

Population dynamics over time

To match the photo-pairs, the repeat photograph was resized so that a distance between two fixed features common to both images was identical. The repeat photograph was then overlain on the historical photograph and rotated and shifted so that landscape features aligned exactly. The matched historical and repeat photographs were then saved independently (Adobe Systems Inc., 2010) and individual trees labelled according to whether these were living, dead or had recruited since the historical photograph was taken (Fig. 2). Mortality and recruitment probabilities were then calculated relative to the digitised original historic image. Age class (juvenile, mature adult and senescent/dead) and degree of rockiness were subjectively scored in both historical and repeat photographs. A juvenile tree was estimated to be generally shorter than 2 m tall with a thin, erect stem and no branching of the primary trunk. A mature adult was usually taller than a juvenile, with a thick primary trunk and a considerable extent of secondary branching, while senescent/dead individuals had only one or two branch tips with foliage or no foliage at all. Depending on the degree of rockiness, photograph sites were classified as either open (<25% rockiness), protected (25% < rockiness < 75%) or well protected (>75% rockiness and

Fig. 2 Example photo-pair. **a** Historical photograph taken by Ken Howes-Howell in 1941 and **b** repeat photograph taken by Timm Hoffman in 2007 at Vogelgesang, Skerpioenspoort. Thirteen living *Widdringtonia cedarbergensis* (*green dots*) and five skeletons (*orange dots*) were recorded in the historical photograph. Only four individuals had survived until 2007, with nine having died (*red dots*) and no new recruits

located within a rocky outcrop). Age class structure was then analysed within three broad time periods, namely 1931–1951 and 1960–1987 (historical photographs), and 2007–2013 (repeat photographs). These time periods, hereafter referred to as 'early', 'middle' and 'recent' respectively, were then compared using Chi squared tests, while the number of trees per photograph was compared for these periods using the nparcomp package in R (version 3.1.0) [22, 23].

Environmental trends
The available fire history for the CWA was accessed via the SANBI Biodiversity GIS portal (http://bgis.sanbi.org), and fire frequency computed for each of the 87 photosites and identified regions [24]. Using GPS co-ordinates for each tree, mean annual temperature (°C) and annual precipitation (mm) values for the period 1950–2000 were extracted from the interpolated Worldclim climate surface [25]. The spatial resolution (30 arc sec) of WorldClim data did not allow for quantifications of temperature and precipitation for each tree. This is similarly the case for fire frequency data, where the temporal resolution did not exactly match the date of all historical photographs, while the myriad micro-habitats that may shelter plants from fires could not be mapped within fire 'polygons'. However, the model variables represent the best available spatial and temporal estimates for climatic conditions presently available. Google Earth was used to determine latitude, longitude and aspect value for each tree [26]. Primary cardinal directions and intercardinal ranges were grouped for ease of analysis (for example: 315° < N ≤ 45°). Further information on the source, resolution and unit of measurement of modelled variables can be found in Additional file 2: Table S2. For climate trend analysis, historical climate data was obtained from the South African Weather Service for Clanwilliam (32°10′52.52″S, 18°53′37.64″E) and the Agricultural Research Council for the Cape Nature Offices at Algeria (32°22′28.65″S, 19° 3′38.34″E) (Fig. 1). Clanwilliam and Algeria were the closest stations with long-term precipitation records (1870–2010 and 1908–2008, respectively). No temperature records exist for Algeria while Clanwilliam has a temperature record from 1963 to 2010. We used the 'segmented' package in R to determine whether regression models of the precipitation and temperature records had unknown break-points [23, 27].

Regional *W. cedarbergensis* population dynamics
All statistical analyses were performed in the R statistical programme [23]. For each region, defined as discrete, geographically separate populations, probability of mortality (Model 1) or recruitment (Model 2) was determined by fitting a generalized linear mixed-effects model

(GLMM) in the 'MASS' package [28]. Habitat, aspect, mean annual temperature (°C) and fire frequency were fitted as fixed effects, while region and photograph site were fitted as random effects. This analysis included both plantations established at the end of the 19th century, as well as natural stands of trees. These models were simply used to describe regional differences in mortality and recruitment between *W. cedarbergensis* populations and not to infer abiotic correlates of population dynamics.

Correlates of *W. cedarbergensis* tree mortality

Only natural stands were used in the analysis to determine environmental and climatic correlates of *W. cedarbergensis* mortality (Model 3). This model exhibited significant positive spatial autocorrelation for spatial lags up to about 5 km (measured using spatial correlograms from the 'ncf' package [29]; see Additional file 3: Figure S1). To account for spatial autocorrelation we included a spatial autocovariate using an exponential correlation structure with latitude and longitude coordinates. The GLMM used to determine correlates of mortality included mortality as the binary response variable, habitat, aspect, mean annual temperature (°C) and fire frequency as fixed effects and region and photograph site as random effects with a binomial family link. We used Wald tests to determine overall significance of the fixed effects [30].

Results

Population dynamics

A total of 1313 living trees were recorded when plantations and natural populations were combined across all historical photographs. Of these, 967 individuals (74%) had died and there were 44 new recruits (Heuningvlei: n = 2, Welbedacht: n = 9, Skerpioenspoort: n = 3 and Middelberg: n = 30), with 3 of these recruits having died in the intervening period, leaving a total of 387 individuals alive. When considering only natural populations at Welbedacht, Skerpioenspoort and Middelberg, 597 individuals (73%) out of 821 had died, leaving only 264 trees alive.

For natural stands, there was no significant change in the average number of trees recorded in each photograph between early (8.65 ± 0.95, n = 51) and middle (6.86 ± 0.83, n = 28) (p = 0.787) periods. However, there was a significant decline (p < 0.001) in trees per photograph for the recent period (2.71 ± 0.34, n = 79). The early period had a significantly greater proportion of juveniles (27%) than the middle (5%) (X^2 = 54.29, df = 1, *p* value < 0.001) and recent period (8%) (X^2 = 36.68, df = 1, p value <0.001). However, no significant change was found in the proportion of juveniles between the middle and recent periods (X^2 = 1.27, df = 1, p value = 0.259).

Regional population dynamics

The plantation at Heuningvlei had the highest probability of mortality (mean ± SD: 0.90 ± 0.04) followed by the natural populations at Skerpioenspoort (0.70 ± 0.08) and Middelberg (0.69 ± 0.13) and then Welbedacht (0.60 ± 0.06) (Fig. 3). Welbedacht had the highest probability of recruitment (0.13 ± 0.08), followed in order by Middelberg (0.06 ± 0.04), Skerpioenspoort (0.04 ± 0.02) and lastly Heuningvlei (0.02 ± 0.01) (Fig. 3).

Environmental trends

The range in fire frequency for all photo-sites was between three and eight fires over the recorded period (1944–2012). Middelberg experienced the highest median fire frequency of five fires, while Welbedacht experienced the lowest median of three fires (Fig. 4). Frequency histograms indicated mixed trends in the lengths of fire return intervals over time at the respective populations. Only Middelberg displayed a generally decreasing fire return interval (*i.e.* more frequent fires) through time (Additional file 4: Figure S2).

Despite high inter-annual variability, precipitation increased significantly from 1908 to 2008 at Algeria (p < 0.001), but showed no significant trend at Clanwilliam only 25 km away (Additional file 5: Figure S3). The ten year mean rainfall amount was also significantly higher (636 ± 51 mm) at Algeria relative to Clanwilliam (177 ± 17 mm) (df = 198, F = 564, p < 0.001). There was a significant increase (p < 0.05) in temperature of approximately 0.6 °C at Clanwilliam over the recording period with segmented regression indicating that most of this increase occurred after 1996 (Fig. 5).

Correlates of *W. cedarbergensis* tree mortality

Fire frequency, aspect, degree of rockiness and mean annual temperature all had a significant effect on *W. cedarbergensis* mortality (p < 0.10) (Table 1: Model 3). Trees in 'open' habitats did not have an increased probability of mortality (Table 1), although a significant difference in the abundance of *W. cedarbergensis* trees in rocky (n = 471), compared to well protected (n = 277) and open (n = 73) sites was observed (df = 2, X^2 = 289.47, p < 0.001). In most cases the model captured the expected direction of relationships between the response and explanatory variables. For example, the probability of mortality increased with an increase in fire frequency (p < 0.05), while an increase in mean annual temperature was almost significant at α = 0.05 (p = 0.07) (Fig. 6). The predicted probability of mortality was greatest on northern, southern and western facing aspects, followed by eastern aspects (Fig. 6). Both elevation (positive) and annual precipitation (negative) had strong co-correlation with mean annual temperature (R^2 = 0.59, p < 0.001 and R^2 = 0.42, p < 0.001, respectively).

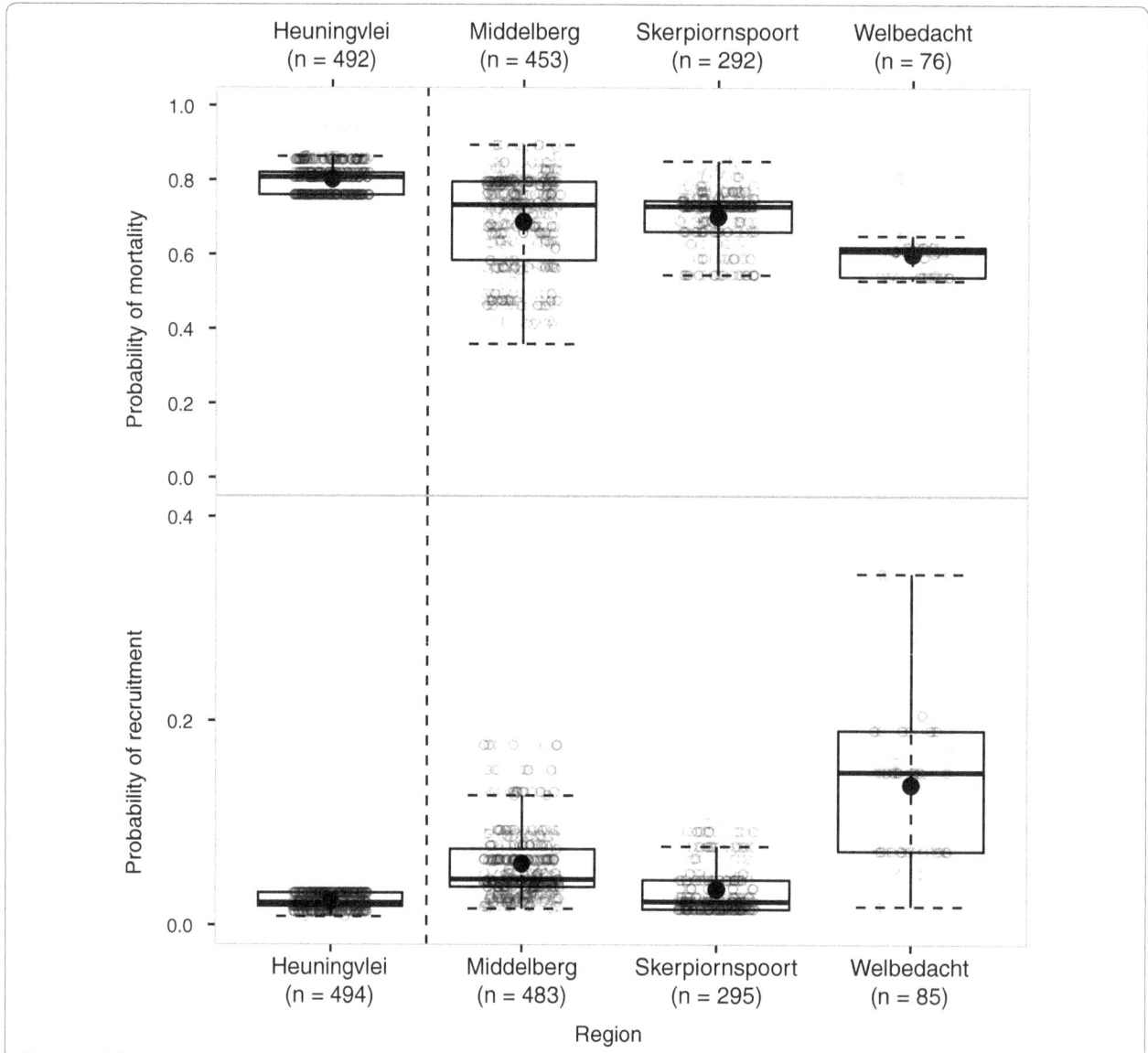

Fig. 3 Modelled probability of *Widdringtonia cedarbergensis* mortality (Model 1) and recruitment (Model 2) at all sampled regions. Raw predicted points with *Tukey boxplots* indicating the median, lower and upper quartiles, and minimum and maximum values with outliers outside the *whiskers*. Means are shown as *filled black dots*

Discussion

The Cederberg was declared a Wilderness Area in 1973 giving it the highest conservation status in South Africa [31]. Grounded in the perception that *W. cedarbergensis* is in decline [6, 7], one of the objectives for the reserve has been the conservation of this iconic tree. Palaeoecological evidence does indeed suggest that *W. cedarbergensis* trees were once more numerous, with a steady decline in fossil pollen deposits since the last glacial maximum [10]. The suggestion is that the decline may be attributed to long-term warming and drying of the West Coast climate in the late Quaternary that may have contributed to

elevated fire frequencies [10]. It has also been proposed that prior to heavy 18th and 19th century logging [5, 6] the species may have occurred in denser communities. Harvesting and more frequent anthropogenic fires led to thinning and reductions in stand size, thereby establishing a positive feedback which promoted further penetration by fires, killing both seedlings and large seed-bearing adults [9].

Without access to accurate demographic records it is very difficult to reconstruct population dynamics before the 20th Century. Our more recent results suggest, however, that population densities were much higher in the

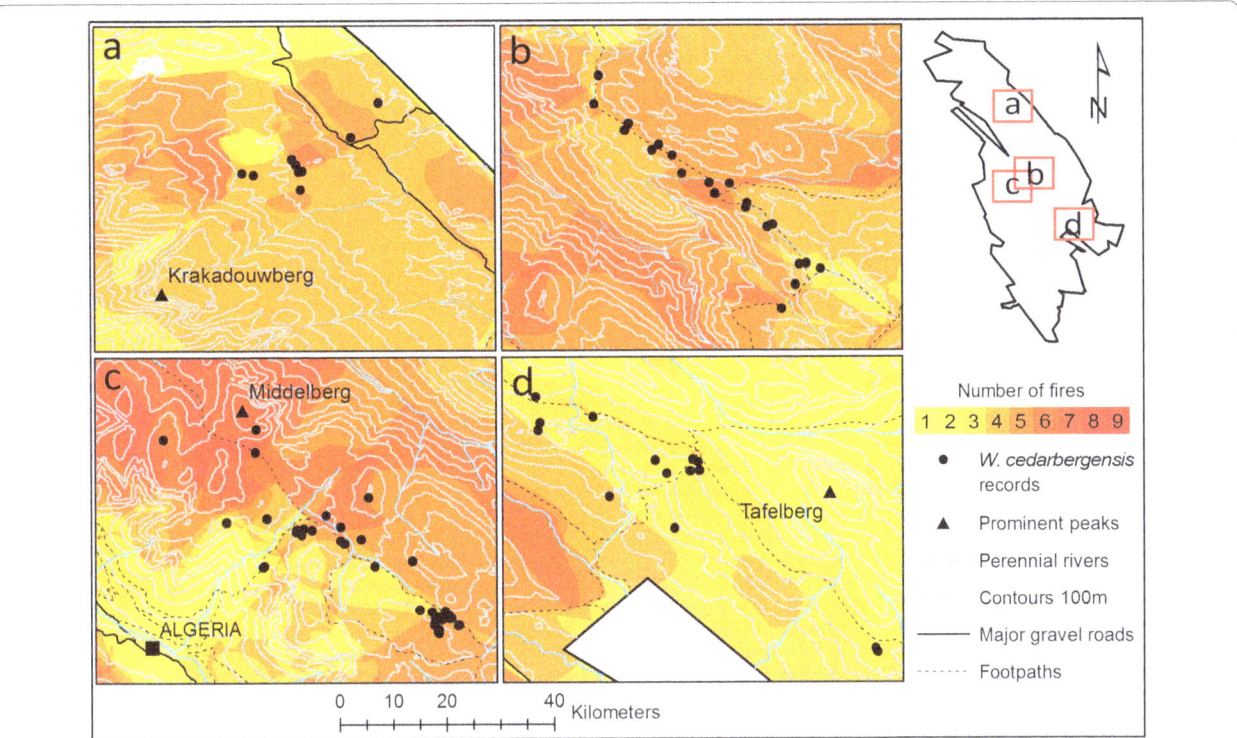

Fig. 4 Fire frequency map for the four studied regions within the Cederberg Wildnerness Area. **a** Heuningvlei, **b** Skerpioenspoort, **c** Middelberg and **d** Welbedacht. The fire record is from 1944 to 2012

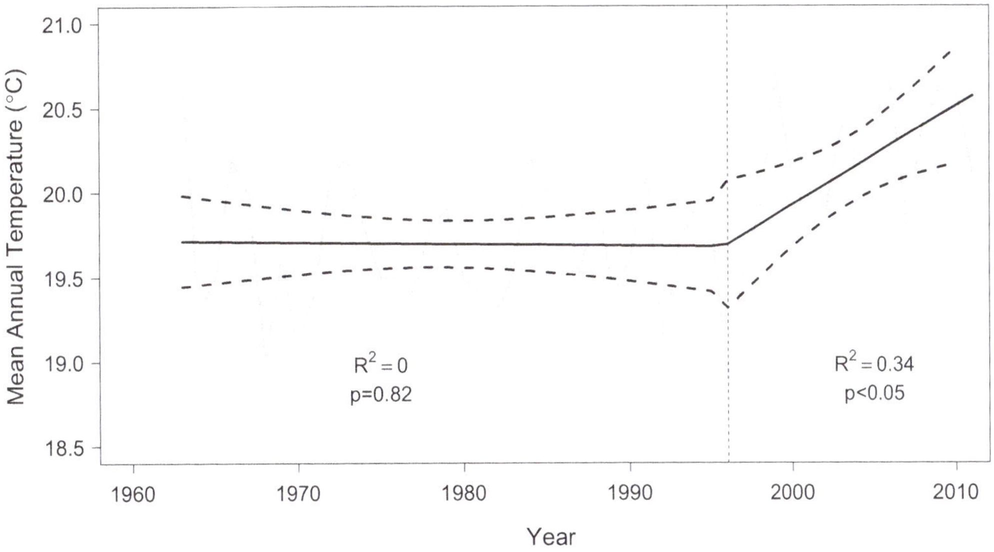

Fig. 5 Mean annual temperature (°C) in Clanwilliam from 1963 to 2010. The *solid line* represents a piecewise regression with an estimated break-point in 1996. The *dashed lines* represent 95% confidence intervals. R^2 values and significance levels are attached

early part of last century than they are today and that the *W. cedarbergensis* decline has continued regardless of any management interventions [13, 32]. In addition, our results show that the greatest declines in trees took place between the periods 1960–1987 and 2007–2013, indicating that *W. cedarbergensis* mortality has increased in the late 20th and early 21st century. Current recruitment rates are also not keeping pace with mortality at any of

Table 1 Fitted coefficients, the standard errors, *P* values of the generalized linear mixed-model for probability of *W. cedarbergensis* mortality across all regions (model 1) and in natural stands (model 3), probability of recruitment across all regions (model 2) and the Wald test of fixed effects

Predictors	Model 1		Model 2		Model 3	
	Estimate	S.E.	Estimate	S.E.	Estimate	S.E.
(a)						
Intercept	−9.14*	2.85	9.23	5.72	−9.96*	4.40
Habitat-open	0.39	0.45	1.19	0.76	0.41	0.61
Habitat-rocky	0.26	0.23	0.38	0.50	0.55*	0.27
Aspect-north	1.45***	0.41	0.00	0.82	1.99***	0.45
Aspect-south	0.49	0.30	0.15	0.69	1.24***	0.34
Aspect-west	0.89**	0.33	0.78	0.63	1.43***	0.41
Fire frequency	0.09	0.12	−0.18	0.23	0.30*	0.14
MAT	0.70**	0.23	−0.96*	0.47	0.64	0.36
Random effects (R.E.)	0.00	0.97	0.00	0.98	1.02	0.90

Predictors	X^2	*p* value	X^2	*p* value	X^2	*p* value
(b)						
Habitat	18.6	0.000	2.5	0.290	9.0	0.029
Aspect	18.8	0.000	8.7	0.068	26.9	0.000
Fire frequency	0.5	0.468	0.6	0.440	4.6	0.033
MAT	9.6	0.002	4.2	0.041	3.2	0.073

R.E.: $n_{region} = 4$; $n_{site} = 86$, R.E.: $n_{region} = 4$; $n_{site} = 86$, R.E.: $n_{region} = 3$, $n_{site} = 80$, $n_{tree} = 1313$, $n_{tree} = 1357$, $n_{tree} = 821$

Habitat well protected and aspect–east set as intercept, *MAT* Mean annual temperature

p values: $1 < 0.10$, $* < 0.05$, $** < 0.01$, $*** < 0.001$

our study sites. Importantly, the spatial consistency of high mortality and low recruitment across geographically isolated populations suggests that the factors driving the decline act over the entire range of the species.

Our model identified increased fire frequency and possibly higher temperatures as the primary drivers for *W. cedarbergensis* mortality. There are more trees still alive at higher elevations with concomitant lower temperatures and greater rainfall. It has been demonstrated that a small increase in temperature can increase vapour pressure deficit (VPD) which increases tree water use resulting in an increase in mortality during severe drought due to xylem cavitation [33]. This increased mortality with increased temperature is also demonstrated by our model showing a significant increase in mortality on warmer north facing slopes.

Although our modelled results suggest *W. cedarbergensis* mortality is not associated with more open habitats, we found that *W. cedarbergensis* trees were positively associated with 'rockier' sites. This suggests a variable not considered in the analysis but weakly correlated to the degree of rockiness such as, for example, available ground water, may be important in *W. cedarbergensis* survival. Previous research has demonstrated that in rocky sites the trees source water from deep pockets between the rocks [11]. This implies a possible affinity for available

water. In addition, directed secondary dispersal [34] and shadier, cooler microhabitats found in rockier environments which are beneficial in seedling establishment [3] are also likely factors that have influenced recruitment and the presence of the trees in such sites.

Temperature-induced mortality has been recognised as a global issue for many tree species [1, 2, 35, 36]. The association of *W. cedarbergensis* trees with sites of lower temperatures at higher elevations and historical trends of increasing temperatures in the Cederberg, should be recognised as possible contributors to mortality. Not even a significant increase in rainfall over the course of the 20th century at the more proximal climate station at Algeria appears to have halted or reversed the negative demographic trend. Climate models for the study area [18] and globally [37] show that temperatures will continue to increase through this century, suggesting that in future favourable habitat for the trees is only likely at higher elevations [38]. Current populations of *W. cedarbergensis* are already at or near the highest locally attainable elevations. In addition, isolated peaks and ridges on which the trees typically occur are often separated from the nearest adjacent higher ground by lower-lying valleys dominated by fire-prone vegetation.

Given the slow rate of *W. cedarbergensis* growth to reproductive maturity, the median fire frequencies at

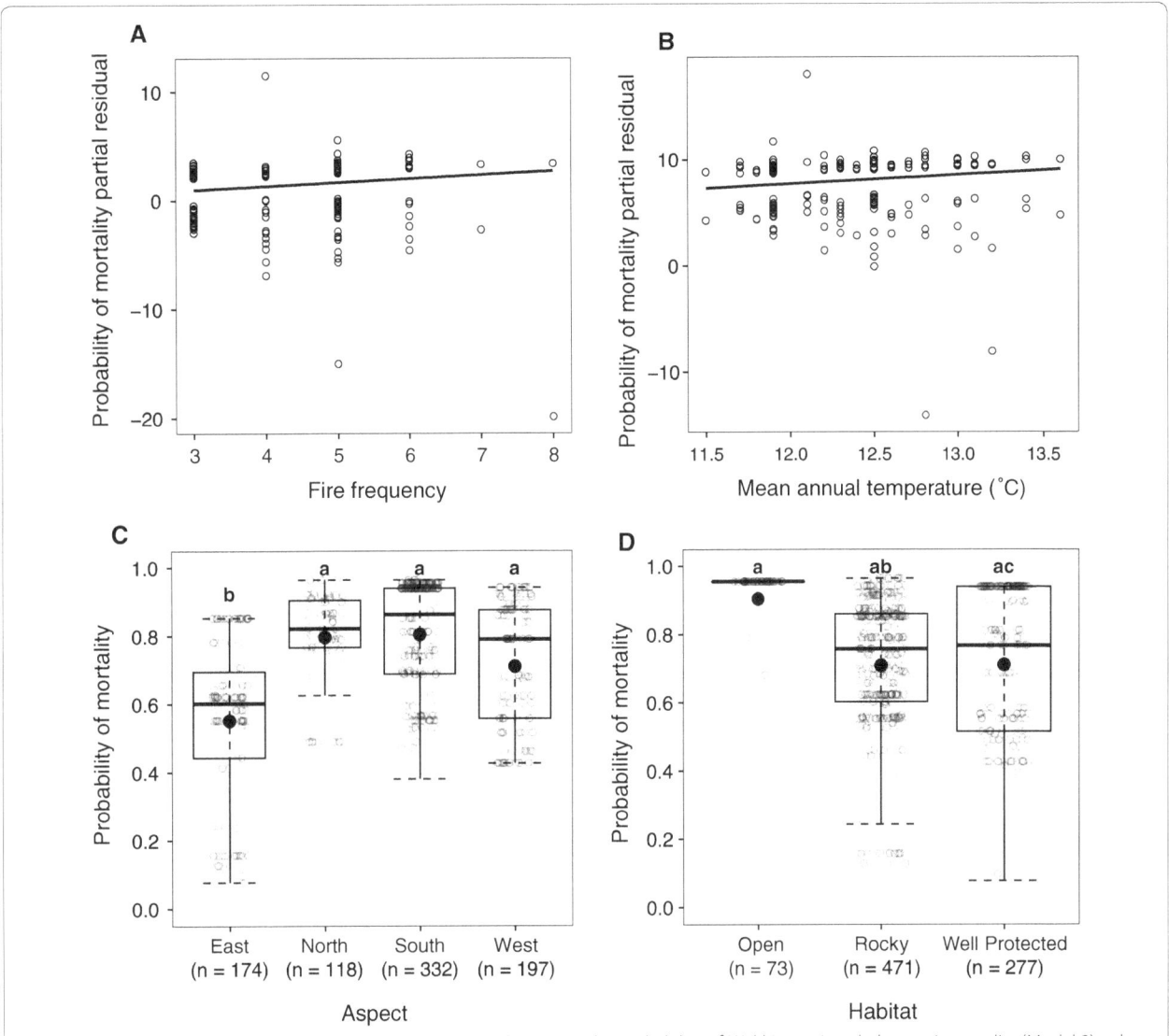

Fig. 6 **A–D** Partial *residual plots* (**A**, **B**) and *predicted points* (**C**, **D**) showing the probability of *Widdringtonia cedarbergensis* mortality (Model 3) only natural populations in relation to environmental and climatic variables. *Regression lines* indicate the partial fit for **A** fire frequency and **B** mean annual temperature; and raw predicted with *Tukey boxplots* for **C** aspect and **D** *habitat* indicate the median, lower and upper quartiles, and minimum and maximum values with outliers outside the whiskers. Means are shown as filled *black dots*. *Different letters* indicate where the modelled probability of mortality was significantly different between aspects or habitat (p < 0.05)

the sampled sites over the 68 year record for the Cederberg suggests that fire return intervals will likely be too short in most regions for the trees to naturally replace themselves [3]. Indeed, in some regions fire return intervals are *decreasing* despite a management policy of a 15–20 year return interval. Our study did not, however, determine the relative impact of less frequent, large, high intensity fires (e.g. in 1959, 1985 and more recently in January 2013), on *W. cedarbergensis* demography and this remains an avenue for further research . Furthermore, future studies may benefit from the use of climatic and environmental variables with a more detailed temporal and spatial resolution, particularly in view of the myriad

micro-habitats and refugia that are available in terrain such as exists in the CWA.

Despite the Cederberg being declared a Wilderness Area in 1973 and one of the management foci being the conservation of the threatened *W. cedarbergensis*, our results show that there has been a significant decline in the number of trees since that time. We suggest that this continued decline is probably due to both natural and anthropogenically mediated increases in fire frequency and temperature. If modelled climate change predictions for increased warming and drying of the Cederberg Mountains [18] are realised, the future survival of the *W. cedarbergensis* is at serious risk.

Conclusions

The outlook for *W. cedarbergensis* remains a concern under the current climate and fire regime. Mortality rates are too high and recruitment rates too low to sustain viable populations in the long-term. The main drivers of mortality appear to be higher temperatures and shorter fire return intervals, while high elevation, rocky, east-facing environments represent refugia that provide fire protection, more reliable groundwater [11] and cooler microhabitats conducive to seedling establishment. Projected drying and temperature increases in the 21st century [18] suggest that the species will be under increasing pressure, both from the impacts on water balance and hydraulic failure under greater water stress, and from elevated fire frequencies.

However, there are interventions that could help sustain current populations or the establishment of new ones. Already established annual seedling out-planting programmes could be improved to increase *W. cedarbergensis* seedling survival. This could be done by incorporating beneficial criteria for *W. cedarbergensis* survival, as outlined in this study, when selecting sites for out-planting [32]. Of the populations which remain, a subset could be selected (based on current demographic profiles, elevation and temperature, degree of rockiness, etc.) for special protection from fire. Lastly, assisted migration [39] to more southern high-lying protected areas should be a serious consideration given the time it would take for viable populations to establish in such areas.

Additional files

Additional file 1: Table S1. Details of photographers for all 87 repeat photograph photo-pairs taken in the Cederberg Wilderness Area between 1931 and 2013.

Additional file 2: Table S2. Source, spatial and temporal resolution, and unit of measurement of environmental variables.

Additional file 3: Figure S1. Moran's *I* plots testing for spatial autocorrelation in model residuals. Spatial autocorrelation in generalized linear mixed-effects model residuals a) without a spatial autocovariate term and b) with a spatial autocovariate term. Significant spatial autocorrelation ($p < 0.05$) at each distance lag is shown by black dots.

Additional file 4: Figure S2. Histograms of fire histories for each region. a) Heuningvlei, b) Skerpioenspoort, c) Middelberg and d) Welbedacht. The fire record is from 1944 to 2012.

Additional file 5: Figure S3. Annual precipitation (mm) in Algeria (1908 to 2008) and Clanwilliam (1870 to 2010). The solid lines represent regression lines. The dashed lines represent 95% confidence intervals. R2 values and significance levels are attached.

Abbreviations

CWA: cederberg wilderness area; GLMM: generalized linear mixed-effects model; SD: standard deviation; RE: random effects; MAT: mean annual temperature.

Authors' contributions

MTH conceived of the study. SLJ, JP, DB, JDMW and MTH carried out repeat photography data collection. JDMW and VV performed the statistical analysis. JDMW and SLJ drafted the manuscript. ECF critically revised the manuscript. All authors read and approved the final manuscript.

Author details

[1] Plant Conservation Unit, University of Cape Town, Cape Town, South Africa. [2] Department of Biological Sciences, University of Cape Town, Cape Town, South Africa. [3] Statistics in Ecology, Environment and Conservation, Department of Statistical Sciences, University of Cape Town, Cape Town, South Africa. [4] African Climate and Development Initiative, University of Cape Town, Cape Town, South Africa.

Acknowledgements

We would like to thank Cape Nature for allowing us to work in the CWA and for providing access to the fire records. In particular we would like to acknowledge the contribution that the late Patrick Lane made to the conservation of the Clanwilliam cedar and all the flora and fauna of the CWA. The South African Weather Service and Agricultural Research Council provided climate data for the area.

Competing interests

The authors declare that they have no competing interests.

References

1. Allen C, Macalady A, Chenchouni H, Bachelet D, McDowell N, Vennetier M, Kitzberger T, Rigling A, Breshears D, Hogg E, Gonzalez P, Fensham R, Zhang Z, Castro J, Demidova N, Lim J-H, Allard G, Running S, Semerci A, Cobb N. A global overview of drought and heat-induced tree mortality reveals emerging climate change risks for forests. For Ecol Manag. 2010;259(4):660–84.
2. McDowell N, Allen CD. Darcy's law predicts widespread forest mortality under climate warming. Nat Clim Change. 2015;5:669–72.
3. Manders PT. An assessment of the current status of the Clanwilliam cedar (*Widdringtonia cedarbergensis*) and the reasons for its decline. S Afr For J. 1986;139:48–53.
4. Farjon A, February E, Higgins S, Fox S, Raimondo D. *Widdringtonia cedarbergensis* The IUCN Red List of Threatened Species 2013. http://dx.doi.org/10.2305/IUCN.UK.2013-1.RLTS.T30365A2793077.en. Accessed 15 Oct 2013.
5. Smith CA. Early 19th century records of the Clanwilliam cedar (Widdringtonia juniperoides Endl). J S Afr For Assoc. 1955;25(1):58–65.
6. Hubbard CS. Observations on the distribution and growth rate of the Clanwilliam cedar *Widdringtonia juniperoides*. S Afr J Sci. 1937;33:572–86.
7. Lückhoff HA. The clanwilliam cedar: its past history and present status. J Mt Club S Afr. 1971;74:33–9.
8. Andrag RH. Studies in die Sederberge (i) Die status van die Clanwilliam sedar (*Widdringtonia cedarbergensis* Marsh) (ii) Buitelugontspanning Unpublished MSc thesis. Stellenbosch: University of Stellenbosch; 1977.
9. Manders PT. A transition matrix model of the population dynamics of the Clanwilliam cedar (*Widdringtomia cedarbergensis*) in natural stands subject to fire. For Ecol Manag. 1987;20:171–86.
10. Meadows M, Sugden J. The History of the Clanwilliam Cedar (*Widdringtonia cedarbergensis*): evidence from Pollen Analysis. S Afr For J. 1991;156:64–71.
11. February E, West A, Newton R. The relationship between rainfall, water source and growth for an endangered tree. Austral Ecol. 2007;32:397–402.

12. Wingfield MJ, von Broembsen SL, Manders PT. A preliminary assessment of the threat of diseases and pests to *Widdringtonia cedarbergensis*. S Afr For J. 1988;147:32–5.

13. Brown PJ, Manders PT, Bands DP, Kruger FJ, Andrag RH. Prescribed burning as a conservation management practice: a case history from the Cederberg mountains, Cape Province, South Africa. Biol Conserv. 1991;56:133–50.

14. Le Maitre DC, Midgley JJ. Plant reproductive ecology. In: Cowling RM, editor. The ecology of fynbos. Cape Town: Oxford University Press; 1992. p. 135–74.

15. Manders PT. The Autecology of *Widdringtonia cedarbergensis* in relation to its conservation management. MSc thesis. Cape Town: University of Cape Town; 1985.

16. Fox SJ. An assessment of the population status and demographic models of *Widdringtonia cedarbergensis*. Unpublished Honours thesis, Department of Botany. Cape Town: University of Cape Town; 2013.

17. Higgins S, February E, Skowno A. Distribution and viability of Clanwilliam cedar (*Widdringtonia cedarbergensis*, Cupressaceae). Final report to WWF-SA table mountain fund. 2001.

18. Midgley GF, Hannah L, Millar D, Rutherford MC, Powrie LW. Assessing the vulnerability of species richness to anthropogenic climate change in a biodiversity hotspot. Glob Ecol Biogeogr. 2002;11:445–51.

19. Coates Palgrave K. Trees of Southern Africa. Cape Town: C. Struik; 1977. p. 59–60

20. Taylor HC. Cederberg: vegetation and flora. Strelitzia 3, Pretoria: National Botanical Institute; 1996.

21. Mucina L, Rutherford MC. The vegetation of South Africa, Lesotho and Swaziland. Strelitzia, 19. Pretoria: South African National Biodivesrity Institute; 2006.

22. Konietschke F, Placzek M, Schaarschmidt F, Hothorn LA. Nparcomp: an R software package for nonparametric multiple comparisons and simultaneous confidence intervals. J Stat Softw. 2015;64(9):1–17.

23. R Core Team. R: A language and environment for statistical computing. Vienna: R foundation for statistical computing. http://www.R-project.org/. 2014.

24. ESRI. ArcGIS desktop: release 10. Redlands: Environ Sys Res Inst; 2011.

25. Hijmans RJ, Cameron SE, Parra JL, Jones PG, Jarvis A. Very high resolution interpolated global terrestrial climate surfaces. Int J Climatol. 2005;25:1965–78.

26. Google Earth Version 7.1.2.2041. http://www.google.com/earth/index. html. Accessed 30 Nov 2015.

27. Muggeo VMR. Estimating regression models with unknown break-points. Stat Med. 2003;22:3055–71.

28. Venables WN, Ripley BD. Modern applied statistics with S. 4th ed. New York: Springer; 2002.

29. Bjornstad ON. ncf: Spatial Nonparametric Covariance Functions. R package version 1.1–7. 2016. http://CRAN.R-project.org/package=ncf. Accessed 14 June 2016.

30. Lesnoff M, Lancelot R. aod: Analysis of Overdispersed Data. R package version 1.3. 2012. http://cran.r-project.org/package=aod. Accessed 14 June 2016.

31. Smith F. Governmental relations in the management of the Cederberg Wilderness Area: some organisational aspects. Politeia. 2002;21(3):58.

32. Mustart PJ, Makua C, Juritz J, Van der Merwe SW, Wessels N. Restoration of the Clanwilliam cedar, *Widdringtonia cedarbergensis*: the importance of monitoring seedlings planted in the Cederberg, South Africa. Biol Conserv. 1995;72:73–6.

33. Will RE, Wilson SM, Zou CB, Hennessey TC. Increased vapor pressure deficit due to higher temperatures leads to greater transpiration and faster mortality during drought for tree seedlings common to the forest-grassland ecotone. N Phytol. 2013;200:366–74.

34. Midgley JJ, Anderson B, Bok A, Fleming T. Scatter-hoarding of Cape Proteaceae nuts by rodents. Evolut Ecol Res. 2002;4:623–6.

35. Suarez ML, Ghermandi L, Kitzberger T. Factors predisposing episodic drought-induced tree mortality in *Nothofagus*—site, climatic sensitivity and growth trends. J Ecol. 2004;92:954–66.

36. Williams AP, Allen CD, Macalady AK, Griffin D, Woodhouse CA, Meko DM, Swetnam TW, Rauscher SA, Seager R, Grissino-Mayer HD, Dean JS, Cook ER, Gangodagamage C, Cai M, McDowell NG. Temperature as a potent driver of regional forest drought stress and tree mortality. Nat Clim Change. 2013;3:292–7.

37. IPCC. Climate change 2013. In: Stocker TF, et al., editors. The physical science basis. Cambridge: Cambridge University Press; 2013.

38. Parmesan C. Ecological and evolutionary responses to recent climate change. Ann Rev Ecol Evol Syst. 2006;37:637–69.

39. McLachlan JS, Hellmann JJ, Schwartz MW. A framework for debate of assisted migration in an era of climate change. Conserv Policy. 2007;21(2):297–302.

Synergistic impacts by an invasive amphipod and an invasive fish explain native gammarid extinction

S. Beggel[1†], J. Brandner[2†], A. F. Cerwenka[1] and J. Geist[1*] ⓘ

Abstract

Background: Worldwide freshwater ecosystems are increasingly affected by invasive alien species. In particular, Ponto-Caspian gobiid fishes and amphipods are suspected to have pronounced effects on aquatic food webs. However, there is a lack of systematic studies mechanistically testing the potential synergistic effects of invasive species on native fauna. In this study we investigated the interrelations between the invasive amphipod *Dikerogammarus villosus* and the invasive fish species *Neogobius melanostomus* in their effects on the native amphipod *Gammarus pulex*. We hypothesized selective predation by the fish as a driver for displacement of native species resulting in potential extinction of *G. pulex*. The survival of *G. pulex* in the presence of *N. melanostomus* in relation to the presence of *D. villosus* and availability of shelter was analyzed in the context of behavioural differences between the amphipod species.

Results: *Gammarus pulex* had a significantly higher susceptibility to predation by *N. melanostomus* compared to *D. villosus* in all experiments, suggesting preferential predation by this fish on native gammarids. Furthermore, the presence of *D. villosus* significantly increased the vulnerability of *G. pulex* to fish predation. Habitat structure was an important factor for swimming activity of amphipods and their mortality, resulting in a threefold decrease in amphipods consumed with shelter habitat structures provided. Behavioral differences in swimming activity were additionally responsible for higher predation rates on *G. pulex*. Intraguild predation could be neglected within short experimental durations.

Conclusions: The results of this study provide evidence for synergistic effects of the two invasive Ponto-Caspian species on the native amphipod as an underlying process of species displacements during invasion processes. Prey behaviour and monotonous habitat structures additionally contribute to the decline of the native gammarid fauna in the upper Danube River and elsewhere.

Keywords: *Dikerogammarus villosus*, *Gammarus pulex*, *Neogobius melanostomus*, Selective predation, Anti-predator behaviour, Species displacement

Background

Worldwide freshwater ecosystems are undergoing major changes in biodiversity, mainly caused by anthropogenic habitat modification and biological invasions [1, 2]. Alteration of habitat and ship traffic are known to be mainly responsible for the dispersal of invasive alien species (IAS) in aquatic ecosystems. Transportation vessels are vectors for introduction of IAS via ballast water or ship-hull transfer from their origins to new areas [3]. In case of successful introduction, the establishment of IAS can result in significant declines of native taxa. Channelized rivers thereby provide both suitable habitat structures [4] as well as migration corridors by interconnecting catchments and enabling dispersal of non-native species [5].

Over the last two decades, ongoing invasions, especially by Ponto-Caspian crustaceans, molluscs and fishes have been reported from the middle and upper sections

*Correspondence: geist@wzw.tum.de
†S. Beggel and J. Brandner contributed equally to this work
[1] Aquatic Systems Biology Unit, School of Life Sciences Weihenstephan, Technical University of Munich, 85354 Freising, Germany
Full list of author information is available at the end of the article

of the Danube River [6–9], the Rhine River [10, 11] and other parts of the world [12, 13].

Among these successful invaders, the amphipod *Dikerogammarus villosus* (Sovinskij 1894) has frequently been proposed to affect native amphipod populations worldwide [5, 14–18]. Corresponding to its first records in the German sections of the Danube River in 1992 and the Rhine River in 1995 [19], significant decreases in abundance and distribution of indigenous amphipods such as *Gammarus pulex* L. 1758 and *Gammarus roeselii* Gervais 1835 have been observed [16, 20, 21], yet these declines have not been mechanistically linked to the simultaneous increase in the abundance of invasive species. Several recent studies identified functional feeding responses and asymmetric mutual predation of *D. villosus* as important mechanisms probably facilitating competitive advantages over other amphipods [21, 22]. Previous studies focused on selective predation of native and invasive amphipods by fishes, e.g., rainbow trout *Oncorhynchus mykiss*, European perch *Perca fluviatilis* [23], and burbot *Lota lota* [24]. Since these fishes only occur in low abundances and in certain areas of the upper Danube River, they can be ruled out as an explanation for the massive declines of native amphipod populations in this river section.

About one decade after the introduction of *D. villosus*, the round goby *Neogobius melanostomus* (Pallas 1814) arrived in the German section of the Danube River in 2004 [25], rapidly spread and displayed high levels of population differentiation [26, 27]. Within few years, this goby species strongly increased in abundance, currently contributing more than three quarters of abundance and about two thirds of biomass of fish in artificial rip-rap bank habitats [7, 8]. Its range expansion and population growth is still ongoing. Extensive samplings during the years 2010 to 2011 [7, 8] were no longer able to detect the formerly abundant native amphipod species, *G. pulex* and *Gammarus roeseli*, in the upper Danube River. Instead they only found the non-native *D. villosus* as the currently most abundant and widely distributed amphipod. Consequently, declines to extinction of native amphipods in the upper Danube River could have been influenced by the invasion of *N. melanostomus*, particularly since amphipods were identified as most important prey for this fish in different field surveys [7, 28, 29].

Since synergistic effects of multiple invasive species can potentially accelerate biodiversity loss and may enforce further homogenization of biological communities, the term "invasional meltdown" has been proposed for such interaction on an ecosystem scale [30]. However, to our knowledge, no study systematically analysed mechanisms and potential sympatric impact of *N. melanostomus* and

D. villosus on indigenous amphipods in experimental trials, to validate species interaction in native species displacement processes.

Invasional processes in general cannot be captured within a short time-frame and the assessment of possible negative consequences for a respective system can be rather complex. However, investigations of species interactions under controlled conditions have the potential to reveal mechanistic relationships that can support the holistic understanding of complex invasion processes.

In this study, selective predation of *N. melanostomus* on native and non-native amphipods was investigated in controlled experiments. The main goals of the present study were (i) to estimate predation preference of *N. melanostomus* towards one of the respective amphipod species, (ii) to determine potential interaction between invasive *D. villosus* and native *G. pulex*, and (iii) to test how potential interactions affect predation by *N. melanostomus*, as synergistic impact in native species displacement.

We hypothesized that (i) there is a feeding preference in *N. melanostomus* towards indigenous amphipod species explaining their massive declines in nature, and that (ii) *D. villosus* can increase the relative predation risk of *G. pulex* by interference competition and a more effective predator avoidance due to the sympatric origin with *N. melanostomus*.

Methods

Test species

Neogobius melanostomus were collected by electrofishing in the upper Danube River (Bad Abbach, Germany: N48°57′11.56″, E11°59′12.53″) under license number 31-7562. The collection of gobies was additionally approved by the local owner of the fisheries rights (Kreisfischereiverein Kelheim e.V.) and the fisheries authority (Fischereifachberatung Niederbayern). Specimens with a mean total length (L_T) of 10.0 ± 2.0 cm were used since this size class is known to preferentially feed on amphipods [7]. After transportation to the laboratory, fish were acclimatised in aerated aquaria (100 × 40 × 50 cm, density of 20 to 30 fish/m², each equipped with 10 clay tubes as shelter) for 1 week prior to the experiments. During the acclimatisation period fish were fed with commercial fish-food (trout chow, Skretting, Norway). Invasive *D. villosus* were collected in the upper Danube River near Kelheim, Germany (N48°54′56.61″, E11°51′43.80″). Since no *G. pulex* were found in the main channel of the Danube River in recent years, we collected them from a small headwater tributary of the Danube River system, the Moosach River (Freising, Germany: N48°23′38.83″, E11°43′26.15″). Accordingly, the native amphipods were completely "naïve" to the other test species. After kick-sampling, amphipods were sorted into two size classes (by sieving with mesh sizes of 2 and 1.5 mm) and transported to the laboratory in an aerated cooler. The smaller size class (size class 1) had a mean $L_T \pm$ SD of 11.8 ± 1.5 mm for G. pulex and 12.0 ± 1.4 mm for D. villosus, (measured on random samples of n = 30 per species and size class).The larger size class (size class 2) in both species had a mean $L_T \pm$ SD of 14.8 ± 1.3 mm for G. pulex and 18.3 ± 2.4 mm for D. villosus. Both amphipod species were acclimatised to laboratory conditions in separated aerated aquaria ($60 \times 30 \times 30$ cm with coarse pebbles as substratum, resembling their natural shelter) for 24–48 h. During the acclimatisation period, amphipods were fed ad libitum with pre-conditioned black alder-leaves (Alnus glutinosa) and fish-food (trout chow, Skretting, Norway). Test species collection and experimental testing was conducted in August and September 2012.

Test conditions

Experimental trials were conducted under constant physicochemical conditions (mean $\pm$ SD; temperature (T)12.3 °C $\pm$ 0.5 °C, dissolved oxygen (DO) 8.9 ± 1.2 mg L^{-1}, electric conductivity (EC, at 25 °C) 1125 ± 5 μS cm^{-1} using local well water. Ionic composition of the water is given in Table 1. Light conditions were 12:12 h dark:light. The test setup consisted of 14 aquaria ($40 \times 25 \times 25$ cm) individually supplied with a constant water flow(-through) of 0.8 L min^{-1}. Adjacent aquaria were shielded from each other to avoid learning effects between the test organisms of different treatments and to avoid mutual disturbance. Each aquarium was filled with 1.5 L of pebbles (16–32 mm) as substratum ("substratum present") or kept without substratum ("substratum absent") and equipped with a clay tube providing shelter for the goby (open side facing back wall). All experimental trials were conducted consequently under daylight conditions to minimize variation due to diel changes in the organisms' behaviour [13, 31]. The experiments were conducted according to German Tierschutzgesetz (§11 TierSchG), approved by the local veterinary board (Landratsamt Freising, license number 32-568) and the animal welfare committee at TUM.

Preliminary experiments

The optimum duration of the feeding trial was determined by preliminary experiments that were not included in the statistical analyses. Duration of 3 h was identified to be most suitable for the quantification of consumption rates (20.7 % $\pm$ 9 SD), based on the criteria that 100 % of the fish were feeding and gut contents were not fully digested to allow prey identification.

Table 1 Water chemistry parameters

Parameter	Concentration [mg L^{-1}]
Sodium (Na^+)	42.3
Potassium (K^+)	9.6
Calcium (Ca^{2+})	123
Magnesium (Mg^{2+})	41.6
Iron (Fe^{3+})	<0.1
Manganese (Mn^{2+})	<0.05
Chloride (Cl^-)	130
Bromide (Br^-)	<0.05
Hydrogen carbonate (HCO_3^-)	311
Sulfate (SO_4^{2-})	58
Nitrate (NO_3^-)	10.4
Ortho-phosphate (PO_4^{3-})	0.09
Ammonium (NH_4^+)	<0.02
Dissolved organic carbon (DOC)	0.94
Total inorganic carbon (TIC)	70.6

Amphipod mortality without predator

Several experimental trials were conducted to test for gammarid mortality without the presence of N. melanostomus caused by random mortality or intraguild predation (IGP, [32]). Each experimental trial consisted of 15 (single species trial) or 30 (mixed species trial) replicates, respectively. Each replicate was stocked with a total of 40 individuals, either 40 specimens of the same species (single species trial), or 20 specimens from both species (mixed species trial). In mixed species trials, two different experimental approaches were conducted: On the one hand, G. pulex and D. villosus of the same size class (size class 1 as described above) were used. In addition, D. villosus individuals of size class 2 vs. G. pulex of size class 1 were exposed to account for the different maximum sizes of the two species. Test duration of each trial was 3 h.

Predation experiments

Experimental trials were conducted in absence and in presence of substratum. Predation experimental trials consisted of 11 (single species trial) or 22 (mixed species trial) replicates without substratum, respectively. In experiments with substratum added, a higher number of replicates was used in the single species trials (G. pulex: 49; D. villosus: 14) due to the greater expected behavioural variability in these treatments with more complex habitat structure. Aquaria were stocked with 40 individuals of the same amphipod species (single species trial) or 20 each from both species (mixed species trial). In mixed species trials, either G. pulex and D. villosus of the same size class (size class 1) were used or size class 2 D. villosus individuals vs. size class 1 G. pulex to account for naturally expected body size-dependent effects. Predation experiments for different size classes were conducted as mixed species trial.

Gobies were not fed 24 h prior to the experiments. Amphipods transferred to the test aquaria were allowed to settle and shelter for 1 h before fish were added. Test duration of each trial was 3 h. Within this timeframe, fish exhibited natural behaviour and consumed amphipods, as expected from preliminary experiments. These short-time experiments allowed for an easy assignment of amphipod species identities and numbers in subsequent gut-content analyses, since prey was nearly undigested. After each feeding trial, fish were caught with a dip-net, euthanized and immediately frozen at -20 °C. The remaining living and dead amphipods in each aquarium were counted. Prior to dissection for stomach content analysis, L_T of the fish (to the nearest 1 mm) and total body mass (W_T to the nearest 0.001 g) were measured. Sex of the fish was determined using the morphology of the urogenital papilla according to Kornis et al. [13] before the experiments and later double-checked during dissection. Intestinal tracts (from the pharyngeal teeth to the anus) were removed, full and empty wet-weight was measured (nearest 0.001 g) to calculate the gut content mass (W_G).

Amphipod swimming behaviour

Experimental trials to quantify species-specific swimming behaviour and potential predator avoidance behaviour in both amphipod species were performed within 30 min intervals. For each amphipod species, 40 individuals were tested per trial with five replicates each. Gobies (n = 10) were not fed 24 h prior to the experiments. Amphipods transferred to the test aquaria were allowed to settle and shelter for 1 h before start of the trial. A single trial consisted of 15 min without plus 15 min with a single *N. melanostomus* present in the aquarium. A photograph was taken every minute to enable standardised counting of amphipods swimming freely in the water column. Experimental trials were conducted with substratum provided in the setup. Since amphipod species could not be differentiated exactly when observing them from outside the tanks, this experiment was not conducted for a mixed species setup.

Data analysis

To quantitatively test for differences in the overall feeding between *N. melanostomus* sexes within the experimental timeframe, the index of stomach fullness (ISF) = $100\ W_G\ {}^*W_T^{-1}$ was calculated, providing a standardized and representative estimate of prey consumption

[7]. The general assumption was tested that females have a higher feed-uptake compared with males, due to a higher energy demand and potential differences in W_T due to variable fish fecundity during the reproductive state in the respective time of the year. Since also single fish and amphipod biomasses were not identical in the experiments, the use of ISF (instead of amphipod counts) to quantify prey consumption enabled higher statistical power, since extremes could be included. For the comparison of differences in feeding on different species and size classes of amphipods, consumption ratios were used as dataset.

Selectivity of goby predation was evaluated using Manly's selectivity index α (discussed by Chesson [33]), calculated according to Eq. 1:

$$\alpha_i = \frac{\ln(\frac{n_{i0}-r_i}{n_{i0}})}{\sum_{j=1}^{m}\ln(n_{j0}-r_j)/n_{j0}}, \quad i = 1, 2, \ldots, m \qquad (1)$$

where n_{i0} is the initial number of prey species i, r_i is the number of prey species i consumed by the goby and m is the number of species, which is two in this study.

In some cases all individuals of the preferred prey species were consumed in the trial, so calculation of α was not possible. To account for that, the approach by Klecka and Boukal [34] was applied, modifying the equation in single cases. Therefore one additional individual was added to the respective n_{i0} and n_{j0}, assuming this additional individual would have survived. This was the case for two trials in which all *G. pulex* were consumed, resulting in a slightly conservative estimate of α. Dead individuals were excluded from this calculation. Statistical analysis were performed on Manly's alpha (α). Differences in selective feeding of the fish were tested by comparing calculated Manly's α values against a hypothetical value of 0.5 (no selectivity) using Wilcoxon's one sample test. For graphical presentation, the alpha values were converted in electivity indices as described in Chesson [33].

$$\varepsilon_i = \frac{m\alpha_i - 1}{(m-2)\alpha + 1}, \quad i = 1, \ldots, m, \qquad (2)$$

Electivity index ε can reach values between -1 and 1 per prey species, where positive values indicate a preferred prey and a value of 0 corresponding to unselective feeding.

All datasets were tested for normal distribution and homogeneity of variance using Shapiro-Wilks and Levene's test, respectively. For comparison of ISF and sex-specific predation, pairwise t tests were used. As assumptions for parametric tests were not met for comparison of mortality rates in control trials without fish, nonparametric Kruskal–Wallis-test was used for multiple comparisons. For post hoc comparisons, Bonferroni-corrected Mann–Whitney-U tests were used. For comparison of amphipod consumption rate of gobies a factorial ANOVA design was chosen, using the factors species (*G. pulex*, *D. villosus*), substratum (absence, presence) and trial (single, mixed). Since the data was skewed, rank transformation of the data was used to overcome violation of ANOVA assumptions. We applied an aligned rank transformation procedure by Wobbrock et al. [35], which is suitable for factorial designs including interactions, since the same main effect, and interaction structure as the original data is retained. Analyses were conducted on partial datasets since including single species and mixed species trials in the same model was not possible, due to the non-independent data. Species were analysed separately using a between subject design with the factors substratum (absence, presence) and trial (single, mixed). *Gammarus pulex* and *D. villosus* were compared with each other separately in single species trials, using a between-subject-design with the factors species (*G. pulex*, *D. villosus*) and substratum (absence, presence). Subsequently, one-way ANOVAs were conducted to compare differences in single vs. mixed trials with and without substratum separately. Mixed species trials for both same and different size-class comparisons were tested using a within-subject-design (repeated measures ANOVA) and are presented separately for each size class comparison.

To test for differences in swimming activity and potential predator avoidance of amphipods, Wilcoxon signed-rank test was applied to compare numbers of free-swimming gammarids with and without fish present (average of 15 observations in 5 tanks). Statistical analyses were performed using the software SPSS 22 (IBM, USA). Significance was accepted at $p < 0.05$.

Results

Amphipod mortality without predator

No significant predation effects (either asymmetric predation by IGP or symmetric by cannibalism) between the tested amphipod species were observed in the trials with the same size class (size class 1) of amphipods, within the test duration of 3 h. Control trials without predator presence revealed similar mortalities in both *G. pulex* and *D. villosus* which remained on average below 2 % (Table 2A). No significant differences were found neither between trials with and without substratum (single trials, Mann–Whitney: *G. pulex*, $z = -0.83$, $p > 0.05$; *D. villosus*, $z = -0.27$, $p > 0.05$; mixed trials, Mann–Whitney: *G. pulex*, $z = -1.76$, $p > 0.05$; *D. villosus*, $z = -2.46$, $p > 0.05$), nor between single- or mixed-species trials (without substratum, Mann–Whitney: *G. pulex*, $z = -1.78$, *D. villosus*, $z = -2.93$, $p > 0.05$; with

Table 2 Relative comparison of amphipod losses found after 3 h under experimental conditions without *N. melanostomus* presence

	Trial	Substratum	Mortality [%] mean (± SD)		N
			G. pulex	*D. villosus*	
A	Single	Presence	1.8 (±3.0)	1.2 (±2.5)	17
	Mixed	Presence	1.3 (±2.7)	1.3 (±3.2)	24
	Single	Absence	0.9 (±2.0)	1.0 (±2.2)	17
	Mixed	Absence	0.3 (±1.3)	0.0 (±0.0)	29
B	Single	Presence	1.8 (±3.0)	1.7 (±2.1)	6
	Mixed	Presence	0.7 (±1.8)	0.7 (±1.8)	16
	Single	Absence	0.9 (±2.0)	1.3 (±1.4)	6
	Mixed	Absence	2.5 (±4.1)	1.9 (±3.1)	15

A G. pulex and *D. villosus* with the same size class, *B D. villosus* larger than *G. pulex*. *Single* only one species per aquarium, *Mixed* both species in aquarium

substratum, Mann–Whitney: *G. pulex*, $z = -0.94$, $p > 0.05$; *D. villosus*, $z = -0.67$, $p > 0.05$). Control trials with different size classes of the two amphipod species (e.g. *G. pulex* size class 1, *D. villosus* size class 2 resulted in different mortality rates in comparison to trials with equal size distribution of both species. Higher mortality rates for both species were recorded in the mixed species trial without the presence of substratum (Table 2B), without being statistically significant. Mortality was highest for *G. pulex* (2.5 %) when both amphipod species were held together in the absence of substratum. As above, no significant differences could be found neither between trials with and without substratum (single trials, Mann–Whitney: *D. villosus*, $z = -0.18$, $p > 0.05$; mixed trials, Mann–Whitney: *G. pulex*, $z = -1.29$, $p > 0.05$; *D. villosus*, $z = -1.19$, $p > 0.05$) nor between mixed- or single-species trials (without substratum, Mann–Whitney: *G. pulex*, $z = -0.83$, *D. villosus*, $z = -0.08$, $p > 0.05$; with substratum, Mann–Whitney: *G. pulex*, $z = -1.39$, $p > 0.05$; *D. villosus*, $z = -1.56$, $p > 0.05$).

Predation experiments

Sex-specific predation

Comparing the ISF (Fig. 1), no significant differences in prey consumption between female (mean ± SD, 4.3 ± 1.5) and male (3.3 ± 1.1) fish were observed in experiments, neither without ($T_{(20)} = 1.552$, $p = 0.136$) nor with substratum ($T_{(20)} = 0.532$, $p = 0.600$). Mean ISF ± SD ranged between 1.50 ± 1.1 for female and 1.52 ± 1.7 for male *N. melanostomus* in presence of substratum. In case of generally reduced availability of amphipods in the open water, i.e. when substratum was provided, ISF was about two times lower than in tests without substratum (Fig. 1). In female *N. melanostomus*, the ISF was significantly ($T_{(24)} = 5.259$, $p < 0.001$)

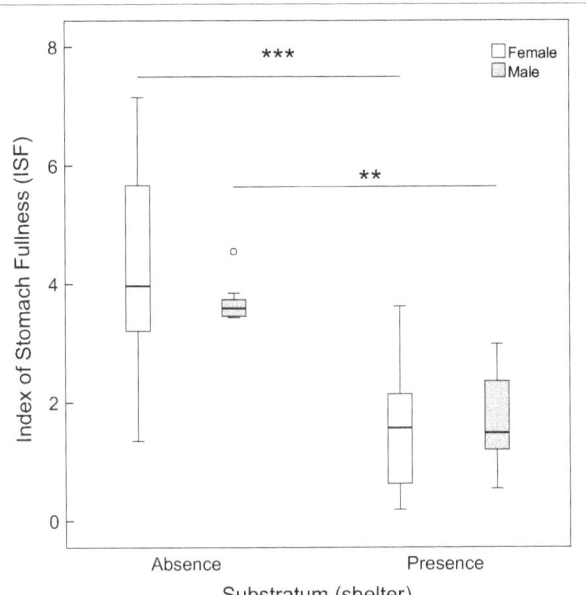

Fig. 1 Influence of fish sex on gammarid predation depending on absence or presence of substratum. Index of stomach fullness (ISF) confirms a generally higher predation rate in absence of substratum. No differences between female (*white*) and male (*grey*) *N. melanostomus* were observed. [*Asterisks* indicate significant differences at p < 0.05 (*), p < 0.01 (**), p < 0.001 (***)]. *Boxplots* represent 25 to 75 % (*boxes*) and 5 to 95 % percentiles (*whiskers*). *Circles* represent outliers (exceeding 1.5 interquartile range). n = 14: female/substratum absent; n = 8: male/substratum absent; n = 12: female/substratum present; n = 10: male/substratum present)

Table 3 Comparison of predation rates of *N. melanostomus* on *Gammarus pulex* based on factorial ANOVA with the factors trial (single species vs. mixed species) and substratum (absence vs. presence)

Factor	SS	df	MS	F	p
Trial					
	7284.869	1	7284.869	2.565	0.112
Error	283960.694	100	2839.607		
Substratum					
	29647.329	1	29647.329	12.256	*0.001*
Error	241893.087	100	2418.931		
Interaction trial x substratum					
	20296.829	1	20296.829	6.031	*0.016*
Error	336567.269	100	3365.673		

Italic values indicate significance of p value (p < 0.05)

SS sum of squares, df degrees of freedom, MS mean square, F F value, p p value of single factors and their interactions, respectively

lower in treatments with substratum than in treatments without substratum. This finding was also significant ($T_{(16)} = 3.538$, p = 0.003) in tests with male *N. melanostomus*.

Between-subject comparison *Gammarus pulex*

The main factor substratum had significant effects on the amount of *G. pulex* consumed by *N. melanostomus* ($F_{(1,100)} = 12.26$, p = 0.001, Table 3). No significant effects of the main factor trial were observed, but a significant interaction between the factors substratum and trial was evident ($F_{(1,100)} = 6.03$, p = 0.03). Separate examination of the datasets with and without substratum showed that, without substratum, no statistically significant difference was evident between single and mixed trials (Fig. 2). *Neogobius melanostomus* consumed 37 ± 14.7 (mean ± SD) and 58.2 ± 24.9 % of *G. pulex*, respectively. In contrast, a significant difference could be observed in the presence of substratum between single and mixed species trials ($F_{(1,\ 69)} = 22.05$, p < 0.001). Gobies consumed 8.6 ± 12.8 (mean ± SD) percent of *G. pulex* in single trials and 27.3 ± 18.2 percent of *G. pulex* when held together with *D. villosus*.

Between-subject comparison *Dikerogammarus villosus*

Similar to the results of *G. pulex*, strongest effects on *D. villosus* predation were observed for the main factor substratum ($F_{(1,65)} = 22.97$, p < 0.001; Table 4). Also the main factor trial showed to be significant ($F_{(1,65)} = 6.88$, p = 0.01), but no significant interaction between the main factors was observed. Separate examination of the datasets without and with substratum showed that, without substratum there were no significant differences in consumed *D. villosus* from single and mixed trials (Fig. 2). *N. melanostomus* consumed 9.5 ± 9.3 (mean ± SD) and 16.4 ± 11.7 percent of *D. villosus*, respectively. In the presence of substratum as shelter single and mixed trials were significantly different ($F_{(1,34)} = 5.12$, p = 0.03). *N. melanostomus* consumed 2.3 ± 3.5 percent of *D. villosus* in single and 6.8 ± 7.6 in mixed trials.

Between-subject comparison—single species trials (*G. pulex* vs. *D. villosus*)

The comparison of predation rates showed that both main factors species and substratum had strong effects of amphipod consumption by *N. melanostomus* (species: $F_{(1,81)} = 59.16$, p < 0.001; substratum: $F_{(1,81)} = 48.91$, p < 0.001; Table 5). The observed feeding pattern in *N. melanostomus* revealed a clear preference for *G. pulex* in the experimental trials with amphipods of the same size-class and a higher susceptibility of this native amphipod compared to non-native *D. villosus* to predation in the absence of substratum (Fig. 2). Also a significant interaction of the factors species and substratum was observed ($F_{(1,81)} = 7.14$, p = 0.009).

Separate examination of the datasets with and without substratum showed significant differences between consumed *G. pulex* and *D. villosus* in the absence

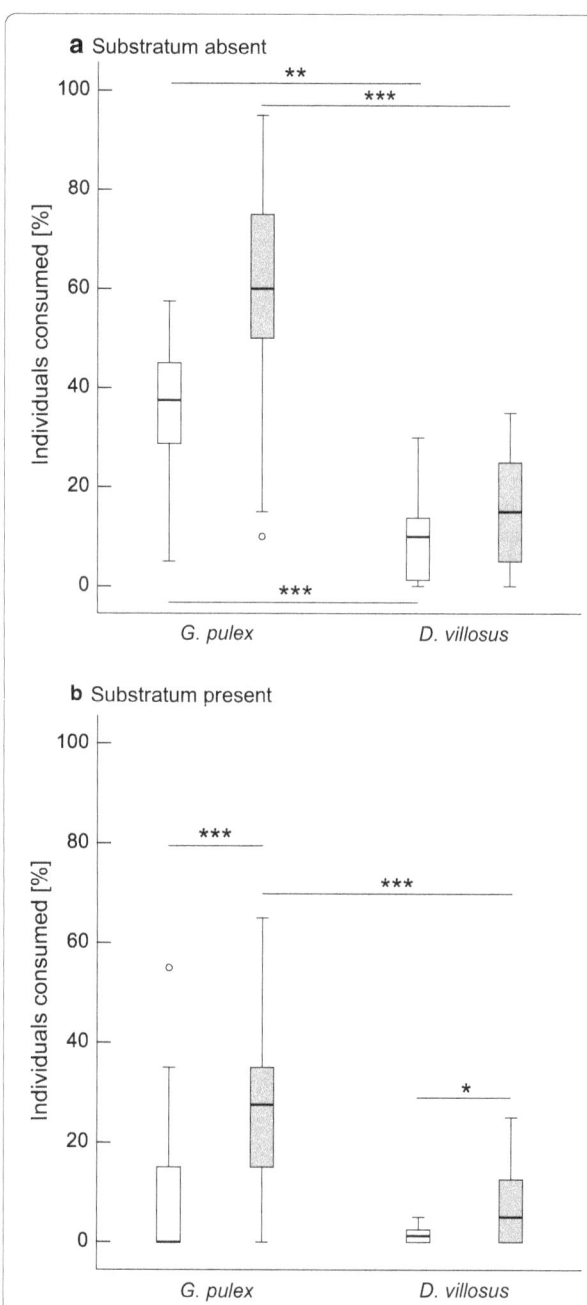

Fig. 2 Predation of *N. melanostomus* on amphipods within a 3 h time-period where either single species (*white*, n = 11) or a combination of both species (*grey*, n = 22) were tested. **a** Substratum absent. **b** Substratum present. [*Asterisks* indicate significant differences at p < 0.05 (*), p < 0.01 (**), p < 0.001 (***)]. *Boxplots* represent 25 to 75 % (*boxes*) and 5 to 95 % percentiles (*whiskers*). *Circles* represent outliers (exceeding 1.5 interquartile range)

Table 4 Comparison of predation rates of *N. melanostomus* on *Dikerogammarus villosus* based on factorial ANOVA with the factors trial (single species vs. mixed species) and substratum (absence vs. presence)

Factor	SS	df	MS	F	p
Trial					
	20544.067	1	20544.067	6.881	*0.011*
Error	194070.532	65	2985.7		
Substratum					
	61621.819	1	61621.819	22.974	*<0.001*
Error	174342.766	65	2418.931		
Interaction trial x substratum					
	5549.039	1	5549.039	1.982	0.164
Error	181939.994	65	2799.077		

Italic values indicate significance of p value (p < 0.05)

SS sum of squares, df degrees of freedom, MS mean square, F F value, p p value of single factors and their interactions, respectively

Table 5 Comparison of predation rates of *N. melanostomus* for single species predation trials based on factorial ANOVA with the factors species (*G. pulex* vs. *D. villosus*) and substratum (absence vs. presence)

Factor	SS	df	MS	F	p
Species					
	59950.019	1	59950.019	59.165	*p < 0.001*
Error	82074.539	81	1013.266		
Substratum					
	94868.898	1	94868.898	48.913	*p < 0.001*
Error	157103.081	81	1939.544		
Interaction species x substratum					
	12026.53	1	12026.53	7.145	*0.009*
Error	136349.47	81	1683.327		

Italic values indicate significance of p value (p < 0.05)

SS sum of squares, df degrees of freedom, MS mean square, F F value, p p value of single factors and their interactions, respectively

(mean ± SD) amphipod specimens per trial (without substratum) and 4 ± 5 specimens (with substratum present), respectively. If *D. villosus* was the only prey species available, 4 ± 4 (mean ± SD) individuals were consumed without and 1 ± 1 with substratum present. In the latter case, several gobies (21 %) with empty guts were recorded.

Within-subject comparison—mixed species trials (G. pulex vs. D. villosus)

In the mixed species trials, i.e. when both amphipod species were available as prey, round gobies clearly preferred *G. pulex* as prey (Fig. 2). This effect was even more pronounced in the presence of substratum, likely

of substratum ($F_{(1,20)}$ = 12.36, p = 0.002). This was even more pronounced in the presence of substratum ($F_{(1,61)}$ = 60.31, p < 0.001). When *G. pulex* was the only food source available, each goby consumed 15 ± 6 due to biological interactions between the both amphipod species within the shelter (Fig. 2). Predation rates were significantly higher on *G. pulex* (percent consumed: median 47.5, mean 42.7, SD 26.6) compared to *D. villosus* (percent consumed: median 10, mean 11.6, SD 10.9), $F_{(1,43)} = 31.34$, p < 0.001, with an effect size of 0.42. Separate examination of the datasets with and without substratum showed that without substratum, in 95.4 % cases more *G. pulex* were consumed and there was a significant higher predation on *G. pulex* (percent consumed: median 60, mean 58.2, SD 24.9) compared to *D. villosus* (percent consumed: median 15, mean 16.4, SD 11.7), $F_{(1,21)} = 9.5$, p = 0.006, with an effect size 0.31). With substratum, in 90.9 % cases more *G. pulex* were consumed and in no case more *D. villosus* were consumed. In 9.1 % of the cases, no difference was found. With substratum, there was a significant higher predation on *G. pulex* (percent consumed: median 27.5, mean 27.3, SD 18.2) compared to *D. villosus* (percent consumed: median 5, mean 6.8, SD 7.6), $F_{(1,21)} = 35.24$, p < 0.001, and the difference was large (effect size 0.63).

Size effects
In the mixed trials with larger *D. villosus* (size class 2) vs. smaller *G. pulex* (size class 1), similar trends were

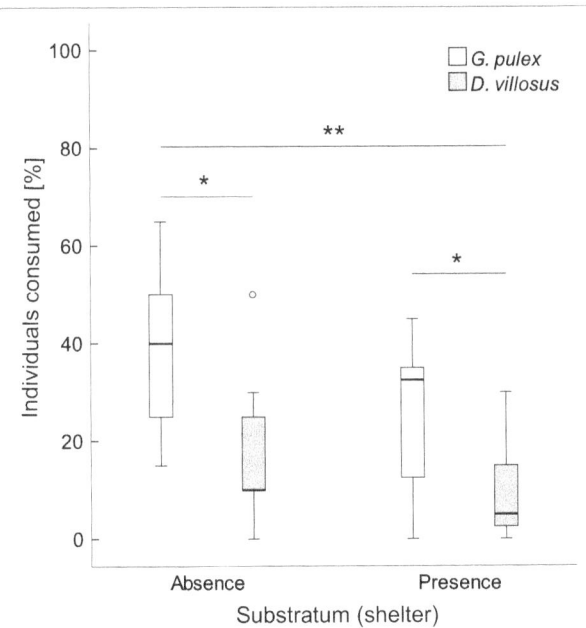

Fig. 3 Predation of *N. melanostomus* within a 3 h time-period on *G. pulex* (*grey*) and *D. villosus* (*white*). *Gammarus pulex* mean size (± SD) 11.8 ± 1.5 mm and *D. villosus* 18.3 ± 2.4 mm. [*Asterisks* indicate significant differences at p < 0.05 (*), p < 0.01 (**), p < 0.001 (***)]. *Boxplots* represent 25 to 75 % (*boxes*) and 5 to 95 % percentiles (*whiskers*). *Circles* represent outliers (exceeding 1.5 interquartile range); n = 12 per substratum trial)

observed as with equally sized amphipods (Fig. 3). Significantly more *G. pulex* were consumed as compared to *D. villosus*, regardless if substratum was absent ($F_{(1,10)} = 8.44$, p = 0.16, effect size 0.46) or present ($F_{(1,11)} = 24.68$, p < 0.001, effect size 0.69). Again, the availability of substratum as shelter resulted in a stronger effect on predation. The fish consumed more *G. pulex* (percent consumed: median 40, mean 38.2, SD 16.9) than *D. villosus* (percent consumed: median 10, mean 16.8, SD 14.7) in 83.3 % of the cases in the absence of shelter. With substratum provided, more *G. pulex* (percent consumed: median 32.5, mean 26.2, SD 14.8) than *D. villosus* (percent consumed: median 5, mean 8.7, SD 8.8) were consumed in 90.9 % of the cases.

Selectivity index
Selectivity analyses using Manly's alpha confirmed a clear preference for *G. pulex* as prey item in mixed species trials (Fig. 4). Alpha values above 0.5 indicate a preference for *G. pulex* and were highest in trials without substratum added and the same size-class of the two amphipod species (0.83 ± 0.18, mean ± SD) compared to trials with presence of substratum (0.81 ± 0.23, mean ± SD). Selectivity was significant for *G. pulex* as preferred prey item (without substratum: z = 3.94, p < 0.001; with substratum z = 3.49, p < 0.001). Interestingly, no significant preference was evident from the selectivity index observed for in the different size-class trials, but higher values in the presence of substratum (0.65 ± 0.37, mean ± SD) were observed compared to trials without substratum (0.50 ± 0.38, mean ± SD).

Amphipod swimming behaviour
Generally, behaviour of the amphipods differed significantly between species: *Gammarus pulex* was about three to four times more active in terms of individuals swimming freely in the water-column during the observation period (Fig. 5). After adding the predator, there was a significant reduction in freely swimming *G. pulex* (mean 2.53, SD 1.67 per min) compared to controls without gobies (mean 0.46, SD 0.27 per min, n = 5), z = −2.02, p = 0.042, and the difference was large (effect size r = −0.90).

Dikerogammarus villosus showed a generally low activity, even in the control group (mean 0.32, SD 0.05 per min, n = 5). After addition of the predator, this effect became more pronounced (mean 0.01, SD 0.03). Wilcoxon comparison revealed a significant difference in activity patterns caused by the presence of the fish (z = −2.03, p = 0.042, effect size r = −0.91). Pairwise comparison between species identified significant differences between the control groups (z = −2.63,

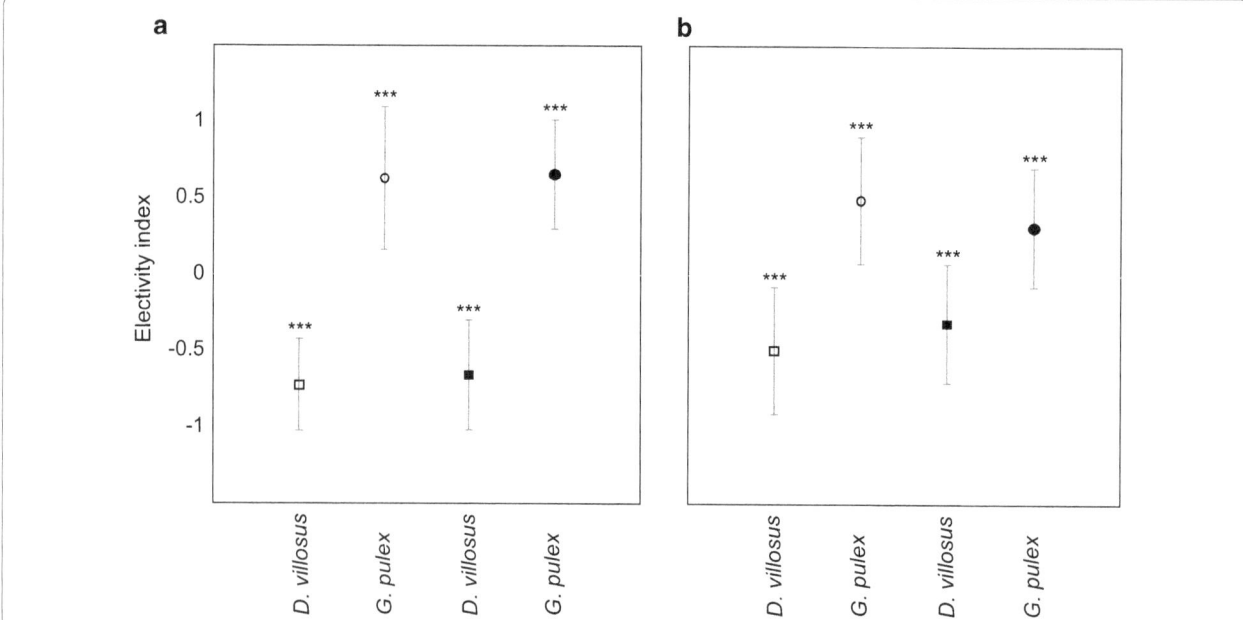

Fig. 4 Prey selectivity of *N. melanostomus*. Mean values ± SE of electivity index are presented. Positive values represent preferred prey species in presence (*open circles* and *squares*) or absence (*black circles* and *squares*) of substratum. *Dashed line* indicates no selectivity. **a** Same size class of *G. pulex* and *D. villosus*. **b** Larger size class of *D. villosus*. Asterisks indicate significant differences to zero at p < 0.05 (repeated measures ANOVA): (*), p < 0.01 (**), p < 0.001 (***)

p = 0.009, effect size r = −0.83) and in the presence of the fish (z = −2.12, p = 0.034, effect size r = −0.67).

Discussion

In several European waterbodies, the arrival of non-native *D. villosus* has simultaneous occurred as the disappearance of native amphipods [16, 36, 37] which is often explained by asymmetric mutual predation [24]. A sympatric and synergistic impact by invasive *D. villosus* and invasive *N. melanostomus* has not yet been considered, but appears likely based on the results of our study.

In our experiments, *D. villosus* faced a lower predation risk against *N. melanostomus* due to a relatively lower swimming activity as compared to *G. pulex*. These findings match obervations in other species pairs such as those by Kinzler and Maier [23] who detected similar results for rainbow trout and European perch. According to van Riel et al. [38], the presence of *D. villosus* can lead to an increased number of *G. pulex* swimming freely in the water column, indicating interference competition. Analogously, we observed a threefold higher predation of *N. melanostomus* on *G. pulex* in the presence of *D. villosus* when shelter was available compared to a single exposure of *G. pulex*. Since substrate structures are often heavily modified in natural habitats for bank erosion protection, this may directly affect predation risk of native gammarids, particularly in anthropogenically modified habitat structures.

The role of IGP

Dikerogammarus villosus is able to prey on other amphipod species without being restricted to vulnerable post-molting stages [36]. On the other hand, the mouthparts of *D. villosus* are not highly specialized just for carnivory and predation, and the species has been described to use a wide spectrum of different food sources [39, 40]. Also gut contents and stable isotope analyses by Koester and Gergs [41] did not provide evidence for *D. villosus* being a carnivorous "killer shrimp". In our study, no significant asymmetric mutual predation between both amphipod species could be confirmed within the short experimental duration. Gammarids are also known to naturally occur at similar densities and the selected experimental setup allowed for isolated analysis of the effects caused by *N. melanostomus* predation. Based upon the results of our experiments, intraguild predation seemingly only plays a minor role and cannot explain the invasion success of *D. villosus* outside of its natural distribution range. In contrast, interactions with *D. villosus* seem to facilitate excessive consumption of the native gammarid by the newly introduced top-predator *N. melanostomus* and appears to play a more important role in the decline of autochthonous amphipods.

Structural diversity of invaded habitats can mitigate potential impact by spatial separation due to different habitat preferences, since it uncouples competition between native and invasive species [42]. Especially in

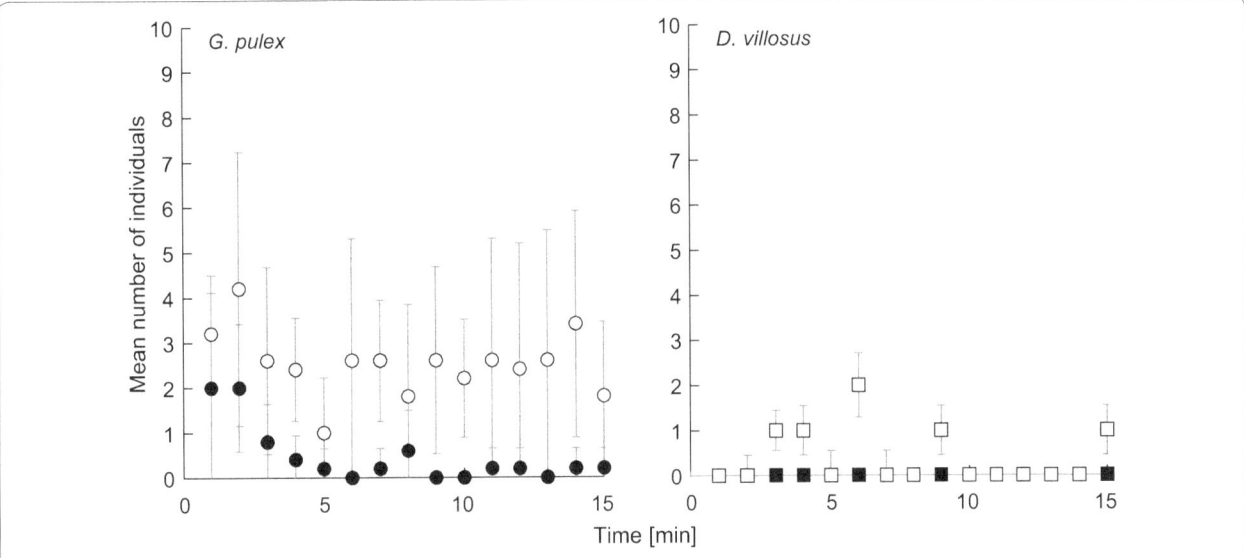

Fig. 5 Mean number (± SD) of individual amphipods swimming actively in the water column within a timeframe of 15 min in absence (*grey*) and 15 min in presence (*black*) of a single *N. melanostomus* specimen. (n = 5 per species)

heavily modified waterbodies such as shipping canals, structural diversity is much lower compared to natural aquatic systems. Technolithal and other monotonic structures within these waterbodies can thus force the respective species towards direct interaction and may consequently increase competitive stress in autochthonous, less generalized species.

Thus, main behavioural traits acting in combination can provide a plausible explanation for the observed eradication of *G. pulex* in natural habitats invaded by *D. villosus* and *N. melanostomus*. In case of *N. melanostomus* and *G. pulex*, preference for one species as a superior food item accompanied by higher susceptibility to predation by different activity patterns in swimming behaviour interacts synergistically and speeds up this process.

Comparing predation of *N. melanostomus* against *G. fasciatus,* an autochthonous amphipod species native to North America, and its Ponto-Caspian invasive counterpart *E. ischnus*, Kestrup and Ricciardi [43] found that competition with *E. ischnus* did not increase the vulnerability of *G. fasciatus* to goby predation. In their experiments *E. ischnus* and *G. fasciatus* appeared to be equally susceptible to goby predation and no preference in feeding of *N. melanostomus* was observed. Kestrup and Ricciardi [43] finally concluded that *N. melanostomus* did not influence the replacement of *G. fasciatus* by *E. ischnus* in the St. Lawrence River. Remarkably, outside laboratory conditions in Lake Erie and Lake Ontario, where *N. melanostomus* continues to spread, *E. ischnus* has replaced *G. fasciatus* as the dominant amphipod on substrates fouled by dreissenids [44]. These amphipods are

mutual predators, too, and dominance of competing amphipods varies with conductivity [45]. Thus, IGP can also be influenced by environmental heterogeneity.

Since *N. melanostomus* used in Kestrup and Ricciardi [43] had a significantly lower (mean = 6.6 cm ± 0.1 SE) L_T compared with our experiments, the experimental setup chosen might have led to misinterpretation: Due to a significantly smaller gape limitation, *N. melanostomus* of this size class preliminary feeds on smaller prey than adult amphipods, such as chelicorophiids, isopods, zooplankton and juvenile amphipods [7]. At a size of about 10–12 cm, depending on time since invasion, *N. melanostomus* is known to mainly feed on amphipods correlating with an ontogenetic diet switch [8]. Thus, in our experiments *N. melanostomus* and amphipods were size-selected with expected maximum impact.

Selective feeding by fish or different vulnerability
Among others, predation by fish is an important factor influencing the composition of amphipod communities. *Neogobius melanostomus* has a rather opportunistic feeding strategy, using the most available prey as compared to other more specialized goby species such as *P. kessleri* [7, 29]. However, *N. melanostomus* shows a clear preference for amphipods over molluscs at early stages of the invasion process when their availability is not limited [8]. Thus, according to our experiments, amphipods appear to be a superior and favourite prey for both sexes, likely facilitating the invasion success of invasive gobies [7, 28, 46]. The choice of amphipods as important prey item in our study appears legitimate since the proportion of

amphipods in *N. melanostomus* prey is highest in all seasons: about two thirds to three quarters to the total food uptake and amphipods were consumed in higher proportions as compared to their availability in environment [7]. Other potentially preferred prey items such as Chironomid larvae have limited availability due to emergence of imagos in summer and autumn. Amphipods can percept predatory fish or injured conspecifics by chemical stimuli and are thus able to react by predator avoidance behaviour (co-evolutionary adaptation) such as reducing time in open water [47], or change of habitat use [48]. Behavioural responses can even be flexible when complex microhabitat structures are available [49]. Since amphipods are known to possess distinct substrate preferences, the presence of preferred microhabitats can mitigate their predation risk [24]. *Dikerogammarus villosus* preferred gravel with a low percentage of sand and stones, whereas *G. pulex* showed no clear substratum preference in a substratum choice experiment [38]. However, in line with our study (see Fig. 5), in presence of invasive *D. villosus*, native *G. pulex* can be found significantly less frequent in gravel [50, 51] independent of *D. villosus* abundance [38, 41]. Such an exclusion from shelter-providing habitats can lead to increased swimming activity in *G. pulex* and consequently to an increased exposure to predatory fish. It might therefore play an important role in the displacement of native amphipods by *D. villosus* [51]. Besides the feeding behaviour, high reproductive potential and rapid growth of *D. villosus* [16, 52, 53] make this species a strong competitor not only for food but also for habitat space [41]. In our study, both sexes of invasive *N. melanostomus* asymmetrically consumed significantly more native *G. pulex* than invasive *D. villosus*, regardless if there was substratum present or not. Calculations of a preferential prey choice index [34] corroborated these findings. On a first glance, these findings are in contrast to Kley et al. [24], who considered that differences in use of spatial niches can permit the co-existence of *D. villosus* and *G. roeseli* in the wild when substrates are diverse. However in their tests, it was autochthonous burbot (*Lota lota*) that did not affect substrate choice or predation risk of *G. roeseli* in presence of *D. villosus*. It is probably due to the smaller prey selectivity of *L. lota* compared to the invasive *N. melanostomus* which can explain these differences. For freshwater amphipods, a predator avoiding mechanism would be to respond to an open water predator by reducing activity and moving towards the sediment, whereas a benthic predator would generally be better avoided by an active escapement through swimming or increased drift escape when predator abundance is low [47, 49]. In case of high or very high numbers of predators, most likely after mass development of invasive alien species, hiding seems to be a better

survival strategy. Proportion of invasive *N. melanostomus* to the total fish fauna can be extremely high, particularly in man-made rip rap habitats where they have been reported to contribute two thirds of all fish counts and about 50 % of total fish biomass [9]. Thus, drift might not work as a perfect escape mechanism. Instead, escaping by drift likely leads to increased predation when predator abundance is high and thus might favour extinction in invaded ecosystems. This theory matches actual field observations from the Danube River, where the breakdown of *G. roeseli* in sections invaded by non-native gobies is reported [7]. Considering these principles, unreduced activity seems to be mostly responsible for higher losses in native *G. pulex*, whereas *D. villosus* appears to be generally less active, consequently being less vulnerable to the benthic ambush predator *N. melanostomus*. According to foraging theories, it therefore appears to be more energy efficient for *N. melanostomus* to feed on native *G. pulex* as an "easy" prey.

The results from our experiments provide a mechanistic explanation for higher feeding rates on *G. pulex* by *N. melanostomus* in absence of suitable substratum which in pristine riverine environments could provide shelter against predation. The observed higher consumption of *G. pulex* by *N. melanostomus* indicates species-specific vulnerability due to gammarid acitivity. Both species similarly respond to a predator, but *D. villosus* is less vulnerable due to its overall lower activity. Consequently, this study provides evidence that a synergistic impact of invasive species with distinct foraging strategies can be greater than the independent effects of the single species. On an ecosystem scale such added effects can pose an important factor in the context of an "invasional meltdown" scenario as defined by Simberloff and Von Holle [30]. Such an invasional meltdown supposedly still seems to occur in the upper Danube River to date [8].

Conclusions

Behavioural traits and interactions between closely related species are important basic mechanisms in understanding species displacement by IAS. The results of this study suggest that both the foraging selectivity of invasive *N. melanostomus*, as well as the behavioural interaction between *D. villosus* and *G. pulex* together result in an increased predation risk for the native gammarid. Previous invasions of *D. villosus* thus likely facilitate the rapid range expansion of *N. melanostomus*, in turn increasing predation on autochthonous amphipods. The results of this study are thus in line with processes described as "invasional meltdown" on the scale of ecosystems. Not only competition for food and habitat resources as well as asymmetric mutual predation between the closely related (amphipod) species can lead to declines of less competitive autochthonous species, but also becoming an easy prey for a new predator enabling excessive resource consumption may lead to the extinction of autochthonous species. Sympatrically, both invasive Ponto-Caspian IAS possess the ability to effectively restructure food-web composition by mutually and synergistically facilitating their invasive potentials, thus promoting further range expansion of both species.

Abbreviations
ANOVA: analysis of variance; DO: dissolved oxygen; EC: electric conductivity; IAS: invasive alien species; IGP: intraguild predation; ISF: index of stomach fullness; L_T: total length; SE: standard error; SD: standard deviation; W_T: total body weight; W_G: gut content weight.

Authors' contributions
The study was conceived and designed by JG and JB. Field sampling of gobies and gammarids, as well as the laboratory work was conducted by SB, JB, and AFC. Data analyses was mostly conducted by SB, with continuous revision by JG and JB. SB, JG and JB jointly interpreted the data and drafted the manuscript which was finalized by SB and JG. All authors read and approved the final manuscript.

Author details
[1] Aquatic Systems Biology Unit, School of Life Sciences Weihenstephan, Technical University of Munich, 85354 Freising, Germany. [2] Wasserwirtschaftsamt Regensburg, Kavalleriestraße 2, 93053 Regensburg, Germany.

Acknowledgements
The authors like to express their special thanks to Lydia Schübel and Heidrun Kerber for their laboratory assistance. This project was partly funded by the Deutsche Forschungsgemeinschaft (DFG), project-numbers GE 2169/1-1(AOBJ: 569812) and DFG SCHL567/5-1.

Competing interests
The authors declare that they have no competing interests.

Animal ethics
This study was conducted in compliance with the international animal care guidelines of the Association for the Study of Animal Behaviour [55] and the ARRIVE guidelines [56]. Gobies were collected by electrofishing in the Danube under license number 31-7562. The collection was additionally approved by the local owner of the fisheries rights (Kreisfischereiverein Kelheim e.V.) and the state fisheries authority (Fischereifachberatung Niederbayern). Laboratory experiments were conducted according to German legislation (German Tierschutzgesetz, §11 TierSchG), license number 32-568), approved by the local veterinary board (Landratsamt Freising, license number 32-568) and the animal welfare committee at TUM.

References
1. Dudgeon D, Arthington AH, Gessner MO, Kawabata Z, Knowler DJ, Lévêque C, Naiman RJ, Prieur-Richard A, Soto D. Stiassny. freshwater biodiversity: importance, threats, status and conservation challenges. Biol Rev. 2006;8:163–82.
2. Keller RP, Geist J, Jeschke JM, Kühn I. Invasive species in Europe: ecology, status and policy. Environ Sci Eur. 2011;23(23):1–17.
3. Wolter C, Röhr F. Distribution history of non-native freshwater fish species in Germany: how invasive are they? J Appl Ichthyol. 2010;26(Suppl. 2):19–27.
4. Sindilariu PD, Freyhof J, Wolter C. Habitat use of juvenile fish in the lower Danube and the Danube Delta: implications for ecotone connectivity. Hydrobiologia. 2006;571:51–61.
5. Bijdevaate A, Jazdzewski K, Ketelaars HAM, Gollasch S, Vander Velde G. Geographical patterns in range extension of Ponto-Caspian macroinvertebrate species in Europe. Can J Fish Aquat Sci. 2002;59:1159–74.
6. Haertl M, Cerwenka AF, Brandner J, Borcherding J, Geist J, Schliewen UK, First record of Babka gymnotrachelus from Germany (Teleostei, Gobiidae, Benthophilinae). Spixiana. 2012;1857(35):155–9.
7. Brandner J, Auerswald K, Cerwenka AF, Schliewen U, Geist J. Comparative feeding ecology of invasive Ponto-Caspian gobies. Hydrobiologia. 2013;703:113–31.
8. Brandner J, Cerwenka AF, Schliewen UK, Geist J. Bigger is better: characteristics of round gobies forming an invasion front in the Danube River. PLoS ONE. 2013;8(9):e73036.
9. Brandner J, Pander J, Mueller M, Cerwenka A, Geist J. Effects of sampling techniques on population assessment of invasive round goby. J Fish Biol. 2013;82:2063–79.
10. Van Riel MC, Vander Velde G, Rajagopal S, Marguillier S, Dehairs F, Bijdevaate A. Trophic relationships in the Rhine food web during invasion and after establishment of the Ponto-Caspian invader Dikerogammarus villosus. Hydrobiologia. 2006;565:39–58.
11. Borcherding J, Staas S, Krüger S, Ondračková M, Šlapanský L, Jurajda P. Non-native Gobiid species in the lower River Rhine (Germany): recent range extensions and densities. J Appl Ichthyol. 2011;27:1–3.
12. Strayer DL. Twenty years of zebra mussels: lessons from the mollusk that made headlines. Front Ecol Environ. 2009;7:135–41.
13. Kornis MS, Mercado-Silva N, Vander Zanden MJ. Twenty years of invasion: a review of round goby Neogobius melanostomus biology, spread and ecological implications. J Fish Biol. 2012;80(2):235–85.
14. Ricciardi A, MacIsaac HJ. Recent mass invasion of the North American Great Lakes by Ponto-Caspian species. Trends Ecol Evol. 2000;15:62–5.
15. Dick JTA, Platvoet D, Kelly DW. Predatory impact of the freshwater invader Dikerogammarus villosus (Crustacea: Amphipoda). Can J Fish Aquat Sci. 2002;59:1078–84.
16. Kley A, Maier G. Reproductive characteristics of invasive gammarids in the Rhine-Main-Danube catchment, South Germany. Limnologica. 2006;36:79–90.
17. Bollache L, Dick JTA, Farnsworth KD, Montgomery WI. Comparison of the functional responses of invasive and native amphipods. Biol Lett. 2008;4:166–9.
18. Bacela-Spychalska K, Van der Velde G. There is more than one 'killer shrimp': trophic positions and predatory abilities of invasive amphipods of Ponto-Caspian origin. Freshw Biol. 2013;58:730–41.
19. Tittizer T, Schöll F, Banning M, Haybach A, Schleuter M. Aquatische Neozoen im Makrozoobenthos der Binnenwasserstraßen Deutschlands. Lauterbornia. 2000;39:1–172.
20. Haas G, Brunke M, Streit B. Fast turnover in dominance of exotic species in the Rhine River determines biodiversity and ecosystem function: an affair between amphipods and mussels. In: Leppäkoski E, Gollasch S, Olenin S, editors. Invasive Aquatic species of Europe. Distribution, impacts and management. Dordrecht: Kluwer Academic Publishers; 2002. p. 426–32.
21. Poeckl M. Success of the invasive Ponto-Caspian amphipod Dikerogammarus villosus by life history traits and reproductive capacity. Biol Invasions. 2009;11:2021–41.
22. Dodd JA, Dick JTA, Alexander ME, Macneil C, Dunn AM, Aldridge DC. Predicting the ecological impacts of a new freshwater invader: functional responses and prey selectivity of the 'killer shrimp', Dikerogammarus villosus, compared to the native Gammarus pulex. Freshw Biol. 2014;59:337–52.
23. Kinzler W, Maier G. Selective predation by fish: a further reason for the decline of native gammarids in the presence of invasives? J Limnol. 2006;65:27–34.
24. Kley A, Kinzler W, Schank Y, Mayer G, Waloszek D, Maier G. Influence of substrate preference and complexity on co-existence of two non-native gammarideans (Crustacea: Amphipoda). Aquat Ecol. 2009;43:1047–59.

25. Paintner S, Seifert K. First record of the round goby, *Neogobius melanostomus* (Gobiidae), in the German Danube. Lauterbornia. 2006;58:101–7.

26. Cerwenka AF, Alibert P, Brandner J, Geist J, Schliewen UK. Phenotypic differentiation of Ponto-Caspian gobies during a contemporary invasion of the upper Danube River. Hydrobiologia. 2014;721:269–84.

27. Cerwenka AF, Brandner J, Geist J, Schliewen UK. Strong versus weak population genetic differentiation after a recent invasion of gobiid fishes (*Neogobius melanostomus* and *Ponticola kessleri*) in the upper Danube. Aquat Invasions. 2014;9:71–86.

28. Borcherding J, Dolina M, Heermann L, Knutzen P, Krüger S, Matern S, van Treeck R, Gertzen S. Feeding and niche differentiation in three invasive gobies in the Lower Rhine. Germany. Limnologica. 2012;43:49–58.

29. Brandner J, Auerswald K, Schaeufele R, Cerwenka AF, Geist J. Isotope evidence for preferential dispersal of fast-spreading invasive gobies along man-made river bank structures. Isot Environ Healt S. 2015;51(1):80–92.

30. Simberloff D, Von Holle B. Positive interactions of nonindigenous species: invasional meltdown? Biol Invasions. 1999;1:21–32.

31. Elliott JM. Day-night changes in the spatial distribution and habitat preferences of freshwater shrimps, *Gammarus pulex*, in a stony stream. Freshw Biol. 2005;50:552–66.

32. Polis GA, Myers CA, Holt RD. The ecology and evolution of intraguild predation: potential competitors that eat each other. Annu Rev Ecol Syst. 1089;20:297–330.

33. Chesson J. The estimation and analysis of preference and its relationship to foraging models. Ecology. 1983;64(4):1297–304.

34. Klecka J, Boukal DS. Who eats whom in a pool? A comparative study of prey selectivity by predatory aquatic insects. PLoS ONE. 2012;6(7):e37741.

35. Wobbrock JO, Findlater L, Gergle D, Higgins JJ. The aligned rank transform for nonparametric factorial analyses using only ANOVA procedures. Proceedings of the ACM Conference on Human Factors in Computing Systems (CHI'11). Vancouver, British Columbia (May 7-12, 2011). New York: ACM Press; 2011: 143–46.

36. Dick JTA, Platvoet D. Invading predatory crustacean *Dikerogammarus villosus* eliminates both native and exotic species. P Roy Soc B. 2000;267:977–83.

37. Kley A, Maier G. Life history characteristics of the invasive freshwater gamarids *Dikerogammarus villosus* and *Echinogammarus ischnus* in the river Main and the Main-Donau canal. Arch Hydrobiol. 2003;156:473–81.

38. Vanriel M, Healy EP, vander Velde G, Bijdevaate A. Interference competition among native and invader amphipods. Acta Oecol. 2007;31:282–9.

39. Mayer G, Maier G, Maas A, Waloszek D. Mouthparts of the Ponto-Caspian invader *Dikerogammarus villosus* (Amphipoda: Pontogammaridae). J Crustacean Biol. 2008;28:1–15.

40. Boeker C, Geist J. Effects of invasive and indigenous amphipods on physico-chemical and microbial properties in freshwater substrates. Aquat Ecol. 2015. doi:10.1007/s10452-015-9539-y.

41. Koester M, Gergs R. No evidence for intraguild predation of *Dikerogammarus villosus* (Sowinsky, 1894) at an invasion front in the Untere Lorze, Switzerland. Aquat Invasions. 2014;9:489–97.

42. Harrison SSC, Bradley DC, Harris IT. Uncoupling Strong predator-prey interactions in streams: the role of marginal macrophytes. Oikos. 2005;108:433–48.

43. Kestrup Å, Ricciardi R. Are interactions among Ponto-Caspian invaders driving amphipod species replacement in the St. Lawrence River? J Great Lakes Res. 2009;35:392–8.

44. Dermott R, Witt J, Um EM, Gonzalez M. Distribution of the Ponto-Caspian amphipod *Echinogammarus ischnus* in the Great Lakes and replacement of native *Gammarus fasciatus*. J Great Lakes Res. 1998;24:442–52.

45. Kestrup Å, Ricciardi R. Environmental heterogeneity limits the local dominance of an invasive freshwater crustacean. Biol Invasions. 2009;11:2095–105.

46. Polačik M, Janáč M, Jurajda P, Adámek Z, Ondračková M, Trichkova T, Vassilev M. Invasive gobies in the Danube: invasion success facilitated by availability and selection of superior food resources. Ecol Freshw Fish. 2009;18:640–9.

47. Wudkevich K, Wisenden BD, Chivers DP, Smith RJF. Reactions of *Gammarus lacustris* to chemical stimuli from natural predators and injured conspecifics. J Chem Ecol. 1997;23:1163–73.

48. Dahl J, Greenberg L. Effects of habitat structure on habitat use by *Gammarus pulex* in artificial streams. Freshw Biol. 1996;36:487–95.

49. Baumgärtner D, Koch U, Rothhaupt KO. Alteration of kairomone-induced antipredator response of the freshwater amphipod *Gammarus roeseli* by sediment type. J Chem Ecol. 2003;29(6):1391–401.

50. MacNeil C, Platvoet D. The predatory impact of the freshwater invader *Dikerogammarus villosus* on native *Gammarus pulex* (Crustacea: Amphipoda); influences of differential microdistribution and food resources. J Zool. 2005;267:31–8.

51. Boets P, Lock K, Messiaen M, Goethals PLM. Combining data-driven methods and lab studies to analyse the ecology of *Dikerogammarus villosus*. Ecol Inform. 2010;5:133–9.

52. Grabowski M, Bacela K, Konopacka A. How to be an invasive gammarid (Amphipoda: Gammaroidea)—comparison of life history traits. Hydrobiologia. 2007;590:75–84.

53. Poeckl M. Strategies of a successful new invader in European fresh waters: fecundity and reproductive potential of the Ponto-Caspian amphipod *Dikerogammarus villosus* in the Austrian Danube, compared with the indigenous *Gammarus fossarum* and *G. roeseli*. Freshw Biol. 2007;52:50–63.

54. Beggel S, Brandner J, Cerwenka A, Geist J. Data from: synergistic impacts by an invasive amphipod and an invasive fish explain native gammarid extinction. Dryad Digital Repository. http://dx.doi.org/10.5061/dryad.c5m0r.

55. Guidelines for the treatment of animals in behavioural research and teaching. Anim Behav. 2012; 83:301–09. doi:10.1016/j.anbehav.2011.10.031.

56. Kilkenny C, Browne WJ, Cuthill IC, Emerson M, Altman DG. Improving bioscience research reporting: the ARRIVE guidelines for reporting animal research. PLoS Biol. 2010;8:e1000412. doi:10.1371/journal.pbio.1000412.

17

Fish with red fluorescent eyes forage more efficiently under dim, blue-green light conditions

Ulrike Katharina Harant[1,2]* and Nicolaas Karel Michiels[1,2]

Abstract

Background: Natural red fluorescence is particularly conspicuous in the eyes of some small, benthic, predatory fishes. Fluorescence also increases in relative efficiency with increasing depth, which has generated speculation about its possible function as a "light organ" to detect cryptic organisms under bluish light. Here we investigate whether foraging success is improved under ambient conditions that make red fluorescence stand out more, using the triplefin *Tripterygion delaisi* as a model system. We repeatedly presented 10 copepods to individual fish ($n = 40$) kept under a narrow blue-green spectrum and compared their performance with that under a broad spectrum with the same overall brightness. The experiment was repeated for two levels of brightness, a shaded one representing 0.4% of the light present at the surface and a heavily shaded one with about 0.01% of the surface brightness.

Results: Fish were 7% more successful at catching copepods under the narrow, fluorescence-friendly spectrum than under the broad spectrum. However, this effect was significant under the heavily shaded light treatment only.

Conclusions: This outcome corroborates previous predictions that fluorescence may be an adaptation to blue-green, heavily shaded environments, which coincides with the opportunistic biology of this species that lives in the transition zone between exposed and heavily shaded microhabitats.

Keywords: Foraging success, Visual ecology, *Tripterygion delaisi*

Background

Fluorescence is a common form of luminescence that can be found throughout the entire biotic world [1]. The functionality of fluorescence for intra-specific communication has already been studied in a variety of organisms within terrestrial as well as aquatic habitats [2–4]. Especially in aquatic environments, where long wavelengths are quickly absorbed, fluorescence allows organisms to restore long-wavelength color patterns by absorbing the abundant photons in the blue-green spectral range and reemitting some of that energy as light at longer wavelengths. This situation applies to fairy wrasses for example, where it has been shown experimentally that the

fluorescence pattern in males plays a role in sexual interactions [5, 6].

Red fluorescence is present in many reef fishes [7, 8]. In small, benthic, predatory fishes, it is often the eyes that fluoresce and they do so more efficiently in deeper water [9, 10]. This depth effect combined with findings that red fluorescence is also phenotypically flexible [11] and becomes more efficient in fish kept in dim environments [12], suggests an optimization to ambient light conditions. Given that several red fluorescent fish can also perceive their own fluorescence [6, 13], we hypothesize that fish with strongly red fluorescent irides may use fluorescence to illuminate and probe their surrounding environment [14]. More specifically, we argue that this form of fluorescence could theoretically be used to induce reflective eyeshine in small prey such as copepods, aiding in their detection. Such active photolocation where prey is illuminated by some kind of private light source

*Correspondence: ulrike.harant@uni-tuebingen.de
[1] Department of Animal Evolutionary Ecology, Institution for Evolution and Ecology, University of Tuebingen, Auf der Morgenstelle 28, 72076 Tuebingen, Germany
Full list of author information is available at the end of the article

has recently been shown in nocturnal flashlight fish [15]. These produce bioluminescent light pulses that might be strong enough to reveal retro-reflection in the eyes of other fish and/or prey nearby. Red fluorescence could be used in a similar way under daylight conditions. This, however, seems more plausible under the heavily shaded, blue-green stenospectral light conditions at depth rather than in shallow, broadly lit euryspectral conditions [9]. We define the euryspectral zone as the depth range close to the surface, with an ambient spectrum that is broader than the visual spectrum of most animals. The stenospectral zone, in contrast, describes the depth range below this, where most of the UV and longer wavelengths have been absorbed by the water column [16, 17], resulting in a spectrum that is narrower than the perception limits of most fish [9, 10]. The transition between the two can be between 5 and 25 m, depending on light conditions and variation in light attenuation by the water column.

Benthic copepods and other micro-crustaceans are a common food source for small fish and their nauplius eyes show strong reflection due to the presence of tapetal cells [18–21]. Own observations and tests (unpubl. data) indicate that copepod eyes reflect incoming light more to the source than elsewhere, similar to a weak retroreflector. Inducing reflective eyeshine in such eyes could therefore be enhanced if the light source (= red fluorescent iris) is close to the pupil, as is also the case for the light organ below the pupil in flashlight fishes [22]. Here, we do not assess whether *Tripterygion delaisi* actually induces and perceives this eyeshine in copepods, but rather examine whether the association between ambient light conditions and foraging behavior is consistent with this hypothesis. More specifically, we test whether fish capture more copepods under bluish light conditions that make fluorescence stand out more compared to broad illumination of the same overall brightness as predicted by our hypothesis. Our model species is the black-faced triplefin *Tripterygion delaisi* (Cadenàt and Blache 1970) [23], a small crypto-benthic fish with strongly red fluorescent irides [12, 24]. Since *T. delaisi* increases the relative efficiency of its fluorescence with decreasing ambient brightness, foraging success was tested under two different brightness levels, mimicking 2° of shading. By doing so, we could assess whether foraging success increases under stenospectral conditions in general, or whether it also requires heavily shaded light conditions.

Methods

We collected *T. delaisi* while SCUBA diving at Stareso (Station de Recherches Sous Marines et Océanographiques) Calvi, Corsica, France in June 2014 and 2015. After transfer to the aquarium facilities at the University Tübingen, Germany, they were held separately in 40, 15 L tanks which were equipped with a living rock as a comfort stone, in a common water recirculation system (20 °C, salinity 34‰, pH 8.2, 12 h light/dark cycle, fed once per day).

Tank illumination

Each aquarium was illuminated with a combined set of 8 LEDs in a single housing covered with a Feno Fe s.soft lt 18 diffuser. The LEDs available were: cold white, UV (395–410 nm), royal blue (450–465 nm), blue (465–485 nm), 2× green (520–535 nm), amber (585–595 nm) and red (620–630 nm). Each LED of each housing could be individually controlled by a DMX standalone unit (Feno fc s.dmx 48d) from off (= 0) to maximum (= 100) allowing spectral shape and brightness to be programmed.

Copepod culture and pilot study on copepod behavior

As a prey model species, we used *Tigriopus californicus* (Baker 1912) a marine harpacticoid copepod that colonizes rock pools from Alaska to Baja California, Mexico [25]. Copepods were cultivated in 1 L tanks (20 °C and 34 ‰ salinity, 12 h light/dark cycle) and fed on a variety of unicellular algae and bacteria. For each of the two experiments, we carried out a pilot experiment in which we tested the preference for the light treatments (stenospectral versus euryspectral) of the copepods. Copepods were inserted into transparent 4 mL cuvettes containing seawater and illuminated each for 2 min with the light treatments used in the main experiment in random order. We then assessed whether the copepods spent significantly more time in the upper or lower half of the cuvette, indicating a preference for a particular light treatment presented. No significant differences were detected (Additional file 1).

First experiment: spectral treatments under shaded conditions

The experimental room in which fish were kept was divided into two benches with 20 aquaria (total $n = 40$). On 10 October 2014 each bench either received a euryspectral or stenospectral treatment with an identical overall irradiance (total irradiance in photons s^{-1} m^{-2}, stenospectral: 2.51×10^{18}, euryspectral: 2.55×10^{18} as in Harant et al. [12]) which represents 3.6% of the total light present just below the water surface on a sunny summer day (Fig. 1). The natural spectrum was measured on the 26 June 2015 at solar noon, close to solar maximum in Corsica, France. These two experimental spectra were designed to mimic the ambient spectra at 5 and 20 m depth, the range in which *T. delaisi* is most abundant. Both benches received a different spectral treatment, alternating every week for 4 weeks. After analyzing the

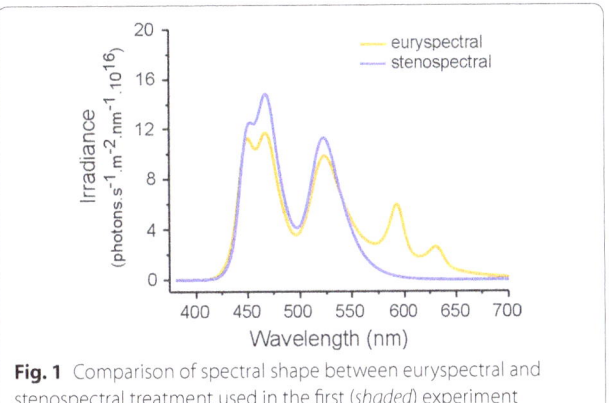

Fig. 1 Comparison of spectral shape between euryspectral and stenospectral treatment used in the first (*shaded*) experiment

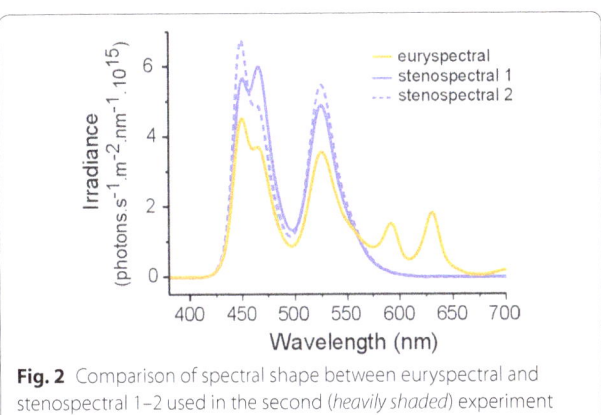

Fig. 2 Comparison of spectral shape between euryspectral and stenospectral 1–2 used in the second (*heavily shaded*) experiment

data, we extended the experimental duration for another 2 weeks to confirm the insignificance of the results (6 weeks total).

Second experiment: same spectra, heavily shaded light conditions

In the second experiment, we used a newly collected set of fish for a test under identical conditions, except that now, brightness was reduced to the lowest level that was manageable to carry out the experiment (about 0.01% of the surface; stenospectral$_1$: 7.05×10^{16}, stenospectral$_2$: 7.05×10^{16}, euryspectral: 7.04×10^{16}). This involved reducing the light produced by the LEDs electronically, but not below a 6% level, where flickering becomes an issue. Since the red LED was already at a low setting in the euryspectral treatment during the first (shaded) experiment, it could not be lowered more. In order to achieve a low light level we therefore added a cap on top of the light diffuser made of 2 layers of neutral density filter (LEE Filters Nr. 210 0.6 ND) which allowed about 6% of the total light intensity to pass.

In *T. delaisi*, brightness perception is mainly mediated by the double cones. According to microspectrophotometric measurements, these peak at 516 and 530 nm. Hence, highest sensitivity in this species lies within the green spectral range [26]. The stenospectral treatments containing more green wavelengths compared to the euryspectral treatment could therefore be perceived as being much brighter regardless of the total brightness. To prevent an increase of foraging success due to this perceived brightness effect, we ran two stenospectral treatments which varied by 10% in the amount of green wavelengths (Fig. 2). However, in order to keep the overall brightness identical, the amount of blue light in the second stenospectral treatment was slightly increased (Fig. 2).

Starting from 26 January 2015, each bench received either a euryspectral or stenospectral light condition, which was swapped after each week for 4 weeks. In the stenospectral treatment the two stenospectral alternatives were changed on a daily basis.

Aquaria experimental setup

The two sides of each aquarium were covered with white non-fluorescing, polypropylene sheets (matt/semi-gloss) sprayed with a fine greyish noise pattern (Hybrid Lack, silver-gray RAL 7001—Additional file 2). This was done to provide a noisy background under the (untested) assumption that it would make it harder for the fish to detect copepods by achromatic contrast alone.

Recording setup

When testing two individual fish in a pair of aquaria, we fitted a GoPro Hero 3+ above each tank, providing a full view of the bottom of the aquarium where the fish move about. In the second experiment, the light was too dim for good recordings, requiring infrared illumination (Versiton SAL-30 IR Illuminator 77 LEDS 30 M (100′) 12VDC 1.5A, peak at 844 nm). For this, the cameras were fitted with a dedicated IR lens (Vision Dimension: 2.97 mm Megapixel M12 × 0.5). In order to minimize the unlikely possibility of sensitivity to strong IR, the IR sources were positioned at ground level in the room, oriented upwards, but not into the tanks. The diffuse reflection from the ceiling and walls was just bright enough to obtain good recordings (Additional file 3).

Fish habituation and testing

Prior to the start of the experiment, fish were familiarized with the pipette that was used to provide copepods: UKH frequently inserted the pipette into the aquaria several times a day during the pre-experimental weeks. In the beginning, the pipettes contained defrosted *Mysis* shrimp (Aki Frost GmbH) which were released into the aquarium (3–4 *Mysis* per insertion). As soon as no flight response was observed anymore, seawater was delivered instead of food. This procedure made sure that the appearance of a pipette triggered positive anticipation without guaranteeing food.

After starting the light treatments, fish were allowed to adapt for 1 week without any other change in maintenance conditions. On the following day (Monday), 5 randomly chosen aquaria pairs ($n = 20$ individuals) in each treatment were tested by injecting 10 copepods in 1 mL seawater 2 cm from the front glass using an automatic pipette. The next day (Tuesday) the remaining 10 aquarium pairs (n = 20 individuals) followed. Upon completing the testing procedure, the spectral treatments were changed to the opposite treatment. The same procedure was then repeated again (Monday/Tuesday) under the second spectral treatment the week after. Each pair of aquaria was tested twice per testing day with one fish first receiving copepods whereas the other only received water from the copepod culture as a control to check for feeding strikes due to odor only. In the second run of the day, the role of positive treatment and control was reversed in each aquarium pair. Between the first and second run in a single aquarium pair, another aquarium pair in the opposite treatment (and bench) was tested to induce a delay between the two successive trials in a single aquarium pair. This 10 min delay allowed fish to go back to their normal routine.

Pre-experimental trials showed that fish usually stop searching for copepods after about 5 min. In addition to that, copepods started to hide within the comfort stone (living rock) after 5–6 min after injection. Hence, if fish were not able to catch all copepods within the first run, the chance of detecting leftover copepods still left in the tank during the control treatment in the second run was small (but see "Results" Section).

Work flow, copepod preparation and video analysis

To enhance video quality, the water inflow of the aquarium was turned off 10 min prior to testing. The copepods were gently taken up by the pipette and released into the aquarium. After insertion of the pipette, videos were recorded for the following 5 min. To prevent observer bias, recorded videos were randomized and transformed to grayscale before analysis. The inserted copepods were too small to be seen on the video, leaving the observer also blind to the copepod treatment and its control. *T. delaisi* shows a saltatory searching behavior [27, 28] which is characterized by approaching prey with small hopping movements, interrupted by scanning of the substrate and a sudden feeding strike once prey is identified. In a pilot study we found that the number of feeding

strikes closely approximated the number of live copepods added to the tank, and never exceeding that number. It confirmed that there are no feeding strikes without copepods, and most or all feeding strikes also resulted in successful prey capture. Only rarely, fish needed two strikes in rapid succession for the same item. Such cases were counted as one strike. Overall, the results show that feeding strikes are a reliable variable for measuring foraging success in *T. delaisi* (Additional file 4). In the main experiments, we used Etholog 2.2.5 [29] to record time (s) for each feeding strike since start of the recording as well as total *n* feeding strikes.

Iris fluorescence of *T. delaisi*

Excitation and emission of iris fluorescence is shown in Fig. 3 with excitation being highest at 550 nm and fluorescence emission peaking at 600 nm [26]. Since *T. delaisi* forages at relatively small distances to prey of a few centimeters only, absorption and scattering is negligible (<1% at 600–650 nm at 4 cm distance, [16, 17]). To calculate the decrease of fluorescence brightness over distance, we fixed an eye of *T. delaisi* on a black stick and illuminated it with a blue Hartenberger Mini Compact LCD divetorch (7×3.5 W 450 nm bulbs) from a distance of 24 cm. Two short pass filters (Thorlabs FD2C subtractive dichroic color short-pass) were attached in front of the torch to cut out longer wavelengths. Since the eyes quickly darken after decapitating a fish due to the dispersal of melanosomes, we treated the eyes with potassium chloride solution [24] for 1 h to reverse this process before using it. The eye was then oriented downwards at an angle of approximately 45° looking at a diffuse white standard (PTFE). A ruler was placed in line with the outer edge of the iris to serve as a reference. Consecutive measurements were taken in 0.5 cm steps using a calibrated PR 740 SpectraScan Spectroradiometer (Photo Research Inc.,) pointed at the diffuse white standard and

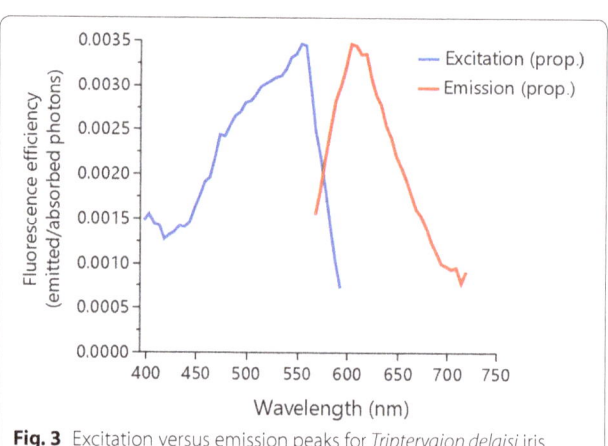

Fig. 3 Excitation versus emission peaks for *Tripterygion delaisi* iris fluorescence

measuring reflected fluorescence starting at 0–2.5 cm distance from the eye. The spectrometer was set to 2 nm bandwidth, an aperture of 0.5, with smart dark enabled at a normal speed with an extended exposure time, and was operated with a calibrated MS-75 lens.

The spectrometer used to take measurements adjusts gain to avoid saturation in the brightest wavelengths resulting in noisy measurements in the longer wavelength range. We therefore used an orange filter (Lee filters, Double C.T. Orange 287) attached in front of the spectrometer lens to suppress the excitation light. Radiance measurements were then corrected for the transmission of the used filter and converted to photon radiance by multiplying the measurements with wavelength*$5.05*10^{15}$ [30]. Photon radiance was then summarized between 600 and 650 nm and averaged among the two measured eyes. Figure 4 displays the exponential loss of iris fluorescence with distance. Note that measurements were taken from a diffuse white standard reflecting all wavelengths equally in a 180° angle. These measurements are therefore very conservative compared to a reflector such as a copepod nauplius eye.

For easier comprehensibility, we provide a demonstration of how red fluorescence behaves with increasing/decreasing proportion of longer wavelengths present in the light environment. Since *T. delaisi* is able to quickly regulate its fluorescence we did not conduct this demonstration with live triplefins but used a special mixture of fluorescent paint which resembles different intensities of fluorescence emission of *T. delaisi* (Additional file 5). We then illuminated the fluorescent patches along with a non-fluorescent red diffuse reflectance standard (Labsphere SCS-RD-010) from a distance of 24 cm with the euryspectral and stenospectral light treatment used during the second experiment. The demonstration shows that with decreasing proportion within the longer wavelength range fluorescence appears more intense while

in direct comparison the non-fluorescent red standard becomes grey.

Statistical analysis of fish behaviour

Data were analysed using a generalized linear mixed model using the lme4 package [31] of R [32]. The response variable *n copepods caught*, was modelled as a binomial (*n copepods caught*/*n copepods missed*) response variable with logit link. In both experiments all initial models contained *light treatment*, *bench*, and *week* as well as all biologically relevant interactions as fixed components. To account for the repeated measurements per fish, fish ID was integrated as a random factor with random slopes. An observation-level random factor (random effect that models extra-Poisson variation of count data, [33]) was added as well to account for overdispersion. By using the Bayesian information criterion (BIC), a backward model selection was performed on random (excluding fish ID) and fixed factors to identify the best fitting model with the fewest predictors. Here, we only present the final models including proxies for the goodness-of-fit of the complete model (conditional R^2) as well as the fixed component (marginal R^2) [34]. Proxies were calculated using the pairwise SEM package for R [35]. Wald z tests were performed to assess the overall significance of fixed effects. All other statistical tests as given, two-way ANOVAs and paired t tests were carried out using JMP 11 (SAS) after confirming normality and homoscedasticity.

All data necessary to reproduce our conclusions are provided in the supplementary files section (Additional files 6, 7).

Pre-results: participation and exclusion criterion

In the first experiment, 31 out of 40 fish participated throughout the entire study whereas the remaining 9 showed no interest and were therefore excluded from the analysis. In the second experiment, 37 of 39 fish successfully participated throughout the whole duration of the experiment. However, 3 males changed to male breeding coloration during the experiment and were excluded from further analysis. Males in breeding coloration increase the content of melanophores in the iris which reduces expressed red fluorescence (unpubl. data). We therefore only considered adults in our analyses that showed their cryptic coloration throughout the entire study.

Pre-results: spectral treatments

In the second experiment there was no detectable difference between the two stenospectral treatments, that differed only slightly in the spectral range (blue-green), which is why these data were pooled together (paired t test: $dF = 33$ $t = 0.25$, $p = 0.81$).

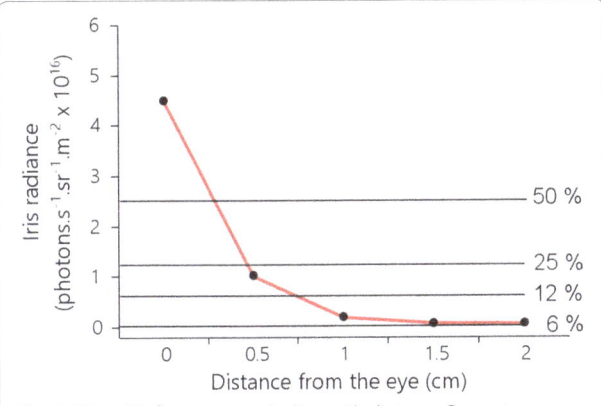

Fig. 4 Mean iris fluorescence decline with distance. Percentages indicate the proportion of light *left* at a given distance

Pre-results: control treatments

Out of 322 recorded control videos (no copepods), only 12 feeding strikes were observed, eleven of which occurred during the second run where fish had previously received copepods. It is therefore likely that these fish caught copepods that were still around from the earlier runs on that day.

Results

Experiment 1: feeding success under shaded conditions

There was no difference in the number of copepods caught under euryspectral versus stenospectral conditions under shaded illumination (Fig. 5; Table 1). Fish caught on average 4.83 ± 1.98 SD copepods in the euryspectral treatment and 4.88 ± 1.82 SD in the stenospectral treatment. Similarly, the time it took until 5 out of 10 copepods were caught ("copepod half-time"), did also not differ between light treatments (Additional file 8).

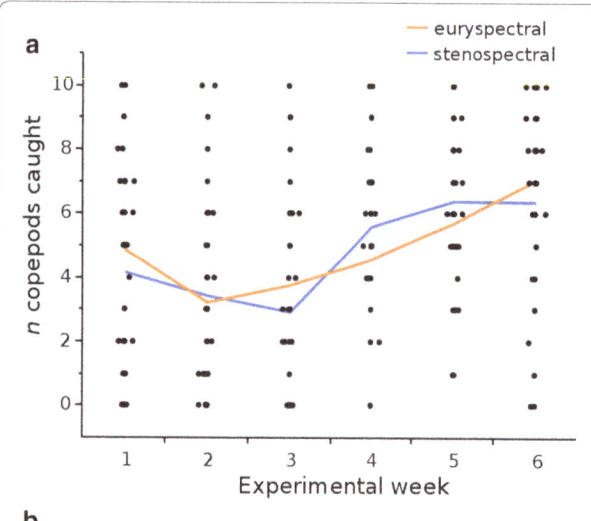

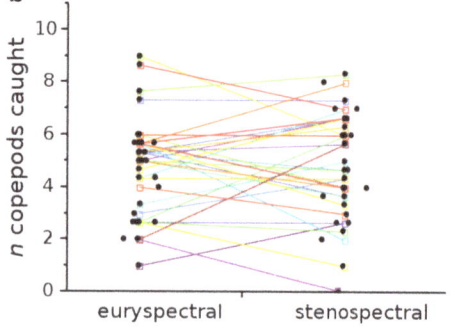

Fig. 5 a Total number of copepods caught during the first (*shaded*) experiment (*n* = 31). *Black dots* represent the total number of copepods caught by individuals per experimental week in the euryspectral or stenospectral light treatment. *Lines* represent mean copepods caught per week in the respective light treatment. **b** Mean copepods caught per fish and spectral treatment. Lines connect mean values for copepods caught of each individual between the two spectral treatments

Interestingly, however, fish significantly increased their foraging success irrespective of the light treatment from 4.5 ± 2.9 SD copepods in week 1 to 6.7 ± 2.8 SD in week 6 (Fig. 3; Table 1), indicating a learning effect.

Experiment 2: foraging success under heavily shaded light conditions

Fish held in the stenospectral treatment caught on average 6.5 ± 1.97 SD copepods while fish tested under euryspectral conditions only caught 5.8 ± 1.63 SD copepods, a significant difference (Table 1; Fig. 6). Hence, the use of red fluorescence under blue-green illumination in deeper *and* more shaded habitats, allows *T. delaisi* to increase its foraging success by an average of 7%. However, fish showed quite some variation. More than one third (44.4%) of the fish for example, increased their foraging success by 15% on average and 29% of all fish increased their efficiency by at least 20% under stenospectral conditions relative to euryspectral conditions. The highest mean increase for any individual was twofold (9 copepods caught compared to 4.5). In contrast, 8 out of 34 fish showed a higher foraging success in the euryspectral treatment compared with the stenospectral treatment (7.3 ± 1 SD compared with 4.9 ± 2 SD copepods). Similar to the first (shaded) experiment, copepod half-time was not affected by treatment (Additional file 9). Improved performance over the course of the experiment could not be confirmed (no effect of week, Table 1), but experiment 1 ran for 6 weeks, experiment 2 for 4 weeks.

When comparing both experiments (shaded and heavily shaded), fish were generally more successful at catching copepods under heavily shaded conditions (average shaded experiment: 4.86 ± 1.68 SD; heavily shaded experiment: 6.29 ± 1.48 SD). This difference, however, needs to be interpreted with care since the two experiments did not run in parallel, but in successive years, with 2 different fish cohorts.

Discussion

Tripterygion delaisi showed on average a 7% higher foraging success under heavily shaded, blue-green light favoring fluorescence compared with broad-spectral or shaded conditions. Although these results do not represent direct evidence that fish use red fluorescence to enhance their prey detection under stenospectral conditions, they are nevertheless consistent with the active photolocation hypothesis. Assuming that red fluorescent irides indeed facilitate prey detection, they probably do so under heavily shaded conditions only, either under rocks or overhangs where the light is dominated by side-welling blue-green scatter from the open water, or at depths, times of day or degrees of cloud cover, where bright and broad-spectral light is lacking.

Table 1 Foraging success and copepod half-time in response to light treatments

Experiment	Parameter	Std-beta coefficient estimate	SE	z	p	Conditional R²	Marginal R²
1: Shaded conditions: foraging success	Intercept	−1.323	0.29	−4.47	<0.001		
	Week	0.34	0.06	5.49	<0.001	0.157	0.058
	Light treatment	0.02	0.21	0.14	0.89		
2: Heavily shaded conditions: foraging success	Intercept	0.73	0.15	5.01	<0.001		
	Light treatment	−0.34	0.17	−1.98	0.047	0.063	0.007

Hence, when fish are hunting in sunlit sites, fluorescence is likely to be of little help for foraging. However, it coincides well with *T. delaisi*'s preference for rocky substrates with crevices and overhangs where brightness transitions are frequent and strong, regardless of depth. Under these conditions, red fluorescence might offer a significant advantage when foraging in the shade.

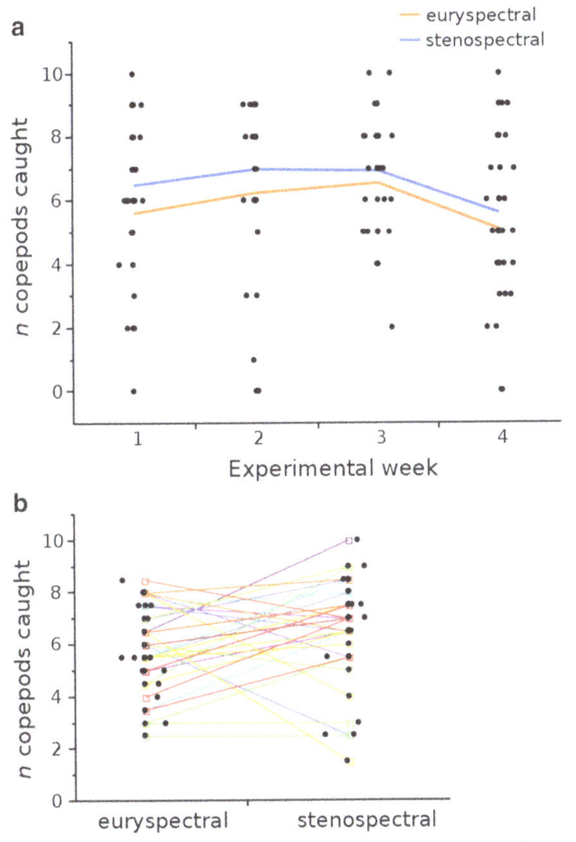

Fig. 6 a Total number of copepods caught during the second (*heavily shaded*) experiment (*n* = 34). *Black dots* represent the total number of copepods caught by individual per experimental week in the euryspectral or stenospectral light treatment. *Lines* represent mean copepods caught per week in the respective light treatment. **b** Mean copepods caught per fish and spectral treatment. Lines connect mean values for copepods caught of each individual between the two spectral treatments

Similar benefits would exist when foraging during dusk and dawn.

Predator–prey interaction

While fish could have theoretically generated a contrast between red illuminated prey and background, explaining the observed increase in foraging success, they could have also used their red fluorescent irides to attract prey. Similar to phototaxis in diurnal vertical migrating invertebrates [36, 37], *T. californicus* could be attracted by certain light cues. This, however, would require copepods to be sensitive to longer wavelengths. Although studies on visual sensitivity of *T. californicus* are rare (but see [20]), our own results obtained from the first pre-experimental study on light preferences of *T. californicus* suggest that such differentiation is absent for the light conditions used here (Additional file 1, red light treatment). However, longer wavelengths are quickly absorbed, implying that red fluorescence can only be effective over very short distances. This is compatible with the saltatory foraging and short-distance strikes typical for *T. delaisi* [27, 28]. Over such short distances, fluorescence could be strong enough to create the proposed contrast. A recent study by Anthes et al. [10] strengthens this hypothesis by showing that red fluorescent irides are a common feature among small benthic predatory fish that predominantly hunt for small invertebrates.

Additionally, wavelengths above 570 nm are rapidly absorbed over larger distances [16, 17, 38]. *T. delaisi* could therefore use red fluorescence to forage more efficiently while remaining hidden from predators nearby.

Individual variation

Fish tested under heavily shaded light showed substantial individual differences despite the fact that fish were given sufficient time to adjust. Such differences were also present in a previous study of a phenotypic response to different light environments [12]. We propose that this degree of variability may represent a form of microhabitat specialization in this very cryptic species. Fish predominantly foraging in exposed sites may show weaker fluorescence because it is less functional and its absence prevents attracting (red-sensitive) visual predators.

Whereas fish that predominantly forage in the shade face the opposite situation. Individual variability also explains the small size of the effect found in the heavily shaded experiment, despite the very strong effects in some individuals. Future work could specifically compare fish collected from exposed sites versus fish collected from overhangs to confirm this view.

Does brightness perception influence foraging success?

In order to keep the overall brightness similar in all spectral treatments, we increased the abundance of blue and green wavelengths within the stenospectral treatment. *T. delaisi* shows highest sensitivity in the green wavelength range [26]. The stenospectral treatment might therefore have been perceived brighter by the fish, regardless of the real overall brightness. We attempted to take care of this by including a second even "greener" stenospectral treatment and comparing foraging success between the two stenospectral treatments. Since foraging success did not differ between these two treatments, a difference in brightness perception between stenospectral and euryspectral treatment alone cannot explain the observed increase in foraging success in the second, heavily shaded experiment. Furthermore, if perceived brightness indeed affected foraging success, a similar effect would have been present in the first, brighter experiment. Although no such effect could be found, we cannot entirely exclude that perceived brightness might still have had a small effect on foraging success in *T. delaisi.*

Conclusions

Summarizing, this study shows that *T. delaisi* forages more efficiently under heavily shaded, blue-green light conditions compared with broad light. Assuming that fish may be using red fluorescent emission to enhance prey detection, this result suggests that the functionality of such a mechanism is more plausible over short distances, under stenospectral, shaded conditions. This offers important clues for the design of future experiments to test active photolocation in this system.

Additional files

Additional file 1. Pilot study: behavior of *Tigriopus californicus* under two different spectra.

Additional file 2. Distraction pattern on polypropylene foil used to cover the walls of the aquaria.

Additional file 3. Video sequence of typical foraging behavior in *T. delaisi* recorded during the experiment.

Additional file 4. Pilot study: feeding strikes in *Tripterygion delaisi.*

Additional file 5. Animated gif illustrating the contrast generated by red fluorescence under the two spectral treatments.

Additional file 6. Foraging success data of the first (shaded) experiment.

Additional file 7. Foraging success data of the second (heavily shaded) experiment.

Additional file 8. R script used to analyze foraging success in the first (shaded) experiment.

Additional file 9. R script used to analyze foraging success in the second (heavily shaded) experiment.

Authors' contributions

UKH and NKM designed the experiments. UKH carried out the collection and analyses of the data, and drafted the manuscript. Both UKH and NKM edited the manuscript. All authors read and approved the final manuscript.

Author details

[1] Department of Animal Evolutionary Ecology, Institution for Evolution and Ecology, University of Tuebingen, Auf der Morgenstelle 28, 72076 Tuebingen, Germany. [2] Department of Biology, Faculty of Science, University of Tuebingen, Auf der Morgenstelle 28, 72076 Tuebingen, Germany.

Acknowledgements

We would also like to thank Martina Hohloch, Gregor Schulte and Oeli Oelkrug for technical support as well as Lukas Miksch for assisting with the video analysis.

Competing interests

The authors declare that they have no competing interests.

Funding

This work was funded by a Reinhart Koselleck Project Grant Mi482/13-1 from the Deutsche Forschungsgemeinschaft to N.K.M. We acknowledge support by Deutsche Forschungsgemeinschaft and Open Access Publishing Fundof University of Tübingen

References

1.	Lagorio MG, Cordon GB, Iriel A. Reviewing the relevance of fluorescence in biological systems. Photochem Photobiol Sci. 2015;14:1538–59.
2.	Arnold KE, Owens IP, Marshall NJ. Fluorescent signaling in parrots. Science. 2002;295:92.
3.	Mazel CH, Cronin TW, Caldwell RL, Marshall NJ. Fluorescent enhancement of signaling in a mantis shrimp. Science. 2004;303:51.
4.	Lim ML, Land MF, Li D. Sex-specific UV and fluorescence signals in jumping spiders. Science. 2007;315:481.
5.	Gerlach T, Theobald J, Hart NS, Collin SP, Michiels NK. Fluorescence characterisation and visual ecology of pseudocheilinid wrasses. Front Zool. 2016;13:13.
6.	Gerlach T, Sprenger D, Michiels NK. Fairy wrasses perceive and respond to their deep red fluorescent coloration. Proc Biol Sci. 2014;281:2014.

7. Michiels NK, Anthes N, Hart NS, Herler J, Meixner AJ, Schleifenbaum F, Schulte G, Siebeck UE, Sprenger D, Wucherer MF. Red fluorescence in reef fish: a novel signalling mechanism? BMC Ecol. 2008;8:16.

8. Sparks JS, Schelly RC, Smith WL, Davis MP, Tchernov D, Pieribone VA, Gruber DF. The covert world of fish biofluorescence: a phylogenetically widespread and phenotypically variable phenomenon. PLoS ONE. 2014;9:e83259.

9. Meadows MG, Anthes N, Dangelmayer S, Alwany MA, Gerlach T, Schulte G, Sprenger D, Theobald J, Michiels NK. Red fluorescence increases with depth in reef fishes, supporting a visual function, not UV protection. Proc Biol Sci. 2014;281:20141211.

10. Anthes N, Theobald J, Gerlach T, Meadows MG, Michiels NK. Diversity and ecological correlates of red fluorescence in marine fishes. Front Ecol Evol. 2016;4:126.

11. Wucherer MF, Michiels NK. A fluorescent chromatophore changes the level of fluorescence in a reef fish. PLoS ONE. 2012;7:e37913.

12. Harant UK, Michiels NK, Anthes N, Meadows MG. The consistent difference in red fluorescence in fishes across a 15 m depth gradient is triggered by ambient brightness, not by ambient spectrum. BMC Res Notes. 2016;9:1.

13. Kalb N, Schneider RF, Sprenger D, Michiels NK. The red-fluorescing marine fish *Tripterygion delaisi* can perceive its own red fluorescent colour. Ethology. 2015;121:566–76.

14. Jack CB. Detecting the detector: a widespread animal sense? 2014. (http://vixra.org/abs/1411.0226). Accessed 22 June 2016.

15. Hellinger J, Jägers P, Donner M, Sutt F, Mark MD, Senen B, Tollrian R, Herlitze S. The flashlight fish *Anomalops katoptron* uses bioluminescent light to detect prey in the dark. PLoS ONE 2017;12:e0170489.

16. Jerlov NG. Optical oceanography. Amsterdam: Elsevier Publishing Company; 1968.

17. Mobley CD. Light and water: radiative transfer in natural waters. Cambridge: Academic press; 1994.

18. Fahrenbach WH. The fine structure of a nauplius eye. Z Zellforsch Mikrosk Anat. 1964;62:182–97.

19. Elofsson R. The ultrastructure of the nauplius eye of Sapphirina (Crustacea: Copepoda). Cell Tissue Res. 1969;100:376–401.

20. Elofsson R. The frontal eyes of crustaceans. Arthropod Struct Dev. 2006;35:275–91.

21. Martin GG, Speekmann C, Beidler S. Photobehavior of the harpacticoid copepod *Tigriopus californicus* and the fine structure of its nauplius eye. Invertebr Biol. 2000;119:110–24.

22. Howland HC, Murphy CJ, McCosker JE. Detection of eyeshine by flashlight fishes of the family Anomalopidae. Vision Res. 1992;32:765–9

23. Cadenat J, Blache J. Description d'une espèce nouvelle, *Tripterygion delaisi* sp. nov., provenant de l'île de Gorée (Sénégal) (Pisces, Clinidae). Bull Mus Natl Hist Nat. 1969;41:1097–105.

24. Wucherer MF, Michiels NK. Regulation of red fluorescent light emission in a cryptic marine fish. Front Zool. 2014;11:1.

25. Haderlie E, Abbott D, Caldwell R. Three other crustaceans: a copepod, a leptostracan, and a stomatopod. In: Morris RH, Abbott DP, Haderlie EC, editors. Intertidal invertebrates of California. Stanford: Standford University Press; 1980. p. 631–5.

26. Bitton PP, Harant UK, Fritsch R, Champ CM, Temple S, Michiels NK. Red fluorescence of the triplefin *Tripterygion delaisi* is increasingly visible against background light with increasing depth. Open Sci. 2017;4(3):161009.

27. O'Brien W, Evans B, Browman H. Flexible search tactics and efficient foraging in saltatory searching animals. Oecologia. 1989;80:100–10.

28. O'Brien WJ, Browman HI, Evans BI. Search strategies of foraging animals. Am Sci. 1990;78:152–60.

29. Ottoni EB. EthoLog 2.2: a tool for the transcription and timing of behavior observation sessions. Behav Res Methods Instrum Comput. 2000;32:446–9.

30. Johnsen S. The optics of life: a biologist's guide to light in nature. Princeton University Press: Princeton; 2011.

31. Bates D, Maechler M, Bolker BM, Walker S. lme4: linear mixed-effects models using Eigen and S4. J Stat Softw. 2014;67:1–48.

32. R Development Core Team. R: a language and environment for statistical computing. R foundation for statistical computing. Vienna; 2013.

33. Harrison XA. Using observation-level random effects to model overdispersion in count data in ecology and evolution. Peer J. 2014;2:e616.

34. Nakagawa S, Schielzeth H. Repeatability for Gaussian and non-Gaussian data: a practical guide for biologists. Biol Rev Camb Philos Soc. 2010;85:935–56.

35. Lefcheck JS. piecewiseSEM: piecewise structural equation modelling in r for ecology, evolution, and systematics. Methods Ecol Evol. 2015;29:1560–8.

36. Ringelberg J. The positively phototactic reaction of *Daphnia magna* Straus: a contribution to the understanding of diurnal vertical migration. Neth J Sea Res. 1964;2:319–406.

37. Longhurst A. Vertical migration. In: Cushing DH, Walsh JJ, editors. The ecology of the seas. Philadelphia: W.B. Saunders Compagny; 1976. p. 116–37.

38. Morel A. Optical properties of pure water and pure sea water. New York: Academic Press; 1974.

18

Wind conditions on migration influence the annual survival of a neotropical migrant, the western yellow-breasted chat (*Icteria virens auricollis*)

Andrew C. Huang[1,2]*, Christine A. Bishop[1,2], René McKibbin[1], Anna Drake[3] and David J. Green[2]

Abstract

Background: Long-distance migratory birds in North America have undergone precipitous declines over the past half-century. Although the trend is clear, for many migrating species underpinning the exact causes poses a challenge to conservation due to the numerous stressors that they encounter. Climate conditions during all phases of their annual cycle can have important consequences for their survival. Here, using 15 years of capture-recapture dataset, we determined the effects of various climate factors during the breeding, wintering, and migrating stages on the annual survival of a western yellow-breasted chat (*Icteria virens auricollis*) population breeding in southwestern Canada.

Results: El Niño effects over the entire annual cycle had little influence on the annual apparent survival of yellow-breasted chats. However, we found evidence that wind conditions during migration, specifically average westerly wind speed or the frequency of storm events, had significant adverse effects on adult annual apparent survival. In comparison, precipitation levels on wintering ground had little to no influence on adult annual apparent survival, whereas growing degree days on the breeding ground had moderate but positive effects.

Conclusions: In the face of climate change and its predicted impacts on climate processes, understanding the influence of weather conditions on the survival of migrating birds can allow appropriate conservation strategies to be adopted for chats and other declining neotropical migrants.

Keywords: Climate conditions, Wind, Storm, Yellow-breasted chats, Neotropical migrants, Climate change

Background

Widespread declines in bird populations are evident across much of North America, with recent findings revealing that an alarming one-third (37%) of North American bird species are of high conservation concern [1]. Long-distance neotropical migrants have declined more steeply than residents and short-distance migrants [1, 2]. Longer journeys mean that these birds encounter more potential impediments along their migratory routes, including collisions with man-made infrastructures, light pollution, increased predation risk, and inadequate food sources [3, 4]. In addition, anthropogenic and environmental threats on the breeding and wintering grounds add to the suite of stressors for neotropical migrants [5, 6]. Unfavourable weather conditions are one of the most detrimental factors that can compromise their survival and reproductive phenology [7–11]. Having a more holistic understanding of the climatic processes and their impacts on migratory birds would allow conservation strategies to be effectively implemented for these species [12].

Migratory birds have endured arduous and perilous journeys between and within North, Central and South Americas for millennia. However, with climate change

*Correspondence: andrew.huang@canada.ca;
andrewhuang220@gmail.com
[1] Environment and Climate Change Canada, Delta, BC, Canada
Full list of author information is available at the end of the article

projections predicting drastic alterations in climatic conditions, uncertainty exists about the future persistence of migratory bird populations [13]. The predicted increase in the frequency, intensity, and duration of extreme weather events—including droughts, intense precipitation, and windstorms—are expected to have adverse effects on avian population dynamics [8, 10, 14, 15]. Neotropical migrants may be particularly susceptible to extreme climate variability, as climate fluctuations and anomalies can result in direct mortality and/or impact food availability during all stages of their annual cycle: breeding, wintering, and migration [16]. Understanding how climatic conditions at different stages of the annual cycle influence the survival of neotropical migrants is critical in evaluating how climate change will impact migratory birds [17, 18].

The objective of this study was to examine how annual survival of a neotropical migrant, western yellow-breasted chat (*Icteria virens auricollis*; hereafter: chat), is influenced by climatic conditions during its breeding, wintering, and spring migratory periods. We developed capture-mark-recapture models constrained with climate variables to evaluate the effects of: (1) El Niño Southern Oscillation (ENSO) during the entire annual cycle; (2) temperature and precipitation conditions on the breeding grounds; (3) precipitation level on the wintering grounds; and finally (4) wind speed, precipitation level, and number of storm events on the spring migration route. Survival was estimated using 15 years (2001–2015) of capture-recapture data from a breeding population of chats in southern British Columbia, Canada. Our study presents the first comprehensive analysis on how adult survivorship of chats in this endangered population [19] in Canada can be influenced by climate factors throughout three major phases of its annual cycle.

Methods

Study species and area

The yellow-breasted chat is a neotropical migratory songbird with a transcontinental distribution extending from southern Canada, the United States, Mexico, to Central America (Fig. 1). The western *auricollis* subspecies has a fragmented breeding distribution in the west, which includes our study area in the south Okanagan valley of British Columbia, Canada, at the northern tip of the geographic range for this subspecies (Fig. 1). Genetic evidence indicates that the western subspecies overwinters on the westcoast of Mexico (southern Baja California, Sinaloa to Oxaca) [20]. Chats in the south Okanagan valley nest in low elevation (>500 m) riparian thickets dominated by wild rose (*Rosa* spp.), snowberry, and other native shrub species [21]. Their diet consists primarily of insects, but also includes fruits and rose petals [22,

23]. We colour-banded, monitored, and resighted birds breeding in the Okanagan valley from Penticton (49° 27′ N, 119° 36′ W) to Osoyoos (49° 1′ N, 119° 26′ W) on the USA border, a distance of 66 km between 2001 and 2015. To ensure that resighted individuals were accurately determined, colour combos were confirmed upon multiple observations and by at least two observers (see McKibbin and Bishop [23, 44] for details). Birds were sexed and aged (second-year [SY] or after second-year [ASY]) based on plumage characteristics and molt limits [24]).

Climate data and models

El Niño Southern Oscillation (ENSO)

ENSO is known to influence a broad spectrum of climate factors in North America including precipitation anomalies in Mexico and southern California [25–27], temperature fluctuations [28], and wind conditions [7, 29]. Climate indices describing ENSO conditions (e.g. El Niño-Southern Oscillation Precipitation Index [ESPI], Southern Oscillation Index [SOI]) have been found to influence the survival rate of several neotropical migrants (e.g. Swainson's Thrush *Catharus ustulatus* [9]; Black-throated Blue Warblers *Setophaga caerulescenes* [18]; Yellow Warblers *Setophaga petechia* [30, 31]). We examined the combined effects of ENSO on the survival rate of chats using the average SOI from May to April ("Model set 1"; Table 1). SOI is a standard index that measures large-scale fluctuations in air pressure occurring between the western and eastern tropical Pacific, and provides a robust index for tracking ENSO phases. We obtained standardized monthly SOI values from the Climate Prediction Center: http://www.cpc.ncep.noaa.gov/data/indices/soi.

Climate conditions on breeding ground

Precipitation and temperature in the Pacific northwest are influenced by ENSO. When ENSO is in its strong phase (i.e. El Niño), jet stream becomes diverted into California, resulting in low precipitation and increased frequency of summer droughts in the Pacific Northwest [6, 32]. Growing degree days (GDD), a measure of heat accumulation, and precipitation level both influence primary productivity and insect emergence [33–36], and in turn food availability for animals in subsequent trophic levels [37–39]. We therefore predicted that precipitation in two periods (October–April and May–July) and GDD prior to start of breeding (January–May) would have positive effects on the survival of chats ("Model set 2"; Table 1). Climate variables on the breeding ground of south Okanagan valley were extracted and averaged from two local weather stations: "Penticton A" (WMO ID: 71889; 49°27′36″ N, 119°36′0″ W; elevation 344.4 m) and "Osoyoos CS" (WMO ID: 71215; 49°1′48″ N, 119°26′24″ W; elevation 282.9 m).

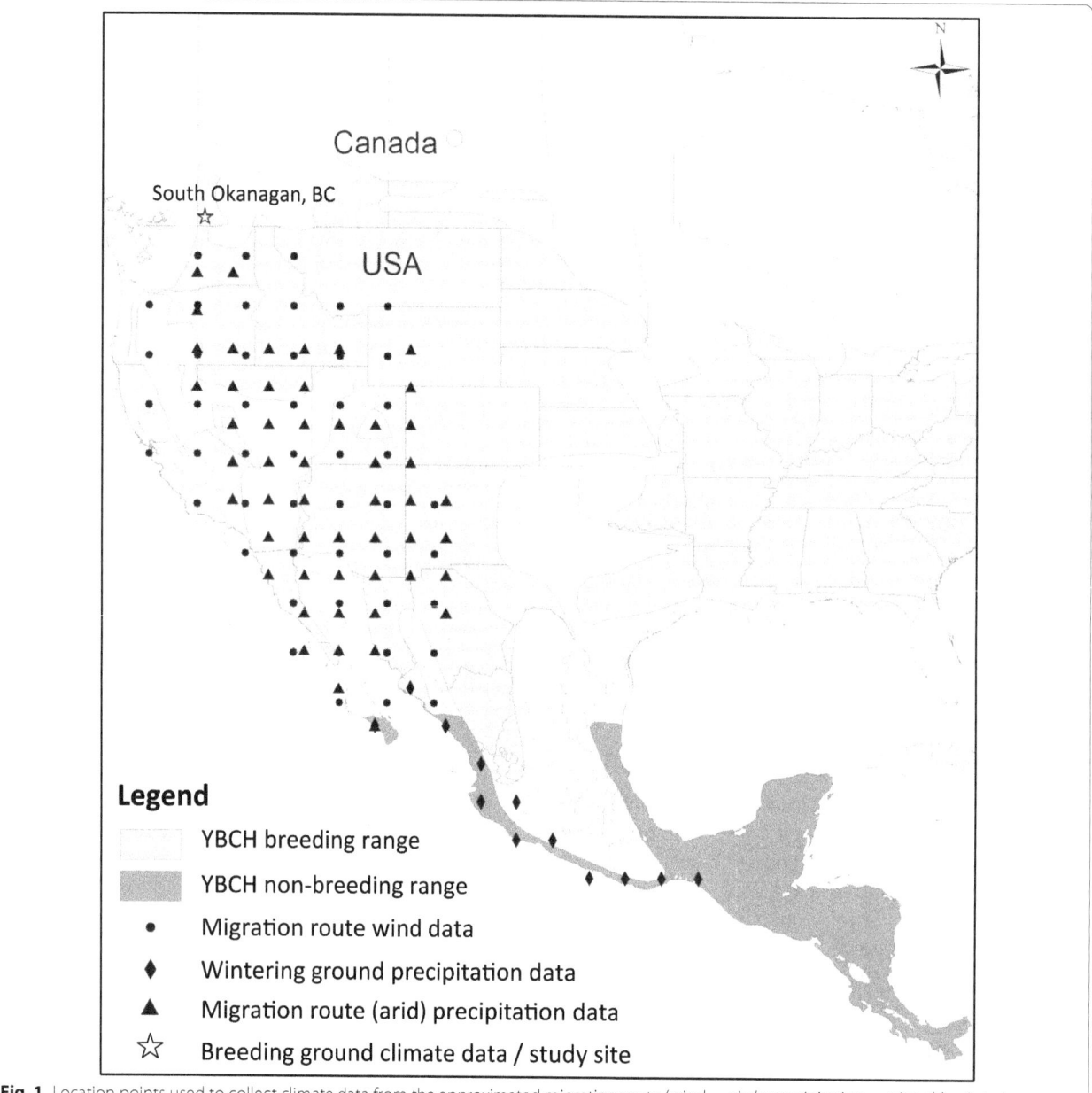

Fig. 1 Location points used to collect climate data from the approximated migration route (wind = *circle*; precipitation = *triangle*), wintering ground (*diamond*), and breeding ground/study site (*star*) of yellow-breasted chats (western subspecies). *Shaded* in *light and dark grey* are the breeding and overwintering ranges of yellow-breasted chats (both subspecies), respectively, adapted from IUCN (April 2016)

Climate conditions on wintering grounds

ENSO has divergent effects on precipitation in western Mexico: El Niño is associated with more rainfall during the winter/spring in northwestern Mexico, whereas La Niña is associated with more summer monsoon rainfall in southwestern Mexico [27]. Rainfall is expected to influence biological productivity on the wintering grounds, and as a result affect the overwinter survival and mass gain of migratory birds prior to spring migration [5, 7]. We therefore predicted that precipitation level on the wintering grounds during the monsoon and/or the late winter would influence the survival of chats ("Model set 3"; Table 1). Modeled rainfall data on the wintering grounds was obtained from the National Center of Environmental Prediction (NCEP)/National Center for Atmospheric Research (NCAR) Reanalysis database, as provided by the NOAA-CIRES Climate Diagnostic Center at Boulder, CO, USA, using R package RNCEP [40]. Precipitation data from the wintering ground was extracted from 12 location points (Fig. 1).

Table 1 Annual and region-specific climate data used to examine climate effects on annual apparent survival of western yellow-breasted chats breeding in the south Okanagan valley, British Columbia, Canada

Model set #	Life cycle stage	Climate variable	Calculated by
1	Annual	Southern Oscillation Index (SOI$_{MAY-APR}$)	Average monthly standardized Southern Oscillation Index (SOI)
2	Breeding ground	Growing degree days (GDD$_{JAN-MAY}$)	Sum positive GDD, where GDD = ([T$_{MAX}$ + T$_{MIN}$]/2)—10
		Precipitation$_{OCT-APR}$ Precipitation$_{MAY-JUL}$	Sum daily rainfall (mm)
3	Wintering ground	Precipitation$_{MAY-NOV}$ Precipitation$_{DEC-APR}$	Sum daily rainfall (mm)
4	Migration route	Precipitation$_{NOV-MAY(ARID)}$	Sum daily rainfall (mm)
		Precipitation$_{NOV-MAY(DESERT)}$	Sum daily rainfall (mm)
		Westerly wind speed (U-wind$_{APR-MAY}$)	Average mean daily U-wind speed (m/s) from 6 p.m.–6 a.m.[b] at 850 and 925 mb[a]
		Southerly wind speed (V-wind$_{APR-MAY}$)	Average mean daily U-wind speed (m/s) from 6 pm – 6am[2] at 850mb and 925mb[a]
		Number of storm nights (Storm$_{APR-MAY}$)	Sum of days with extreme (>95 percentile) U or V wind speeds

[a] Within the altitudinal range of migrant songbirds [67, 68]

[b] Includes 6 p.m., 12 and 6 a.m. (12 pm was excluded) as most migrants begin migration after dusk, peak at midnight, and end before dawn

Climate conditions on spring migration route

Wind conditions and storm events during the migration period are expected to influence flight costs and survival during flight [7, 8]. In addition, habitat quality at stopover sites can also be affected by precipitation that falls in the winter and spring (November–May). Rainfall in arid and/or desert regions within the flyway may have a disproportionate effect on survival, and variation in precipitation in the desert region of the western flyway was associated with variation in the annual survival of Swainson's Thrush [9]. We predicted that westerly (U) and southerly (V) wind speeds and the number of storm nights on the migratory flyway during the migratory period (April–May) would have negative effects, and that precipitation in either the desert region or arid region of the flyway would have positive effects on the annual survival of chats ("Model set 4"; Table 1). Modeled climate data on the migration route was also obtained from the NCEP/NCAR Reanalysis database using RNCEP [40]. Precipitation data were extracted from "arid" (N = 52) and "desert" (N = 23) zones within the western flyway [41], whereas wind speed data was extracted from the entire flyway (N = 49; Fig. 1).

Analysis

We estimated apparent annual adult survival from 2001 to 2015 using the Cormack-Jolly-Seber model. We calculated the probability of an adult returning to the study site (ϕ) after controlling for the probability that banded individuals were resighted or recaptured, hereafter described as resighting probability (ρ), using program MARK 5.1 [42, 43]. Probability of return (ϕ) reflects both survival and emigration, and our apparent annual

survival estimates therefore underestimate annual survival. The global model that allowed adult survival to vary as a function of gender, age and year and the resighting probability to vary as a function of gender and year fit the data well and showed no evidence of overdispersion (median procedure, ĉ < 1).

We first determined the best model structure for the resighting rate, and then modeled survival rates with candidate models containing gender, age, year, and all possible interactions (Table 1). We then developed a series of candidate model sets to examine whether annual apparent survival varied with annual ENSO (Model set 1), breeding ground conditions (Model set 2), winter conditions (Model set 3), or migration conditions (Model set 4). Model set 2 included models with both Precipitation$_{OCT-APR}$, Precipitation$_{MAY-JUL}$, and GDD, the precipitation variables combined or alone, and GDD alone, in addition to the best model in the earlier temporal analysis (n = 6 models). Model set 3 included models with Precipitation$_{MAY-NOV}$ or Precipitation$_{DEC-APR}$ and the best model in the earlier temporal analysis (n = 3 models). Both variables were not included because they were highly correlated. Model set 4 included models with both U + V-wind, U and V-wind alone, the number of storm nights in April and May, Precipitation$_{NOV-MAY}$ on either the desert or arid region of the migratory flyway, and the best model from the earlier temporal analysis (n = 7 models). Finally, we asked whether overall ENSO conditions or conditions on the breeding grounds, wintering grounds or on migration best described variation in annual apparent survival. In this candidate set we included the top model in each of Model sets 1–4 and the best model in the earlier temporal analysis. We used a hierarchical modelling approach so that we can evaluate climate effects that operate in a single season allowing comparison with other studies focused on one season alone, and then compete the best models from within each season to assess which period plays the most important role in explaining variation in annual survival. With the exception of the annual SOI, all climate variables were standardized by subtracting the mean and dividing the standard deviation allowing estimated effect sizes to be compared directly. At each stage of the analysis we used Akaike's Information Criterion to rank competing models. Model weights and beta estimates of effect sizes were used to compare models and assess the importance of individual climate variables.

Results

We colour-banded a total of 313 chats (118 females, 195 males) between 2001 and 2015. Ninety-four birds were resighted or recaptured at least once a year or more after banding with one male resighted for 6 consecutive years after being banded as an SY bird in 2005. Out of the 148 re-encounters, 93 (63%) were resighted but not recaptured; the remaining 55 (37%) were predominantly both recaptured and resighted, with a few individuals being recaptured but not resighted.

Apparent survival rate

The best resighting model indicated that resighting varied with gender (Table 2); males were more likely to be resighted on the breeding grounds than females (male = 0.59 ± 0.05; female = 0.39 ± 0.09). This model received marginally more support than a model where resighting of males and females did not differ and substantially more support than a model where resighting varied with both gender and year.

The best model examining temporal variation in the annual apparent survival of chats suggested that annual apparent survival was relatively constant across the 15 years of the study. The top (i.e. the null model) received slightly more support than models indicating that survival varied with gender and/or age, and substantially more support than models indicating that survival varied across years (Table 2). The estimated annual apparent survival from the top model was 0.57 ± 0.03, similar to the estimate of 0.65 from a previous study [44]. See Additional file 1: Table S1 for estimates of annual survival from the simple temporal model.

Climate variables predicting survival rates
Relationship between climate variables

ENSO conditions over the course of the year, measured using the average SOI from May to April ($SOI_{MAY-APR}$),

Table 2 Summary of models examining gender (g), age (a), and temporal (t) variation in (a) resighting probability, and (b) apparent annual survival of the western yellow-breasted chats breeding in south Okanagan valley, British Columbia (n = 313 individuals, 461 encounters

Model	AIC_c	ΔAIC_c	Weight	K
(a) Resighting models				
Phi(g*a*t) p(g)	774.28	0	0.51	58
Phi(g*a*t) p(.)	774.30	0.02	0.49	57
Phi(g*a*t) p(g + t)	794.37	20.09	0	71
Phi(g*a*t) p(g*t)	830.29	56.01	0	84
(b) Apparent annual survival models				
Phi(.)	708.67	0	0.24	3
Phi(g)	709.22	0.55	0.18	4
Phi(a)	709.42	0.75	0.17	4
Phi(g + a)	709.47	0.80	0.16	5
Phi(g + a+ga)	709.57	0.90	0.16	6
Phi(g + a+t)	713.22	4.55	0.02	18
Phi(g + a+ga + t)	713.32	4.67	0.02	19
Phi(g + t)	713.97	5.30	0.02	17
Phi(a + t)	714.68	6.01	0.01	17
Phi(t)	714.97	6.30	0.01	16
Phi(g*a*t)	774.28	65.61	0	58

We show all models in the resighting candidate set and all models with ΔAIC_c < 10 and the global Phi model in the apparent annual survival candidate set. In all apparent annual survival models resighting probability varies with gender

Phi(.) null model, *Phi(g*a*t)* global model

was postively correlated with average westerly wind speeds and the frequency of storms in April and May ($U-wind_{APR-MAY}$, r = 0.605, p = 0.02; $Storm_{APR-MAY}$, r = 0.522, p = 0.06). $SOI_{MAY-APR}$ was not significantly correlated with any of the other climate variables (Additional file 1: Table S2). Breeding season climate variables ($GDD_{JAN-MAY}$, $Precipitation_{OCT-APR}$, $Precipitation_{MAY-JUL}$) were not significantly intercorrelated (all r < 0.45, Additional file 1: Table S2). Precipitation on the wintering grounds from May to November was correlated with precipitation on the wintering grounds from December to April (r = 0.816, p = 0.004). Unsurprisingly, average westerly wind speeds on the migration flyway in April and May increased as the frequency of storms increased (r = 0.841, p = 0.002), and precipitation in desert regions of the migration flyway from November to May was highly correlated with precipitation in arid regions of the flyway (r = 0.927, p < 0.001). Across periods the only climate variables that were significantly intercorrelated were precipitation on the breeding grounds from May to July and the frequency of storms on migration (r = 0.615, p = 0.02, Additional file 1: Table S2).

El Niño Southern Oscillation (ENSO)

Annual apparent survival of chats was not related to the average ENSO conditions experienced over the course of the year. The model that included the $SOI_{MAY-APR}$ term received less support than the model indicating that annual apparent survival was constant, and the beta estimate for the $SOI_{MAY-APR}$ term had 95% CI intervals that spanned zero (-0.28 ± 0.20, -0.68 to 0.12).

Climate conditions on breeding ground

The number of growing degree days was positively associated, and precipitation on the breeding grounds between May and July was negatively associated, with the annual apparent survival of chats. The top model in Model set 2, that examined breeding ground effects on annual survival, included both $GDD_{JAN-MAY}$ and $Precipitation_{MAY-JUL}$. This model received slightly more support than the model with only the $GDD_{JAN-MAY}$ term and 15 times the support of the null model (Table 3). Beta estimates for the $GDD_{JAN-MAY}$ term had 95% CI that did not span zero (0.30 ± 0.13, $0.06-0.55$), whereas those

for $Precipitation_{MAY-JUL}$ were lower and had 95% CI that spanned zero (-0.23 ± 0.16, -0.53 to 0.08).

Climate conditions on wintering ground

Precipitation on the wintering grounds was not associated with variation in the annual apparent survival of chats. Models containing the $Precipitation_{DEC-APR}$ and/ or the $Precipitation_{MAY-NOV}$ terms received less support than the null model in Model set 3 (Table 3).

Climate conditions on spring migration route

Wind speed on the migration flyway during April and May was negatively associated with the annual apparent survival of chats (Fig. 2). Three models in Model set 4 that examined migration effects on annual survival received strong support, and all three models included terms associated with wind speed and/or frequency of storm events (Table 3). The top model in Model set 4, that included only the westerly wind speed term ($U\text{-}wind_{APR-MAY}$), received over 20 times the support of the null model. Beta estimates for the U-wind term had confidence

Table 3 Models examining the relationship between climate variables on the breeding grounds, wintering grounds, and on migration and the apparent annual survival of western yellow-breasted chats breeding in the Okanagan valley, British Columbia, Canada (n = 313 individuals, 461 encounters)

Variables	AIC_C	ΔAIC_C	Weight	K
a. Breeding ground				
$GDD_{JAN-MAY}$ + $precipitation_{MAY-JUL}$	704.09	0	0.30	5
$GDD_{JAN-MAY}$	704.34	0.25	0.26	4
$GDD_{JAN-MAY}$ + $precipitation_{OCT-APR}$	705.14	1.05	0.17	5
$GDD_{JAN-MAY}$ + $precipitation_{OCT-APR}$ + $precipitation_{May-JUL}$	705.85	1.76	0.12	6
$Precipitation_{OCT-APR}$	707.78	3.69	0.05	4
$Precipitation_{MAY-JUL}$	708.24	4.15	0.04	4
$Precipitation_{OCT-APR}$ + $precipitation_{MAY-JUL}$	708.52	4.44	0.03	5
Phi(.)	708.67	4.58	0.02	3
Phi(g*a*t)	774.28	70.19	0	58
b. Wintering ground				
Phi(.)	708.67	0	0.44	3
$Precipitation_{DEC-APR}$	709.10	0.42	0.36	4
$Precipitation_{MAY-NOV}$	710.28	1.61	0.20	4
Phi(g*a*t)	774.28	65.61	0	58
c. Migration route				
$U\text{-}wind_{APR-MAY}$	702.88	0	0.43	4
$Storm_{APR-MAY}$	703.43	0.55	0.33	4
$U\text{-}wind_{APR-MAY}$ + $V\text{-}wind_{APR-MAY}$	704.78	1.90	0.17	5
$Precipitation_{NOV-MAY(ARID)}$	708.57	5.69	0.03	4
Phi(.)	708.67	5.79	0.02	3
$Precipitation_{NOV-MAY(DESERT)}$	710.06	7.19	0.01	4
$V\text{-}wind_{APR-MAY}$	710.31	7.43	0.01	4
Phi(g*a*t)	774.28	71.4	0	58

K number of parameters, GDD growing degree days, Phi(.) null model, Phi(g*a*t) global model

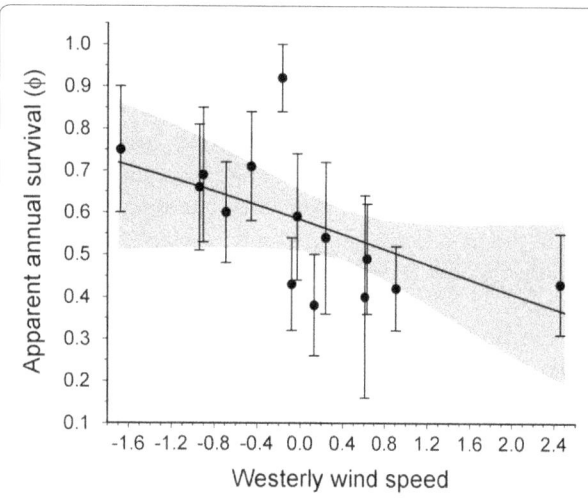

Fig. 2 Annual apparent survival (±SE) of adult western yellow-breasted chats in the south Okanagan valley, British Columbia, Canada from 2001 to 2015 in relation to standardized westerly wind speed during migration. *Solid lines* and *shaded area* represent predicted apparent annual survival ±95% CI from the top model

Table 4 Comparison of best-supported climate models from the candidate sets examining apparent annual survival of western yellow-breasted chats during each stage of the annual cycle (migration, breeding, wintering), the annual ENSO model and the null model (n = 313 individuals, 461 encounters)

Stage	Model	AIC$_c$	ΔAIC$_c$	Weight	K
Migration	U-wind$_{APR-MAY}$	702.88	0	0.59	4
Breeding	GDD$_{JAN-MAY}$ + precipitation$_{MAY-JUL}$	704.09	1.21	0.32	5
Annual	Phi(.)	708.68	5.8	0.03	3
Annual	SOI$_{MAY-APR}$	708.86	5.98	0.03	4
Wintering	Precipitation$_{DEC-APR}$	709.097	6.22	0.03	4
Annual	Phi(g*a*t)	774.28	75.4	0	58

K number of parameters, *GDD* growing degree days, SOI Southern Oscillation Index, *Phi(.)* null model, *Phi(g*a*t)* global model

intervals that did not span zero (−0.36 ± 0.13, −0.61 to −0.1). The second ranked model that received a similar level of support contained the Storm$_{APR-MAY}$ term, and the third ranked- model included both the U-wind$_{APR-MAY}$ and the V-wind$_{APR-MAY}$ terms. Models with terms associated with precipitation at stopover habitat in desert or arid regions of the migration flyway received negligible support (Table 3).

Strongest predicting climate conditions on annual survival
When competing models examining different hypotheses for climate effects on annual apparent survival were tested, the top model indicated that annual apparent annual survival rates were best predicted by westerly wind on the migratory flyway during April and May. This model received nearly twice the support of the model linking annual survival and conditions on the breeding grounds that contained the Precipitation$_{MAY-JUL}$ + GDD$_{JAN-MAY}$ terms (Table 4). See Additional file 1: Table S3 for beta estimates, standard errors and 95% confidence intervals for logit link function parameters in climate and null models in Table 4.

Discussion

Large-scale climatic phenomenon such as ENSO have profound impacts on regional weather conditions, including temperature, rainfall patterns, and wind conditions. These climate regimes influence the survival and breeding phenology of neotropical migratory birds [7, 9, 31, 45]. Temperature and precipitation have major implications on foliage productivity and insect abundance,

which in turn can affect the survival of migrating songbirds [33–39]. Adverse wind conditions and extreme storm events, in comparison, can either cause direct mortality or result in higher energetic cost for migrating individuals [7, 8]. In this study, we found evidence that the annual adult survival rate of a neotropical migrant, the western yellow-breasted chat, was negatively and most strongly associated with westerly wind speed during the spring migration from the central west coast of Mexico to southwestern Canada. Further, the frequency of storm events in their spring migration route had a negative effect on their annual survival, whereas GDD from January to May on their breeding ground had a positive effect.

The negative effect of wind conditions was either as a result of higher average westerly wind speed or the high frequency of storm events on the migration route of yellow breasted chats. These two climate indices were correlated, and both described variation in annual apparent survival, making it hard to distinguish between the effects of extended periods with high crosswinds and the effects of less frequent extreme events. Previous studies have similarly shown that wind conditions during migration negatively influenced the annual apparent survival of other migratory birds [46], including yellow warblers (*Setophaga petechial* [7]), and chimney swifts (*Chaetura pelagica* [10]). Our 15 years of long-term study corroborated the findings of other shorter term studies that varied from 3 to 9 years [7, 10]. Favourable tailwind conditions facilitate migratory flight, and thereby allow birds to expend less energy per unit distance. On the other hand, turbulences and strong winds against the direction of their flight path can result in less efficient migratory flights, leading to greater energy expenditure [47, 48]. Adverse wind conditions deplete their energy reserves, potentially rendering individuals to either die of exhaustion or become more susceptible to depredation risks. Although most migrants stay grounded until windstorms have abated, individuals in the midst of migrating are at most risk of perishing or becoming displaced from these extreme storm events. The effects of windstorms are likely even more detrimental to the survival of migrants when they are flying across long stretches of landmass or waterbodies without suitable stopover sites [9]. In the case of chats, the Great Basin, the Sonoran Desert, and the Gulf of Mexico may all act as migratory barriers.

An alternative explanation for the negative association between westerly wind speed and annual apparent survival is that strong westerly crosswinds cause migrating chats to stray off course and breed elsewhere, likely in the eastern or southern parts of their breeding range [7, 49]. When faced with prevailing crosswinds, migrants may be pushed off their flight path, or compensate by reorienting their flight to offset the drift in order to remain philopatric to their breeding site at the cost of higher energy expenditure [49–53]. Migrants are more likely to opt for the latter strategy in situations where being blown off course means encountering inhospitable terrains such as oceans and deserts [49]. In our study, chats that experienced strong westerly crosswinds on the pacific flyway could potentially afford to deviate eastward and still encounter potentially suitable stopover or breeding sites. Alternatively, chats may have also settled for more southern breeding grounds when unable to overcome the strong winds or storms, leading to individuals from the Okanagan (northern edge population) to shift further south in years where migratory conditions were unfavourable. This potential population dynamic shift can be confirmed by conducting similar survival estimates in the other populations across its range. Recent technological advances allowing tracking of small birds (e.g. MOTUS towers [54]) may provide an opportunity to assess how chats and other species on different flyways respond to variation in wind conditions during flight [11].

Our study provided evidence that GDD on the breeding ground prior to the breeding season had a positive effect on the annual survival of chats. GDD has strong influences on primary productivity and insect biomass [33, 34], and has been shown to be a key driver of avian distribution and diversity [55, 56]. Plant productivity positively predicts insect abundance [57], which in turn, affects the breeding phenology and success of insectivorous passerines, including dusky flycatchers (*Empidonax oberholseri* [58]), tit species (*Parus* spp. [59]), and horned larks (*Eremophila alpestris* [60]). Further, increased plant growth means denser foliage, potentially allowing for nests to be better concealed from predators and harsh weather [58, 60, 61]. In years where GDD promoted ample food

source and possibly better nest concealment in our study area, adult chats may have perceived the breeding habitat as high quality, and as a result demonstrated higher site fidelity.

Abundant rainfall facilitates plant growth and insect biomass, and thereby has the potential to increase the survival of birds that use these areas for breeding, overwintering, or refueling during migration [9, 18, 30, 31]. The productivity of neotropical migrants breeding in the Pacific northwest of North America was higher in El Niño years (−ve SOI values), which are associated with wetter springtime weather along the Pacific slope from southern California to central Mexico [31, 62]. Contrary to those studies, we found little to suggest that ENSO or precipitation levels contributed to the the annual survival of chats. Similarly, other studies also found no evidence for ENSO or rainfall effects on the survival of American redstarts (*Setophaga ruticilla* [63]) and yellow warblers [7] (but see LaManna et al. [9]). An explanation for this lack of evidence could be that in years where primary productivity was poor along the migratory route, chats made more punctuated migratory bouts, enabling them to require lower fat store accumulation at stopover sites, as opposed to opting for the long-jump strategy which would demand greater metabolic expense. On its wintering ground, sites with declining Enhanced Vegetation Index (a proxy for plant productivity) had reportedly a positive impact on the within-winter survival of chats [64]. This suggests that chats have adapted well to wintering habitat types characterized by relatively low plant productivity (e.g. lowland shrub-steppe and scrub habitat), and that while strong ENSO phases and greater precipitation may promote vegetation growth in these habitats, the cascading impact on chat survival is not significant.

Conclusions

Climate change forecasts indicate an increase in the frequency and intensity of extreme weather, including storm events, droughts, and prolonged precipitation [65]. With stronger climate fluctuations and anomalies against a backdrop of anthropogenic changes to the landscape, these stressors present a challenge for long-distant neotropical migrants during all stages of their annual cycle [8, 10, 15–18]. Our study showed that storm events were more frequent during weaker ENSO events (i.e. higher SOI values), and that westerly wind speed within the western flyway was also positively correlated with SOI values [7]. However, the direction and degree to which ENSO is affected by climate change are unclear [66]. We found that the frequency of storm events was negatively associated with chat survival, whereas GDD had a positive association. Both of those climate indices are expected to increase with climate change; whether the effects of one factor would offset the effects of another is unknown. Therefore, given such uncertainties, we are currently unable to predict with confidence the mechanistic effects of climate change on the population dynamics of chats. The interplay between these climate factors within the context of climate change will need to be further explored to elucidate how neotropical migrant bird survival will respond to a changing climate. We also recommend conducting a similar study for multiple chat populations across its range, allowing for a more complete and broader picture of the population dynamics with respect to climate conditions. Furthermore, understanding the cumulative effects of climate change and other anthropogenic threats such as habitat loss and man-made migration hazards will be critical to prioritize appropriate strategies for chats and other neotropical migratory passerines.

Authors' contributions
ACH, DJG, and CAB designed the study. ACH and CAB conducted the field work. ACH and AD managed and compiled the data. DJG conducted most of the data analysis. ACH wrote the manuscript with contributions from all co-authors. All authors read and approved the final manuscript.

Author details
[1] Environment and Climate Change Canada, Delta, BC, Canada. [2] Department of Biological Sciences, Center for Wildlife Ecology, Simon Fraser University, Burnaby, BC, Canada. [3] Department of Forest and Conservation Sciences, University of British Columbia, Vancouver, BC, Canada.

Acknowledgements
We thank Michael Bezener, Richard Armstrong, Tiinesha Begaye, Tim Forrester, Natasha Barlow, Jenn Baici, Jon Ruddy, Hilary Lefort, Julien St-Amand and many others for their contributions in the field. We also thank the En'owkin Centre, the Osoyoos Indian Band, and many other private land owners for access to their lands.

Competing interests
The authors declare that they have no competing interests.

Funding
The study was funded by Environment and Climate Change Canada, the En'owkin Centre, and the Science Horizons Program (Government of Canada).

References
1. North American Bird Conservation Initiative. The State of North America's birds 2016. Environment and climate change Canada: Ottawa; 2016. http://www.stateofthebirds.org.
2. Nebel S, Mills A, McCracken JD, Taylor PD. Declines of aerial insectivores in north America follow a geographic gradient. Avian Conserv Ecol. 2010;5:1.
3. Calvert AM, Bishop CA, Elliot RD, Krebs EA, Kydd TM, Machtans CS, Robertson GJ. A synthesis of human-related avian mortality in Canada. Avian Conserv Ecol. 2013;18:11.
4. Faaborg J, Levey DJ, Johnson DH, Holmes RT, Anders AD, Bildstein KL, et al. Conserving migratory land birds in the New World: do we know enough? Ecol Appl. 2010;20:398–418.
5. Ockendon N, Johnston A, Baillie SR. Rainfall on wintering grounds affects population change in many species of Afro-Palaearctic migrants. J Ornithol. 2014;155:905–17.

6. Brown DP, Comrie AC. A winter precipitation 'dipole' in the western United States associated with multidecadal ENSO variability. Geophys Res Lett. 2004;31:L09203.
7. Drake A, Rock CA, Quinlan SP, Martin M, Green DJ. Wind speed during migration influences the survival, timing of breeding, and productivity of a neogropical migrant, Setophaga petechia. PLoS ONE. 2014;9:e97152.
8. Newton I. Can conditions experienced during migration limit the population levels of birds? J Ornithol. 2006;147:146–66.
9. LaManna JA, George TL, Saracco JF, Nott MP, DeSante DF. El Nino-southern oscillation influences annual survival of a migratory songbird at a regional scale. Auk. 2012;129:734–43.
10. Dionne M, Maurice C, Gauthier J, Shaffer F. Impact of Hurricane Wilma on migrating bird: case of the chimney swift. Wilson J. Ornithol. 2008;120:784–92.
11. Dossman BC, Mitchell GW, Norris DR, Taylor PD, Guglielmo CG, Matthews SN, Rodewald PG. The effects of wind and fuel stores on stopover departure behavior across a migratory barrier. Behav Ecol. 2016;27:567–74.
12. Faaborg J, Holmes RT, Anders AD, Bildstein KL, Dugger KM, et al. Recent advances in understanding migration systems of New World land birds. Ecol Monogr. 2010;80:3–48.
13. Huber DG, Gulledge J. Extreme weather and climate change: understanding the link, managing the risk. Arlington: Pew Center on Global Climate Change; 2011.
14. Fisher RJ, Wellicome TI, Bayne EM, Poulin RG, Todd LD, Ford AT. Extreme precipitation reduces reproductive output of an endangered raptor. J Appl Ecol. 2015;52:1500–8.
15. Studds CE, Marra PP. Fluctuations in rainfall to nonbreeding performance in a long-distance migratory bird, Setophaga ruticilla. Clim Res. 2007;35:115–22.
16. Huntley B, Collingham YC, Green RE, Hilton GM, Rahbek C, Willie SG. Potential impacts of climatic change upon geographical distributions of birds. Ibis. 2006;148:8–28.
17. Klaassen M, Hoye BJ, Nolet BA, Buttemer WA. Ecophysiology of avian migration in the face of current global hazards. Philos Trans R Soc B. 2012;367:1719–32.
18. Sillett TS, Holmes RT, Sherry TW. Impacts of a global climate cycle on population dynamics of a migratory songbird. Science. 2000;288:2040–2.
19. Environment and Climate Change Canada. Recovery strategy for the yellow-breasted Chat auricollis subspecies (Icteria virens auricollis) (Southern Mountain population) in Canada. Species at risk act recovery strategy series. Environment and Climate Change Canada, Ottawa; 2016.
20. Lovette IJ, Clegg SM, Smith TB. Limited utility of mtDNA markers for determining connectivity among breeding and overwintering locations in three neotropical migrant birds. Conserv Biol. 2004;18:156–66.
21. McKibbin R, Bishop CA. Habitat characterization of breeding territories of the western Yellow-breasted Chat in the south Okanagan, British Columbia, Canada. Northwest Nat. 2010;91:145–56.
22. Eckerle KP, Thompson CF. Yellow-breasted Chat (Icteria virens). In The Birds of North America, no. 575. In: Poole A, Gill F, editords. Academy of Natural Sciences, Philadelphia. Washington, DC: American Ornithologists' Union; 2001.
23. McKibbin R, Bishop CA. Feeding observations of the western yellow-breasted chat in the south Okanagan valley, British Columbia, Canada during a 7 year study period. BC Birds. 2007;18:24–5.
24. Pyle P. Identification guide to North American birds-part 1. Bolinas: Slate Creek Press; 1997.
25. Rauscher SA, Giorgi F, Diffenbaugh NS, Seth A. Extension and intensification of the Meso-American mid-summer drought in the twenty-first century. Clim Dyn. 2008;31:551–71.
26. Tourigny E, Jones CG. An analysis of regional climate model performance over the tropical Americas. Part I: simulating seasonal variability of precipitation associated with ENSO forcing. Tellus. 2009;61:323–42.
27. Caso MC, Gonzalez-Abraham C, Ezcurra E. Divergent ecological effects of oceanographic anomalies on terrestrial ecosystems of the Mexican Pacific coast. Proc Natl Acad Sci USA. 2007;104:10530–5.
28. Smith CA, Sardeshmukh P. The effect of ENSO on the intraseasonal variance of surface temperature in winter. Int J Climatol. 2000;20:1543–57.
29. Finley J, Raphael M. The relationship between El Nino and the duration and frequency of the Santa Ana winds of Southern California. Prof Geogr. 2007;59:184–92.

30. Mazerolle DF, Dufour KW, Hobson KA, denHann HE. Effects of large-scale climatic fluctuations on survival and production of young in Neotropical migrant songbird, the yellow Warbler Dendroica petechial. J Avian Biol. 2005;36:155–63.

31. Nott MP, Desante DF, Siegel RB, Pyle P. Influences of the El Niño/Southern Oscillation and the North Atlantic Oscillation on avian productivity in forests of the Pacific Northwest of North America. Glob Ecol Biogeogr. 2002;11:333–42.

32. Gershunov A, Barnett TP. Interdecadal modulation of ENSO teleconnections. Bull Am Meteorol Soc. 1998;79:2715–26.

33. Phillimore AB, Proios K, O'Mahony N, Bernard R, Lord AM, Atkinson S, Smithers RJ. Inferring local processes from macro-scale phonological pattern: a comparison of two methods. J Ecol. 2013;101:774–83.

34. Hodgson JA, Thomas CD, Oliver TH, Anderson BJ, Brereton TM, Crones EE. Predicting insect phenology across space and time. Glob Change Biol. 2011;17:1289–300.

35. Polis GA, Hurd SD, Jackson CT, Pinero FS. El Nino effects on the dynamics and control of an island ecosystem in the Gulf of California. Ecol. 1884;1997(78):1897.

36. Van Bael SA, Aiello A, Valderrama A, Medianero E, Samaniego M, Wright SJ. General herbivore outbreak following en El Nino-related drought in a lowland Panamanian forest. J Trop Ecol. 2004;20:625–33.

37. Moyes K, Nussey DH, Clements MN, Guinness FE, Morris A, Morris S, Pemberton JM, Kruuk LE, Clutton-Brock TH. Advancing breeding phenology in response to environmental change in a wild red deer population. Glob Change Biol. 2011;17:2455–69.

38. Manjujano S. Preliminary evidence of the importance of ENSO in modifying food availability for white-tailed deer in a Mexican tropical dry forest. Biotropica. 2006;38:695–9.

39. Rotenberry JT, Wiens JA. Weather and reproductive variation in shrub-steppe sparrows: a hierarchical analysis. Ecology. 1991;72:1325–35.

40. Kemp MU, Emiel van Loon E, Shamoun-Baranes J, Bouten W. RNCEP: global weather and climate data at your fingertips. Methods Ecol Evol. 2012;3:65–70.

41. Kottek M, Grieser J, Beck C, Rudolf B, Rubel F. World map of the Köppen-Geiger climate classification updated. Meteorol Z. 2006;15:259–63.

42. Lebreton J-D, Burnham KP, Clobert J, Anderson DR. Modeling survival and testing biological hypotheses using marked animals: a unified approach with case studies. Ecol Monogr. 1992;62:67–118. doi:10.2307/2937171.

43. White GC, Burnham KP. Program MARK: survival estimation from populations of marked animals. Bird Study. 1999;46:S120–39. doi:10.1080/00063659909477239.

44. McKibbin R, Bishop CA. Site fidelity and annual survival of the western yellow-breasted Chat (Icteria virens auricollis) at the northern edge of its range. Can Field Nat. 2012;126:135–42.

45. Macmynowski DP, Root TL, Ballard G, Geupel GR. Changes in spring arrival of nearctic-neotropical migrants attributed to multiscalar climate. Glob Change Biol. 2007;13:2239–51.

46. Erni B, Liechti F, Bruderer B. The role of wind in passerine autumn migration between Europe and Africa. Behav Ecol. 2005;16:732–40.

47. Mitchell GW, Woodworth BK, Taylor PD, Norris DR. Automated telemetry reveals age specific differences in flight duration and speed are driven by wind conditions in a migratory songbird. Mov Ecol. 2015;3:19.

48. Liechti F. Birds: blowin' by the wind? J Ornithol. 2006;147:202–11.

49. Horton KG, Van Doren BM, Stepanian PM, Hochachka WM, Farnsworth A, Kelly JF. Nocturnally migrating songbirds drift when they can and compensate when they must. Sci Rep. 2016;6:21249.

50. McLaren JD, Shamoun-Baranes J, Bouten W. Wind selectivity and partial compensation for wind drift among nocturnally migrating passerines. Behav Ecol. 2012;23:1089–101.

51. Zehnder S, Åkesson S, Liechti F, Bruderer B. Nocturnal autumn bird migration at Falsterbo, south Sweden. J Avian Biol. 2001;32:239–48.

52. Peterson AC, Niemi GJ, Johnson DH. Patterns in diurnal airspace use by migratory landbirds along an ecological barrier. Ecol Appl. 2014;25:673–84.

53. Able KP. The orientation of passerine nocturnal migrants following offshore drift. Auk. 1975;94:320–30.

54. Taylor PD, Crewe TL, Mackenzie SA, Lepage D, Aubry Y, et al. The Motus wildlife tracking system: a collaborative research network to enhance the understanding of wildlife movement. Avian Conserv Ecol. 2017;12:8.

55. Honkanen M, Roberge J-M, Rajasärkkä A, Mönkkönen M. Disentangling the effects of area, energy and habitat heterogeneity on boreal forest bird species richness in protected areas. Glob Ecol Biogeogr. 2010;19:61–71.

56. DesGranges J, LeBlanc M. The influence of summer climate on avian community composition in the eastern boreal forest of Canada. Avian Conserv Ecol. 2012;7:2.

57. Haddad NM, Tilman D, Haarstad J, Ritchie M, Knops JM. Contrasting effects of plant richness and composition on insect communities: a field experiment. Am Nat. 2001;158:17–35.

58. Borgmann KL, Conway CJ, Morrison ML. Breeding phenology of birds: mechanisms underlying seasonal declines in the risk of nest predation. PLoS ONE. 2013;8:e65909.

59. Cole EF, Long PR, Zelazowski P, Szulkin M, Sheldon BC. Predicting bird phenology from space: satellite-derived vegetation green-up signal uncovers spatial variation in phenological synchrony between birds and their environment. Ecol Evol. 2015;5:5057–74.

60. Du B, Liu C, Yang M, Bao S, Guan M, Liu N. Horned larks on the Tibetan Plateau adjust the breeding strategy according to the seasonal changes in the risk of nest predation and food availability. J Avian Biol. 2014;45:466–74.

61. Burhans DE, Thompson FR. Relationship of songbird nest concealment to nest fate and flushing behavior of adults. Auk. 2001;118:237–42.

62. Swetnam TW, Betancourt JL. Mesoscale disturbance and ecological response to decadal climatic variability in the American Southwest. J Climate. 1998;11:3128–47.

63. Wilson S, Ladeau SL, Tøttrup AP, Marra PP. Range-wide effects of breeding- and nonbreeding-season climate on the abundance of a Neotropical migrant songbird. Ecology. 2011;92:1789–98.

64. Saracco JF, Ruiz-Gutierrez V. Development of restoration and scaling options for songbirds. Point Reyes Station: The Institute for Bird Populations; 2012.

65. Easterling DR, Meehl GA, Parmesan C, Changnon SA, Karl TA, Mearns LO. Climate extremes: observations, modeling, and impacts. Science. 2009;289:2068–74.

66. Collins M, An S, Cai W, Ganachaud A, et al. The impact of global warming on the tropical Pacific Ocean and El Nino. Nat Geosci. 2010;3:391–7.

67. Alerstam T, Chapman JW, Ba¨ckman J, Smith AD, Karlsson H, et al. Convergent patterns of long-distance nocturnal migration in noctuid moths and passerine birds. Proc R Soc B Biol Sci. 2011;278:3074–80.

68. Felix RK Jr, Diehl RH, Ruth JM. Seasonal passerine migratory movements over the arid southwest. Stud Avian Biol. 2008;37:126–37.

19

Adaptation to new nutritional environments: larval performance, foraging decisions, and adult oviposition choices in *Drosophila suzukii*

Nuno F. Silva-Soares[1*], A. Nogueira-Alves[1], P. Beldade[1,2] and Christen Kerry Mirth[1,3*]

Abstract

Background: Understanding how species adapt to new niches is a central issue in evolutionary ecology. Nutrition is vital for the survival of all organisms and impacts species fitness and distribution. While most *Drosophila* species exploit rotting plant parts, some species have diversified to use ripe fruit, allowing earlier colonization. The decomposition of plant material is facilitated by yeast colonization and proliferation. These yeasts serve as the main protein source for *Drosophila* larvae. This dynamic rotting process entails changes in the nutritional composition of the food and other properties, and animals feeding on material at different stages of decay are expected to have behavioural and nutritional adaptations.

Results: We compared larval performance, feeding behaviour and adult oviposition site choice between the ripe fruit colonizer and invasive pest *Drosophila suzukii*, and a closely-related rotting fruit colonizer, *Drosophila biarmipes*. Through the manipulation of protein:carbohydrate ratios in artificial diets, we found that *D. suzukii* larvae perform better at lower protein concentrations and consume less protein rich diets relative to *D. biarmipes*. For adult oviposition, these species differed in preference for substrate hardness, but not for the substrate nutritional composition.

Conclusions: Our findings highlight that rather than being an exclusive specialist on ripe fruit, *D. suzukii*'s adaptation to use ripening fruit allow it to colonize a wider range of food substrates than *D. biarmipes*, which is limited to soft foods with higher protein concentrations. Our results underscore the importance of nutritional performance and feeding behaviours in the colonization of new food niches.

Keywords: *Drosophila suzukii*, *Drosophila biarmipes*, Foraging, Niche, Nutrition, Nutritional geometry

Background

The food substrates animals exploit play an important role in defining their ecological niche, determining their geographic distribution and abundance [1]. The possibility of exploiting new food substrates provides the opportunity for species to expand their distribution range and can contribute to diversification. To colonize new foods, animals need to adapt to the nutritional, physical, and chemical properties of the food [2]. Many studies have shown how this adaptation often involves acquiring the ability to exploit foods of differing physical properties or tolerate new, frequently toxic, chemical compositions [3, 4]. For example, the fruit fly *Drosophila sechellia* and tobacco hornworm *Manduca sexta* have evolved tolerance to toxic compounds of their food plants, *Morinda* and *Nicotiana*, respectively [3, 4]. Thus, the range of potential food types a species can exploit will define its ability to expand its nutritional niche [2]. In addition, to expand their nutritional niche species need to adapt to differences in the nutrient content of the new foods [2].

*Correspondence: nsoares@igc.gulbenkian.pt; christen.mirth@monash.edu
[1] Instituto Gulbenkian de Ciência, Rua da Quinta Grande nº6, 2780-156 Oeiras, Portugal
[3] School Biological Sciences, Monash University, 25 Rainforest Walk, Melbourne, VIC 3800, Australia
Full list of author information is available at the end of the article

The nutritional composition of food substrates impacts a wide range of life-history and other traits [5–8]. Studies in grasshoppers [9] and caterpillars [10, 11] have shown that niche breadth greatly affects and is affected by an animal's nutritional requirements and foraging strategies. Colonization of new nutritional substrates entails behavioural, physiological, and morphological adaptations that allow individuals to find, choose, and use that substrate. As such, a comprehensive understanding of how animals adapt to new foods requires combining analyses of how animals perform on different diets with analyses of the range of physical and non-nutritional properties animals can exploit [2]. Comparing the impact of diet composition on life history traits and food preferences between closely-related species with divergent nutritional niches can further our understanding of adaption to new nutritional resources and new habitats.

Species of the genus *Drosophila* offer a powerful system for an integrated analysis of the diversification of niche breath. *Drosophila* larvae explore a wide variety of food types and have a diverse range of foraging strategies [12]. The *melanogaster* species group, in particular, contains both generalist species, like *Drosophila melanogaster*, capable of colonizing several different kinds of rotting fruits and fungi, and specialized species, such as *D. sechellia*, that preferentially lays eggs in *Morinda* fruits, which are toxic to most other species [3]. *D. sechellia's* colonization of this new niche has been accompanied by a number of physiological and behavioural adaptations, including changes in the olfactory system that resulted in aversive odors becoming attractive [13], increased tolerance to the toxins in *Morinda* [3], and reduced dopamine synthesis as the species relies heavily on food-derived dopamine [14]. Although *D. sechellia* is an elegant example of an adaption to a new food niche, it is difficult to disentangle adaptations to the toxins from adaptations to the new nutritional environment.

Drosophila species differ not only in the breadth of plant and fungal material, but also in the stage of substrate decay used for oviposition. Decaying organic matter like rotting fruits are dynamic environments that change in nutrient composition (notably, protein and carbohydrate content), pH, and microbial communities throughout the rotting process [15, 16]. Yeasts, crucial to the rotting process and the main source of protein for Drosophilid flies [17–19], colonize decaying organic matter in a species-specific sequential manner [16–18] and increase in concentration as fruits rot. Because differences in oviposition preference dictate the environment in which larvae will develop [20], species-specific preferences in the order of colonization provide an opportunity to explore how species partition nutritional resources.

The Southeast-Asian *D. suzukii* (Matsumura) is unusual among *Drosophila* in its preference for colonizing ripe, rather than rotting, soft-skinned fruits such as strawberries [21]. In recent years, *D. suzukii* has become an invasive pest species, both in Europe and North America, causing severe agricultural damage to several fruit growing industries [21–23]. Studies of *D. suzukii's* adaptation to ripe fruit have focused on the adult females' large and serrated ovipositor that can pierce and lay eggs under the ripe fruit's skin [24]. Despite its economical relevance and potential threat to the environment, little is known about how *D. suzukii* larvae cope with the low protein nutritional environment of ripe fruit. The closely-related *Drosophila biarmipes* (Malloch), unable to pierce ripe fruit skin and limited to colonizing rotting fruit [24], provides a good reference for comparison to *D. suzukii* to explore how adaptation to a new temporal niche reflects nutritional requirements, larval feeding behaviour, and oviposition strategies.

Here, we make use of *D. suzukii* and *D. biarmipes* to understand how differences in their nutritional niche affect the response of life history traits to the macronutrient composition of the larval diet, and how that correlates with larval feeding behaviour and adult oviposition preferences. We use Nutritional Geometry [5, 25, 26] to test the hypothesis that *D. suzukii* larvae perform better in diets with lower protein content when compared to *D. biarmipes*. This approach uses artificial diets varying the quantities of two nutrients to determine how various life history traits map to the nutrient space, and has been applied to animals across many taxa [7–9, 11, 27–31]. Then, we use behavioural assays to test how larvae choose their feeding substrates and how adults choose oviposition sites. We find that differences in the response of larval life history traits and feeding choices reflect divergence in the macronutrient requirements between species, whereas oviposition choice reflects differences in the range of substrates a species can exploit.

Methods

Fly stocks

We used the L19 strain of *D. suzukii* generously provided by Vincent Debat (French Natural History Museum, Paris), and the 14023-0361.11 strain of *D. biarmipes* obtained from the *Drosophila* Species Stock Center (San Diego). *D. suzukii* were maintained at 22.5 °C, and *D. biarmipes* at 25 °C both on standard laboratory fly food: 45 g of molasses, 75 g of sucrose, 70 g of cornmeal, 10 g of agar, 1100 ml of water, and 25 ml of a 10% Nipagin solution per litre. For the maintenance of *D. suzukii*, we added strips of autoclaved paper as perching sites, and dry yeast to increase egg production.

Protein and sugar quantification in strawberries

To quantify the protein and sugar content in decaying strawberries, we placed single ripe strawberries into 11 plastic cups and inserted these cups into fly population cages (11 × 20.5 × 27 cm) at 25 °C. We replicated this three times. To simulate natural yeast inoculation, we introduced 50 males (25 *D. suzukii* and 25 *D. biarmipes)* inside each cage where a 6 cm diameter Petri dish containing standard fly food and live-yeast paste (Baker's yeast) provided a source of yeast during the first 2 days of the experiment. From each replicate, we froze (−20 °C) one strawberry per day on days 1–10 and on day 13 after the start of the yeast inoculation. Samples were later thawed and lysed using glass beads in a Qiagen Tissue-Lyser (10 min at maximum speed). After centrifugation (10 min at 6000×*g*), we measured protein (Pierce BCA Protein assay kit; Thermo Scientific #23227) and glucose/sucrose (Glucose and Sucrose colourimetric/fluorimetric assay kit; Sigma #MAK013) concentrations in the supernatant, following each manufacturer's instructions but using half-size reactions.

To characterize changes in macronutrient content of strawberries as they decayed, we log (x+1) transformed the P and C concentration data and used a linear mixed effect model, with replicate as the random effect and day as fixed effect as in [30].

Larval performance assays

Using nutritional geometry, we raised larvae of *D. biarmipes* and *D. suzukii* on 24 different diets varying in protein, carbohydrate, and caloric content. For each diet, we mixed solutions of yeast (Lesaffre SAF-Instant Red) and sucrose (Sidul, Santa Iria de Azóia, Portugal) that were one of four concentrations (45, 90, 180, and 360 mg/ml) with 0.5% agar corresponding to generate diets of four different caloric concentrations (0.18, 0.36, 0.72 and 1.44 kcal/ml, respectively). For each caloric concentration, we mixed the solutions to produce six different protein to carbohydrate (P:C) ratios: 1.5:1 (corresponding to pure yeast solution), 1:1, 1:2, 1:4, 1:8, and 1:16. All foods were autoclaved and we added 1% of propionic acid (Acros organics, Geel, Belgium) and 1% solution of nipagin (10% p-hydroxy benzoic acid methyl ester in 95% ethanol) to prevent fungal and bacterial contamination. This experiment was conducted at 25 °C and replicated four times for each study species. We measured five life-history traits: survival from first instar to pupariation, larval development time (checked twice per day, at 10:00 and 18:00), female and male pharate adult weight, and ovariole number. Pharate adult weight and ovariole number were measured as described in [8].

Larval foraging assays

We performed two types of larval foraging assays: a two-choice assay where larvae were offered the choice between two P:C ratios, and a no-choice assay where larvae fed on a single diet. Prior to the assays, larvae were reared on standard fly food in densities of approximately 200 individuals in 6 cm Petri dishes filled with standard fly food. We collected L3 (third larval instar) larvae 0–5 h post ecdysis for all assays.

For the two-choice assays, we used 9.2 cm split Petri dishes with each half filled with one food type. We placed ten L3 larvae on each side of the plate (total of 20 per plate) and allowed them to forage for either 2 or 4 h at 25 °C in the dark. We offered larvae one of three choice pairs of foods equal in caloric content (0.72 kcal/ml) but differing in P:C ratios: 1.5:1 vs 1:8; 1:1 vs 1:8; 1:1 vs 1:16. Each choice pair and time point was replicated 10 times. To distinguish between the diets offered, we dyed foods with either red or blue food dye (1% Rayner, Billingshurst, UK), switching the colours between replicate assays to control for colour preference. At the end of the assay, we collected larvae and scored gut colour content of individual larvae followed by the spectrophotometric analysis of the larval pooled gut contents of each replicate. All larvae from each replicate were transferred into one 2 ml Eppendorf and frozen at −20 °C. We later extracted the dye from the sample by adding 80 μl of ice-cold methanol, homogenizing the sample with Qiagen TissueLyser (1 min at maximum speed), and centrifuging at 13,000×*g* for 10 min at 4 °C. We measured the absorbance of 60 μl of the supernatant from each sample at 450 nm for the red dye and at 600 nm for the blue using a Victor3 multi-label plate reader (Perkin Elmer, Waltham, USA).

The no-choice assay design was similar, except that larvae were left to forage on a single P:C ratio (either 1.5:1, 1:1, 1:2, 1:4, 1:8 or 1:16). All foods were isocaloric (0.72 kcal/ml) and dyed blue for the subsequent quantification of the amount of food ingested by spectrophotometric analysis.

Adult oviposition assays

We performed two types of oviposition preference assays: choice between food P:C ratios and choice between food hardness. From each species, we selected 20 newly-eclosed females and 10 newly-eclosed males from density controlled bottles. To transfer *D. suzukii* flies between vials, we are required by regulation to anesthetise them with CO_2. To minimize the effects of exposure to CO_2 on oviposition behavior for both species, we sorted twenty virgin females and ten males into vials and allowed them to mate for 7–9 days. This resulted in the death of some of the females, mainly due to the animals getting stuck on the food. As a result, we had a variable number of females in each vial for the oviposition assay. To adjust for this variability, we report the number of eggs laid normalized by number of females in the vial. Further, we used only vials that contained a minimum of 10 surviving females and 5 surviving males to run 25 replicates of each preference assay. Flies from individual vials were transferred to assay arenas containing three 15 ml Falcon tube lids with food, glued onto a Petri dish (6 cm) capped by a 200 ml plastic cup. All foods were isocaloric (0.72 kcal/ml), but were either of one of three P:C ratios (1:1, 1:4, and 1:8) or of different agar concentrations (1, 2, and 3%). Assays were run at 25 °C for 6 h in complete darkness. At the end of the assay, we froze all flies and assessed oviposition preference by counting the number of eggs in each food patch and dividing it by the number of females in the assay arena.

Statistical analyses

All statistical analyses were done in R (version 3.0.2, R Development Core Team 2013, https://www.r-project.org/). All datasets and scripts are publically available from Figshare (doi: 10.4225/03/58ca18ae80d1a).

From the nutritional geometry experiment, we estimated the response of each life history trait to the protein, carbohydrate, and caloric content of the food based on the methods described in [6, 8, 30]. For survival, we fit a generalized linear model with a quasi-binomial distribution, to account for the overdispersion of the data, and with a logit link function. For the remaining traits, we fit linear mixed effect models, with replicate as the random effect. The effect of both the linear and quadratic components of carbohydrate and protein, as well as their cross-product, was included in our models. To visualize each trait's response on the macronutrient space defined by our panel of diets, we used non-parametric thin plate splines. To compare response shapes between traits and between species, we standardized the dependent variable values to a mean of zero and a unit standard deviation, and compared responses using partial F tests.

The two-choice larval assays provided two types of data: number of larvae with blue and/or red colour in their guts (scored by eye), and quantification of each colour in pooled larvae (determined by spectrophotometer absorbance). For the dataset of larval gut colour (scored by eye) we calculated the proportion of larvae in each replicate that ate protein-rich food only, protein-poor food only, both foods, and no food. We then tested for differences for proportion of individuals in these categories between: (1) choice pairs; (2) species; (3) time points; and (4) possible interactions; by fitting a generalized linear model with a quasi-poisson distribution to account for overdispersion of the data.

For the spectrophotometer absorbance data, we calculated the proportions of protein-rich versus protein-poor diet consumed based on the proportion of measured absorbance values for each dye. We then tested for differences in the proportion of protein-rich food using: choice pairs; species; time points; and possible interactions as factors; by fitting a generalized linear model with a quasi-poisson distribution. For the proportion of protein-rich food in the gut for each treatment and each species, we then tested for significant departure from the null hypothesis (no preferences for either protein rich or protein poor foods: $\mu = 0.5$) using Wilcoxon signed rank test. We also compared the amounts of protein-rich food ingested between species for each treatment by pair-wise comparison of least squared means. We adjusted for multiple comparisons using sequential Bonferroni (Holm) correction of the p values ($\alpha = 0.05$).

The no-choice larval assays allow us to use the quantity of food larvae ingest in each diet to assess how they regulate macronutrient intake [5]. To do this, we calculated quantity of ingested protein and carbohydrate in each diet, by measuring the absorbance values for the blue dye extracted from the gut. By comparing protein and carbohydrate intake across diets, we can determine how larvae prioritise their macronutrient intake. To linearize the data, we applied the log(x+1) transformation. To compare macronutrient intake between treatments and between species, we first normalized protein and carbohydrate intake by species by subtracting the species-specific median from each sample [26]. Thus, a value of zero indicates there is no nutritional offset relative to the median, while positive and negative offsets represent excess or deficits in the intake, respectively (as in [32]). We then analysed the effect of species, P:C ratio, and assay time on each macronutrient offset using linear models. We chose to pool the data of both time points, even if time had a significant effect for protein consumption, because the main macronutrient differences between species were relative to carbohydrates (Additional file 4: Tables S1, Additional file 5: Table S2). We compared the different macronutrient offsets slopes by comparison of the least squared trends, adjusting for multiple comparisons with sequential Bonferroni (Holm) correction.

Finally, we assessed adult oviposition preference in relation to substrate macronutrient composition and substrate hardness in two separate assays. For each assay, to test for differences in the proportion of eggs laid (1) between each food type, (2) between species, and (3) respective possible interactions, we used generalized linear model, with a quasi-binomial distribution to account for overdispersion. We compared the proportion of eggs laid in each substrate against a null distribution of 33%

(no preference between the three substrates offered) by comparing least squared means, adjusting for multiple comparisons using sequential Bonferroni (Holm) correction.

Results

Larval nutritional performance matches each species use of fruit decay stage

We started by measuring protein and sugar content of strawberries, one of the preferred oviposition substrates of *D. suzukii* [33], over the course of 14 days to verify our assumption that P:C ratio is lower in ripe fruits and increases with the rotting process (Additional file 1: Figure S1). Similarly to what had been shown for figs [30], we found that protein concentrations significantly increased with rotting (linear mixed effect model, p value = 0.045, R^2 = 0.191), while sugar concentrations significantly decreased (p value <0.001, R^2 = 0.533). This corresponded to significant increases in P:C ratio with decay (p value <0.001, R^2 = 0.460).

D. suzukii prefers to lay its eggs on ripe fruits while *D. biarmipes* colonizes fruits at later stages of decay. Given the changes in P:C during fruit decay, we hypothesized that, relative to *D. biarmipes*, *D. suzukii* would perform better at lower P:C ratios and/or have higher tolerance to high concentrations of carbohydrates/low concentrations of protein. To test this, we quantified the response of five larval life history traits (survival, developmental time, male/female body mass, and ovariole number) across 24 diets varying in protein, carbohydrate, and caloric composition. We equated better performance with the highest values for survival, body mass and ovariole number, and lowest values for development time. We analysed the effects of protein and carbohydrate content on each trait, and compared response surfaces between traits and species.

For both species, few individuals reached the pupal stage on 1:16 P:C foods (regardless of the caloric value), with survival below 2% for *D. suzukii* and below 7% for *D. biarmipes*. Also for both species, we found that the highest mean survival was obtained on 1.5:1 P:C foods but at different caloric contents: 0.36 kcal/ml for *D. suzukii* (77% survival) and 1.44 kcal/ml for *D. biarmipes* (96% survival). A summary of the trait value ranges can be found in Tables S3 and S4 (Additional files 6, 7). On average, *D. biarmipes* larvae developed faster (105 h), had more ovarioles (26.3 ovarioles), and were larger (1.13 mg male and 1.42 mg females) when raised on the highest protein concentrations (Additional file 7: Table S4). For *D. suzukii*, larvae developed fastest (131 h), had the most ovarioles (24.5), and grew into largest flies (1.47 mg male and 1.82 mg female) when raised in intermediate protein concentrations (Additional file 6: Table S3). When testing

the effect of macronutrient composition on the different traits, we found that protein content, both linear and quadratic components, contributed significantly to all life history traits in both species (Table 1). On the other hand, carbohydrate content only played a significant role for *D. biarmipes* male body mass (linear and quadratic components) and female body mass (in interaction with protein content).

Figure 1 represents the response of the different life history traits to our macronutrient landscape (male mass in Additional file 2: Figure S2). For *D. suzukii*, we found similar response surfaces for all traits (Additional file 8: Table S5), with performance generally increasing between low and intermediate protein levels and plateauing at intermediate protein levels and high P:C ratios. For *D. biarmipes*, all traits were maximized at high protein level and high P:C ratios, but differed in the shape of the response to macronutrient variation. We found three groups of significantly different responses in *D. biarmipes* traits (Additional file 9: Table S6). In Fig. 1 this is particularly clear in relation to how the traits responded to increasing protein levels: survival increased steadily until reaching a plateau, developmental time decreased first steeply and later more shallowly until its minimum, and body mass (both male and female) increased without ever reaching a plateau within our macronutrient panel. The response for ovariole number was statistically indistinguishable from that of both survival and developmental time. When we compared trait responses between species, we found differences for all traits, with the exception of survival (Table 2).

When compared to *D. suzukii*, *D. biarmipes* larvae prefer and consume more protein

We ran two types of behavioural assays to test the hypotheses that macronutrient regulation differs between *D. suzukii* and *D. biarmipes* larvae (no-choice assay) and that larvae choose and consume foods that maximize their performance (two-choice assay).

In our no-choice experiment (Fig. 2A, B), where larvae were left to feed on only one of six P:C ratios (1.5:1; 1:1; 1:2; 1:4; 1:8; and 1:16), we observed that *D. biarmipes* had higher median protein (0.022 mg) and carbohydrate (0.082 mg) intake than *D. suzukii* (0.015 mg protein, 0.044 mg carbohydrate). We then compared the two species in terms of the intake offset for each macronutrient from its normalized median of 0 (Fig. 2B, for mean consumption values see Additional file 10: Table S7). Briefly, an offset value of 0 represents no deviation from the species-specific median across all foods for the respective macronutrient. Conversely, positive or negative offset values represent higher or lower consumption, respectively, of that macronutrient relative to

Table 1 The linear and quadratic effects of carbohydrate (C) and protein (P), and their cross product, in the larval diet on five life history traits: survival, developmental time, female and male adult mass and ovariole number in *Drosophila suzukii* and *Drosophila biarmipes*

Trait	C	p	C²	p²	C × p	R²
D. suzukii						
Survival						
β	−0.002	*0.107*	<−0.001	<−0.001	<−0.001	–
t value	−0.315	*11.287****	−1.401	−10.979***	−0.26	
Dev. time						
β	0.095	*−2.122*	<−0.001	*0.009*	0.001	0.54
t value	0.816	*−11.77****	−0.341	*9.833****	1.074	
Female mass						
β	0.003	*0.019*	<−0.001	*<−0.001*	<−0.001	0.36
t value	1.983	*8.58****	−1.365	*−7.377****	−1.355	
Male mass						
β	0.001	*0.013*	<−0.001	*<−0.001*	<−0.001	0.41
t value	1.284	*8.19****	−1.262	*−7.288****	−0.457	
Ovariole no.						
β	0.016	*0.107*	<−0.001	*<−0.001*	<−0.001	0.09
t value	1.172	*5.017****	−0.718	*−3.84****	−1.61	
D. biarmipes						
Survival						
β	−0.004	*0.126*	<−0.001	*<−0.001*	<−0.001	–
t value	−1.051	*13.236****	−1.24	*−12.438****	−0.074	
Dev. time						
β	0.184	*−3.617*	<0.001	*0.016*	<−0.001	0.74
t value	1.26	*−14.58****	1.669	*12.09****	−0.581	
Female mass						
β	0.001	*0.117*	<−0.001	*<−0.001*	*<−0.001*	0.66
t value	1.9	*13.273****	−1.756	*−9.049****	*−2.512**	
Male mass						
β	*0.001*	*0.007*	<−0.001	*<−0.001*	<−0.001	0.54
t value	*2.491**	*8.972****	*−2.801***	*−5.977****	−0.707	
Ovariole no.						
β	−0.002	*0.186*	<−0.001	*<−0.001*	<0.001	0.43
t value	−0.188	*10.849****	−1.52	*−9.486****	0.498	

For all traits with exception of survival, the models were linear mixed-effects models fit by maximum likelihood. Survival data was analysed with a generalized linear model, assuming a quasi-binomial distribution of survival probabilities and a logit link. Significant coefficients are in italics

* p < 0.05, ** p < 0.01, *** p < 0.001

the median (see "Methods"). Low nutrient offsets and respective variation across treatments indicate a strong nutrient intake regulation and the same logic applies to the reverse situation. Our analysis revealed that both *D.* *suzukii* and *D. biarmipes* regulate protein consumption over carbohydrate consumption; protein intake deviates significantly less from the median than carbohydrate intake (Fig. 2A), absolute consumption values in Fig. 2A).

(See figure on next page.)
Fig. 1 The effects of protein and carbohydrate content of the larval diet on four larval life-history traits of *D. suzukii* (*left column*) and *D. biarmipes* (*right column*). The fitted response surfaces of the effects of 24 different diets varying in protein, carbohydrate, and caloric composition for: (*first row*) proportion of larvae surviving from first instar larvae to pupae; (*second row*) developmental time from first instar larvae to pupae; (*third row*) female pharate weight; and (*fourth row*) total number of ovarioles of adult females. *Dashed black lines* represent the P:C ratios. We replicated four times each block of 24 diets per species using 30 first instar larvae per diet. *Filled black circles* represent the respective nutritional coordinates of each of the 24 diets used (if a dot is absent not enough larvae survived that treatment to measure the trait)

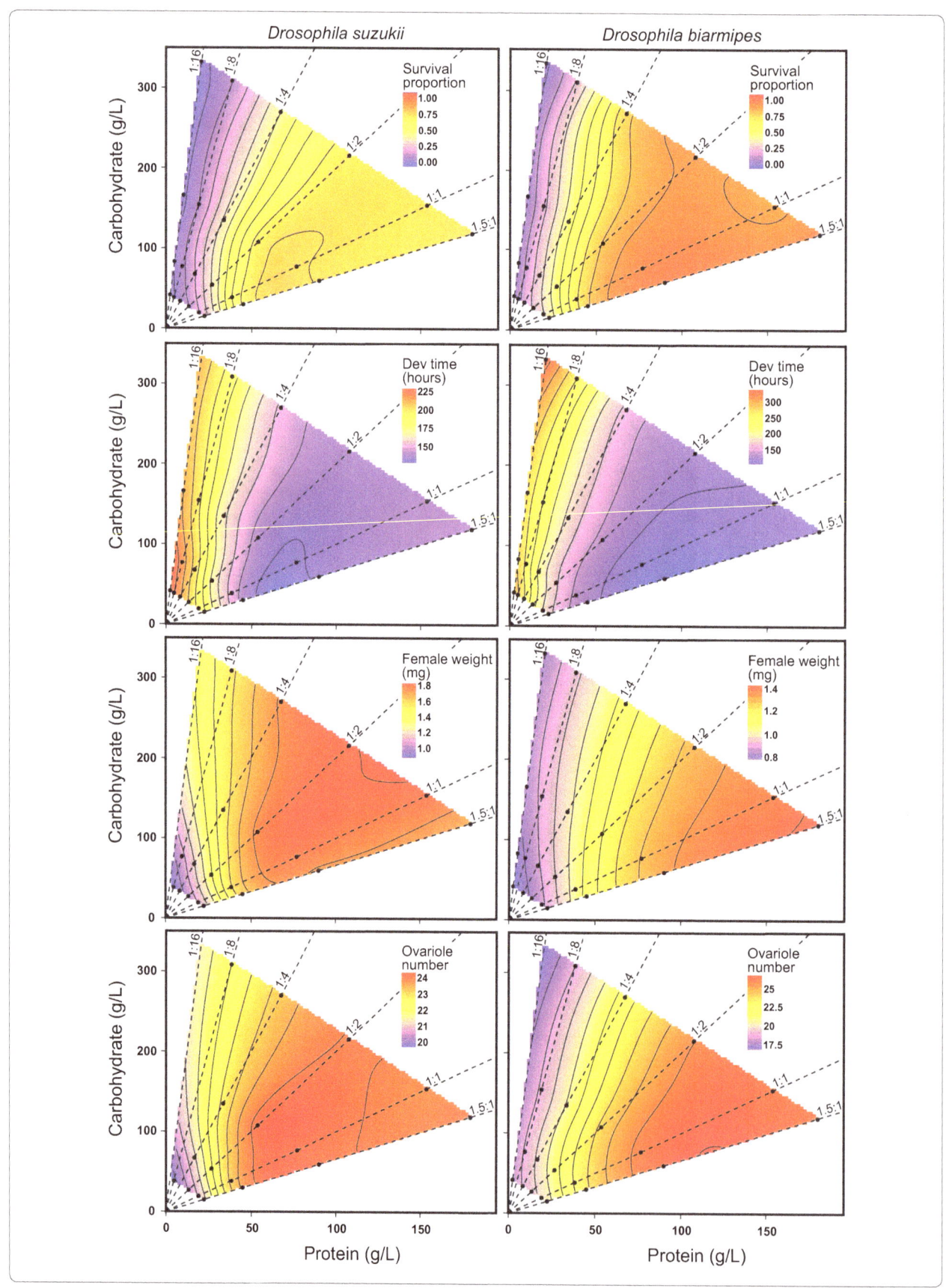

Table 2 Differences in the response surfaces between *D. suzukii* and *D. biarmipes* for each trait

Trait	D. freedom	L ratio	p value
Survival	5	1.793	0.877
Dev. time	5	*58.999*	*<0.001*
Female weight	5	*72.124*	*<0.001*
Male weight	5	*21.968*	*<0.001*
Ovariole no.	5	*32.213*	*<0.001*

Using partial F tests, we compared the response surfaces generated from linear mixed effects models on the scaled parameter values. Response surfaces that show significant differences are highlighted in italics

We also observed that deviations in protein and carbohydrate consumption differed between the two species: *D. biarmipes* had steeper slope for the offset of carbohydrates consumed than *D. suzukii* (Fig. 2A), suggesting a stronger response to P:C variation and higher food consumption rate (Fig. 2B).

We then assessed whether *D. suzukii* and *D. biarmipes* larvae chose differently when given a choice between a protein rich and a protein poor diet of equal caloric value (1.5:1 vs 1:8; 1:1 vs 1:8; and 1:1 vs 1:16—Fig. 2C, D). Our data on the colour inside the gut of individual larvae revealed that, for both species, larvae tended to not mix

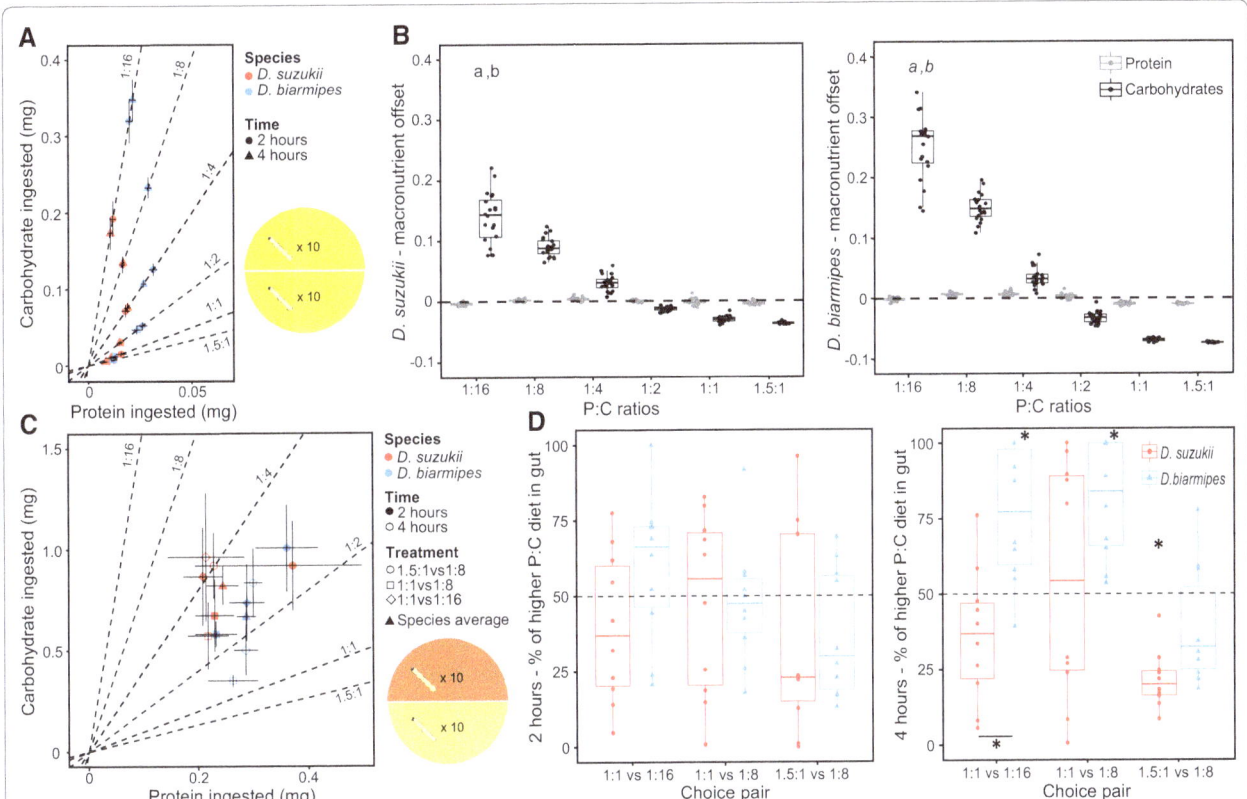

Fig. 2 *D. suzukii* and *D. biarmipes* differ in the quantity of food eaten depending on the diet. **A** Amount of protein and carbohydrate ingested in the no-choice assay. Twenty larvae were offered one of six P:C ratios and were able to forage for 2 or 4 h. *Each dot* represents the mean value of 10 replicates and the *error bars* are 95% confidence intervals of the means. **B** Differences between protein and carbohydrates offsets in *D. biarmipes* and *D. suzukii*. Each condition was replicated ten times (both time points were pooled). *Dashed line* represents the normalized median for each macronutrient (0 = no macronutrient offset). Differences in font type (*regular* versus *italic*) between letters represent significant differences across least squared trends for *a* protein and *b* carbohydrates between the two species. **C** Amount of protein and carbohydrate ingested in the two-choice assay. *Each dot* represents the mean value of 10 replicates, except the *triangles*, which represent the average intake target for both species calculated from the pooled data of all treatments/time points. *Error bars* are 95% confidence intervals of the means. Twenty larvae were offered a choice between two protein to carbohydrate (P:C) ratios in three different combinations. Larvae were able to forage for 2 or 4 h and each time point was replicated ten times. The quantity of food in the larval gut was determined by spectrophotometer. **D** The figures show the percentage of the total amount of food ingested that corresponded to the higher protein food (1:1 for the first and second food pairs, and 1.5:1 for the third food pair) found in the guts of *D. suzukii* and *D. biarmipes* larvae in the two-choice assay. *Black asterisks* represent significant differences to no choice (50%-*dashed black line*—see Additional file 13: Table S10) and *grey asterisks* represent significant differences between species for the same diet (Least squared means comparison, see Additional file 14: Table S11)

foods and fed preferentially on the protein rich food, with a higher proportion of larvae doing so in *D. biarmipes* relative to *D. suzukii* (Additional file 3: Figure S3; Additional file 11: Table S8). This result agrees with our no-choice assay results where larvae prioritized protein over carbohydrate regulation. Our spectrophotometric dataset, allowing the quantification of protein and carbohydrate ingested in pooled larva, showed significant differences between the two species after 4 h of choice assay (Fig. 2D, and absolute consumption values in Fig. 2C). With the exception of when given a choice between 1.5:1 and 1:8 foods, *D. biarmipes* larvae always consumed significantly more of the protein-rich food (Fig. 2D, see Additional file 12: Table S9, Additional file 13: Table S10). On the other hand, *D. suzukii* larvae consumed as much protein-rich as protein-poor food (Fig. 2D), showing no preference between available P:C ratios. The exception was when given a choice between 1.5:1 and 1:8 foods: in this case *D. suzukii* larvae avoided 1.5:1 (pure yeast) food.

Adult oviposition preference correlates with substrate hardness rather than substrate nutritional status

The nutritional response surfaces for life history traits (Fig. 1) and nutritional preference of L3 larvae (Fig. 2D) suggested that *D. suzukii* both perform better on lower protein concentrations and ingest less protein than *D. biarmipes*. We next hypothesized that adult females would choose to lay their eggs in food substrates more suitable for their larvae. To test this hypothesis, we assessed oviposition site preference by letting flies choose between three isocaloric P:C ratios for oviposition: 1:1, 1:4, and 1:8. Our prediction was that *D. biarmipes* would prefer to lay eggs in the protein-rich food (1:1) with P:C ratios closer to those of rotting fruit, while *D. suzukii* would prefer oviposition substrates with lower P:C ratios, closer to those of ripe fruit. However, we found that *D. biarmipes* and *D. suzukii* did not differ in preference. They both preferred the lowest P:C ratio (1:8) for oviposition, and laid the fewest eggs on the highest P:C ratio (Fig. 3A; Table 3; Additional file 14: Table S11).

We next turned our attention to another property that differs between ripe and rotting fruit. *D. suzukii* females have piercing ovipositors that are unique among *Drosophila* flies and allow them to penetrate the harder skin of ripe fruits [24]. This lead us to hypothesize that substrate hardness might underlie differences in oviposition site preferences between the two species. To test if *D. biarmipes* and *D. suzukii* make oviposition decisions based on substrate hardness, we gave females the choice between 1:8 P:C ratio foods differing in agar concentration: 1, 2, and 3%. Our prediction was that *D. suzukii* would prefer higher agar concentrations relative to *D. biarmipes*. We found that, while *D. suzukii* females did

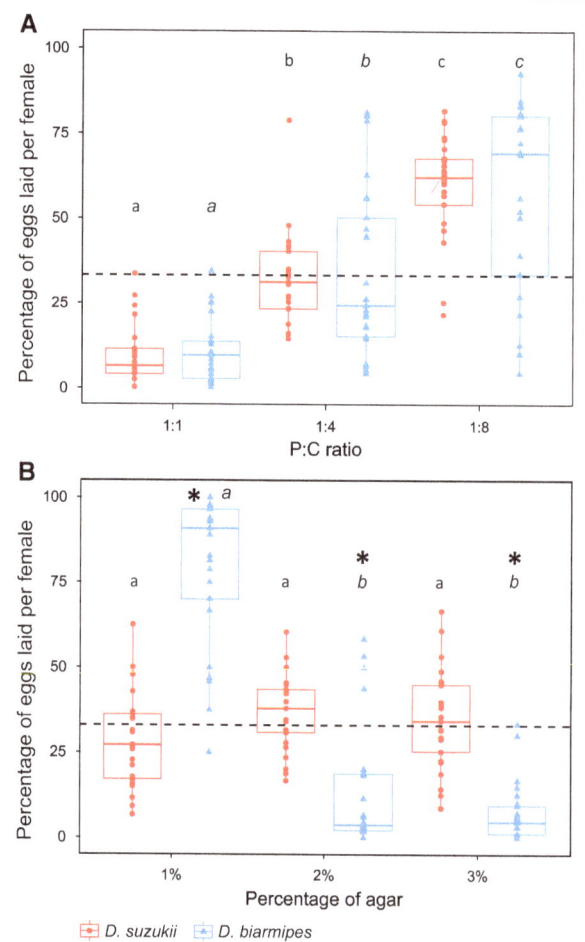

Fig. 3 *D. suzukii* and *D. biarmipes* have the same oviposition preference for P:C ratios, but differ for substrate hardness. The oviposition site preference was estimated by the percentage of eggs laid in each diet. The *letters* (*red* for *D. suzukii*, *blue italics* for *D. biarmipes*) indicate significant differences in the proportion of eggs laid between different diets within a species, with significant differences marked by *different letters*, as determined by least squared means assuming no-choice value of 33% (*dashed line* in *all panels*—see Additional file 14: Table S11, Additional file 15: Table S12). Each treatment was replicated 25 times. *Stars* represent significant differences from the null hypothesis of 33%. **A** Females were offered a choice between three diets differing in their P:C ratios for oviposition. **B** Females were offered a choice between three diets differing in their hardness (agar concentration) for oviposition

not show any preference between food patches, laying equal proportions of eggs in each of the three, *D. biarmipes* showed a clear preference for the softest food (1% agar) (Fig. 3B; Table 3; Additional file 15: Table S12).

Discussion

In this study, we aimed to understand how new nutritional niches impact nutritional performance and foraging behaviour. We studied these in two closely related

Table 3 *D. suzukii* and *D. biarmipes* show significant preferences in oviposition for P:C ratios and react differently to hardness

	D. freedom	Pr(chi)
Nutritional preference		
P:C ratios	2	*<0.001***
Species	1	1.000
Food*species	2	0.828
Hardness preference		
Agar %	2	*<0.001***
Species	1	1.000
Food*species	2	*<0.001***

We fit the data with generalized linear models using a quasi-binomial distribution to account for the overdispersion of the data

Significant differences are highlighted in italics: * p < 0.05, ** p < 0.01, *** p < 0.001

species of *Drosophila* that differ in the order in which they colonize fruit; *D. suzukii* that colonizes ripe fruit and *D. biarmipes* that colonizes rotting fruit. Given this difference and the changes in protein and carbohydrate composition as fruit decays (Additional file 1: Figure S1), we hypothesized that *D. suzukii* larvae would have better performance in and prefer, lower P:C ratios when compared to *D. biarmipes* larvae.

Effects of dietary macronutrient composition on larval life-history traits

Supporting our hypothesis, we found that *D. suzukii* maximizes performance for all life history traits examined at lower protein concentrations when compared to *D. biarmipes*. *D. suzukii* responded similarly across all traits, surviving better, developing faster and into larger flies with more ovarioles at intermediate protein concentrations. This is in agreement with a previous study showing absence of trade-offs for *D. suzukii* life history traits for larvae on protein-poor diets [34]. In contrast, *D. biarmipes* larvae survived better, and developed faster and into larger flies with more ovarioles at the highest protein concentrations. Similar differences in the response of life history traits to diet composition have previously been observed between *Drosophila simulans* and *Zaprionus indianus,* two species that co-inhabit fig plantations in southern Brazil but colonize figs at different stages of decay [30]. This presents the interesting possibility that the order of fruit colonization correlates with larval nutritional performance across *Drosophila* species, with early colonizers performing better at lower P:C ratios when compared to late colonizers.

Fruits, either ripe or rotten, are complex substrates with several macro and micronutrients varying across fruit types/species and decaying stage. In our study, we changed protein concentrations by manipulating

yeast content of our artificial diets. However, as in rotting fruits, yeast contributes more than protein to the diet. Because of this, we cannot disentangle the effect of changes in protein from changes in other nutrients provided by the yeast. For future studies, it will prove valuable to use a synthetic medium, such as developed by [35], to distinguish effects of changes in protein and carbohydrates from changes in other nutrients. Also, analyses of the effect of dietary P:C ratio variation on other life-history traits, such as egg production rate and lifetime fecundity, might uncover relevant trade-offs for the ecology of these fruit flies. Lastly, even though nutritional geometry is a powerful analytical tool, it cannot fully recreate the changes in dietary conditions of rotting fruit. It is, thus, possible that in our artificial setting we either diminish or enhance differences in performance between species. Further studies exploring effects of other properties of rotting fruit, such as changes in concentration of chemical by-products of the rotting process and in microbial composition, as well as interspecific competition would certainly provide a more detailed characterization of the differences in how *D. suzukii* and *D. biarmipes* use food resources.

Larval nutritional preferences and macronutrient compromise

We expected the differences in the response of life history traits to nutrition to be reflected in the larval foraging preferences. While larvae have limited mobility and usually are restricted to forage on the fruit where females laid their eggs, rotting fruits are not homogeneous and larvae might be able to choose between patches with different nutritional composition [16, 20]. Indeed, we found that, when given the choice between two foods with different P:C ratios, the rotting fruit colonizer *D. biarmipes* consumed more of the protein rich diets (in two of three choices provided), while ripe fruit colonizer *D. suzukii* consumed indistinguishable amounts of protein-rich and protein-poor foods (Fig. 2D). In contrast, when we scored individual larval preferences (Additional file 3: Figure S3), we found that, for both species, more larvae chose protein rich food than protein poor food, and rarely mixed the diets. Taken together, these datasets imply that these species use different foraging strategies when offered a choice between food types. While *D. biarmipes* larvae select between food types, in *D. suzukii* larvae the preference for protein rich food was weaker and rather then choosing the protein rich food they compensated by altering their food intake when feeding on protein poor diets. The latter observation contrasts with results from *D. melanogaster,* whose larvae mix between unbalanced foods to achieve an intermediate P:C ratio [8]. This suggests interesting differences in foraging strategies between these three species, with *D. suzukii* and *D. biarmipes* preferring to forage in a single food and regulating their macronutrient intake through the volume of food ingested.

We also found that both *D. suzukii* and *D. biarmipes* were more affected by the protein than by the carbohydrate concentrations (Table 1). This was reflected in their foraging strategies on single foods: both species prioritized protein intake at the expense of consuming excess carbohydrates (Fig. 2A). However, compared to *D. biarmipes* larvae, *D. suzukii* consumed less protein (Fig. 2A, B). When examining normalized macronutrient intakes (Fig. 2A), we found that the offset for carbohydrate intake showed steeper responses to changes in diet P:C ratio in *D. biarmipes* when compared to *D. suzukii* (Fig. 2A). This suggests that rather than simply differing in ingestion rates, *D. suzukii* and *D. biarmipes* differ in macronutrient intake target [36].

Adult oviposition preferences and niche exploration

For *Drosophila* females, yeast represents an important cue for oviposition substrate selection [37]. This selection is crucial as it defines the environment for the next generation of larvae. We expected *D. suzukii* and *D. biarmipes* females to make oviposition choices that reflected the differences in larval nutritional performance and macronutrient preferences. However, we found that these species did not differ in their preference for P:C ratios for oviposition (Fig. 3A). This is in contrast to what had been described for another fly species pair that differ in order of fruit colonization, *Zaprionus indianus* (earlier colonizer of figs) and *D. simulans* (later colonizer) [30]. *Z. indianus* preferred to oviposit on low P:C ratios and *D. simulans* preferred higher P:C ratios but only when competing with *Z. indianus*. This suggests that *D. suzukii* and *D. biarmipes* use different cues to distinguish their order of colonization than do *Z. indianus* and *D. simulans*.

Both *D. suzukii* and *D. biarmipes* preferentially laid their eggs on substrates with low P:C ratios (1:8), similar to what had been previously described for *D. melanogaster* [8, 20]. These ratios are suboptimal for larval performance for all three species [8, 20]. However, because fruits continue to change as they ripen and then rot, the P:C ratio larvae will be eating is likely to be higher from that chosen for oviposition [8, 20]. It seems likely that both *D. biarmipes* and *D. suzukii* are following a similar strategy as *D. melanogaster*: choosing oviposition sites that will be beneficial to larvae with time.

Despite the absence of ovipositional differences between *D. suzukii* and *D. biarmipes* relative to the nutritional composition of the substrate (Fig. 3A), they still differed in oviposition preference relative to another relevant physical property of decaying fruits, substrate hardness (Fig. 3B). *D. biarmipes* preferred laying their eggs in the softest medium while *D. suzukii* females showed no preference for any hardness (Fig. 3B). Thus, it seems that the unique ovipositor of *D. suzukii* widens the range of potential substrates it can exploit allowing this species to use resources unavailable to other species of *Drosophila* [24]. This is also consistent with the natural spatial distribution for *D. suzukii* female flies that exploit oviposition sites on tree crowns, contrasting with other species that only use fruit that has fallen on the ground [38]. Furthermore, this illustrates two alternative strategies for interspecific competition during oviposition connected to different dimensions of the nutritional niche [2]; based on macronutrient content between *Z. indianus* and *D. simulans* [30] and determined by substrate hardness in *D. suzukii* and *D. biarmipes* (this study).

Conclusions

Thanks to its modified ovipositor, *D. suzukii* has the unique capability of colonizing ripe fruit, inaccessible to other *Drosophila* species like *D. biarmipes*, which are restricted to rotting fruit. Using nutritional geometry, we reveal differences between these species going beyond the ovipositor and extending to differences in the nutritional responses of life history traits, larval foraging behaviour, and adult oviposition preference. Our findings highlight that rather than being an exclusive specialist on ripe fruit, *D. suzukii* has a generalist profile relative to food exploitation, being able to colonize a wider range of food substrates than *D. biarmipes*, which is limited to higher P:C ratios and soft substrates (i.e. rotting fruits). Altogether, these results match well with the predictions proposed by Machovsky-Capusta et al. [2] for a successful invasive species [2]. *D. suzukii* not only has a flexible performance relative to the dimensions of macronutrient performance and food composition, but also excels at food exploitation among *Drosophila* flies. For the future, it would be interesting to extend this analysis not only to other *Drosophila* species, but also to other taxa. The *Drosophila* clade provides significant molecular resources with *D. melanogaster* and has other species with interesting ecological contexts, such as the specialist *D. sechellia*, allowing us to address both evolutionary and mechanistic questions relative to nutritional adaptation. By analyzing other taxa including other invasive species, our capacity to predict/control the damage of biological invasions will improve. These studies ultimately would provide more profound insight into phenotypic adaptation to new foods.

Additional files

Additional file 1: Figure S1. The macronutrient composition of straw-berries changes with the stage of decay. The plots show the log transfor-mations of protein (top left) and sugar (sucrose and glucose) amounts in ug per μl (top right) and protein to sugar ratio (bottom left) over the course of 14 days in rotting strawberries for three replicates. Black lines indicate the regression estimates from linear models and the grey shaded areas represent 95% confidence intervals.

Additional file 2: Figure S2. The effects of protein and carbohydrate content of the larval diet on male adult mass of *D. suzukii* (left) and *D. biarmipes* (right). The fitted response surfaces of the effects of 24 different diets varying in protein, carbohydrate, and caloric composition for male pharate weight as proxy for male adult mass. Dashed black lines represent the P:C ratios. Filled black circles represent the respective nutritional coor-dinates of each of the 24 diets used (if a dot is absent, not enough larvae survived that treatment to measure the trait).

Additional file 3: Figure S3. *D. suzukii and D. biarmipes* choose protein rich food over a period of 2 and 4 hours. Twenty larvae were offered a choice between two protein to carbohydrate (P:C) ratios in three different combinations (1:1/1:16, 1:1/1:8 and 1.5:1/1:8). Twenty larvae were left to forage for two or four hours and each treatment was replicated ten times per time point. Larval food preference was first assessed by eye to deter-mine the colour of the larval gut. Each larva was assigned to one of four possible categories: protein rich food; carbohydrate rich food; both foods and none. (Top row) The percentage of *D. suzukii* and *D. biarmipes* larvae assigned to each category after two hours of foraging (Bottom row).

Additional file 4: Table S1. Effects of diet, time, larval species and pos-sible interactions in the amount of protein that third instar larvae ingested for the no-choice larval assays.

Additional file 5: Table S2. Effects of diet, time, larval species and pos-sible interactions in the amount of carbohydrates that third instar larvae ingested for the no-choice larval assays.

Additional file 6: Table S3. Mean values and standard deviation (StDev) for each trait for each diet of the nutritional geometry for *D. suzukii.*

Additional file 7: Table S4. Mean values and standard deviation (StDev) for each trait for each diet of the nutritional geometry for *D. biarmipes.*

Additional file 8: Table S5. Pairwise comparisons between the response surfaces of the five life history traits in *Drosophila suzukii.*

Additional file 9: Table S6. Pairwise comparisons between the response surfaces of the five life history traits in *Drosophila biarmipes.*

Additional file 10: Table S7. Mean values (mg) and standard deviation (StDev) for protein and carbohydrate intake for each P:C ratio of the no-choice experiment.

Additional file 11: Table S8. Effects of pair of diets presented, time, lar-val species and possible interactions for the amount of larvae that chose: Protein rich food; Carbohydrate rich food; Both foods; and None. We analyzed our data with generalized linear models using a quasi-possion distribution (ANOVA type II).

Additional file 12: Table S9. Effects of pair of diets presented, time, larval species and possible interactions in the amount of protein rich food that third instar larvae chose to consume. We analyzed our data with a generalized linear model using a quasi-poisson distribution (ANOVA type II).

Additional file 13. Table S10. Differences of consumption of each diet for each treatment in *D. suzukii* and *D. biarmipes.* We used Wilcoxon signed ranked test each to compare the percentage of protein-rich diet present in the gut to 50% (no-choice). Treatments that show significant differ-ences are highlighted in bold.

Additional file 14: Table S11. Least squared means (Lsmean), standard errors (St. error), and groups for *D. suzukii* and *D. biarmipes* nutritional ovi-position assays, with significant differences denoted by different numbers in the group column (adjusting p-values using the Bonferroni method for a significance level of 0.05).

Additional file 15. Table S12. Least squared means (Lsmean), standard errors (St. error), and groups for *D. suzukii* and *D. biarmipes* hardness ovipo-sition assays, with significant differences denoted by different numbers in the group column (adjusting p-values using the Bonferroni method for a significance level of 0.05).

Authors' contributions
NSS, PB and CKM conceived the study and designed all the experiments. NSS conducted the nutritional geometry experiments. NSS and ANA conducted the larval preference and oviposition assays. NSS and CKM completed the statistical analysis. NSS, PB and CKM wrote the manuscript. All authors read and approved the final manuscript.

Author details
[1] Instituto Gulbenkian de Ciência, Rua da Quinta Grande nº6, 2780-156 Oei-ras, Portugal. [2] Université Toulouse III Paul Sabatier, Bâtiment 4R1, 118 Route de Narbonne, 31062 Toulouse Cedex 9, France. [3] School Biological Sciences, Monash University, 25 Rainforest Walk, Melbourne, VIC 3800, Australia.

Acknowledgements
The authors would like to thank IGC's Fly facility for providing fly food for stock maintenance and also Vincent Debat for providing the *D. suzukii* strain used in this work. The authors would like to thank Carla Srgò for her critical reading of the manuscript.

Competing interests
The authors declare that they have no competing interests.

Funding
This work was funded by the Fundação para a Ciência e Tecnologia (FCT) in the form of a PhD fellowship to Nuno Filipe da Silva Soares (SFRH/BD/91391/2012) and supported by Instituto Gulbenkian de Ciência (IGC) for providing facilities and equipment essential for this study.

References
1. Hutchinson GE. Concluding remarks. Cold Spring Harb Symp Quant Biol. 1957;22:415–27.
2. Machovsky-Capuska GE, Senior AM, Simpson SJ, Raubenheimer D. The multidimensional nutritional niche. Trends Ecol Evol. 2016;31:355–65.
3. R'Kha S, Capy P, David JR. Host-plant specialization in the Drosophila melanogaster species complex: a physiological, behavioral, and genetical analysis. Proc Natl Acad Sci USA. 1991;88:1835–9.
4. Appel HM, Martin MM. Significance of metabolic load in the evolution of host specificity of *Manduca Sexta.* Ecology. 1992;73:216–28.
5. Simpson SJ, Raubenheimer D. A multi-level analysis of feeding behaviour: the geometry of nutritional decisions. Philos Trans R Soc B. 1993;342:381–402.
6. Lee KP, Simpson SJ, Clissold FJ, Brooks R, Ballard JWO, Taylor PW, et al. Lifespan and reproduction in Drosophila: new insights from nutritional geometry. Proc Natl Acad Sci USA. 2008;105:2498–503.
7. Fanson BG, Taylor PW. Protein:carbohydrate ratios explain life span pat-terns found in Queensland fruit fly on diets varying in yeast:sugar ratios. Age. 2012;34:1361–8.

8. Rodrigues MA, Martins NE, Balancé LF, Broom LN, Dias AJS, Fernandes ASD, et al. Drosophila melanogaster larvae make nutritional choices that minimize developmental time. J Insect Physiol. 2015;81:69–80.

9. Raubenheimer D, Simpson SJ. Nutrient balancing in grasshoppers: behavioural and physiological correlates of dietary breadth. J Exp Biol. 2003;206:1669–81.

10. Lee KP, Raubenheimer D, Behmer ST, Simpson SJ. A correlation between macronutrient balancing and insect host-plant range: evidence from the specialist caterpillar Spodoptera exempta (Walker). J Insect Physiol. 2003;49:1161–71.

11. Lee KP, Behmer ST, Simpson SJ. Nutrient regulation in relation to diet breadth: a comparison of Heliothis sister species and a hybrid. J Exp Biol. 2006;209:2076–84.

12. Markow TA, O'Grady P. Reproductive ecology of Drosophila. Funct Ecol. 2008;22:747–59.

13. Dekker T, Ibba I, Siju KP, Stensmyr MC, Hansson BS. Olfactory shifts parallel superspecialism for toxic fruit in Drosophila melanogaster sibling D. sechellia. Curr Biol. 2006;16:101–9.

14. Lavista-Llanos S, Svatoš A, Kai M, Riemensperger T, Birman S, Stensmyr MC, et al. Dopamine drives Drosophila sechellia adaptation to its toxic host. Elife. 2014;3:1–17.

15. Pesis E, Fuchs Y, Zauberman G. Starch content and amylase activity in avocado fruit pulp. J Am Soc Hortic Sci. 1978;103:673–6.

16. Morais PB, Martins MB, Klaczko LB, Mendonca-Hagler LC, Hagler AN. Yeast succession in the amazon fruit Parahancornia amapa as resource partitioning among Drosophila spp. Appl Environ Microbiol. 1995;61:4251–7.

17. Fogleman JC, Starmer WT, Heed WB. Larval selectivity for yeast species by Drosophila mojavensis in natural substrates. Proc Natl Acad Sci USA. 1981;78:4435–9.

18. Starmer WT, Fogleman JC. Coadaptation of Drosophila and yeasts in their natural habitat. J Chem Ecol. 1986;12:1037–55.

19. Buser CC, Newcomb RD, Gaskett AC, Goddard MR. Niche construction initiates the evolution of mutualistic interactions. Ecol Lett. 2014;17:1257–64.

20. Lihoreau M, Poissonnier L-A, Isabel G, Dussutour A. Drosophila females trade off good nutrition with high quality oviposition sites when choosing foods. J Exp Biol. 2016. doi:10.1242/jeb.142257.

21. Lee JC, Bruck DJ, Curry H, Edwards D, Haviland DR, Van Steenwyk RA, et al. The susceptibility of small fruits and cherries to the spotted-wing drosophila Drosophila suzukii. Pest Manag Sci. 2011;67:1358–67.

22. Burrack HJ, Smith JP, Pfeiffer DG, Koeher G, Laforest J. Using volunteer-based networks to track Drosophila suzukii (Diptera: Drosophilidae) an invasive pest of fruit crops. J Integr Pest Manag. 2012;4:1–5.

23. Cini A, Ioriatti C, Anfora G. A review of the invasion of Drosophila suzukii in Europe and a draft research agenda for integrated pest management. Bull Insectol. 2012;65:149–60.

24. Atallah J, Teixeira L, Salazar R, Zaragoza G, Kopp A. The making of a pest: the evolution of a fruit-penetrating ovipositor in Drosophila suzukii and related species. Proc R Soc B. 2014;281:20132840.

25. Raubenheimer D, Simpson SJ. Integrative models of nutrient balancing: application to insects and vertebrates. Nutr Res Rev. 1997;10:151–79.

26. Simpson SJ, Raubenheimer D. Assuaging nutritional complexity: a geometrical approach. Proc Nutr Soc. 1999;58:779–89.

27. Lee KP, Cory JS, Wilson K, Raubenheimer D, Simpson SJ. Flexible diet choice offsets protein costs of pathogen resistance in a caterpillar. Proc R Soc B. 2006;273:823–9.

28. Salomon M, Mayntz D, Lubin Y. Colony nutrition skews reproduction in a social spider. Behav Ecol. 2008;19:605–11.

29. Dussutour A, Latty T, Beekman M, Simpson SJ. Amoeboid organism solves complex nutritional challenges. Proc Natl Acad Sci USA. 2010;107:4607–11.

30. Matavelli C, Carvalho MJA, Martins NE, Mirth CK. Differences in larval nutritional requirements and female oviposition preference reflect the order of fruit colonization of Zaprionus indianus and Drosophila simulans. J Insect Physiol. 2015;82:66–74.

31. Aryal A, Coogan SCP, Ji W, Rothman JM, Raubenheimer D. Foods, macronutrients and fibre in the diet of blue sheep (Psuedois nayaur) in the Annapurna Conservation Area of Nepal. Ecol Evol. 2015;5:4006–17.

32. Almeida de Carvalho MJ, Mirth CK. Food intake and food choice are altered by the developmental transition at critical weight in Drosophila melanogaster. Anim Behav. 2017;126:195–208.

33. Poyet M, Le Roux V, Gibert P, Meirland A, Prévost G, Eslin P, et al. The wide potential trophic niche of the Asiaticc fruit fly Drosophila suzukii: the key of its invasion success in temperate Europe? PLoS ONE. 2015;10:1–26.

34. Jaramillo SL, Mehlferber E, Moore PJ. Life-history trade-offs under different larval diets in Drosophila suzukii (Diptera: Drosophilidae). Physiol Entomol. 2015;40:2–9.

35. Piper MDW, Blanc E, Leitão-Gonçalves R, Yang M, He X, Linford NJ, et al. A holidic medium for Drosophila melanogaster. Nat Methods. 2014;11:100–5.

36. Raubenheimer D, Simpson SJ. The geometry of compensatory feeding in the locust. Anim Behav. 1993;45:953–64.

37. Becher PG, Flick G, Rozpedowska E, Schmidt A, Hagman A, Lebreton S, et al. Yeast, not fruit volatiles mediate Drosophila melanogaster attraction, oviposition and development. Funct Ecol. 2012;26:822–8.

38. Poyet M, Eslin P, Héraude M, Le Roux V, Prévost G, Gibert P, et al. Invasive host for invasive pest: when the Asiatic cherry fly (Drosophila suzukii) meets the American black cherry (Prunus serotina) in Europe. Agric For Entomol. 2014;16:251–9.

Permissions

The contributors of this book come from diverse backgrounds, making this book a truly international effort. This book will bring forth new frontiers with its revolutionizing research information and detailed analysis of the nascent developments around the world.

We would like to thank all the contributing authors for lending their expertise to make the book truly unique. They have played a crucial role in the development of this book. Without their invaluable contributions this book wouldn't have been possible. They have made vital efforts to compile up to date information on the varied aspects of this subject to make this book a valuable addition to the collection of many professionals and students.

This book was conceptualized with the vision of imparting up-to-date information and advanced data in this field. To ensure the same, a matchless editorial board was set up. Every individual on the board went through rigorous rounds of assessment to prove their worth. After which they invested a large part of their time researching and compiling the most relevant data for our readers.

The editorial board has been involved in producing this book since its inception. They have spent rigorous hours researching and exploring the diverse topics which have resulted in the successful publishing of this book. They have passed on their knowledge of decades through this book. To expedite this challenging task, the publisher supported the team at every step. A small team of assistant editors was also appointed to further simplify the editing procedure and attain best results for the readers.

Apart from the editorial board, the designing team has also invested a significant amount of their time in understanding the subject and creating the most relevant covers. They scrutinized every image to scout for the most suitable representation of the subject and create an appropriate cover for the book.

The publishing team has been an ardent support to the editorial, designing and production team. Their endless efforts to recruit the best for this project, has resulted in the accomplishment of this book. They are a veteran in the field of academics and their pool of knowledge is as vast as their experience in printing. Their expertise and guidance has proved useful at every step. Their uncompromising quality standards have made this book an exceptional effort. Their encouragement from time to time has been an inspiration for everyone.

The publisher and the editorial board hope that this book will prove to be a valuable piece of knowledge for researchers, students, practitioners and scholars across the globe.

List of Contributors

Hannah W. Grooms, Elizabeth M. Velazquez, Joshua Traub and Brian J. Kennedy
Department of Civil and Environmental Engineering, Northwestern University, 2145 Sheridan Road, Evanston, IL 60208, USA

Mark W. Westneat
Department of Zoology, Field Museum of Natural History, 1400 South Lake Shore Drive, Chicago, IL 60605, USA

Timothy D. Swain, Emily DuBois, Jillian Henss, Michelle E. Wagner and Luisa A. Marcelino
Department of Civil and Environmental Engineering, Northwestern University, 2145 Sheridan Road, Evanston, IL 60208, USA
Department of Zoology, Field Museum of Natural History, 1400 South Lake Shore Drive, Chicago, IL 60605, USA

Andrew Gomes, Valentina P. Stoyneva, Andrew J. Radosevich, Justin Derbas Jeremy D. Rogers and Vadim Backman
Department of Biomedical Engineering, Northwestern University, 2145 Sheridan Road, Evanston, IL 60208, USA

Arabela A. Grigorescu
Keck Biophysics Facility, Northwestern University, 633 Clark Street, Evanston, IL 60208, USA

Kevin Sanborn, Shoshana Levine, Mark Schick and George Parsons
Fishes Department, John G. Shedd Aquarium, 1200 South Lake Shore Drive, Chicago, IL 60605, USA

Brendan C. Biggs
Division of Water Resource Management, Florida Department of Environmental Protection, 2600 Blair Stone Road, Tallahassee 32399, USA

Peter Schausberger
Department of Behavioural Biology, University of Vienna, Vienna, Austria
Group of Arthropod Ecology and Behavior, Department of Crop Sciences, University of Natural Resources and Life Sciences, Vienna, Austria

Stefan Peneder
Group of Arthropod Ecology and Behavior, Department of Crop Sciences, University of Natural Resources and Life Sciences, Vienna, Austria

Volker Nehring
Centre for Social Evolution, University of Copenhagen, Copenhagen, Denmark
Department for Ecology and Evolution, Biology I, Freiburg University, Hauptstr. 1, 79104 Freiburg, Germany

Francesca R. Dani
Centro di Servizi di Spettrometria di Massa (CISM), University of Florence, Florence, Italy
Dipartimento di Biologia, University of Florence, Florence, Italy

Luca Calamai
Centro di Servizi di Spettrometria di Massa (CISM), University of Florence, Florence, Italy
Dipartimento di Scienza del Suolo e Nutrizione della Pianta, University of Florence, Florence, Italy

Stefano Turillazzi
Department for Ecology and Evolution, Biology I, Freiburg University, Hauptstr. 1, 79104 Freiburg, Germany
Centro di Servizi di Spettrometria di Massa (CISM), University of Florence, Florence, Italy

Horst Bohn
Zoologische Staatssammlung München, Munich, Germany

Klaus-Dieter Klass
Senckenberg Naturhistorische Sammlungen Dresden, Dresden, Germany

Patrizia d'Ettorre
Centre for Social Evolution, University of Copenhagen, Copenhagen, Denmark
Laboratoire d'Ethologie Expérimentale et Comparée (LEEC), Université Paris 13, Sorbonne Paris Cité, Villetaneuse, France

Chunyan Yi, Chunyan Zheng, Ling Zeng and Yijuan Xu
Department of Entomology, South China Agricultural University, Guangzhou 510640, China

Stephen J. Jacquemin
Department of Biological Sciences, Dwyer Hall, Wright State University-Lake Campus, Celina, OH 45822, USA

Mark Pyron
Department of Biology, Aquatic Biology and Fisheries Center, Ball State University, Muncie, IN 47306, USA

Peter W. Shearer and Preston H. Brown
Mid-Columbia Agricultural Research and Extension Center, Oregon State University, 3005 Experiment Station Drive, Hood River, OR 97331, USA

Jessica D. West ad Joanna C. Chiu
Department of Entomology and Nematology, University of California, Davis, CA 95616, USA

Vaughn M. Walton
Department of Horticulture, Oregon State University, Corvallis, OR 97331, USA

Nicolas Svetec
Department of Evolution and Ecology, University of California, Davis, CA 95616, USA

Sylke Wohlrab
Department of Ecological Chemistry, Alfred Wegener Institute, Helmholtz Centre for Polar and Marine Research, 27570 Bremerhaven, Germany

Erik Selander
Department of Biological and Environmental Sciences, University of Gothenburg, Vasaparken, 40530 Gothenburg, Sweden

U. John
Department of Ecological Chemistry, Alfred Wegener Institute, Helmholtz Centre for Polar and Marine Research, 27570 Bremerhaven, Germany Helmholtz Institute for Functional Marine Biodiversity (HIFMB), 26111 Oldenburg, Germany

Andrew N. Gherlenda, Ben D. Moore, Scott N. Johnson and Markus Riegler
Hawkesbury Institute for the Environment, Western Sydney University, Locked Bag 1797, Penrith, NSW 2751, Australia

Anthony M. Haigh
School of Science and Health, Western Sydney University, Locked Bag 1797, Penrith, NSW 2751, Australia

Martins Briedis
Department of Zoology, Palacký University, tř. 17. listopadu 50, 77146 Olomouc, Czech Republic

Peter Adamík
Department of Zoology, Palacký University, tř. 17. listopadu 50, 77146 Olomouc, Czech Republic Museum of Natural History, nám. Republiky 5, 77173 Olomouc, Czech Republic

Steffen Hahn
Department of Bird Migration, Swiss Ornithological Institute, Seerose 1, 6204 Sempach, Switzerland

Virginie Ravigné, Frédéric Chiroleu and Serge Quilici
CIRAD, UMR PVBMT, 97410 Saint Pierre, France

Brahim Chermiti
Institut Supérieur Agronomique de Chott-Mariem, Laboratoire d'Entomologie et de Lutte Biologique, Université de Sousse, 4042 Sousse, Tunisia

Abir Hafsi
CIRAD, UMR PVBMT, 97410 Saint Pierre, France. Institut Supérieur Agronomique de Chott-Mariem, Laboratoire d'Entomologie et de Lutte Biologique, Université de Sousse, 4042 Sousse, Tunisia

Benoit Facon
CIRAD, UMR PVBMT, 97410 Saint Pierre, France UMR « Centre de Biologie pour la Gestion des Populations », INRA-SPE, 755 avenue du Campus, Agropolis, CS 30016, 34988 Montferrier sur Lez, Cedex, France

Pierre-François Duyck
CIRAD, UMR PVBMT, 97410 Saint Pierre, France UMR « Peuplements Végétaux et Bio-agresseurs en Milieu Tropical », CIRAD Pôle de Protection des Plantes, 7 chemin de l'Irat, 97410 Saint Pierre, La Réunion, France

Topi K. Lehtonen
Department of Biosciences, Åbo Akademi University, Tykistökatu 6, 20520 Turku, Finland Section of Ecology, Department of Biology, University of Turku, 20014 Turku, Finland

Department of Biological and Environmental Science, University of Gothenburg, 40530 Gothenburg, Sweden
School of Biological Sciences, Monash University, Melbourne, VIC 3800, Australia

Charlotta Kvarnemo
Department of Biological and Environmental Science, University of Gothenburg, 40530 Gothenburg, Sweden

Bob B. M. Wong
School of Biological Sciences, Monash University, Melbourne, VIC 3800, Australia

Daniel Johansson, Ricardo T. Pereyra and Kerstin Johannesson
Department of Marine Sciences, University of Gothenburg, Tjärnö, Strömstad, Sweden
Centre for Marine Evolutionary Biology, University of Gothenburg, Tjärnö, Strömstad, Sweden

Marina Rafajlović
Centre for Marine Evolutionary Biology, University of Gothenburg, Tjärnö, Strömstad, Sweden
Department of Physics, University of Gothenburg, Gothenburg, Sweden

Marjorie C. Sorensen
Department of Zoology, University of Cambridge, Cambridge CB2 3EJ, UK

Graham D. Fairhurst
Department of Biology, University of Saskatchewan, Saskatoon S7N 5E2, Canada

Susanne Jenni-Eiermann
Swiss Ornithological Institute, Sempach, Switzerland

Jason Newton
NERC Life Sciences Mass Spectrometry Facility, Scottish Universities Environmental Research Centre, Rankine Avenue, East Kilbride G75 0QF, UK

Elizabeth Yohannes
Limnological Institute, University of Konstanz, Mainaustrasse 252, 78464 Constance, Germany

Claire N. Spottiswoode
Department of Zoology, University of Cambridge, Cambridge CB2 3EJ, UK
DST-NRF Centre of Excellence at the FitzPatrick Institute, University of Cape Town, Cape Town, South Africa

Gilberto Pasinelli
Swiss Ornithological Institute, Sempach, Switzerland
Department of Evolutionary Biology and Environmental Studies, University of Zurich, Zurich, Switzerland

Alex Grendelmeier
Swiss Ornithological Institute, Sempach, Switzerland
Division of Conservation Biology, Institute of Ecology and Evolution, University of Bern, Bern, Switzerland

Michael Gerber
Department of Evolutionary Biology and Environmental Studies, University of Zurich, Zurich, Switzerland
Schweizer Vogelschutz SVS/BirdLife Schweiz, Zurich, Switzerland

Raphaël Arlettaz
Division of Conservation Biology, Institute of Ecology and Evolution, University of Bern, Bern, Switzerland
Swiss Ornithological Institute, Valais Field Station, Sion, Switzerland

J. D. M. White, S. L. Jack, M. T. Hoffman, J. Puttick and D. Bonora
Plant Conservation Unit, University of Cape Town, Cape Town, South Africa
Department of Biological Sciences, University of Cape Town, Cape Town, South Africa

E. C. February
Department of Biological Sciences, University of Cape Town, Cape Town, South Africa

V. Visser
Statistics in Ecology, Environment and Conservation, Department of Statistical Sciences, University of Cape Town, Cape Town, South Africa
African Climate and Development Initiative, University of Cape Town, Cape Town, South Africa

S. Beggel, A. F. Cerwenka and J. Geist
Aquatic Systems Biology Unit, School of Life Sciences Weihenstephan, Technical University of Munich, 85354 Freising, Germany

J. Brandner
Wasserwirtschaftsamt Regensburg, Kavalleriestraße 2, 93053 Regensburg, Germany

Ulrike Katharina Harant and Nicolaas Karel Michiels
Department of Animal Evolutionary Ecology, Institution for Evolution and Ecology, University of Tuebingen, Auf der Morgenstelle 28, 72076 Tuebingen, Germany
Department of Biology, Faculty of Science, University of Tuebingen, Auf der Morgenstelle 28, 72076 Tuebingen, Germany

Andrew C. Huang and Christine A. Bishop
Environment and Climate Change Canada, Delta, BC, Canada
Department of Biological Sciences, Center for Wildlife Ecology, Simon Fraser University, Burnaby, BC, Canada

René McKibbin
Environment and Climate Change Canada, Delta, BC, Canada

David J. Green
Department of Biological Sciences, Center for Wildlife Ecology, Simon Fraser University, Burnaby, BC, Canada

Anna Drake
Department of Forest and Conservation Sciences, University of British Columbia, Vancouver, BC, Canada

Nuno F. Silva-Soares and A. Nogueira-Alves
Instituto Gulbenkian de Ciência, Rua da Quinta Grande nº6, 2780-156 Oeiras, Portugal

P. Beldade
Instituto Gulbenkian de Ciência, Rua da Quinta Grande nº6, 2780-156 Oeiras, Portugal
Université Toulouse III Paul Sabatier, Bâtiment 4R1, 118 Route de Narbonne, 31062 Toulouse Cedex 9, France

Christen Kerry Mirth
Instituto Gulbenkian de Ciência, Rua da Quinta Grande nº6, 2780-156 Oeiras, Portugal
School Biological Sciences, Monash University, 25 Rainforest Walk, Melbourne, VIC 3800, Australia

Index

www.ingramcontent.com/pod-product-compliance
Lightning Source LLC
Chambersburg PA
CBHW080408190526
45161CB00003B/164